EOS Annual Meeting (EOSAM 2020)

EPJ Web of Conferences Volume 238 (2020)

Online
7 - 11 September 2020

Editors:

Humberto Michinel
Manuel F. Costa
Orlando Frazao

ISBN: 978-1-7138-1795-6

Printed from e-media with permission by:

Curran Associates, Inc.
57 Morehouse Lane
Red Hook, NY 12571

Some format issues inherent in the e-media version may also appear in this print version.

For additional information, please contact EDP Sciences – Web of Conferences
at the address below.

EDP Sciences – Web of Conferences
17, Avenue du Hoggar
Parc d'Activité de Courtabœuf
BP 112
F-91944 Les Ulis Cedex A
France

Phone: +33 (0) 1 69 18 75 75
Fax: +33 (0) 1 69 28 84 91

contact-edps@webofconferences.org

Additional copies of this publication are available from:

Curran Associates, Inc.
57 Morehouse Lane
Red Hook, NY 12571 USA
Phone: 845-758-0400
Fax: 845-758-2633
Email: curran@proceedings.com
Web: www.proceedings.com

TABLE OF CONTENTS

edp sciences

Web of Conferences

Statement of Peer review

In submitting conference proceedings to *Web of Conferences*, the editors of the proceedings certify to the Publisher that

1. They adhere to its **Policy on Publishing Integrity** in order to safeguard good scientific practice in publishing.
2. All articles have been subjected to peer review administered by the proceedings editors.
3. Reviews have been conducted by expert referees, who have been requested to provide unbiased and constructive comments aimed, whenever possible, at improving the work.
4. Proceedings editors have taken all reasonable steps to ensure the quality of the materials they publish and their decision to accept or reject a paper for publication has been based only on the merits of the work and the relevance to the journal.

Title, date and place of the conference EOSAM2020, September 7 to 11, 2020, Porto, Portugal

Proceedings editor(s): Manuel Filipe P. C. M. Costa

Date and editor's signature June 25, 2020

Statement of Peer review

In submitting conference proceedings to *Web of Conferences*, the editors of the proceedings certify to the Publisher that

1. They adhere to its **Policy on Publishing Integrity** in order to safeguard good scientific practice in publishing.
2. All articles have been subjected to peer review administered by the proceedings editors.
3. Reviews have been conducted by expert referees, who have been requested to provide unbiased and constructive comments aimed, whenever possible, at improving the work.
4. Proceedings editors have taken all reasonable steps to ensure the quality of the materials they publish and their decision to accept or reject a paper for publication has been based only on the merits of the work and the relevance to the journal.

Title, date and place of the conference

Proceedings editor(s):

Orlando Frazao

Date and editor's signature

17, avenue du Hoggar - PA de Courtabœuf – BP 112 - 91944 Les Ulis Cedex A (France)
Tél. : 33 (0)1 69 18 75 75 - Fax : 33(0)1 69 07 45 17 – www.edpsciences.org

edp sciences

Web of Conferences

Statement of Peer review

In submitting conference proceedings to *Web of Conferences*, the editors of the proceedings certify to the Publisher that

1. They adhere to its **Policy on Publishing Integrity** in order to safeguard good scientific practice in publishing.
2. All articles have been subjected to peer review administered by the proceedings editors.
3. Reviews have been conducted by expert referees, who have been requested to provide unbiased and constructive comments aimed, whenever possible, at improving the work.
4. Proceedings editors have taken all reasonable steps to ensure the quality of the materials they publish and their decision to accept or reject a paper for publication has been based only on the merits of the work and the relevance to the journal.

Title, date and place of the conference

Proceedings editor(s):

Date and editor's signature

Humberto Michinel

17, avenue du Hoggar - PA de Courtabœuf – BP 112 - 91944 Les Ulis Cedex A (France)
Tél. : 33 (0)1 69 18 75 75 - Fax : 33(0)1 69 07 45 17 – www.edpsciences.org

Preface

The 2020 annual congress of the European Optical Society (EOS) has been celebrated online for the first time in its history. This has been a bold movement, forced by the exceptional circumstances of the global pandemic faced by the World. Despite the hard circumstances, we consider that the response from the European optical community has risen to the occasion as it shown by the number and quality of the communications presented that are published in this volume.

Therefore, the conference has been extremely successful in its contribution to the main mission of the EOS: bring together and encourage the cooperation in Europe of all with an interest in optics, photonics and related scientific fields, to make practical use of research results, and to support the industrial exploitation of optics and photonics.

More than a hundred of talks and posters were presented and all the national optical societies in Europe were represented as well as international partners like the International Commission for Optics, the European Physical Society, the Japan Society of Applied Physics, the Optical Society of Korea, the Optical Society of Japan, the Chinese Optical Society, the Taiwan Photonics Society and the Optical Iberoamerican Network (RIAO).

The four-day online conference included five plenary talks given by outstanding scientists working in the most important areas of optics and photonics. Yuri Kivshar form the Australian National University spoke about the main advances in photonic properties of metasurfaces, Luis Plaja from the University of Salamanca presented the latest results on ultra-high power lasers, Hatice Altug from the Ecole Polytechnique Federale de Lausanne lectured on nanophotonic biosensors and José Luis Santos from the University of Porto gave a talk on new paths in optical sensing.

Each day, in addition to plenary talks, parallel video sessions were held in combination with life chat interactions between the congress attendees and the authors. In addition, a call was made for submission of selected papers for publication in JEOS:RP, the journal of the European Optical Society, covering the latest advances in both theory and applications of optics and photonics grouped on the following TOMs (Topical Meetings):

TOM 1- Silicon Photonics and Guided-Wave Optics

TOM 2- Computational, Adaptive and Freeform Optics

TOM 3- Optical System Design, Tolerancing and Manufacturing

TOM 4- Bio-Medical Optics

TOM 5- Resonant Nanophotonics: fundamentals and applications

TOM 6- Frontiers in Optical Metrology

TOM 7- Organic & Hybrid Semiconductor Materials and Devices

TOM 9 - Tapered optical fibers, from fundamental to applications

TOM 10- Optical communications

https://doi.org/10.1051/epjconf/202023800001

TOM 11- Thermal radiation and energy management

TOM 12- Nonlinear Optics: materials and applications

TOM 13- Advances and Applications of Optics and Photonics

We would like to express our gratitude to the Board of the European Optical Society, the EOS Executive Director, the EOS Office staff and the chairs of the different TOMs for their support and help in the organization of the Conference. We finally want to acknowledge all the authors that have submitted their work to be included in this volume and very specially thank all those collaborating persons who acted as reviewers of the papers included in this volume for their excellent work.

The editors:

Humberto Michinel (Universidade de Vigo)

Manuel F. Costa (Universidade do Minho)

Orlando Frazao (Universidade de Porto)

Laser processed semiconductors for integrated photonic devices -INVITED

Anna C. Peacock[1,*], *Stuart J.* MacFarquhar[1], *Yohann* Franz[1], *Antoine F. J.* Runge[1], *Sakellaris* Mailis[1], *Swe Z.* Oo[1,2], *Vinita* Mittal[2], *Harold M. H.* Chong[2], and *Ozan* Aktas[1]

[1]Optoelectronics Research Centre, University of Southampton, Southampton, UK
[2]School of Electronics and Computer Science, University of Southampton, Southampton, UK

Abstract. We report results of laser processing of amorphous silicon and silicon-germanium semiconductor materials for the production of integrated photonic platforms. As the materials are deposited and processed at low temperatures, they are flexible, low cost, and suitable for multi-layer integration with other photonic or electronic layers. We demonstrate the formation of waveguides via crystallization of pre-patterned silicon components and functional microstructures through crystallization and compositional tuning of silicon-germanium alloy films. These results open a route for the fabrication of high density, multi-functional integrated optoelectronic chips.

1 Introduction

Group IV semiconductor materials such as silicon (Si) and germanium (Ge) are excellent materials for integrated photonic devices as they are CMOS compatible, have high refractive indices, which allows for small device footprints, and they offer the potential for integration with electronics [1]. Currently, most of the focus on these materials is on single-crystal platforms, owing to the superior optical and electronic properties over their amorphous and polycrystalline counterparts. However, single-crystal platforms still face significant integration challenges, for example in multi-layer and multi-material systems, due to fabrication and growth constraints. Thus, more recently there has been increased interest in Group IV materials that can be deposited using cheap and flexible methods, followed by annealing to achieve high quality, i.e., close to single-crystal, polycrystalline components [2].

Here we report results on two different Group IV material platforms that can be produced via deposition followed by laser crystallization. Compared to conventional thermal annealing treatments, laser crystallization offers a key advantage in that it is highly localized, so that the heating and crystallization can be confined to individual components, reducing the overall thermal budget. In our first demonstration, we show that it is possible to produce high quality poly-Si waveguides by laser processing pre-patterned amorphous Si (a-Si) components that have been deposited at temperatures < 400 °C [3]. For the second demonstration, we focus on the crystallization and compositional tuning of amorphous SiGe (a-SiGe) alloy films deposited at ~200 °C. Here, the added advantage of composition tuning is critical as it allows for direct laser writing of graded index microstripes, which could be exploited for the production of optoelectronic circuits and components (e.g., waveguides, gratings, modulators) with tunable functionality [4].

2 Materials deposition

The fabrication of the a-Si components begins with the formation of a 4.6 µm thick thermally grown buried oxide layer on top of a c-Si substrate. A 480 nm thick film of a-Si is then grown using a hot-wire chemical vapor deposition (HWCVD) technique, with silane (SiH_4) as the only precursor, at a temperature of 320 °C. Following the deposition, e-beam lithography and plasma etching were used to pattern a series of straight waveguides with widths of 0.5 µm, 1 µm, 1.5 µm and 2 µm in the a-Si film.

The a-SiGe films were directly fabricated on c-Si wafers, which were dipped in buffered hydrofluoric acid (HF) for 3 minutes to remove the native oxide. The 400 nm thick SiGe films were deposited by plasma enhanced chemical vapour deposition (PECVD) using SiH_4 (5 sccm) and GeH_4 (50 sccm) precursors, with an RF power of 15 W at a pressure of 300 mTorr and a temperature of 200 °C. The initial atomic content of Ge is 60%.

3 Laser processing

Laser processing of the a-Si and a-SiGe samples was carried out with the setup shown in Fig. 1. The light source was an Argon ion laser emitting continuous wave (CW) radiation at 488 nm with a maximum power of 350 mW. The setup included 3D motorized stages capable of programmed movements at speeds ranging from 0.01 mm/s up to 100 mm/s. The power was adjusted using a

* Corresponding author: acp@orc.soton.ac.uk

polarization cube and a half wave plate. The beam was focused on the top surface of the samples using either a 10x or 20x objective lens to produce spot diameters of 4.7 μm and 2.5 μm, respectively. A pellicle beam splitter, a CCD camera and a white light source were used to image the surface of the samples. In both cases the samples were heated with the laser until melting of the semiconductor occurred.

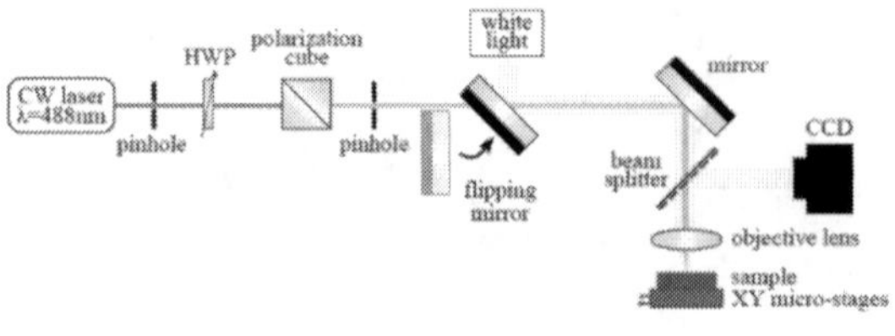

Fig. 1. Laser writing setup. HWP is half-wave plate.

4 Results and discussion

4.1 Laser crystallization of a-Si waveguides

For processing of the a-Si components, the waveguides were heated until complete melting. As a consequence, the waveguides are reshaped while in a liquid state by surface tension that acts on the liquid-air interface. Therefore, the initially rectangular cross section of the a-Si waveguide forms a semi-circular shape in the poly-Si form, as shown in Fig. 2. Here, the SEM micrograph in (a) is an un-processed 2 μm wide a-Si waveguide, and in (b) is a 2 μm wide poly-Si waveguide, processed with a power of 200 mW at a speed of 0.1 mm/s.

Fig. 2. SEM micrographs showing cross-sections of a 2μm wide waveguide before (a) and after (b) laser processing.

Micro-Raman spectroscopy, Secco wet etching and X-ray diffraction measurements reveal the high crystalline quality of the processed waveguides with the formation of millimeter long crystal grains. Optical losses as low as 5.3 dB/cm have been measured, indicating their suitability for the development of integrated circuits [3].

4.2 Laser crystallization and compositional tuning of a-SiGe films

Laser assisted melting of SiGe alloy films induces phase segregation of the Si and Ge atoms, producing Si-rich and Ge-rich regions. Our results show that the spatial profile and amount of phase-segregation in the SiGe thin films can be engineered by controlling the scan speed of the laser. Depending on the speed, different spatial redistributions of the alloy components can be achieved, as shown in Fig. 3.

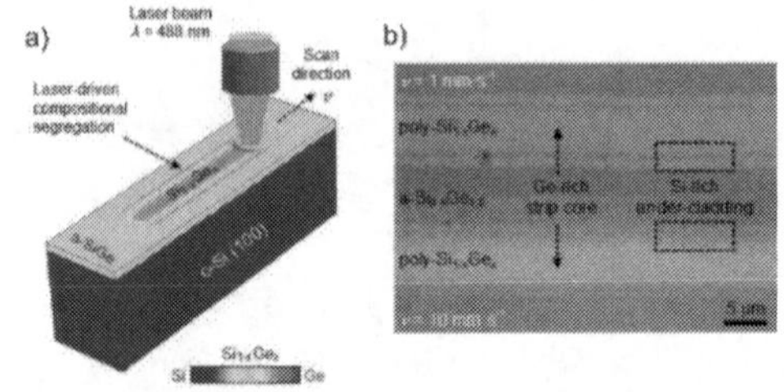

Fig. 3. a) Schematic of compositional tuning to produce crystalline microstripes. b) Optical microscope image shows two polycrystalline stripes, written with difference speeds, and hence Ge concentrations.

Speeds below a threshold of 5 mm/s result in a small compositional segregation within the solidified region, shown by the darker central stripe in Fig. 3b. Note, the Ge content on the top surface is slightly higher than the initial composition. However, above this threshold, a Ge-rich region is obtained at the centre between two lower index Si-rich lateral regions, which can help to promote optical waveguiding in the Ge-rich core. Moreover, higher Ge content can be obtained by using higher scan speeds. By changing the spot size and power, we have control over the size, composition gradient, and shape of the poly-SiGe microstructures written by the laser. Importantly, compositional tuning allows for modifying the refractive index and bandgap of the material, which is useful in the construction of gratings, modulators and detectors that can operate over different wavelength ranges [4].

5 Conclusion

We have demonstrated laser processing of amorphous silicon waveguides and silicon-germanium films for the construction of integrated photonic devices. A key feature of our technique is that the materials are produced using low cost, low temperature and flexible deposition methods, making the components compatible with multi-layer and multi-material architectures.

The authors thank the Engineering and Physical Sciences Research Council (EPSRC) for funding.

References

1. R. Soref, D. Buca, S.Q. Yu, OPN, January Issue, 34 (2016)
2. J. Orcutt, S.D. Tang, S. Kramer, K. Mehta, H. Li, V. Stojanovic, Opt. Express **20**, 7243 (2012)
3. Y. Franz A.F.J. Runge, S.Z. Oo, G. Jimenez-Martinez, N. Healy, A. Khokhar, A. Tarazona, H.M.H. Chong, S. Mailis, A.C. Peacock, Opt. Express **27**, 4462, (2019)
4. O. Aktas, S.Z. Oo S.J. MacFarquhar, V. Mittal, H.M.H. Chong, A.C. Peacock, ACS Appl. Mater. Interfaces 12, 9457 (2020)

Ultra-broadband silicon photonics devices based on subwavelength metamaterials -INVITED

A. V. Velasco[1,*], D. González-Andrade[1], A. Herrero-Bermello[2], J. M. Luque-González[3], R. Halir[3], J. G. Wangüemert-Pérez[3], A. Ortega-Moñux[3], A. Dias[2], I. Molina-Fernández[3], and P. Cheben[4]

[1]Spanish National Research Council (CSIC), Institute of Optics, 28006 Madrid, Spain
[2]Alcyon Photonics S.L, 28006 Madrid, Spain
[3]Universidad de Málaga, ETSI Telecomunicación, Departamento de Ingeniería de Comunicaciones 29071 Málaga, Spain
[4]National Research Council of Canada, Ottawa, Ontario, K1A 0R6, Canada

Abstract. Subwavelength structured waveguides provide tailorable optical properties that can be leveraged to overcome bandwidth limitations in a wide range of photonic devices. In this invited talk, we present an overview of recent developments on subwavelength engineered building blocks, including phase shifters, mode multiplexers, polarization beam splitters and zero-birefringence waveguides.

Subwavelength grating (SWG) metamaterials are becoming established as a fundamental tool for the development of a new generation of silicon photonics building blocks with engineered optical properties [1, 2]. In this work, we present the fundamentals of this technology and its application to the design of several high-performance photonic devices. Firstly, we report on an ultra-broadband phase shifter [3], and on a mode converter and multiplexer/demultiplexer based on a subwavelength engineered multimode interference coupler [4]. Then, we introduce tilted SWG metamaterials, which provide an additional degree of freedom in the design by enabling to tailor the anisotropic response of the optical medium [5]. Finally, we present the application of tilted SWGs to the design of a polarization splitter [6,7] and a zero-birefringence waveguide [8].

Acknowledgments

We acknowledge funding through Ministerio de Economía y Competitividad (TEC2016-80718-R, CDTI SNEO-20181232, RTI2018-097957-B-C33), Ministerio de Educación, Cultura y Deporte (FPU16/06762,), Proyecto I+D+i en el marco del Programa Operativo FEDER Andalucía 2014-2020 (UMA18-FEDERJA-219), Universidad de Málaga, Alcyon Photonics (8.06/5.59.5045), the Community of Madrid – FEDER funds (S2018/NMT-4326), and the Horizon 2020 research and innovation program under the Marie Sklodowska-Curie grant No. 734331.

References

1. P. Cheben et al., "Subwavelength integrated photonics," Nature 560 (2018).
2. R. Halir et al., "Subwavelength-grating metamaterial structures for silicon photonic devices," Proc. IEEE 106 (2018).
3. D. González-Andrade et al., "Ultra-broadband nanophotonic phase shifter based on subwavelength metamaterial waveguides," Photon. Res. 8 (2020).
4. D. González-Andrade et al., "Experimental demonstration of a broadband mode converter and multiplexer based on subwavelength grating waveguides," Opt. Laser Technol. 129 (2020).
5. J. M. Luque-González et al., "Tilted subwavelength gratings: controlling anisotropy in metamaterial nanophotonic waveguides," Opt. Lett. 43 (2018).
6. A. Herrero-Bermello et al., "Design of a broadband polarization splitter based on anisotropy-engineered tilted subwavelength gratings," IEEE Photonics J. 11 (2019).
7. A. Herrero-Bermello et al., "Experimental demonstration of metamaterial anisotropy engineering for broadband on-chip polarization beam splitting," Opt. Express 28 (2020).
8. A. Herrero-Bermello et al., "Zero-birefringence silicon waveguides based on tilted subwavelength metamaterials," IEEE Photonics J. 11 (2019).

* Corresponding author: a.villafranca@csic.es

Quantum dot multi-wavelength lasers for Tbit/s coherent communications and 5G wireless networks -INVITED

Z.G. Lu, J.R. Liu, Y.X. Mao, K. Zeb, G.C. Liu, J. Webber, M. Rahim, G. Pakulski, C.Y. Song, M. Vachon, P. Barrios, D. Poitras, S.R. Wang, W.H. Jiang, *and P. Poole*
Advanced Electronics and Photonics (AEP) Research Centre, National Research Council (NRC) Canada, Ottawa, ON, Canada

Abstract. We will present the design, growth, fabrication, electronic control and packaging of the InAs/InP quantum dot (QD) multi-wavelength lasers (MWLs) developed by AEP of NRC in Canada. Their key technical specifications include L-I-V curves, optical and RF beating spectra, relative intensity noise (RIN), and optical phase noise of each individual wavelength channel, as well as timing jitter are investigated. Data bandwidth transmission capacity of 5.376 Tbit/s and 10.3 Tbit/s respectively in the PAM-4 and 16-QAM modulation formats are demonstrated by only using a single QD MWL chip. We have also developed a novel monolithic InAs/InP QD dual-wavelength (DW) DFB laser as a compact optical beat source to generate millimeter-wave (MMW) signals. Because of using a common cavity, highly coherent and correlated optical modes with optical linewidth as low as 15.83 kHz, spectrally pure MMW signals around 46.8 GHz with a linewidth down to 26.1 kHz were experimentally demonstrated. By using this QD DW-DFB laser, a 1Gbaud (2Gbps) MMW over-fiber transmission link is demonstrated with PAM-4 signals. The results show that the demonstrated device is suitable for high speed high capacity MMW fiber-wireless integrated fronthaul of 5G networks.

1. Introduction

As worldwide data traffic continues to grow, it becomes urgent to utilize communication resources optimally, facilitating capacity growth at reduced energy consumption. Quantum dot (QD) semiconductor multi-wavelength lasers (MWLs) are promising light sources for Terabit/s dense-WDM optical coherent networks, optical signal processing and millimeter wave (MMW) generation because of many unique advantages as compared with quantum well (QW) and bulk semiconductor materials [1]. Such monolithic QD-based semiconductor MWLs solve the obvious cost, power consumption and packaging problems by replacing many separate lasers for each channel by only a single semiconductor laser chip. Other advantages include compact size, low power consumption, simple fabrication, and the ability for hybrid integration with silicon substrates. In recent years we have successfully developed InAs/InP QD MWLs with repetition rates from 10 GHz to 437 GHz and a total output power up to 50 mW per facet at room temperature [1-5]. In this paper we give an overview of the performance of the QD MWLs and their applications in Terabit/s optical communication links and 5G wireless networks.

2. QD materials and QD MWLs for Tbit/s networks

The InAs/InP QD material was grown by chemical beam epitaxy (CBE) on exactly (001)-oriented *n*-type InP substrates [6]. A 355 nm thick InGaAsP waveguide core contained five stacked layers of InAs QDs with $In_{0.816}Ga_{0.184}As_{0.392}P_{0.608}$ (1.15Q) barriers as the gain medium, surrounded by *n*- and *p*- type InP cladding layers. Single lateral mode ridge waveguide QD-MWLs were fabricated with a stripe width of 2.0 μm and a Fabry-Perot (F-P) laser cavity via standard photolithography and combination of dry-, wet-etching and contact metallization techniques. Fig. 1(a) shows a scanning electron microscope (SEM) image and a corresponding schematic cross-section is shown in Fig. 1 (b).

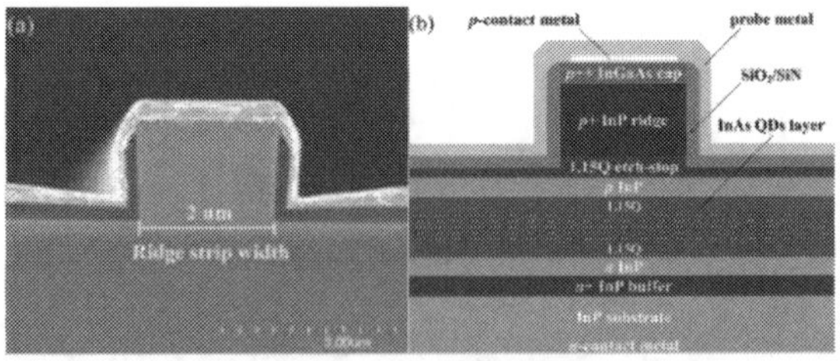

Fig. 1. (a) Cross-sectional SEM image showing the facet of a fabricated F-P InAs/InP QD laser; (b) Schematic cross-sectional diagram of a shallow-etched ridge-waveguide InAs/InP QD laser

Fig. 2(a) shows the light output of the lasers as a function of drive current with blue dots for 11 GHz, red dashes for 25 GHz, and green solid for 34.5 GHz lasers at 20°C. Fig. 2 (b) shows the lasing spectra at the indicated drive current for the three lasers. We have investigated optical linewidth for individually filtered longitudinal lasing modes by using an OE4000 automated laser phase noise measurement system and their timing jitters based on their RF beating signals.

https://doi.org/10.1051/epjconf/202023801003

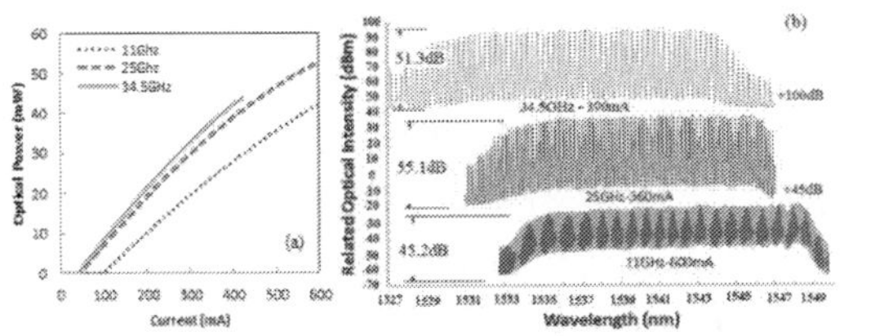

Fig. 2 (a) Light-current characteristics (b) Optical spectra of 11, 25 and 34.5 GHz QD MWLs at 0.01-nm resolution.

To demonstrate the performance of these lasers in a Tbit/s system we have used both PAM-4 and 16-QAM modulation schemes. Fig. 3(a) shows the experimental setup for PAM-4 and 16-QAM data transmission system where the optical modulation is performed by a dual polarization QAM transmitter. A PAM-4 signal is created using the arbitrary waveform generator (AWG) for a NRZ PRBS with a pattern length of 2^{15}-1 bits at a symbol rate of 28 GBaud on two uncorrelated channels (*IY and IX*). Fig. 3(b) shows the Bit Error Rate (BER) as a function of received power for back-to-back (B2B) and after 25 km of SMF transmission using a single filtered wavelength mode at 1547.855 nm (one of the 48 channels) in the 34.5 GHz laser. The inset shows the measured eye diagram after 25 km SMF showing open eyes. The potential aggregate transmission capacity for the 48 channels available using the 34.5 GHz laser is 5.376 Tbit/s (PAM-4 48 × 28 GBaud PDM) over 25 km of a SMF transmission. Fig. 3(c) shows a 16-QAM constellation diagram at a symbol rate of 23 GBaud obtained from a filtered single laser mode at 1544.16 nm for the 25 GHz laser. An error vector magnitude (EVM) of 7.74% and BER of 6.15×10^{-6} were obtained. We obtained 16-QAM results at the base rate of 23 GBaud over 56 individual channels, which corresponds to an aggregate transmission capacity of 10.3 Tbit/s (16QAM 56x23 GBaud PDM) B2B.

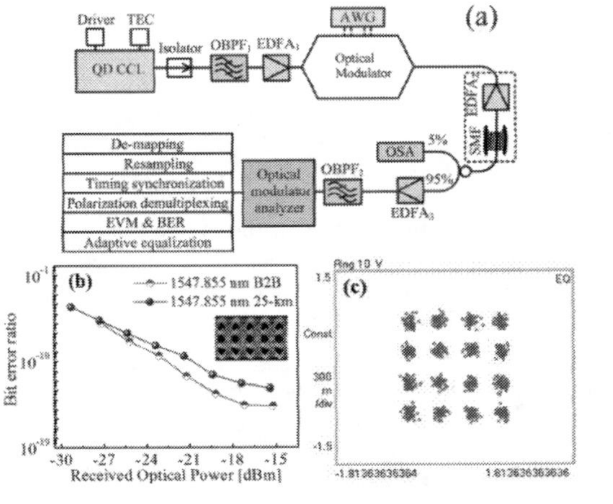

Fig. 3 (a) A schematic of PAM 4 and QAM data format system. (b) Measured BER versus received power for PAM 4 B2B and after 25-km of SMF transmission using a comb line of 1547.855 nm of 34.5 GHz QD MWL. The inset shows corresponding eye diagram for 25-km of SMF transmission. (c) A constellation eye diagram obtained from filtered channel of 1544.16 nm for the 25 GHz QD MWL in the double polarization 16-QAM data format at 23 GBaud B2B transmission.

3. QD DW DFB lasers for 5G wireless networks

The device was implemented in a p-n blocked buried heterostructure distributed feedback (BH DFB) laser with a novel synthesized aperiodic non-uniform diffraction grating layer placed below the InAs/InP QD active layer, which provides the distributed feedback such that two longitudinal modes lase simultaneously [7]. Fig. 4 shows its L-I curve, the insect shows its optical spectrum at 360 mA and their generated MMW signal of 46.8 GHz. The phase noise and RIN were measured experimentally for the two dominant optical modes. At 360 mA, optical linewidths of 25.9 kHz and 29.4 kHz were measured for the two optical modes at λ_1 =1539.522 nm and λ_2=1539.896 nm, respectively. Average RIN was measured as -150.8 dB/Hz for λ_1 and -151.4 dB/Hz for λ_2 over the frequency range from 10 MHz to 20 GHz.

For MMW over-fiber transmission link, λ_2 of the QD DW-DFB laser was modulated using a 40 GHz Mach-Zehnder intensity modulator (MZ-IM) with 1Gbaud (2Gbps) PAM-4 signals generated through AWG and λ_1 was used as beat note signal for heterodyne MMW carrier generation. A clear eye diagram of PAM-4 signals of 1Gbaud transmission bandwidth along with transmitted/received signals were measured using Tektronix digital phosphor real-time oscilloscope. These results indicate the promising applications of the demonstrated QD DW-DFB laser as a compact mm-wave optical beat source in heterodyne mm-wave communication systems of 5G and beyond.

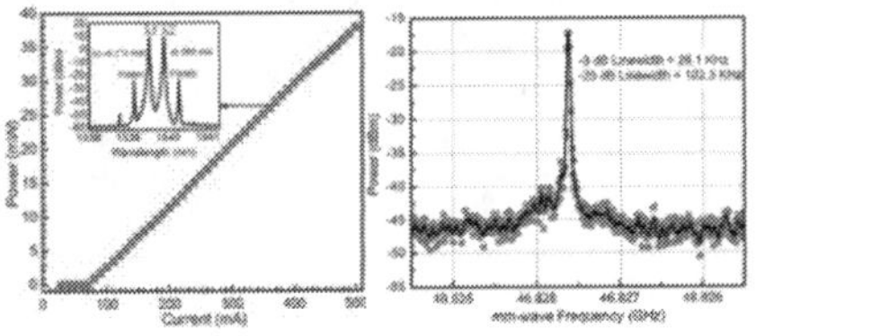

Fig. 4. L-I characteristics (inset shows the optical spectrum at 360 mA) and their generated MMW signal.

References

1. Z.G. Lu "Quantum dot coherent comb lasers for terabit optical networking systems", (Invited paper) In Proc. Photonics West, San Francisco, CA, USA, Vol. 10921, 109210N (4 March 2019).
2. Z.G. Lu, et al. "312-fs pulse generation from a passive C-band InAs/InP quantum dot mode-locked laser," Opt. Express, **16**, 10835-10840 (2008).
3. Z.G. Lu, et al. "An L-band monolithic InAs/InP quantum dot mode-locked laser with femtosecond pulses," Opt. Express, **17**, 13609–13614 (2009).
4. Z.G. Lu, et al. "Ultra-high repetition rate InAs/InP quantum dot mode-locked lasers," Optics Communications, Vol. 284, No. 9, pp. 2323-2326 (May 2011).
5. Z.G. Lu, et al. "High performance InAs/InP quantum dot 34.462-GHz C-band coherent comb laser module," Opt. Express, Vol. 26, No. 2, 2160-2167(2018).
6. P. J. Poole, et al. "Growth of InAs/InP-based quantum dots for 1.55 μm laser applications," J. Cryst. Growth 311(6), 1482–1486 (2009).
7. M. Rahim, et al. "Monolithic InAs/InP quantum dash dual-wavelength DFB laser with ultra-low noise common cavity modes for millimeter-wave applications", Optics Express 27 (24), 35368-35375 (November 25, 2019).

Kerr Micro-combs for Radio Frequency Photonics -INVITED

Xingyuan Xu,[1†] Mengxi Tan,[1] Jiayang Wu,[1] Sai T. Chu,[3] Brent E. Little,[4] Roberto Morandotti,[5] Thach Nguyen[2], Arnan Mitchell,[2] and David J. Moss[1,*]

[1]*Optical Sciences Centre, Swinburne University of Technology, Hawthorn, VIC 3122 Australia*
[2]*School of Engineering, RMIT University, Melbourne, VIC 3000, Australia*
[3]*Department of Physics, City University of Hong Kong, Hong Kong, China.*
[4]*Xi'an Institute of Optics and Precision Mechanics Precision Mechanics of CAS, Xi'an, China.*
[5]*INRS –Énergie, Matériaux et Télécommunications, Varennes, Québec, J3X 1S2, Canada.*
[†]*Current Address: Electro-Photonics Laboratory, Dept. of Electrical and Computer Systems Engineering, Monash University, Clayton, 3800 VIC, Australia*
**dmoss@swin.edu.au*

Abstract— **We review applications of Kerr micro-combs in RF photonic systems including fractional differentiators, Hilbert Transformers and many other functions.**

Radio frequency (RF) and microwave photonics, which bring together the worlds of radiofrequency engineering and optoelectronics [1], exploit the potential of optical technologies and benefit RF systems in many respects, including high speed, broad operation bandwidth, low loss, and strong immunity to electromagnetic interference [2-4]. A diverse range of photonic approaches to RF signal generation, transmission, processing, and sensing have been proposed and widely employed in RF systems and communication networks [5-13]. Nevertheless, most RF systems are composed of discrete components, which impose certain drawbacks in terms of cost, power consumption and reliability, thus holding RF photonic systems from reaching maturity and replacing traditional RF solutions [14, 15]. Meanwhile, advances in integrated photonics [16-21], driven by the compelling economics of ever smaller footprint and lower power consumption, have created new possibilities and opportunities for RF photonics. Commercialized wafer scale fabrication of III-V, dielectrics, elemental semiconductor and nonlinear crystals have solved key challenges for the co-integration of lasers, modulators, photodetectors, and passive components, and have paved the road for integrated RF photonics to bring it closer towards commercial applications.

We review our recent work on RF and microwave photonic applications of integrated micro-combs, including a reconfigurable RF photonic intensity differentiator, RF channelizer and others. By employing an on-chip nonlinear micro-ring resonator (MRR), we generate a broadband Kerr comb based on soliton crystals, with a record low FSR of 49GHz, generating a large number of comb lines and use it as a high-quality multi-wavelength source for a transversal differentiator. By programming and shaping the power of individual comb lines according to corresponding tap weights, reconfigurable intensity differentiators with variable differentiation orders can be achieved. Detailed analyses of the operation principle and experimental demonstrations of fractional, 1st, 2nd, and 3rd order differentiations are performed.

As one of the most powerful tools in RF photonic systems, optical frequency combs can serve as multi-wavelength sources and establish multiple RF channels, and thus can greatly increase the capacity for transmission and performance for transversal processers [22-26]. Unfortunately, traditional approaches like discrete laser arrays, mode-locked lasers, or cascaded modulators all have limitations of one form or another, such as the cost, ability to be integrated or the number of available wavelengths, and thus pose challenges for integrated RF photonic systems.

Micro-comb sources, particularly those based on novel CMOS-compatible platforms [27, 28], offer new possibilities for integrated RF photonics. In 2008-10, new platforms for nonlinear optics, including Hydex [27, 29-36] and silicon nitride [37], were introduced that exhibit negligible nonlinear absorption in the telecom band, a moderate nonlinear parameter and extremely high nonlinear figure of merit, which are ideal for micro-comb generation. Following the first report of Kerr frequency comb sources in 2007 [38], the first integrated CMOS compatible integrated optical parametric oscillators were reported in 2010 [27, 37], and since then this field has exploded. Many cutting-edge applications have been demonstrated based on CMOS-compatible micro-combs, ranging from filter-driven mode-locked lasers [39-42] to quantum physics [43-48]. Meanwhile, for RF photonics, many new applications have been investigated with the fundamental advantages of micro-combs demonstrated [49-51].

As compared with conventional intensity differentiators based on laser diode arrays, the cost, size and complexity can be greatly reduced. Our scheme enables a high degree of reconfigurability in terms of processing functions and operation bandwidth, offering a reconfigurable platform for diverse microwave photonic computing functions. By programming the waveshaper's tap coefficients, our scheme can also apply to other computing functions such as Hilbert transforms. The operation bandwidth is fundamentally limited by the Nyquist zone, which is determined by the comb spacing. In our case, the frequency spacing of the Kerr comb generated by the nonlinear MRR reaches 200 GHz, thus leading to a potential operation bandwidth of over 100 GHz.

[1] J. Capmany, and D. Novak, "Microwave photonics combines two worlds," Nature Photonics, **1**(6), 319-330 (2007).

[2] A. J. Seeds, "Microwave photonics," IEEE Transactions on Microwave Theory and Techniques, **50** (3), 877-887 (2002).

[3] R. C. Williamson, and R. D. Esman, "RF photonics," IEEE Journal of Lightwave Technology, **26** (9-12), 1145-1153 (2008).

[4] J. P. Yao, "Microwave Photonics," IEEE Journal of Lightwave Technology, **27** (1-4), 314-335 (2009).

[5] Q. Z. Cen, *et al.*, "Rapidly and continuously frequency-scanning opto-electronic oscillator," Optics Express, **25** (2), 635-643 (2017).

[6] J. Dai, X. Y. Xu, J. L. Ke *et al.*, "Self-Oscillating Triangular Pulse Generator Based on 90 degrees Photonic-Assisted Phase Shifter," IEEE Photonics Technology Letters, **29** (3), 271-274 (2017).

[7] J. Dai, *et al.*, "Self-oscillating optical frequency comb generator based on an optoelectronic oscillator," Opt Express, **23** (23), 30014 (2015).

[8] K. Xu, *et al.*, "Microwave photonics: radio-over-fiber links, systems, and applications," Photonics Research, **2** (4), B54-B63 (2014).

[9] X. Y. Xu, J. Dai, Y. T. Dai *et al.*, "Broadband and wide-range feedback tuning scheme for phase-locked loop stabilization of tunable optoelectronic oscillators," Optics Letters, **40** (24), 5858-5861 (2015).

[10] A. X. Zhang, *et al.*, "Stable RF delivery by lambda dispersion-induced optical tunable delay," Opt Lett, **38** (14), 2419 (2013).

[11] S. Mansoori, and A. Mitchell, "RF transversal filter using an AOTF," IEEE Photonics Technology Letters, **16** (3), 879-881 (2004).

[12] H. Emami, *et al.*, "Wideband RF photonic in-phase and quadrature-phase generation," Optics letters, **33** (2), 98-100 (2008).

[13] S. L. Pan, and J. P. Yao, "Optical generation of polarity- and shape-switchable ultrawideband pulses using a chirped intensity modulator and a 1st order MZ interferometer," Optics Letters, **34** (9), 1312-1314 (2009).

[14] D. Marpaung, *et al.*, "Nonlinear Integrated Microwave Photonics," J. of Lightwave Technology, **32** (20), 3421-3427 (2014).

[15] D. Marpaung *et al.*, "Integrated microwave photonics," Laser & Photonics Reviews, **7** (4), 506-538 (2013).

[16] J. Wu, *et al.*, "Compact tunable Si photonic differential-equation solver for general linear time-invariant systems," Opt Express, **22** (21), 26254 (2014).

[17] J. Y. Wu, *et al.*, "Nested Configuration of Silicon Microring Resonator With Multiple Coupling Regimes," IEEE Photonics Technology Letters, **25** (6), 580-583 (2013).

[18] J. Y. Wu, P. Cao, T. Pan *et al.*, "Compact on-chip 1 x 2 wavelength selective switch based on silicon microring resonator with nested pairs of subrings," Photonics Research, **3** (1), 9-14 (2015).

[19] J. Y. Wu, *et al.*, "On-Chip Tunable Second-Order Differential-Equation Solver Based on a Silicon Photonic Mode-Split Microresonator," J. of Lightwave Technology, **33** (17), 3542-3549 (2015).

[20] J. Y. Wu, *et al.*, "Passive silicon photonic devices for microwave photonic signal processing," Optics Communications, **373**, 44-52 (2016).

[21] J.Wu, et al., "Micro-ring resonator quality factor enhancement via an integrated Fabry-Perot cavity", APL Photonics, **2** 056103 (2017).

[22] A. M. Weiner, "Microwave Photonic Filters Based on Optical Frequency Combs," 2012 Ieee Photonics Conference (IPC), 296-297 (2012).

[23] O. F. Yilmaz, *et al.*, "True time delays using conversion/dispersion with flat magnitude response for wideband analog RF signals," Optics Express, **20** (8), 8219-8227 (2012).

[24] R. Wu, *et al.*, "Supercontinuum-based 10-GHz flat-topped optical frequency comb generation," Optics Express, **21** (5), 6045-6052 (2013).

[25] A. J. Metcalf, *et al.*, "High-Power Broadly Tunable Electrooptic Frequency Comb Generator," IEEE J. Selected Topics in Quantum Electronics, **19** (6), (2013).

[26] E. Hamidi, et al.,"Tunable Programmable Microwave Photonic Filters based on an Optical Frequency Comb," IEEE Trans. on Microwave Theory, **58** (11), 3269-3278 (2010).

[27] L. Razzari, *et al.*, "CMOS-compatible integrated optical hyper-parametric oscillator," Nature Photonics, **4** (1), 41-45 (2010).

[28] D. J. Moss, *et al.*, "New CMOS-compatible platforms based on silicon nitride and Hydex for nonlinear optics," Nature Photonics, **7** (8), 597-607 (2013).

[29] M. Ferrera, *et al.*, "Low-power continuous-wave nonlinear optics in doped silica glass integrated waveguide structures," Nature Photonics, **2** (12), 737-740 (2008).

[30] D. Duchesne, *et al.*, "Efficient self-phase modulation in low loss, high index doped silica glass integrated waveguides," Optics Express, **17** (3), 1865-1870 (2009).

[31] M. Ferrera, *et al.*, "Low power four wave mixing in an integrated, micro-ring resonator with Q=1.2 million," Optics Express, **17** (16), 14098-14103 (2009).

[32] D. Duchesne, *et al.*, "Supercontinuum generation in a high index doped silica glass spiral waveguide," Optics Express, **18** (2), 923-930 (2010).

[33] M. Ferrera, et al., "Ultra-Fast Integrated All-Optical Integrator", Nature Communications, **1** Article 29 (2010). DOI:10.1038/ncomms1028

[34] A. Pasquazi, *et al.*, "All-optical wavelength conversion in an integrated ring resonator," Optics Express, **18** (4), 3858-3863 (2010).

[35] A. Pasquazi, *et al.*, "Efficient wavelength conversion and net parametric gain via FWM in a high index doped silica waveguide," Optics Express, **18** (8), 7634-7641 (2010).

[36] M. Peccianti, *et al.*, "Subpicosecond optical pulse compression via an integrated nonlinear chirper," Optics Express, **18** (8), 7625-7633 (2010).

[37] J. S. Levy, *et al.*, "CMOS-compatible multiple-wavelength oscillator for on-chip optical interconnects," Nature Photonics, **4** (1), 37-40 (2010).

[38] P. Del'Haye, et al., "Optical frequency comb generation from a monolithic microresonator," *Nature* **450**, (7173) 1214-1217 (2007).

[39] M.Peccianti, *et al.*, "Demonstration of an ultrafast nonlinear microcavity modelocked laser", Nature Communications, **3** 765 (2012).

[40] M.Kues, et. al., "Passively modelocked laser with an ultra-narrow spectral width", Nature Photonics, **11** (3) 159 (2017).

[41] A. Pasquazi, *et al.*, "Self-locked optical parametric oscillation in a CMOS compatible microring resonator: a route to robust optical frequency comb generation on a chip," Optics Express, **21** (11), 13333-13341 (2013).

[42] A. Pasquazi, *et al.*, "Stable, dual mode, high repetition rate mode-locked laser based on a microring resonator," Optics Express, **20** (24), 27355-27362 (2012).

[43] C. Reimer, *et al.*, "Integrated frequency comb source of heralded single photons," Optics Express, **22** (6), 6535-6546 (2014).

[44] C.Reimer, et al., "Cross-polarized photon-pair generation and bi-chromatically pumped optical parametric oscillation on a chip", Nature Communications, **6** Article 8236 (2015).

[45] L. Caspani, *et al.*, "Multifrequency sources of quantum correlated photon pairs on-chip: a path toward integrated Quantum Frequency Combs," Nanophotonics, **5** (2), 351-362 (2016).

[46] C. Reimer, *et al.*, "Generation of multiphoton entangled quantum states by means of integrated frequency combs," Science, **351** (6278), 1176-1180 (2016).

[47] M.Kues, et al., "On-chip generation of high-dimensional entangled quantum states and their coherent control", Nature, **546** (7660) 622 (2017).

[48] P. Roztocki, M. Kues, C. Reimer *et al.*, "Practical system for the generation of pulsed quantum frequency combs," Optics Express, **25** (16), 18940-18949 (2017).

[49] T. G. Nguyen, *et al.*, "Integrated frequency comb source based Hilbert transformer for wideband microwave photonic phase analysis," Optics Express, **23** (17), 22087-22097 (2015).

[50] X.Xu, et al., "Microwave Photonic All-optical Differentiator based on an Integrated Frequency Comb Source", APL Photonics, **2** (9) 096104 (2017).

[51] X. X. Xue, *et al.*, "Programmable Single-Bandpass Photonic RF Filter Based on Kerr Comb from a Microring," IEEE J. of Lightwave Technology, **32** (20), 3557-3565 (2014).

EPJ Web of Conferences **238**, 01005 (2020) https://doi.org/10.1051/epjconf/202023801005
EOSAM 2020

Silicon photonics for high data rate applications -INVITED

David J. Thomson[1,*], *Weiwei* Zhang[1], *Ke* Li[1], *Kapil* Debnath[1], *Shenghao* Liu[1], *Bigeng* Chen[1], *Muhammad* K. Husain[2], *Fanfan* Meng[1], *Ali* Z. Khokhar[1], *James* Byers[2], *Martin* Ebert[1], *Jamie* D. Reynolds[1], *Mehdi* Banakar[1], *Lorenzo* Mastronardi[1], *Callum* G. Littlejohns[1], *Frederic* Y. Gardes[1], *Goran* Z. Mashanovich[1], *Shinichi* Saito[2] and *Graham* T. Reed[1].

[1]Optoelectronics Research Centre, University of Southampton, SO17 1BJ, UK
[2]Electronics and Computer Science, University of Southampton, SO17 1BJ, UK

Abstract. The high speed conversion of signals between the optical and electrical domains is crucial for many key applications of silicon photonics. Electro-optic modulators integrated with electronic drive amplifiers are typically used to convert an electrical signal to the optical domain. Design of these individual elements is important to achieve high performance, however a true optimisation requires careful co-design of the photonic and electronic components considering the properties of each other. Here we present our recent results in this area together with a MOSCAP type modulator with the potential for high speed, high efficiency and highly linear modulation.

1 Introduction

Many applications of silicon photonics require operation in the 10's of GHz range, from data communication to radio over fibre and LIDAR. High speed modulation in silicon has been demonstrated for over a decade now [1] and typically devices demonstrated are based upon the depletion of free carriers in a waveguide based, reversed biased PN diode. Integration of such devices with the required driving electronics usually causes a degradation of the operating speed due to the parasitics from the integration process and/or the driver performance itself. In order to produce an optical signal modulated at such high frequencies and operating at low power both the design and implementation of both the modulator and drive electronics therefore require careful design. Here we present our work on the co-design and integration of a silicon photonics modulator and CMOS electronic driver using both wire bonding and flip-chip bonding approaches.

The above mentioned carrier depletion based devices typically have low modulation efficiencies owing to the limited depletion widths and tolerable doping levels. Devices require interaction lengths on the order of milimetres in order to keep to CMOS compatible driving voltages. For high speed operation on such device lengths travelling wave electrodes are required with impedance matched terminations that result in large power consumption. Another issue with depletion based devices is the linearity of the phase shift versus voltage which can have implications when producing higher order modulation formats, in radio over fibre applications and LIDAR. MOSCAP modulators which are based upon the accumulation of high densities of electrons and holes around a thin dielectric layer in the waveguide have a far superior linearity together with the potential for an order of magnitude improvement in efficiency. In this case devices lengths are in the 100's of micrometre range and can be driven as a lumped element, which typically requires a lower drive power. Here we also present our recent work on this type of modulator.

2 MOSCAP modulator

A deterring factor to the use of the MOSCAP modulator is difficulty of the fabrication as compared to the carrier depletion approach. The main challenge lies in the realisation of the thin dielectric layer in the waveguide and keeping the optical and electrical properties of the silicon either side of it sufficiently good so as to maintain a high device performance. In typical devices from the literature the dielectric layer is positioned on top of high quality single crystal silicon with polycrystalline silicon deposited on top of that. The properties of the polycrystalline silicon are however inferior optically and electrically to single crystal silicon and therefore impact the device performance.

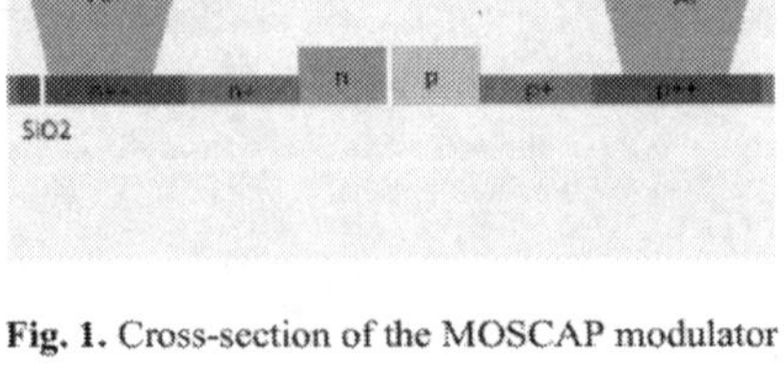

Fig. 1. Cross-section of the MOSCAP modulator

In our work the dielectric layer runs vertically in the waveguide (as shown in figure 1) and a regrowth process

* Corresponding author: d.thomson@soton.ac.uk

is used to transform deposited amorphous silicon into single crystal silicon. The resulting device therefore has high quality silicon on both sides of the dielectric layer. Results from a proof of concept batch of devices have demonstrated efficiencies on the order of 1.8V.cm with modulation in the 10's of Gbit/s (figure 2).

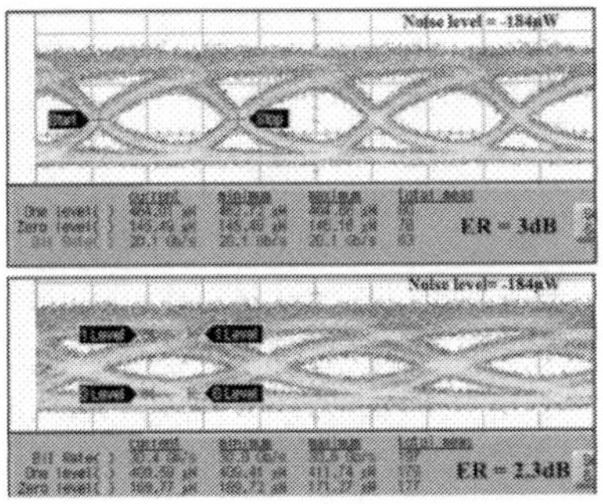

Fig. 2. Optical eye diagrams from the MOSCAP modulator

3 U-shape modulator and CMOS driver

Recently with have performed work on the close integration of the design of both silicon optical modulators and CMOS drive electronics to produce high performance, power efficient modulation units. One implementation of this work uses a carrier depletion based optical modulator designed in a U-shape Mach Zehnder modulator (MZM) configuration to allow access to both input and termination pads of the MZM on one side of the chip (figure 3). Both CMOS driver and MZM were designed with common-centre topology, making either wire-bonding or flip-chip bonding solutions possible. By terminating the modulator electrodes on the CMOS chip dynamic control of the termination impedance can be performed allowing fine tuning of the performance.

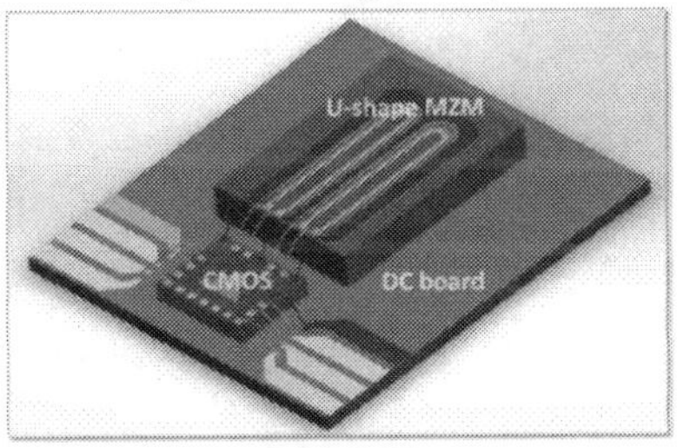

Fig. 3. Schematic of CMOS driver – U shape modulator integration with wire bonds.

In the design process the properties of the modulator and parasitics of the integration approach are incorporated into the CMOS driver design. The design of either individual component can be altered and the overall unit performance analysed in order to perform a full optimisation. Wire bond and flip-chip bond integration approaches have been followed and the results compared (figure 4) [2].

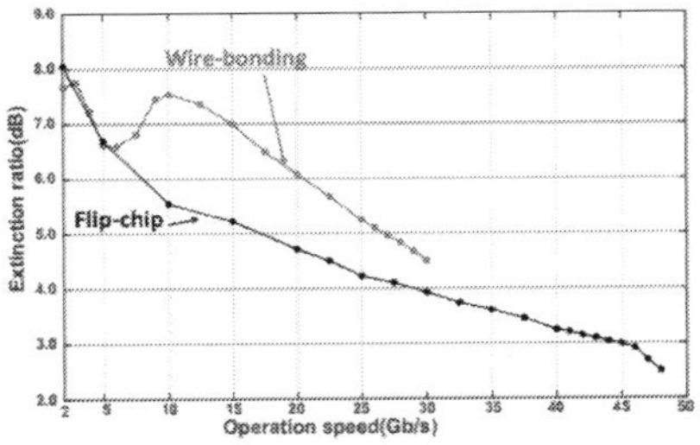

Fig. 4. Eye extinction ratio versus data rate for wire bond and flip chip bond integration approaches.

Figure 4 shows the achievable eye extinction ratio at different data rates with the two integration techniques. At lower frequencies the wire bonding approach yields a better performance in terms of extinction ratio. This is due to inductive peaking from the parasitics of the bond wires in the 10-12Gbit/s range. As the modulation rate is increased the extinction ratio of the flip-chip bonded approach roles off at a slower rate allowing modulation to 40Gbit/s. With wire bonding the eye is closed at around 25Gbit/s.

Summary

An efficient, linear optical modulator based upon the accumulation of free carriers around a thin, vertical dielectric layer in a silicon waveguide has been presented together with a modulation efficiency around 1.8V.cm and operating speed in the 10's of Gbit/s. Work around the co-design of CMOS electronic driver and silicon optical modulator has compared integration through wire bonding and flip-chip bonding. Wire bonding gives a better performance at lower data rates due to a peaking effect from the bond wires, however at data rates of 25Gbit/s and beyond the eye is closed Whereas for flip-chip integration modulation at 40Gbit/s is possible.

Acknowledgements

This work is supported by funding from Rockley Photonics/EPSRC (EP/R003076/1), European Commission Horizon 2020 project PICTURE (780930), EPSRC projects (EP/M009416/1), (EP/M008975/1), (EP/N013247/1) and Hitachi. G. T. Reed is grateful to the Royal Society/Wolfson Foundation for funding. D. J. Thomson acknowledges funding from the Royal Society for his University Research Fellowship.

References

1. G. T. Reed et al. "Silicon optical modulators," Nature Photonics, **4**, 518-526, (2010).
2. Ke Li et al. "Co-design of electronics and photonics components for Silicon Photonics transmitters," Proceedings of ECOC (2018)

https://doi.org/10.1051/epjconf/202023801006

Deformed microcavities with very high Q-factors and directional farfield emission

Arne Behrens[1,*], *Martí* Bosch[2], *Martina* Hentschel[3], and *Stefan* Sinzinger[1]

[1]FG Technische OptikInstitut für Mikro- und Nanotechnologien MacroNano ®, TU Ilmenau
[2]FG Computational Physics, TU Ilmenau
[3]Intstitut für Physik, Technische Universität Chemnitz

Abstract. We report the design and optimized fabrication of deformed whispering gallery mode resonators in silica with solely ICP-RIE. This allows us to control the morphology of the resonators more freely and results in low surface roughness. The light was coupled into the resonator using a state of the art tapered fiber approach and we determined the Q-factor in the range of 10^5

1 Introduction

Optical microcavities are structures that confine light to a defined volume using the mechanisms of reflection and resonance. A special type of modes inside a circular optical microcavity is called whispering gallery mode (wgm) because they are confined by the total internal reflection occurring at the cavity boundary. One main property of a resonator is the Q-factor, which describes the ratio of stored energy inside the resonator to the loss energy per cycle. A high Q-factor, therefore, means long photon lifetime inside the cavity and this increases the light-matter interaction drastically. This is the main reason why high-Q resonators are a research platform for several fundamental investigations and experimental applications [1]. In the recent time high-Q resonators have been used for e.g. quantum electro dynamics [2] ,telecommunications [3], optomechanics, nonlinear optics [4] and sensing [5].

2 Asymmetric resonant cavities

One intriguing field of study are asymmetric resonant wgm cavities. While circular cavities exhibit a strongly isotropic radiation and propagation pattern, asymmetric resonant cavities show a more complex behavior. Much theoretical work was dedicated to the exploration of optical microcavities which produce a strongly directional farfield and to combine it with the property of high Q factors and therefore low loss [6].

Depending on the refractive indices of microresonstoars with specific deformations generate a directional far-field while still exhibiting high Q-factors. The limacon is such a morphology for refractive index larger than 3 [7]. Since most of the optical media with very low absorption have a refractive index close to 1.5 investigations

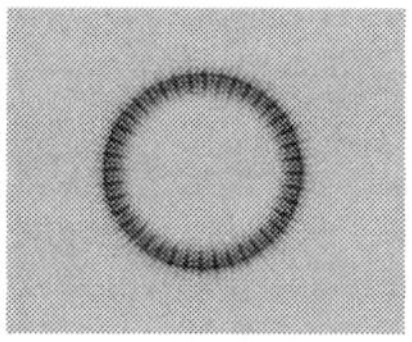

(a) Circle

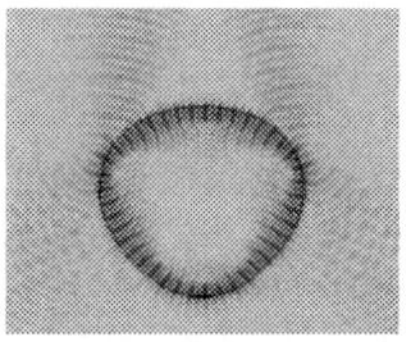

(b) Shortegg

Figure 1: Comparison of the mode structure

showed that a suitable morphology for this regime is the Shortegg, with a shape following the equation2.

$$r(\phi) = R_0(1 + 0.16\cos(\phi) - 0.022\cos(2\phi) - 0.05\cos(3\phi))$$

3 Fabrication

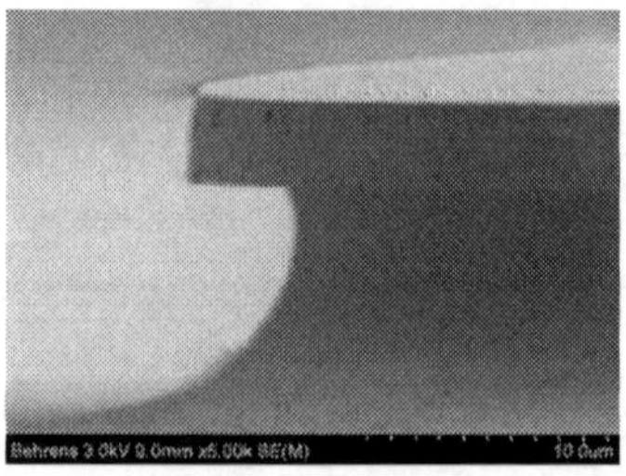

Figure 2: SEM image of a microdisk resonator

To investigate the behavior and potential applications of deformed cavities establish a fabrication process that allows us to fabricate resonators with very precise control of the deformation. The state of the art fabrication process

*e-mail: arne.behrens@tu-ilmenau.de

for UHQ resonator was shown by [8]. They use buffered HF for the etching of the silica layer and underetch the silica disk with XeF_2 vapour. After the structuring they use a reflow process to create a smooth boundary which results in a shrinkage of the resonator. First, we need to etch the 2 μm thick silica layer, which is fabricated by thermal oxidation. For this, we use an inductively coupled plasma (ICP)-RIE with CHF_3/Ar chemistry. By fine-tuning the etch gases and the ion energy it is even possible to adjust the slope angle of our resonator which allows us to confine mode in the z-direction. The following silicon underetching is achieved by using a cryogenic SF_6/O_2 ICP RIE which exhibits very high selectivity to silica which is critical for achieving a low surface roughness. We achieved a surface roughness of $R_Q = 2$ nm.

4 Experimental measurement

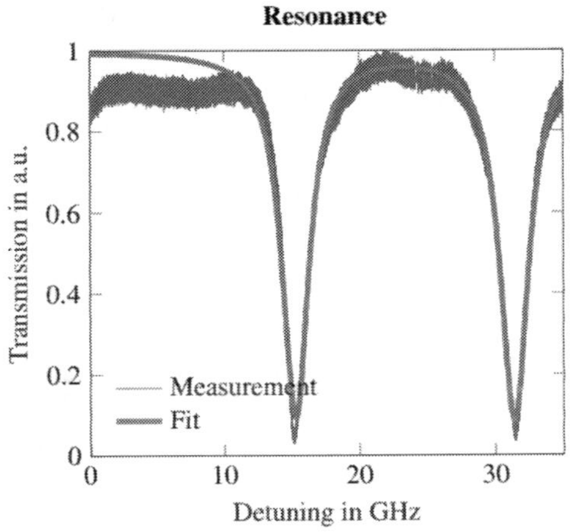

Figure 3: Fine scan of the resonance compared to a Lorentzian fit with a spectral window of 35 GHz

2cm

For coupling light into the resonator, the most prominent method is the use of a so-called tapered fiber.

The shown measurements were conducted using a shortegg resonator with a diameter of 100 μm or $kR = 200$. The highest Q-factor measured was 10^5 at a wavelength of about 1529 nm shown in 3. The measurement data shows a good match with the simulated mode behavior. The divergence on the left side can be attributed to some low q modes. The fitted model consists of two Lorentzian peaks with the position 15.3 GHz and 31.4 GHz with an individual width of 2.3 GHz and 2.5 GHz. This indicates the presence of mode splitting and mode broadening which could be caused by several pertubations like surface roughness, microparticles which are stuck to the surface or the presence of the tapered fiber itself.

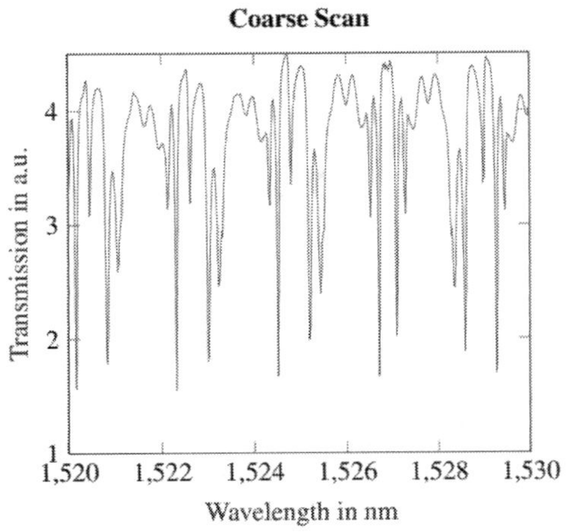

Figure 4: Measured resonances by monitoring the transmission of a tapered single mode fiber

5 Simulation

To evaluate the experimental measurements we designed a wave optical simulation in comsol. There we computed the eigenmodes of a 100 μm Shortegg to compare them to our measurements. They resulted in a maximum Q-factor of $3 * 10^6$. However this is the maxium intrinsic Q-factor which could be achieved in the eigenmode evaluations.

6 Conclusion

We demonstrated the fabrication of High-Q using ICP-RIE which resulted in measured Q-factors in the range of 10^5 in deformed cavities with a diamter of 100 μm.

Acknowledgements

I want to thank Prof. Sfle Nic Chormaic for the possibily to visit the Light-Matter-Interaction Unit at OIST and to conduct measurements at their laboratory.

References

[1] K.J. Vahala, nature **424**, 839 (2003)

[2] S. Spillane, T. Kippenberg, K. Vahala, K. Goh, E. Wilcut, H. Kimble, Physical Review A **71**, 013817 (2005)

[3] T.A. Ibrahim, R. Grover, L. Kuo, S. Kanakaraju, L. Calhoun, P. Ho, IEEE Photonics Technology Letters **15**, 1422 (2003)

[4] T.J. Kippenberg, R. Holzwarth, S.A. Diddams, science **332**, 555 (2011)

[5] M.D. Baaske, M.R. Foreman, F. Vollmer, Nature nanotechnology **9**, 933 (2014)

[6] J. Wiersig, M. Hentschel, Physical review letters **100**, 033901 (2008)

[7] J. Kreismann, S. Sinzinger, M. Hentschel, Physical Review A **95**, 011801 (2017)

Enhanced efficiency thermo-optic phase-shifter by using multi-mode-interference device

Bharat Pant[1], *Weiwei* Zhang[1], *Denh* Tran[1], *Mehdi* Banakar[1], *Han* Du[1], *Xingzhao* Yan[1], *Callum G.* Littlejohns[1], *Graham T.* Reed[1], *David J.* Thomson[1].

[1]Optoelectronic Research Centre, University of Southampton, England

Abstract. We demonstrate a method for power reduction in thermo-optic based Mach-Zehnder interferometer (MZI) by using the multimode region of a 2x2 Multi-Mode-Interferometer (MMI) as the modulation region. The light is circulated through the same multimode region twice and therefore utilizes the already present change in temperature leading to additional phase change, and an increase in efficiency. Power saving of 29.6% compared to a conventional thermo-optic phase shifter using single mode waveguides has been experimentally demonstrated. The reported devices show minimal insertion loss penalty compared to generic devices and do not add any additional fabrication complexity. Such an approach could also be applied to higher speed devices, for example those employing free carrier effects.

1 Introduction

Thermo-optic phase-shifters have found wide application such as in switching, advance communications [1], and neural networks [2]. More complex devices including Mach-Zehnder modulator arrays, MZI [3] and ring resonator modulators [4] are now being implemented on Silicon-on-Insulator (SOI) platform due to the compatibility and cost benefits [5,6]. Various thermo-optic phase-shifter based interferometric devices are summarized in [7,8] with different approaches to minimize $P_\pi * \tau$, where P_π is the power required to for temperature change and τ is the $1/e$ limiting time constant of heat diffusion.

Tuneability in silicon MMI have been recently demonstrated [9,10]. In [11], reduction in power consumption was demonstrated by utilizing dense packaging of waveguides, arranged in double spiral geometry to increase the overlap of induced temperature distribution. In this work we present an alternative technique for power reduction by passing light through the same modulation region twice thus making more efficient use of the modulation effect.

2 Device design

In our approach, we designed a 2x2 port MMI modulator device with heaters on top of the multi-mode region acting as the modulation region. Fig. 1 shows top view of the device where multimode region length is chosen as the first self-imaging length or commonly known as cross configuration. In the work of [9] the MMI structure itself performs the switching function where the mode interference is modified by the thermally induced refractive index change, causing the light to switch between output ports. In this work, the MMI length is short and the heating more uniformly distributed over the MMI region and therefore the power outputted from each port does not significantly change during modulation.

The light enters the MMI device from top left port and exits from the bottom right port, which is then reintroduced into the device via in-port 2 (top right) and using a 180-degree bend. The final output is available from bottom left port.

Since light passes through the same modulation region twice, power applied to cause the change in refractive index of the multi-mode waveguide can be utilized for phase change on second pass as well, effectively reduce the power to 50% of the power required compared to the single pass through a multimode case.

The MMI width was chosen as 6 μm, with the ports tapering from 0.45 μm to 1.5 μm and positioned at a gap of 0.53 μm symmetrically along the width. The MMI length was simulated to be 89.60 μm for cross configuration, and the heater length and width chosen to be 80 μm and 2 μm respectively. The MMI modulator was placed in both arms of an imbalanced MZI and one of the arms was actuated to observe the change in intensity from the imbalanced MZI output. Similar devices with multi-mode region length 179.20 μm, corresponding to bar configuration, and heater length and width of 160 μm and 2 μm were also fabricated. Normalization devices with exact heater dimensions (as used in both cross and bar MMI MZI configurations) were fabricated for comparison.

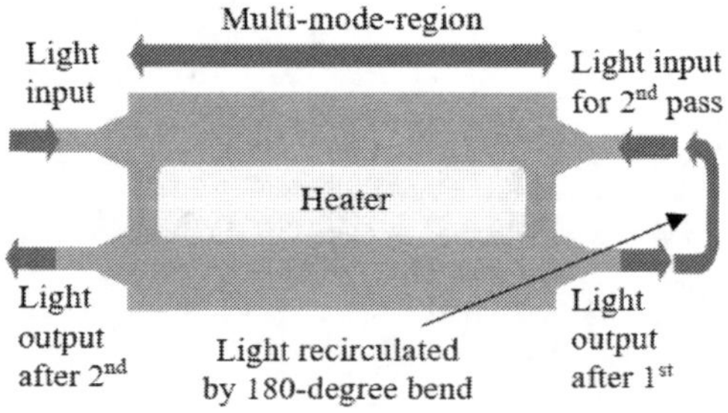

Fig. 1. Top view of the 2x2 MMI used as phase modulator.

* Corresponding author: bharat.pant@soton.ac.uk

2.1. Fabrication

Fabrication of the devices was performed within a CORNERSTONE MPW run on 220 nm SOI, within the nanofabrication cleanroom facilities at the University of Southampton. The first two steps involved defining the grating couplers and RIB waveguides with 100 nm Silicon slab by DUV lithography and ICP etching. This is followed by depositing a 1μm thick PECVD SiO_2 cladding followed by two lift-off steps for the heater filament and contacts pads. The heater filament is made of TiN and has a thickness of 150 nm and the contact pads consist of a stack of Ti (30 nm) and Au (200 nm).

2.2 Device characterisation

Light from a tuneable laser source was coupled into and out of the devices by using fibres aligned with chip grating couplers. A Keithley power supply was used to power the heaters. The MMI MZ devices recorded an additional insertion loss lower than 0.5 dB compared to the normalization MZ device.

The phase shift was calculated using the formulae

$$\Delta\varphi = \frac{2\pi\,\Delta\lambda}{FSR} \tag{1}$$

, where $\Delta\varphi$ is the change in phase compared to when no voltage is applied, $\Delta\lambda$ is the change in the spectral dip due to the applied voltage and FSR is the free spectral range of the imbalanced MZI.

Fig.2 shows the power vs phase shift comparison between normalization MZ devices and identical MMI based devices for both fabricated heater length. For devices with heater length 80 μm, normalization MZI show a value of 24.23 mW per π phase shift whereas the MMI MZI devices recorded 17.05 mW per π phase shift. This translates to a power saving of 7.18 mW or 29.6 % compared to the normalization MZI.

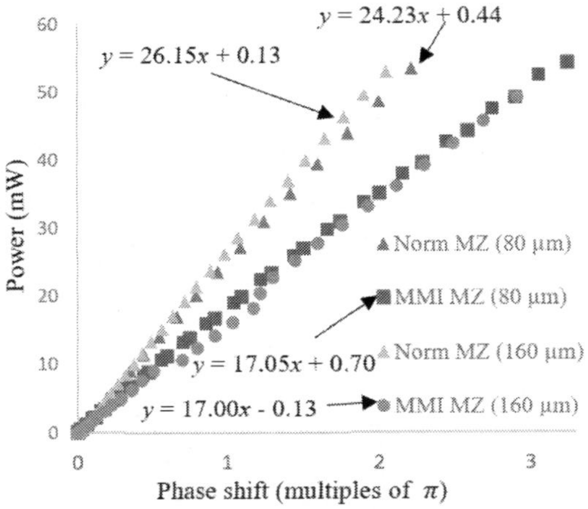

Fig. 2. Power vs phase shift for devices with heater lengths of 80 μm and 160 μm. Normalization MZI are represented by triangles, and MMI based MZI devices are represented as blue squares (80 μm) and green circles (160 μm).

Similarly, for devices with heater length of 160 μm, the normalization MZ device showed a value of 26.15 mW per π phase change and the MMI MZI devices recorded a value of 17 mW leading to and power saving of 9.15 mW or 35%.

3 Conclusion

A proof of concept for enhancing the efficiency of thermo-optic phase shifters has been demonstrated by utilizing the multimode region of a 2x2 MMI as the modulation region and traversing the light twice through it. The experimentally demonstrated devices show a power reduction of in excess of 29.6% compared to thermo-optic phase shifters using single mode waveguides.

The author would like to acknowledge CORNERSTONE (EP/L021129/1) for their contribution in fabrication. D. J. Thomson acknowledges funding from the Royal Society for his University Research Fellowship.

References

1. D. J. Thomson, C. G. Littlejohns, S. Stanković, M. Nedeljkovic, and S. A. Reynolds, in *Wiley Encyclopedia of Electrical and Electronics Engineering* (2015).
2. A. N. Tait, T. F. de Lima, E. Zhou, A. X. Wu, M. A. Nahmias, B. J. Shastri, and P. R. Prucnal, Scientific Reports **7** (1), 7430 (2017).
3. A. Samani, V. Veerasubramanian, E. El-Fiky, D. Patel, and D. V. Plant, IEEE Photonics Journal **8** (1), 1 (2016).
4. W. Bogaerts, P. De Heyn, T. Van Vaerenbergh, K. De Vos, S. Kumar Selvaraja, T. Claes, P. Dumon, P. Bienstman, D. Van Thourhout, and R. Baets, Laser & Photonics Reviews **6** (1), 47 (2012).
5. C. Doerr, Frontiers in Physics 3, 37 (2015).
6. D. Thomson, A. Zilkie, J. E. Bowers, T. Komljenovic, G. T. Reed, L. Vivien, D. Marris-Morini, E. Cassan, L. Virot, and J.-M. Fédéli, Journal of Optics **18** (7), 073003 (2016).
7. M. Jacques, A. Samani, E. El-Fiky, D. Patel, Z. Xing, and D. V. Plant, Opt. Express **27** (8), 10456 (2019).
8. G. Coppola, L. Sirleto, I. Rendina, and M. Iodice, Optical Engineering **50** (7), 071112 (2011).
9. A. Rosa, A. Gutierrez, A. Brimont, A. Griol, and P. Sanchis, Opt. Express **24**, 191 (2016).
10. R. Á, A. Brimont, A. Griol, and P. Sanchis, presented at the 11th International Conference on Group IV Photonics (GFP), 2014 (unpublished).
11. A. Densmore, S. Janz, R. Ma, J. H. Schmid, D.-X. Xu, A. Delâge, J. Lapointe, M. Vachon, and P. Cheben, Opt. Express **17** (13), 10457 (2009).

Efficient modeling techniques for plasmonic and photonic devices

Amgad Abdrabou[1] and *S. S. A.* Obayya[2,*]

[1]Department of Mathematics, City University of Hong Kong, Kowloon, Hong Kong, China
[2]Center for Photonics and Smart Materials, Zewail City of Science and Technology, Giza, 12578, Egypt

Abstract. Plasmonics plays a vital role in realizing nanophotonic devices for integrated optics due to its strong light localization into subwavelength dimensions beyond the diffraction limit. Therefore, plasmonics has a wide range of applications such as sensing, solar cells, microscopy, etc. Plasmonics modelling techniques are necessary for understanding the underlying physics of plasmonic devices. However, correct modelling of these devices is still an obstacle facing some of existing modeling techniques. In this paper, we discuss the shortcomings of the existing tools for analysing plasmonic devices, highlight their main features and suggest some powerful remedies.

1 Introduction

Nowadays, plasmonics is playing a vital role in the realizing nanophotonic devices for integrated optics, where light can be confined into subwavelength dimensions beyond the diffraction limit [1]. This featured plasmoincs in a wide range of applications such as sensing, solar cells, etc [1–3]. To understand the underlying physics of plasmonics and for design purposes, correct modelling techniques are essential tools. Unfortunately, many of the conventional techniques either fail [4], or show incurable instabilities [5, 6] due to strong discontinuities at metal-dielectric interfaces and the inherent difficulties in imposing the interface boundary conditions. Many trials have been made to overcome such problem by developing efficient and accurate tools specifically for plasmonics [6–9].

In this paper, we introduce an efficient method based on the pseudo-spectral method [10–12], for plasmonic modelling with a proper application of physical boundary conditions at interfaces of discontinuity. The pseudo-spectral method can incorporate maps which redefine the basis functions over a desired domain.

2 Mathematical Formulation

For multi-layered structures invariant along both x and z directions as shown in figure 1, the 1D Helmholtz equation for TE polarization is given by

$$\frac{\partial^2 u}{\partial y^2} + {k_0}^2 \varepsilon(y)\, u = \beta^2 u, \quad y \in (-\infty, \infty) \tag{1}$$

while for the TM case we have

$$\varepsilon(y)\frac{\partial}{\partial y}\left[\frac{1}{\varepsilon(y)}\frac{\partial \phi}{\partial y}\right] + {k_0}^2 \varepsilon(y)\, u = \beta^2 u, \quad y \in (-\infty, \infty) \tag{2}$$

*e-mail: sobayya@zewailcity.edu.eg

where $E_x = u(y)e^{i\beta z}$ for the TE case, and $H_x = u(y)e^{i\beta z}$ for the TM case, k_0 is the freespace wavenumber and β is the propagation constant.

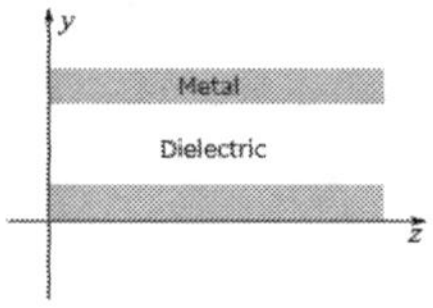

Figure 1. Schematic for a typical multi-layered plasmonic structure.

3 Multi-domain pseudo-spectra method

The eigenvalue problems (1) for TE and (2) for TM polarization can be solved numerically using the Chebyshev pseudo-spectral method [10, 19]. By referring to Fig. 1, we can assume that the points of discontinuity of $\varepsilon(y)$ are located at $0 = y_0 < y_1 < y_2 < \cdots < y_N = b$, where $b > 0$. For $s = 1, 2, \cdots, N$, the sth-subdomain is defined by $y_{s-1} < y < y_s$. The Chebyshev collocation points discretize y on each subdomain using $N_s + 1$ points as

$$\xi_{s,k} = y_{s-1} + \frac{y_s - y_{s-1}}{2}\left[1 - \cos(k\pi/N_s)\right], \quad 0 \le k \le N_s.$$

Over each subdomain s, the method approximates the first derivative of $u(y)$ by a differentiation matrix $\boldsymbol{D}^{(s)}$ [10] such that

$$\begin{bmatrix} u'(y_{s-1}^+) \\ \boldsymbol{u}_s' \\ u'(y_s^-) \end{bmatrix} \approx \boldsymbol{D}^{(s)} \begin{bmatrix} u(y_{s-1}^+) \\ \mathbf{u}_s \\ u(y_s^-) \end{bmatrix}, \tag{3}$$

where $\boldsymbol{u}_s$ is a column vector of length $N_s - 1$ whose entries are u at the interior discretization points, i.e.,

$$\boldsymbol{u}_s = \left[u(\xi_{s,1}), u(\xi_{s,2}), \cdots, u(\xi_{s,N_s-1})\right]^T, \tag{4}$$

https://doi.org/10.1051/epjconf/202023801008

the prime denotes the derivative with respect to y. The matrix $\boldsymbol{D}^{(s)}$ can be partitioned as

$$\boldsymbol{D}^{(s)} = \left[\begin{array}{c|c|c} d_1^{(s)} & \mathbf{d}_2^{(s)} & d_3^{(s)} \\ \hline \mathbf{d}_4^{(s)} & \mathbf{D}_5^{(s)} & \mathbf{d}_6^{(s)} \\ \hline d_7^{(s)} & \mathbf{d}_8^{(s)} & d_9^{(s)} \end{array}\right]. \quad (5)$$

This leads to the approximations [19]

$$u'(y_{s-1}^+) = d_1^{(s)} u(y_{s-1}) + \mathbf{d}_2^{(s)} \boldsymbol{u}_s + d_3^{(s)} u(y_s), \quad (6)$$

$$u'(y_s^-) = d_7^{(s)} u(y_{s-1}) + \mathbf{d}_8^{(s)} \boldsymbol{u}_s + d_9^{(s)} u(y_s), \quad (7)$$

for the derivatives at the points of discontinuity, and

$$\boldsymbol{u}' = \mathbf{d}_4^{(s)} u(y_{s-1}) + \mathbf{D}_5^{(s)} \boldsymbol{u}_s + \mathbf{d}_6^{(s)} u(y_s), \quad (8)$$

for the derivative at the interior points. The above expressions can be used to impose the boundary conditions efficiently at discontinuities. In addition, it is straight forward to use basis function composed with maps to handle semi-infinite domains [11].

4 Conclusion

The paper addresses an efficient pseudo-spectral method for modelling plasmonic devices. Appropriate conformal maps are used to redefine the basis functions over finite or semi-infinite subdomains, eliminating the need for PMLs. Besides, a precise application of the interface boundary condition is presented. This work can be extended to structures with 2D cross-section or 3D structures having a discontinuity along the z axis.

References

[1] E. Ozbay, Science **311**, 189 (2006)

[2] G. Veronis, S. Fan, Optics Express **15**, 1211 (2007)

[3] A.M. Heikal, F.F.K. Hussain, M.F.O. Hameed, S.S. Obayya, Journal of Lightwave Technology **33**, 2868 (2015)

[4] M. Koshiba, K. Hayata, M. Suzuki, IEEE transactions on microwave theory and techniques **32**, 587 (1984)

[5] K.P. Prokopidis, D.C. Zografopoulos, IEEE Microwave and Wireless Components Letters **24**, 659 (2014)

[6] A.M. Said, A. Heikal, N.F. Areed, S. Obayya, IEEE Photonics Journal **8**, 1 (2016)

[7] K. Stannigel, M. König, J. Niegemann, K. Busch, Optics Express **17**, 14934 (2009)

[8] W. Shin, Ph.D. thesis, Citeseer (2013)

[9] K.S. Atia, A. Heikal, S. Obayya, Optics express **23**, 22199 (2015)

[10] L.N. Trefethen, *Spectral methods in MATLAB*, Vol. 10 (Siam, 2000)

[11] A. Abdrabou, A. Heikal, S. Obayya, Optics express **24**, 10495 (2016)

[12] C.C. Huang, Optics express **18**, 23711 (2010)

[13] M. Koshiba, Y. Tsuji, Photonics Technology Letters, IEEE **8**, 1208 (1996)

[14] H. Rao, R. Scarmozzino, R. Osgood, Photonics Technology Letters, IEEE **11**, 830 (1999)

[15] H. El-Refaei, D. Yevick, I. Betty, Photonics Technology Letters, IEEE **12**, 389 (2000)

[16] P.L. Ho, Y.Y. Lu, Photonics Technology Letters, IEEE **13**, 1316 (2001)

[17] H. Zhang, J. Mu, W.P. Huang, J. Lightwave Technol. **26**, 600 (2008)

[18] K.S. Atia, A. Said, A. Heikal, S. Obayya, JOSA B **36**, 1402 (2019)

[19] D. Song, L. Yuan, Y.Y. Lu, J. Opt. Soc. Am. A **28**, 613 (2011)

Numerically solving generated Jacobian equations in freeform optical design

Lotte B. Romijn[1,*], *Martijn J.H.* Anthonissen[1,**], *Jan H.M.* ten Thije Boonkkamp[1,***], and *Wilbert L.* IJzerman[1,2,****]

[1]Eindhoven University of Technology (TU/e), Department of Mathematics and Computer Science, PO Box 513, 5600 MB Eindhoven, The Netherlands

[2]Signify Research, High Tech Campus 7, 5656 AE Eindhoven, The Netherlands

Abstract. We present an efficient numerical algorithm that can be used to solve the generalized Monge-Ampère equations for a single freeform reflector and lens surface. These equations are instances of so-called 'generated Jacobian equations' which are characterized by associated generating functions. The algorithm has a wide applicability to any optical system that can be described by a smooth generating function.

1 Introduction

We consider two optical systems: (System 1) a lens surface for a point source and far-field target intensity, and (System 2) a reflector surface for a parallel incoming beam and near-field target illuminance. For System 1 the source emits light radially outward and we consider a source emittance in spherical coordinates $f_1(\phi, \theta)$ in [lm/sr]. The target intensity distribution $g_1(\psi, \chi)$ in [lm/sr] in the far field is expressed with respect to a different set of spherical coordinates (ψ, χ) with origin the lens surface approximated as a point in space (far-field approximation). The first surface of the lens is spherical and the second surface is freeform. For System 2 we are given an emittance of the source domain $f_2(\mathrm{x})$ in [lm/m^2] and a target illuminance in the near field given by $g_2(\mathrm{y})$ in [lm/m^2], where x and y are the local Cartesian coordinates of the source and target planes, respectively. In figure 1 a schematic representation is given for System 2. Our goal is to compute the optical surfaces that define the mappings $\boldsymbol{m}_1$ and $\boldsymbol{m}_2$ transforming f_1 into g_1 and f_2 into g_2, respectively.

2 Mathematical model

In previous work on System 1 [1], we used optimal transport theory to derive a relation of the form $u_1(\boldsymbol{x}) + u_2(\boldsymbol{y}) = c(\boldsymbol{x}, \boldsymbol{y})$, where u_1 specifies the location of the freeform optical surface and $c(\boldsymbol{x}, \boldsymbol{y})$ is a logarithmic cost function in optimal transport theory with $(\boldsymbol{x}, \boldsymbol{y})$ denoting the stereographic coordinates obtained from performing coordinate transformations on the source and target domains. Subsequently, we could derive an optical map $\boldsymbol{m}$ and derive the generalized Monge-Ampère equation as

$$\det(\mathrm{D}\boldsymbol{m}) = \frac{\det(\boldsymbol{P}(\boldsymbol{x}))}{\det(\boldsymbol{C}(\boldsymbol{x}, \boldsymbol{m}(\boldsymbol{x})))} = \frac{f_1(\boldsymbol{x})\,(1 + |\boldsymbol{m}(\boldsymbol{x})|^2)^2}{g_1(\boldsymbol{m}(\boldsymbol{x}))\,(1 + |\boldsymbol{x}|^2)^2},$$

where $\mathrm{D}\boldsymbol{m}$ is the Jacobi matrix of $\boldsymbol{m}$, $\boldsymbol{P}$ is a symmetric positive-definite (SPD) matrix and $\boldsymbol{C} = \mathrm{D}_{\boldsymbol{xy}}c = (c_{x_i y_j})$ is the matrix of mixed second-order partial derivatives with respect to the stereographic coordinates of the source and target.

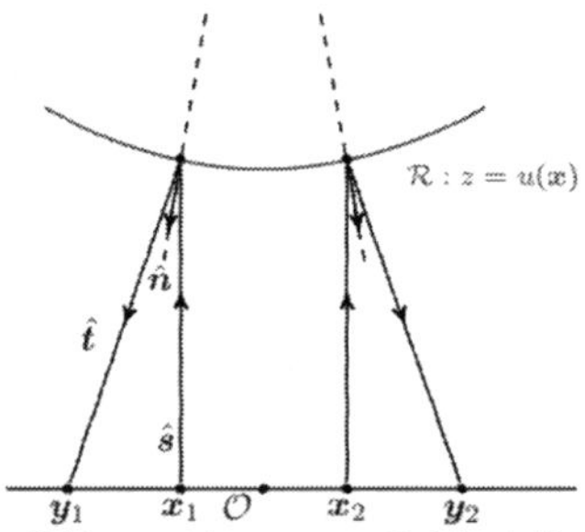

Figure 1: Schematic representation of a parallel beam of light rays, reflected by the reflector surface onto a target in the near field.

However, for some optical systems such as System 2 a relation of the form $u_1(\boldsymbol{x}) + u_2(\boldsymbol{y}) = c(\boldsymbol{x}, \boldsymbol{y})$ does not exist in any coordinate system. In this paper, we generalize the cost-function formulation in optimal transport theory to a generating-function formulation by defining $u(\boldsymbol{x}) = G(\boldsymbol{x}, \boldsymbol{y}, z(\boldsymbol{y}))$ and $z(\boldsymbol{y}) = H(\boldsymbol{x}, \boldsymbol{y}, u(\boldsymbol{x}))$, where u is the unknown location of the optical surface, G and H are smooth functions and $G(\boldsymbol{x}, \boldsymbol{y}, \cdot)$ and $H(\boldsymbol{x}, \boldsymbol{y}, \cdot)$ are each other's inverses. For both systems, we can reformulate the generalized Monge-Ampère equation as a generated Jaco-

*e-mail: l.b.romijn@tue.nl
**e-mail: m.j.h.anthonissen@tue.nl
***e-mail: j.h.m.tenthijeboonkkamp@tue.nl
****e-mail: wilbert.ijzerman@signify.com

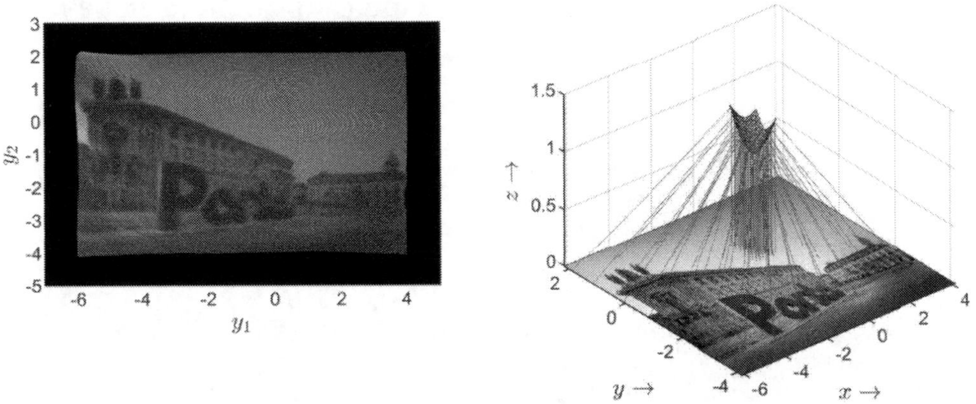

Figure 2: (Left) Ray traced image of Porto projected on a screen in the near field. (Right) Cartoon of a single freeform reflector that converts a small parallel beam (in red) into an output beam (in blue) that produces a picture of Porto on a screen in the near field. Only a small subset of the total rays traced (approximately 100.000) has been plotted.

bian equation [2]

$$\det(\mathrm{D}\boldsymbol{m}) = \frac{\det(\boldsymbol{P}(\boldsymbol{x}))}{\det(\boldsymbol{C}(\boldsymbol{x}, \boldsymbol{m}(\boldsymbol{x}), u(\boldsymbol{x})))} = \frac{F(\boldsymbol{x})}{G(\boldsymbol{m}(\boldsymbol{x}))},$$

where $\boldsymbol{P}$ is a symmetric positive-definite (SPD) matrix and we redefine $\boldsymbol{C}$ as $\boldsymbol{C} = \mathrm{D}_{\boldsymbol{x}\boldsymbol{y}}H(\boldsymbol{x}, \boldsymbol{m}(\boldsymbol{x}), u(\boldsymbol{x})) = (H_{x_i y_j})$. Note the added dependency of $\boldsymbol{C}$ on the surface u. For System 1, we consider the stereographic coordinates $(\boldsymbol{x}, \boldsymbol{y})$ and have $F(\boldsymbol{x}) = f_1(\boldsymbol{x})/(1 + |\boldsymbol{x}|^2)^2$ and $G(\boldsymbol{m}(\boldsymbol{x})) = g_1(\boldsymbol{m}(\boldsymbol{x}))/(1 + |\boldsymbol{m}(\boldsymbol{x})|^2)^2$. For System 2, we take Cartesian coordinates $\boldsymbol{x} = \mathrm{x}$ and $\boldsymbol{y} = \mathrm{y}$, and derive $F(\boldsymbol{x}) = f_2(\mathrm{x})$ and $G(\boldsymbol{m}(\boldsymbol{x})) = g_2(\boldsymbol{m}(\mathrm{x}))$.

For System 1, we use Hamilton's angular characteristic and derive $H(\boldsymbol{x}, \boldsymbol{y}, u(x)) = u(x)\left(n - 1 + \frac{2|\boldsymbol{x}-\boldsymbol{y}|^2}{(1+|\boldsymbol{x}|^2)(1+|\boldsymbol{y}|^2)}\right)$. Inverting this equation gives the generating function

$$G(\boldsymbol{x}, \boldsymbol{y}, z(\boldsymbol{y})) = z(\boldsymbol{y}) \left(n - 1 + \frac{2\,|\boldsymbol{x} - \boldsymbol{y}|^2}{(1 + |\boldsymbol{x}|^2)\,(1 + |\boldsymbol{y}|^2)}\right)^{-1}.$$

For System 2, the function H is Hamilton's point characteristic $H(\boldsymbol{x}, \boldsymbol{y}, u(\boldsymbol{x})) = u(\boldsymbol{x}) + \sqrt{|\boldsymbol{y} - \boldsymbol{x}|^2 + u(\boldsymbol{x})^2}$ which is equal to the optical path length from a point $\boldsymbol{x}$ on the source domain to a point $\boldsymbol{y}$ on the target domain, see figure 1. Inverting this equation gives the generating function $G(\boldsymbol{x}, \boldsymbol{y}, z(\boldsymbol{y})) = u$ with $z(\boldsymbol{y}) = 1/H$ as derived in [2, 3]

$$G(\boldsymbol{x}, \boldsymbol{y}, z) = \frac{1}{2\,z} - \frac{z}{2}|\boldsymbol{x} - \boldsymbol{y}|^2.$$

3 Numerical method

Using a generalized least-squares approach [4] we iteratively update the mapping $\boldsymbol{m}$. We write the generalized Monge-Ampère equation as the matrix equation $\boldsymbol{C}\mathrm{D}\boldsymbol{m}(\boldsymbol{x}) = \boldsymbol{P}(\boldsymbol{x})$, where $\boldsymbol{P}(\boldsymbol{x})$ satisfies $\det(\boldsymbol{P}(\boldsymbol{x})) = F(\boldsymbol{x})/G(\boldsymbol{m}(\boldsymbol{x}))$, cf. [5]. An iterative procedure is used to find the mapping, which involves an efficient procedure to find the numerical solution of a constrained minimization problem for each grid point to compute $\boldsymbol{P}$ and the solution of a linear elliptic boundary value problem to compute $\boldsymbol{m}$. In order to update the matrix $\boldsymbol{C}$ during the iterative procedure, the location of the optical surface is calculated from the mapping also in a least-squares sense.

The numerical method can be used to compute reflector and lens surfaces for various examples. For instance, we can compute the reflector surface that transforms light of a parallel beam into a near-field picture of Porto, as shown in figure 2. For System 1 we will compare the results of the cost-function approach previously presented in [1] to the new generating-function approach.

By using a generating-function approach we are able to extend the applicability of the least-squares procedure originally presented in [5] to any optical system that can be described by a smooth generating function.

References

[1] L. Romijn, J. ten Thije Boonkkamp, W. IJzerman, JOSA A **36**, 1926 (2019)

[2] N.S. Trudinger, Discrete Contin. Dyn. Syst. **34**, 1663 (2012)

[3] F. Jiang, N. Trudinger, N.S. Arch. Ration. Mech. Anal **229**, 547 (2018)

[4] L.B. Romijn, J.H. ten Thije Boonkkamp, W.L. IJzerman, J. Comput. Phys. **408**, 109283 (2020)

[5] C. Prins, Ph.D. thesis, Eindhoven University of Technology, Eindhoven (2014)

Symmetric spatio-temporal focusing of ultrashort pulses

A. Treffer[1], **D. Kühn**[2], **F. Wyrowski**[2], **R. Grunwald**[1]

[1]Max Born Institute for Nonlinear Optics and Short-Pulse Spectroscopy, 12489 Berlin, Germany
[2]Institute of Applied Physics, Friedrich-Schiller University, Albert-Einstein-Straße 15, 07745 Jena, Germany
email: treffer@mbi-berlin.de

Summary

Simultaneous spatio-temporal focusing of few-cycle pulses is usually performed by stretcher-compressor configurations causing pulse-front tilt. By shaping and superimposing two reciprocal sub-beams, opposite tilts can compensate each other. Pulse propagation for asymmetric and symmetric focusing of near infrared 5-fs-pulses is studied by numerical simulations and proof-of-principle experiments.

Introduction

The propagation of ultrashort pulses in optical media is often limited by nonlinear effects and damage because of high peak intensities. These constraints can be circumvented by stretching and compressing pulses by dispersive components. A particular stretcher-compressor concept is known as simultaneous spatio-temporal focusing (SSTF) [1-2]. The main goal of SSTF is to provide a focus with minimum pulse duration at the position of interest. Theoretical studies demonstrated the capability of the approach [3]. Simultaneous localization in space and time, along with robust temporal transfer, enables to investigate remote nonlinear processes behind dispersive elements even with low-pulse-energy Ti:sapphire oscillators. Therefore, SSTF is of increasing interest for applications in materials processing [4], microscopy [5], plasma physics [6], or nonlinear optics. Previously, a single-channel SSTF system was studied by numerical simulations [3]. It was shown that dispersion causes travel time differences that lead to pulse-front tilt (PFT) with characteristic signatures in spatial, spectral and temporal domain. Here we report on an alternative approach based on two SSTF channels with reciprocal grating configuration that allows for compensating PFT and thus to form symmetric pulse fronts. Results of numerical studies as well as first proof-of-principle experiments are presented.

Symmetric stretcher-compressor approach

The schematic drawings in Fig. 1a and Fig. 1b show a conventional single-path two-grating setup for SSTF and a two-path system, respectively. The reciprocal grating arrangement (Fig. 1b) is capable to symmetrize the focus and thus to partially compensate for dispersion-induced pulse-front tilt (PFT). If the length of both channels is properly adjusted, the phase difference in the center is minimized. The degree of compensation of PFT depends on the superposition angle and the temporal coherence of the pulse. Physical optics simulations with VirtualLab Fusion software were used to calculate the pulse-front tilt of the traditional and the two-path system. Figs. 2a-f show the field amplitude at the detector plane of the system depicted in Fig. 1a, the same system mirrored at its symmetry axis, and the superposition of both, respectively (Fig 1b) with and without dispersion. Predictions were confirmed by proof-of-principle experiments with a Ti:sapphire laser oscillator

(Laser Quantum, central wavelength 700 nm, minimum pulse duration 9 fs, pre-chirped) stretched by a prism pair (1200 l/mm) and lens focusing (f = 75.6 mm). Pulse characterization was performed with a GRENOUILLE setup. Currently, experimental conditions are confined to reduce parasitic chirp effects.

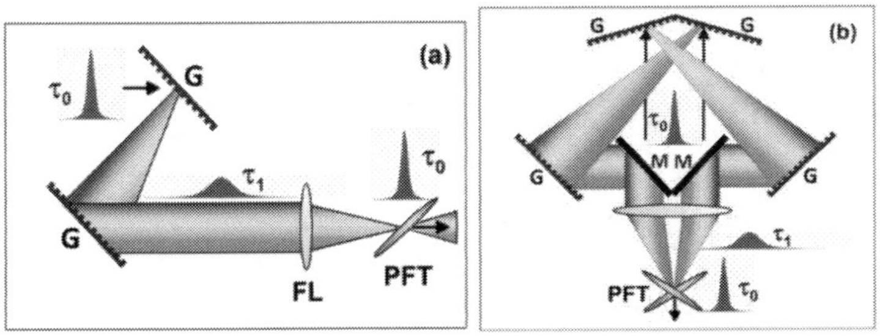

Fig. 1. Configurations for spatio-temporal focusing: (a) double-grating setup with asymmetric focusing, (b) 4-grating setup with symmetric focusing (PFT = pulse-front tilt, G = grating, FL = ocusing lens, τ_0, τ_1 = pulse durations with and without compression, M = mirror, schematically).

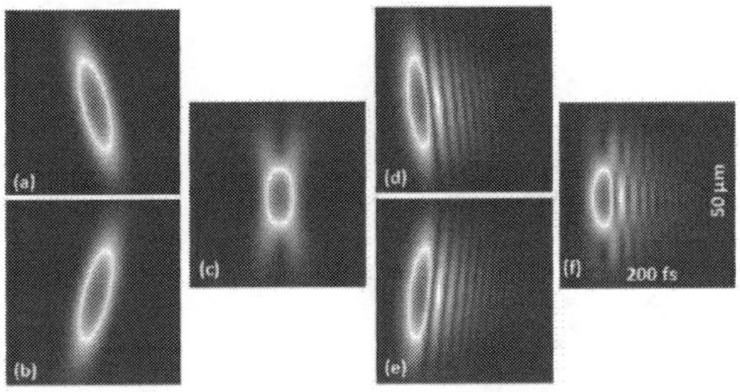

Fig. 2. Field distribution of single and dual channel configuration in the focal region simulated for a Gaussian pulse (center wavelength 619 nm, FWHM bandwidth 40 nm, initial FWHM pulse duration 14.2 fs, GDD = 9600 fs2); perfect compensation: (a), (b) left and right arm, (c) superposition of both; with realistic chirp: (d), (e), (f), respectively (normalized). The grating orientation determines the direction of pulse front tilt. Therefore, the superposition creates a cross-shaped field (see c,f).

Conclusions

A double-stretcher-compressor scheme for simultaneous spatio-temporal focusing was proposed to symmetrize the pulse front. Proof-of-principle experiments confirm the approach. To enable a tuning of the superposition angle, adaptive configurations on the basis of MEMS components will be studied in a next step.

The work was supported by DFG (Gr 1782/16-1).

References

[1] G. Zhu, J. v. Howe, M. Durst et al., *Opt. Express* **13**, 2153–2159 (2005)
[2] D. Oron, E. Tal, and Y. Silberberg, *Opt. Express* **13**, 1468–1476 (2005)
[3] S. Zhang, F. Wyrowski, R. Kammel and S. Nolte, *Proc. SPIE* **8972**, 89721B (2014)
[4] B. Sun, P. S. Salter, C. Roider et al., *Light: Sci. & Appl.* **7**, 17117 (2018)
[5] G. Song, A. Nakamura, K. Hirosawa et al., *Rev. Sci. Instrum.* **86**, 083701 (2015)
[6] Z. Wang, B. Teng, G. Li et al., *Opt. Lett.* **40**, 5726–5729 (2015)

Modelling Light Scattering in the Context of Freeform Optical Design

Simon Kronberg[1,*], *Martijn J.H.* Anthonissen[1,**] *Jan H.M.* ten Thije Boonkkamp[1,***], and *Wilbert L.* IJzerman[1,2,****]

[1]Eindhoven University of Technology (TU/e), Department of Mathematics and Computer Science, PO Box 513, 5600 MB Eindhoven, The Netherlands
[2]Signify Research, High Tech Campus 7, 5656 AE Eindhoven, The Netherlands

Abstract. We present an initial attempt to model scattering in freeform optical design. Scattering is modelled as a convolution of the unperturbed specular distribution and a spreading function. Deconvolution is used to recover the equivalent specular distribution, for which design procedures are well established.

1 Introduction

Most freeform optical design relies on geometrical optics, where light is represented by rays which travel in straight lines in homogeneous media and adhere to the law of reflection and Snell's law of refraction at optical interfaces. This approach, whilst mathematically appealing, neglects light scattering, which could negatively affect the performance of the optical element. We present an initial attempt to include these effects via a convolution of an unperturbed specular distribution and a spreading function. The design problem would thus entail prescribing a desired distribution, deconvolving it with the spreading function to get an "equivalent specular problem" and using previously developed tools in the rest of the design process.

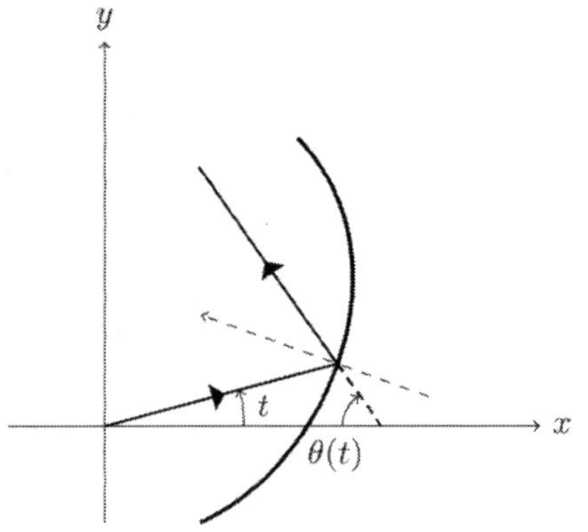

Figure 1. Illustration of the law of reflection.

2 Mathematical model

Our model is currently restricted to two-dimensional problems, but it may readily be generalised to three dimensions. We propose to model the scattering phenomena as a two-step process, whereby the incident ray first experiences a specular reflection, followed by a scattering event. The specular reflection may be described by a map $m\colon t \to \theta$, [1], where t and $\theta(t)$ are the directions of the incident and reflected ray, see Fig. 1. Similarly, the inverse specular mapping reads $t = m^{-1}(\theta)$.

Fix t, such that the energy balance in the associated specular direction $\theta = \theta(t)$ becomes

$$\tilde{\mathcal{G}}(\theta) = \mathcal{I}(m^{-1}(\theta))\,\frac{\mathrm{d}m^{-1}}{\mathrm{d}\theta}(\theta), \tag{1}$$

where $\tilde{\mathcal{G}}(\theta)$ is the specular intensity distribution as governed by the mapping m and the initial source distribution $\mathcal{I}(t)$. In reaching (1), we utilised the inverse specular map and its first derivative whilst considering how a small pencil of rays centred about t gets mapped by m.

The scattering event essentially creates a second map

$$s_\alpha\colon \theta \to \beta, \quad \beta \equiv \theta + \alpha,$$

where α is a stochastic variable representing the angular deviation from the specular direction. By again fixing t for the case with scattering included, and considering the integral over the source distribution $\mathcal{I}(t)$, we get

$$\int_{\mathcal{T}} \mathcal{I}(t)\,\mathrm{d}t = \int_{\mathcal{B}}\int_{\mathcal{A}} p(\alpha)\,\mathcal{I}(m^{-1}(\beta-\alpha))\,\frac{\mathrm{d}m^{-1}}{\mathrm{d}\theta}(\beta-\alpha)\,\mathrm{d}\alpha\,\mathrm{d}\beta, \tag{2}$$

where the spreading function $p(\alpha)$ is a probability density function, and the limits of the integrals are taken to be the physically relevant directions.

Next, we state the global energy balance as

$$\int_{\mathcal{T}} \mathcal{I}(t)\,\mathrm{d}t = \int_{\mathcal{B}} \mathcal{G}(\beta)\,\mathrm{d}\beta, \tag{3}$$

*e-mail: s.kronberg@tue.nl
**e-mail: m.j.h.anthonissen@tue.nl
***e-mail: j.h.m.tenthijeboonkkamp@tue.nl
****e-mail: wilbert.ijzerman@signify.com

 https://doi.org/10.1051/epjconf/202023802003

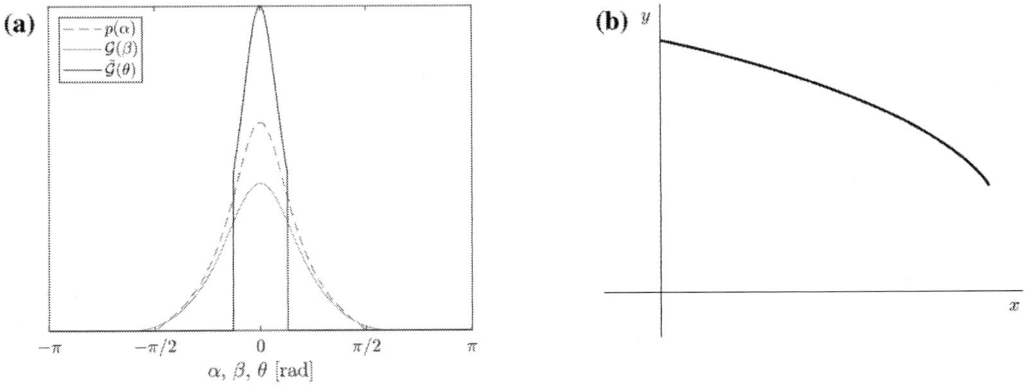

Figure 2. Two-dimensional example of reflector design using the proposed method; **(a)** the given intensity distribution $\mathcal{G}(\beta)$ and the spreading density function $p(\alpha)$ are deconvolved to get $\tilde{\mathcal{G}}(\theta)$; **(b)** the corresponding specular reflector yielding $\tilde{\mathcal{G}}(\theta)$ in the far field when $\mathcal{I}(t)$ is taken to be unity for all relevant values of t.

where $\mathcal{G}(\beta)$ is a prescribed intensity distribution after reflection and scattering. Comparing (2) and (3), it is clear that

$$\begin{aligned}\mathcal{G}(\beta) &= \int_{\mathcal{A}} p(\alpha)\,\mathcal{I}(m^{-1}(\beta-\alpha))\,\frac{\mathrm{d}m^{-1}}{\mathrm{d}\theta}(\beta-\alpha)\,\mathrm{d}\alpha \\ &= \int_{\mathcal{A}} p(\alpha)\,\tilde{\mathcal{G}}(\beta-\alpha)\,\mathrm{d}\alpha, \qquad (4)\end{aligned}$$

where

$$\tilde{\mathcal{G}}(\beta-\alpha) := \mathcal{I}(m^{-1}(\beta-\alpha))\,\frac{\mathrm{d}m^{-1}}{\mathrm{d}\theta}(\beta-\alpha)$$

is a specular distribution without the effect of scattering. To reach this expression, recall that $\beta-\alpha=\theta$, and compare the above expression to that in (1). The two are clearly identical.

3 Solution algorithm

The proposed solution algorithm may be outlined as follows. Given $\mathcal{I}(t)$, $\mathcal{G}(\beta)$ and $p(\alpha)$:

1) Compute the deconvolution from (4) to get $\tilde{\mathcal{G}}(\theta)$.
2) Design the specular reflector with target distribution $\tilde{\mathcal{G}}(\theta)$ using contemporary tools and techniques, such a Monge-Ampère solver [2].

The critical step for the present discussion naturally becomes computing the deconvolution, where physical constraints such as positivity are crucial.

4 Two-dimensional example

An example of the proposed method is outlined in Fig. 2, where the equivalent specular intensity distribution $\tilde{\mathcal{G}}(\theta)$ is found by deconvolving the spreading function $p(\alpha)$ from the desired intensity distribution $\mathcal{G}(\beta)$. In this case, the deconvolution can be carried out via Matlab's \ operator, but this is not generally the case as the matrix representation of $p(\alpha)$ often is nearly singular [3]. Then a simple design algorithm from Maes [1] based on conservation of energy is applied to obtain the specular reflectors in **(b)**.

In particular, the desired intensity distribution $\mathcal{G}(\beta)$ is shown in Fig. 2 **(a)**, together with a normalised bidirectional transmission distribution function measured from a 1.5 mm thick polycarbonate sample with 0.3% (weight) silicon scatterer particles acting as $p(\alpha)$. The equivalent specular distribution $\tilde{G}(\theta) = 4\pi/(\pi^2 + 64\,\theta^2)$ has compact support $-\pi/8 \le \theta \le \pi/8$. $\mathcal{I}(t)$ is taken as unity for $\pi/4 \le t \le \pi/2$ and zero for all other t.

References

[1] M. Maes, *Mathematical Methods for Reflector Design* (CIP-gegevens Koninklijke Bibliotheek, Den Haag, 1997), ISBN 90-74445-36-5
[2] L.B. Romijn, J.H.M. ten Thije Boonkkamp, W.L. IJzerman, J. Opt. Soc. Am. A **36**, 1926 (2019)
[3] P. Jansson, *Deconvolution of Images and Spectra* (Academic Press, 1997), ISBN 9780123802224

Phase-shift mask fabrication at micrometric scale by ion-exchange in glass for astronomical wavefront sensors

Xesús Prieto-Blanco[1,*], *Carlos* Montero-Orille[1], *Héctor* González-Núñez[1], *Vicente* Moreno[1], *Manuel P.* Cagigal[2], and *Jesús* Liñares[1]

[1]Quantum Materials and Photonics Group, Department of Applied Physics, Faculty of Physics / Faculty of Optics and Optometry, University of Santiago de Compostela, Campus Vida s/n, E-15782, Santiago de Compostela, Galicia, Spain.
[2]Department of Applied Physics, University of Cantabria, 39005 Santander, Spain.

Abstract. Photolithography combined with ion-exchange in glass is a well-known technology that can be applied to develop many different optical devices. In this work, we present the complete procedure to generate small circular phase-shift masks with diameters of only a few microns and high control in the phase change produced. It is a strategic element in applications such as optical astronomy.

1 Introduction

Binary phase-shift masks elements have multiple applications in different fields and present remarkable advantages with respect to diffractive amplitude elements. They can be used in complex optical tasks (sometimes combined with binary amplitude masks) such as apodization, super-resolution imaging, coronography or (de)multiplexing in optical communications among others [1–3]. They are also at the core of a new quaternary adaptive optical system for astronomical telescopes based on a point diffraction interferometer (PDI) [4]. Its diffracting element is a $\pi/2$-phase disk a few micrometers in diameter. Phase-shift masks usually generate the phase difference by changing the optical thickness of a transparent medium. However, it is also possible generate a phase change through the modification of the refractive index in different areas of the medium. The technique of ion-exchange in glass can be used to perform this task [5, 6]. In this work, a phase-shift element to be used in a PDI is designed, fabricated by silver-sodium ion-exchange in glass and characterized by the m-lines method for guided optical modes and by differential interference contrast (DIC) microscopy. The rest of the work is presented as follows: Section 2 outlines the materials and methods used; experimental results are presented and analysed in Section 3; and finally Section 4 summarizes the conclusions of this work.

2 Materials and Methods

Soda-lime microscope slide glasses of 1 mm thickness (*Gold Seal*^{TM}) are used as a material substrate, since they present enough regularity in their surface and thickness. The glasses are thoroughly cleaned and aluminized (about 0.25 μm thick layer) in a vacuum chamber (*Balzers BAE 250 Coating System*). Next, the positive photoresist *MICROPOSIT*^{TM}*S1813 G2* is deposited above the aluminium layer. A binary mask printed on paper and transferred to the photoresist by using a Schneider Xenon-Sapphire 4.5/90 lens, which is optimized for a 14:1 reduction. The object is illuminated with four 125 W Hg lamps for several hours. Then, the sample is developed, etched in acid and cleaned in acetone. The result is an intermediate mask that acts as an object for an old photomicrographic lens Zeiss S-Planar 1.1/68 that performs a second reduction step (1:5). We built a transmission lighting system that consists of a 75 W Hg lamp, a diaphragm, an interferential filter (436 nm) and lenses that conjugate the diaphragm in the input pupil of the Zeiss lens. The final aluminium mask on glass is obtained similarly to the intermediate one but the exposure time is reduced to a few minutes.

The ion-exchange process takes place in molten salt bath inside a oven with stabilized temperature. We choose the eutectic mixture $Ag_{0.05}Na_{0.475}K_{0.475}NO_3$ at 300 °C in order to obtain a slow penetration and a high index change in the unmasked region. A carefully calibration of the diffusion time is required to adjust the exact phase change. After the ion-exchange process, the aluminium mask is removed although some regions are left as references to locate the phase-shift mask.

The exchanged layer normally also forms a several mode waveguide. We measured its effective indexes with a *Metricon 2010/M* by the prism coupling technique. From them, the gradient index profile can be recovered and integrated to obtain the cumulative phase difference between masked and unmasked regions when the element is perpendicularly illuminated [3]. This procedure provides the phase-shift at 632.8 nm, but we can extend it to a different wavelength taking into account the exchanged-glass dispersion [6]. Finally, the spatial characteristics of the mask

*e-mail: xesus.prieto.blanco@usc.es

https://doi.org/10.1051/epjconf/202023802004

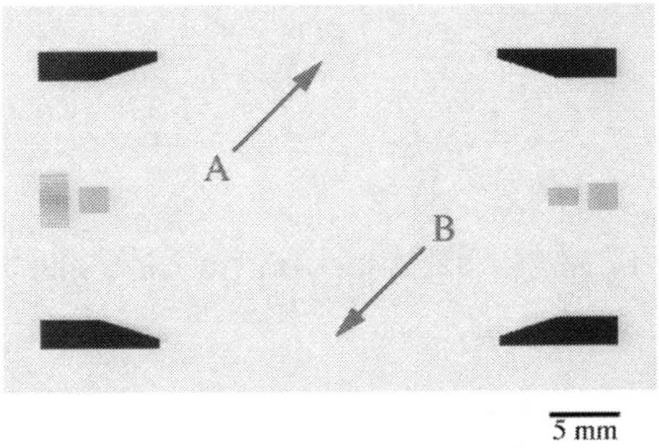

Figure 1. Picture of the intermediate mask made of aluminium on a 75 x 50 mm glass.

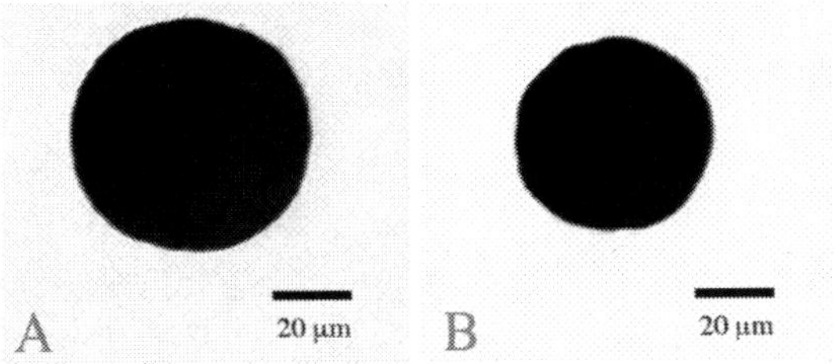

Figure 2. Images of the circular dots of the intermediate mask shown in Figure 1.

were investigated with a commercial high-resolution DIC microscope (*Nikon Eclipse Ni-U*).

3 Experimental Results

Next, we show the performance of the technique in the fabrication of a circular binary mask with a diameter in the range 5–10 μm and a phase-shift difference of $\pi/2$ rad for 632.8 nm. We must stress that this is a strategic element designed for astronomy applications (such as shown in ref. [4]). The phase pattern can be made from any of the following masks: an aluminium layer on the entire surface except for a hole, or only an aluminium dot on a clean surface. We use the second one to avoid phase defects arising from digs, scratches, pores. . . in the aluminium layer.

Figures 1 and 2 show the intermediate mask which includes some reference patterns to help the localization of the two dots defining the desired binary mask. Figure 3 shows the final phase elements as viewed through a DIC microscope where the brightness is related to the partial derivative of the phase along a diagonal. The inner diameters of these masks are (A) 6.1 μm and (B) 4.1 μm. The phase shift achieved is 0.47π rad after a diffusion time of 10 min 45 s. Therefore, the phase-shift error is less than $\lambda/66$ for λ=632.8 nm. We could even increase the accuracy with a better diffusion time adjustment.

As shown in the pictures, the mask is not purely binary. The side diffusion generates a ramp of approximate 2 μm for this phase shift. Normally, this does not affect for certain applications although it imposes a limit on the minimum spatial size of the mask that can be fabricated.

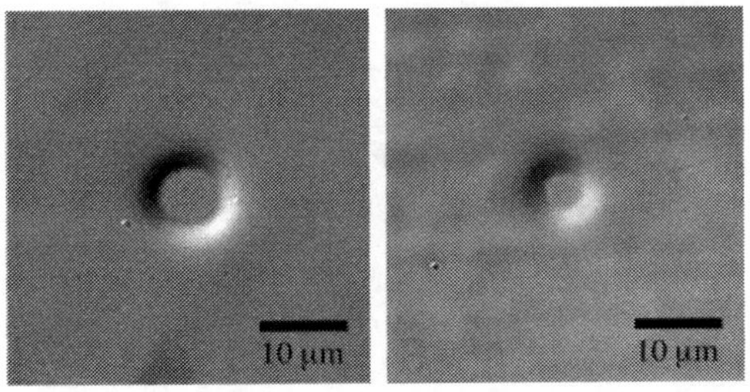

Figure 3. Pictures of the final two circular phase-shift masks obtained in the fabricated element. The images were obtained with a DIC microscope in a transmission configuration with a Nikon TU Plan Fluor 100×/0.9 objective.

4 Conclusions

High quality micrometric-size phase-shift masks can be produced with the technique described in this work. These elements can be used in numerous applications where the need for a very accurate phase-shift is required. These phase-shift masks made by ion-exchange also presents several advantages over other techniques based on thickness modifications, as for example, a more robust element because it is inside the glass substrate. The main limitation of this technique is the side diffusion which prevents to achieve sub-micrometric patterns.

Funding. This research was funded by Xunta de Galicia, Consellería de Educación, Universidades e FP, Grant GRC N. ED431C2018/11; Ministerio de Economía y Competitividad, Central Government of Spain, Grant N. AYA2016-78773-C2-2-P and the Fondo Europeo de Desenvolvemento Rexional (FEDER).

References

[1] X. Gao, F. Gan, W. Xu, Opt. Laser Tech. **39**, 1074-1080 (2007)

[2] F. Patru, P. Baudoz, R. Galicher, Q. Cao, K. Wang, L. Xing, F. Boussaha, J. Firminy, M. Bonafous, Opt. Express **26**, 10007-10012 (2018)

[3] J. Liñares, C. Montero-Orille, V. Moreno, D. Mouriz, M.C. Nistal, X. Prieto-Blanco, Appl. Opt. **56**, 7099-7106 (2017)

[4] M.P. Cagigal, A. Fuentes, M.A. Cagigas, P.J. Valle, X. Prieto-Blanco, V.F. Canales, Opt. Express **27**, 24524-24537 (2019)

[5] J. Liñares, X. Prieto-Blanco, V. Moreno, C. Montero-Orille, D. Mouriz, M.C. Nistal, D. Barral, Opt. Express **25**, 10925-10938 (2017)

[6] X. Prieto-Blanco, C. Montero-Orille, V. Moreno, E.F. Mateo, J. Liñares, Appl. Opt. **54**, 3308-3314 (2015)

An energy conservative *hp*-scheme for light propagation using Liouville's equation for geometrical optics

Robert A.M. van Gestel[1,*], *Martijn J.H.* Anthonissen[1], *Jan H.M.* ten Thije Boonkkamp[1], and *Wilbert L.* IJzerman[1,2]

[1]Eindhoven University of Technology, PO Box 513, 5600 MB Eindhoven, the Netherlands
[2]Signify - High Tech Campus 7, 5656 AE, Eindhoven, the Netherlands

Abstract. In this contribution an alternative method to standard forward ray-tracing is briefly outlined. The method is based on a phase-space description of light propagating through an optical system. The propagation of light rays are governed by Hamilton's equations. Conservation of energy and étendue for a beam of light, allow us to derive a Liouville's equation for the energy propagation through an optical system. Liouville's equation is solved numerically using an *hp*-adaptive scheme, which for a smooth refractive index field is energy conservative. A proper treatment of optical interfaces ensures that the scheme is energy conservative over the full domain.

1 Introduction

Standard forward ray-tracing methods for geometrical optics, based on Monte Carlo methods with bin counting, are in general not energy conservative and exhibit a slow convergence. A recently new approach is based on a phase-space description of light, where each light ray propagating through an optical system evolves according to an optical Hamiltonian [1, 2]. The 'time' parameter for Hamilton's equations can be the arc length of the ray, or in our case it will be the distance along a chosen optical axis, denoted by z. A light ray in phase-space is described by position $\boldsymbol{q} \in \mathbb{R}^d$ and momentum $\boldsymbol{p} \in \mathbb{R}^d$ coordinates, where the momentum is related to the direction of a light ray. For two-dimensional optics $d = 1$, while for three-dimensional optics $d = 2$. Furthermore, the momentum is restricted to Descartes' disk, by $|\boldsymbol{p}| \leq n(z, \boldsymbol{q})$ where n is the refractive index field.

The Hamiltonian description of light allows us to apply Liouville's theorem, where we now consider a differential element of phase-space. In the context of optics such elements in phase-space are called étendue. It describes a beam of light instead of individual light rays. Moreover, in a lossless optical system, energy must be conserved, implying that the luminous flux of a beam must be conserved when moving through an optical system. Furthermore étendue is conserved. Étendue $\mathrm{d}\mathcal{U}$ and luminous flux $\mathrm{d}\Phi$ are related by another invariant, known as basic-luminance ρ which is defined by

$$\rho = \frac{\mathrm{d}\Phi}{\mathrm{d}\mathcal{U}}. \tag{1}$$

The basic-luminance is measured in lumen per projected area per steradian, and is a function of the distance along the optical axis z, the position coordinates $\boldsymbol{q}$ and momentum coordinates $\boldsymbol{p}$, i.e., $\rho = \rho(z, \boldsymbol{q}, \boldsymbol{p})$. The invariance of ρ implies

$$\rho(z + \Delta z, \boldsymbol{q}(z + \Delta z), \boldsymbol{p}(z + \Delta z)) = \rho(z, \boldsymbol{q}(z), \boldsymbol{p}(z)). \tag{2}$$

This allows us to derive a Liouville's equation for the basic-luminance, which reads

$$\frac{\partial \rho}{\partial z} + \frac{\partial h}{\partial \boldsymbol{p}} \cdot \frac{\partial \rho}{\partial \boldsymbol{q}} - \frac{\partial h}{\partial \boldsymbol{q}} \cdot \frac{\partial \rho}{\partial \boldsymbol{p}} = 0, \tag{3}$$

where $h = h(z, \boldsymbol{q}, \boldsymbol{p}) = -\sqrt{n(z, \boldsymbol{q})^2 - |\boldsymbol{p}|^2}$ is the optical Hamiltonian. Liouville's equation holds where h is sufficiently smooth. Near an optical interface h will be discontinuous and we have to apply Snell's law or the law of specular reflection [2]. A proper treatment of the optical interfaces is needed, since Snell's law and the law of specular reflection cause for non-local boundary conditions in phase-space, at which the momentum $\boldsymbol{p}$ is discontinuous.

2 Numerical method and results

As a proof of concept we restrict ourselves to two-dimensional optics ($d = 1$). In general, Liouville's equation cannot be solved for complex systems, hence a numerical method is required. For spatial (phase-space) discretisation we apply the discontinuous Galerkin-spectral element method (DG-SEM) by Kopriva [3], while for z-integration we use a low-storage fourth order explicit Runge-Kutta [4] method. The spatial discretisation is *hp*-adaptive, meaning we can increase the accuracy either by using more elements via *h*-refinement, or by increasing the order of accuracy via *p*-refinement. For light moving towards an optical interface, there is no condition on ρ, whilst for light moving away from an optical interface

*e-mail: r.a.m.v.gestel@tue.nl

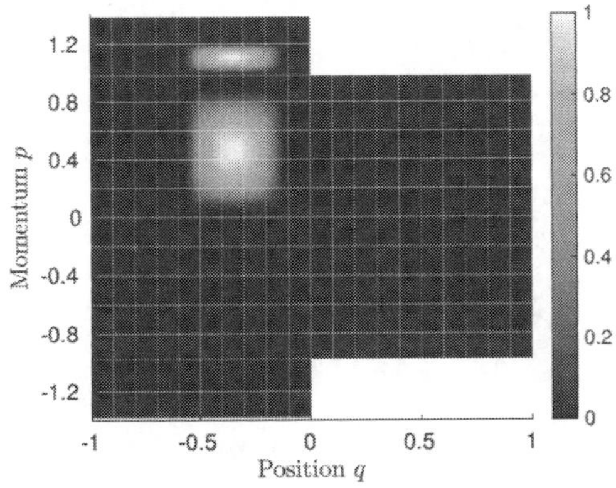

Figure 1. Initial basic-luminance distribution $\rho(z, q, p)$ at $z = 0$.

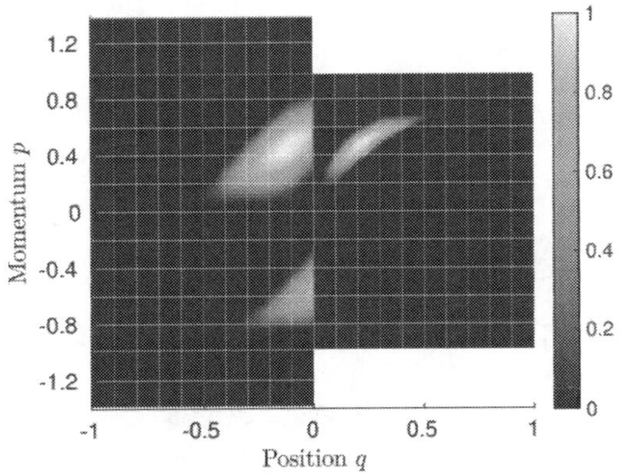

Figure 2. Basic-luminance distribution $\rho(z, q, p)$ at $z = 0.7$.

we have a Dirichlet boundary condition where the value is to be determined from the values on the other side of the interface, according to Snell's law or the law of specular reflection. However, simply taking the values from the other side does not suffice. A special approach related to the treatment of non-conforming geometries [5] is required, with an additional constraint for luminous flux conservation.

Test case: As a proof of concept we will apply the strategy outlined above to a test-case dubbed the 'bucket of water' from [6]. First, Liouville's equation is made dimensionless by some appropriate scaling, such that $q \in [-1, +1]$ and that the maximum value of the initial condition is equal to 1. The test-case consists of a simple jump in the refractive index, i.e, $n(q) = 1.4$ if $q \leq 0$ and $n(q) = 1$ if $q > 0$. Note that the optical axis is parallel to the optical interface. As an initial condition, we take the phase-space distribution shown in Figure 1, and after solving Liouville's equation till $z = 0.7$ we obtain the distribution displayed in Figure 2. Note that the distribution is perfectly discontinuous along the optical interface. The distribution involves three different segments, the first one is the source part in $\{q \leq 0, p \geq 0\}$, which has only propagated through the medium with refractive index n_1. The second one is in the so-called reflected region $\{q \leq 0, p < 0\}$, where a part of the distribution ends up due to reflection. The third segment is in the refracted region $\{q > 0, p \geq 0\}$, where a part of the distribution ends after being refracted at the optical interface. Moreover, to prove energy conservation, the relative error in the total luminous flux as a function of z is plotted in Figure 3. We observe energy conservation up to machine precision.

3 Conclusion

The described variable order method for light propagation through an optical system based on Liouville's equation is promising. A special treatment of the optical interfaces ensures that the scheme is fully energy conservative. Future research will be conducted to extend the method to three-dimensional optics where we have to solve Liouville's equation on a four-dimensional phase-space.

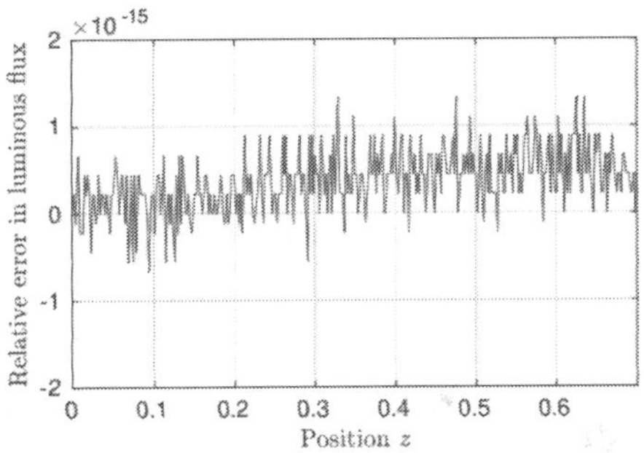

Figure 3. Relative error in the total luminous flux.

References

[1] K.B. Wolf, *Geometric optics on phase space* (Springer Science & Business Media, 2004)

[2] B.S. van Lith, J.H.M. ten Thije Boonkkamp, W.L. IJzerman, Journal of Computational Physics: X **3**, 100030 (2019)

[3] D.A. Kopriva, *Implementing spectral methods for partial differential equations: Algorithms for scientists and engineers* (Springer Science & Business Media, 2009)

[4] D.W. Zingg, T.T. Chisholm, Applied Numerical Mathematics **31**, 227 (1999)

[5] D.A. Kopriva, S.L. Woodruff, M.Y. Hussaini, International journal for numerical methods in engineering **53**, 105 (2002)

[6] B.S. van Lith, J.H.M. ten Thije Boonkkamp, W.L. IJzerman, T.W. Tukker, Journal of Scientific Computing **68**, 739 (2016)

Light shaping from a physical-optics point of view

Liangxin Yang[1,*], *Irfan* Badar[1,2], *Christian* Hellmann[2,3], and *Frank* Wyrowski[1]

[1]Institute of Applied Physics, Friedrich Schiller University Jena, Jena, Germany
[2]Wyrowski Photonics UG, Jena, Germany
[3]LightTrans International UG, Jena, Germany

Abstract. In the design of optical element for light shaping, a geometric-optics assumption is usually used, where the validity of the assumption is rarely discussed in literature. In this work, the field tracing techniques for modeling light-shaping systems are presented, which reveals the optical element resulted from those geometric-base algorithm is not always accurate enough for the design task. An example is demonstrated with the functional embodiment of the element. The simulation result shows that diffraction effect may occur, especially in paraxial situation. However, the designed result start with the assumption is well-introduced initial guess for further optimization with the iterative Fourier transform algorithm (IFTA).

1 Introduction

The design of a light-shaping system is aimed at redistributing the energy of the light source for certain applications. In literature, different algorithms have been proposed, attempt to find a suitable optical element to fulfill the task. One type of algorithms is based on the geometric assumption, where a one-to-one map between the source energy distribution and the target one is assumed, and a local relationship between their coordinate is established. By using the assumption, either a phase-only holography optical element(HOE) [1, 2], or a freeform surface can be designed [3, 4].

Typically, for analyzing the design result, a ray-tracing simulation is performed, which gives a result corresponding exactly to the target, and is usually given as proof of the validity of the design. However, using a simulation technique which is also under the geometric-optics assumption is not enough for a full analysis of the accuracy of the design, since any eventual physical-optics effects would go unnoticed, both in the design and in the subsequent simulation. Therefore, to answer the question of whether the geometric-optics assumption is accurate enough for the design algorithm, the designed result has to be investigated via physical-optics simulation.

2 Field tracing techniques

In this work, we present a field tracing technique under physical optics for modeling the light-shaping system. The modeling methods are sequentially applied in the light path and illustrated in a field tracing diagram as shown in Fig. 1. The field tracing diagram builds up an algorithm demonstrates how the target field is calculated from the input,

$$\boldsymbol{E}^{\text{out}}(\boldsymbol{\rho}')\big|_{\text{ptar}} = \mathcal{F}^{-1}\left\{\tilde{\mathcal{P}}\mathcal{F}\left\{\mathcal{B}\,\boldsymbol{E}^{\text{in}}(\boldsymbol{\rho})\big|_{\text{pin}}\right\}\right\}. \tag{1}$$

where all the $\boldsymbol{E}$ are the vectorial electric fields, defined either in the space domain, with $\boldsymbol{\rho} = (x, y)$ or the spatial-frequency domain, with $\boldsymbol{\kappa} = (k_x, k_y)$. $\mathcal{B}$ is an operator in general, which acts on the field in the $\boldsymbol{\rho}$ domain. $\tilde{\mathcal{P}}$ denotes the free space propagation operator in the $\boldsymbol{\kappa}$ domain. And $\mathcal{F}$ and $\mathcal{F}^{-1}$ are the direct and inverse Fourier transform operators for the fields between two domains.

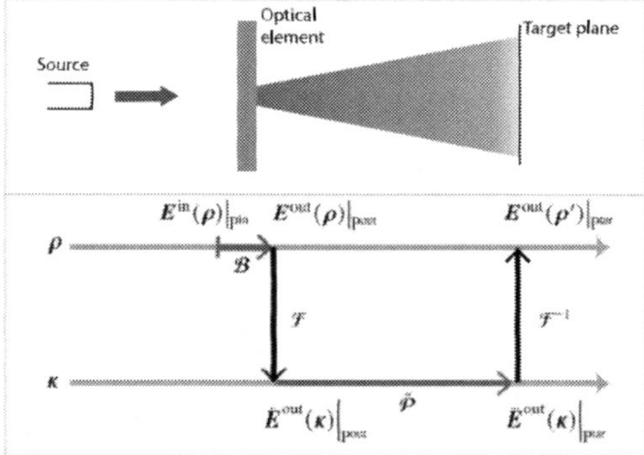

Figure 1. A light-shaping system and its field tracing diagram.

Numerically, there are several algorithms available to compute the Fourier transform in Eq. 1 [5, 6], e.g. Fast Fourier Transform (FFT), semi-analytical Fourier Transform (SFT), or Homeomorphic Fourier Transform (HFT). HFT and inverse HFT are approximated pointwise Fourier transforms, which produce a one-to-one map between the coordinates of the fields expressed in the $\boldsymbol{\rho}$ domain and in the $\boldsymbol{\kappa}$ domain [6].

*e-mail: liangxin.yang@uni-jena.de

 https://doi.org/10.1051/epjconf/202023802006

In the field tracing diagram, $\tilde{\mathcal{P}}$ is a pointwise operator. Therefore, if the $\mathcal{B}$ is also a pointwise operator, and all the direct Fourier transforms in the field tracing can be approximated by the HFT, the whole field tracing simulation of the light-shaping system gives a pointwise calculation, which forms the homeomorphism between $\boldsymbol{E}^{\text{in}}(\rho)\big|_{p_{\text{in}}}$ and $\boldsymbol{E}^{\text{out}}(\rho')\big|_{p_{\text{tar}}}$, i.e. $\rho'(\rho)$ is a bijective mapping function. And that is the typical assumption usually made in freeform surface design algorithms. However, in certain design task, the assumption is not always satisfied, especially the accuracy of the HFT is not always sufficient.

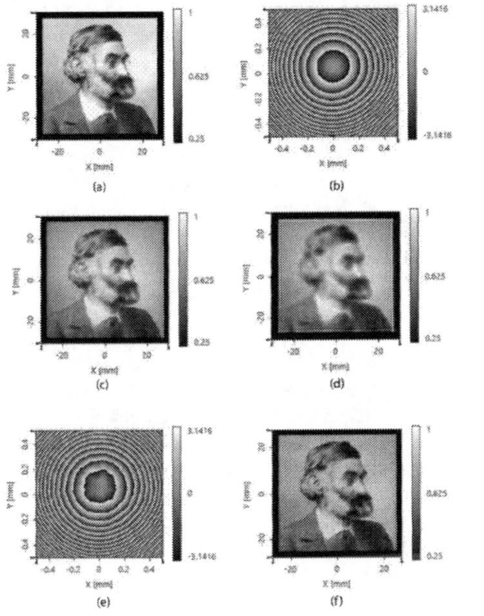

Figure 2. (a) Target irradiance distribution (normalized). (b)The wrapped phase response function designed with geometric assumption (Unit: rad). (c) Irradiance: simulation result, with the HFT used. (d) Irradiance: simulation result, with the FFT used. (e)The optimized phase response function by IFTA (Unit: rad). (f) Irradiance: simulation result with the optimized phase response function.

3 Example and discussion

For investigating the assumption in the design algorithm, an example is illustrated in Fig. 2 to demonstrate the effect. A target irradiance pattern is shown in Fig. 2 (a). The design task is to shape an input plane wave for achieving the target irradiance distribution in the far field zone. The input beam size is 1 mm and the target plane is set perpendicular to the optical axis and located 1 m away from the component.

The functional embodiment of the optical element, which is considered as a phase response function, is designed by using the geometric-optics assumption where a local mapping between the irradiance of the input and the target is established. The mathematical model of "optimal mass transport (OMT)" [7] is applied to calculate the mapping. When the mapping is obtained, the phase response function can be concluded from geometric relationship [2]. The designed result is shown in Fig. 2 (b).

With the phase response function, simulation is performed by the field tracing techniques. $\mathcal{F}$ and $\mathcal{F}^{-1}$ are selected as HFT to start with. Therefore, the assumption in the design algorithm is also used in the simulation. The irradiance at the target plane is measured in the detector, and shown in Fig. 2 (c). For further investigation, the $\mathcal{F}$ and $\mathcal{F}^{-1}$ are switched to rigorous FFT. The result is shown in Fig. 2 (d). The result in Fig. 2 (c) coincides with the target in Fig. 2 (a), however, the result in Fig. 2 (d) shows obvious diffraction effect and low resolution, which proves that in this system, the HFT is not accurate enough for the calculation of the Fourier transform, so that the one-to-one map between the input and target does not exist, i.e. the design from the algorithm with geometric-optics assumption is not valid.

For optimization of the result, the iterative Fourier transform algorithm (IFTA) is performed, starting with an initial guess given by previous result. The optimized phase response function is shown in Fig. 2 (e). Comparing the results, one can see that the phase response function after IFTA is quite closed to the initial one, so it is reasonable to conclude the result from the mapping type algorithm as a good option to start with. Fig 2 (f) shows the simulation result with the optimized function, which compared to the irradiance distribution in Fig. 2 (d), the diffraction fringe has been removed and the resolution is improved.

All the simulation are done via the software VirtualLab Fusion [8].

References

[1] H. Aagedal, M. Schmid, S. Egner, J. Müller-Quade, T. Beth, F. Wyrowski, J. Opt. Soc. Am. A **14**, 1549 (1997)

[2] A. Hermerschmidt, H.J. Eichler, S. Teiwes, J. Schwartz, Proc. SPIE **3291**, 40 (1998)

[3] Z. Feng, L. Huang, G. Jin, M. Gong, Opt. Express **21**, 28693 (2013)

[4] R. Wu, Z. Feng, Z. Zheng, R. Liang, P. Benítez, J.C. Miñano, F. Duerr, Laser & Photonics Reviews **12**, 1700310 (2018)

[5] Z. Wang, S. Zhang, O. Baladron-Zorita, C. Hellmann, F. Wyrowski, Opt. Express **27**, 15335 (2019)

[6] Z. Wang, O. Baladron-Zorita, C. Hellmann, F. Wyrowski, Opt. Express **28**, 10552 (2020)

[7] C. Prins, R. Beltman, J. ten Thije Boonkkamp, W. IJzerman, T. Tukker, SIAM Journal on Scientific Computing **37**, B937 (2015)

[8] Physical optics simulation and design software "Wyrowski VirtualLab Fusion", developed by Wyrowski Photonics GmbH, distributed by LightTrans International UG, Jena, Germany.

Multimode fibre probe calibration

Petr Jákl[1,*], *Martin* Šiler[1], *Jan* Ježek[1], *Johanna* Trägårdh[1], *Tomáš* Čižmár[1,2], *Pavel* Zemánek[1]

[1]Institute of Scientific Instruments of the CAS, v.v.i., Královopolská 147, 61264 Brno, Czech republic
[2]Leibniz Institute of Photonic Technology, 07745 Jena, Germany

Abstract. Multimode fibres (MMF) used in endoscopy have advantage of small diameter and flexibility, thus causing less damage to living animals. However, the imaging requires wavefront shaping techniques to obtain a sharp image despite the mode dispersion in the waveguide. We suggest version of transmission matrix calibration which uses internal modes of the waveguide and, thus, lessens requirements on the endoscopy apparatus removing the external reference path.

1 Endoscopy imaging

The prevalence of neurodegenerative diseases connected with an increasing age of the population requires preclinical studies of animal models to develop a knowledge base of neuronal processes and blood flow deep in the brain. Especially experiments *in vivo* are of immense importance for the future of the research of Alzheimer's, Parkinson's and other diseases in order to translate the results to human medicine [1]. Measurements in brain tissue require access to the deep layers, but the penetration of microscopy techniques do not go deeper than approximately 1.5 mm. Endoscopy probes come in different sizes and imaging quality. From several millimetres in diameter (GRIN lenses) [2], over fibre bundles [3] to multimode optical fibres (MMF) [4]. Especially the latter are of increasing popularity for their narrow footprint and flexibility. Obtaining an image using MMF probe is not straightforward, though. Due to different group velocities of the fibre modes, the phases are mixed and resulting image is an apparently chaotic distribution of speckles.

To characterize the fiber and achieve successful imaging, it is necessary to map the system response described by a transmission matrix (TM) [5,6]. The calibration process requires a highly homogeneous reference beam in order to extract modes' phases using phase-shifting interferometry. Although a high quality matrix (as well as resulting endoscopic images) is obtained, the experimental setup is somewhat complicated by the necessity of introduction of a reference beam in the calibration part of the apparatus and the overall phase stabilization of the optical paths not to introduce artefacts in the measurement.

A suggested alternative calibration technique uses internal modes of the endoscopic probe as a reference for TM calibration. However, the intensity distribution of the reference beam in the sample plane is a speckle pattern which results in blind spots in the measured TM due to lack of interference signal in the reference mode intensity minima. One of the previous ideas was to run the calibration procedure second time with another mode set as a reference field. Since the speckle patterns are different, it is possible to use the second measurement to cover the holes in TM [7]. Our approach is a little bit different. Running several procedures, each with different single internal reference mode, we obtain N "holey" TMs. Every output mode is defined as a superposition of all the input modes with phase and amplitude measured with reference R_1 in the first measurement. The same output mode measured with reference R_n is reconstructed from input modes with a constant phase shift $\Delta\varphi=\varphi_{R1}-\varphi_{Rn}$; the value of $\Delta\varphi$ is revealed by dot product of the input mode vectors. Combined TM is then created by adding all the particular TMs to TM_1 where the added columns are shifted by $\Delta\varphi$ and negated.

Although the calibration procedure is time consuming due to repetition of the calibration task with different references, it results in a high quality homogeneous image.

2 Experimental results

The experimental setup requires a device for wavefront shaping (a spatial light modulator SLM in our case), transformation optics to project the Fourier plane of SLM to the MMF probe input facet and an imaging path to observe the output modes of the probe. We have used

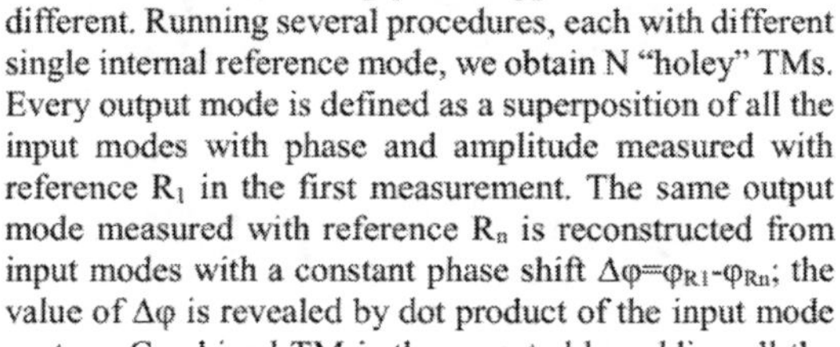

Fig. 1 : Experimental apparatus with spatial light modulator (SLM), quarter waveplates (QWP) and microscope objectives (MO).

* Corresponding author: jakl@isibrno.cz

circularly polarized light (one polarization state only), as it is less sensitive to crosstalk with the other polarization state. An infrared laser (λ=1070 nm) and an optical fibre with NA=0.22, d=50 µm was used. The external reference beam is introduced to setup for comparison and it is not used during calibration with internal references.

To estimate the performance of the internal calibration technique, we have measured the intensity of all the output modes addressed one by one. The values were put together to form an image of output fibre facet for quick visual check and their uniformity was expressed using the interference contrast formula. We have characterized the results from calibration with randomly chosen references. Obviously, the more internal

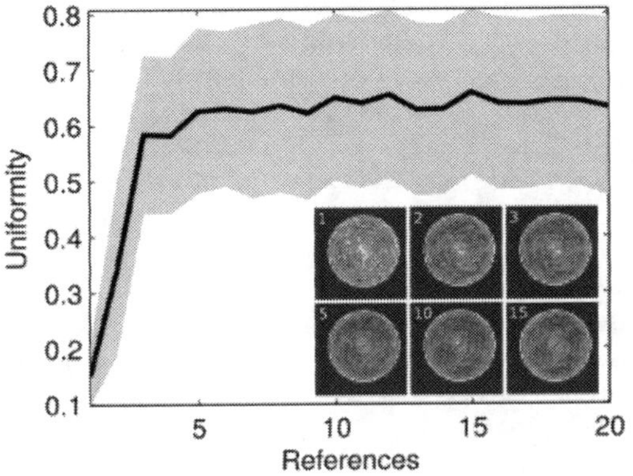

Fig. 2 :Uniformity measurement for 1-20 randomly chosen internal references. Inset shows intensity of all output modes across the fiber facet to visually compare homogenity.

references we used, the less blind spots or fluctuations in the output mode intensity we observed. Surprisingly, it was saturated quickly, as more than five references rarely had any effect on the uniformity.

We have used the calibrated endoscope to record images of 1951 USAF resolution test prepared by electron beam lithography in a chromium layer. The target was placed approximately 10 µm behind the output facet, the output modes were addressed one by one and the detector further down the optical path collected light passing through. This way, we have simulated the endoscopy

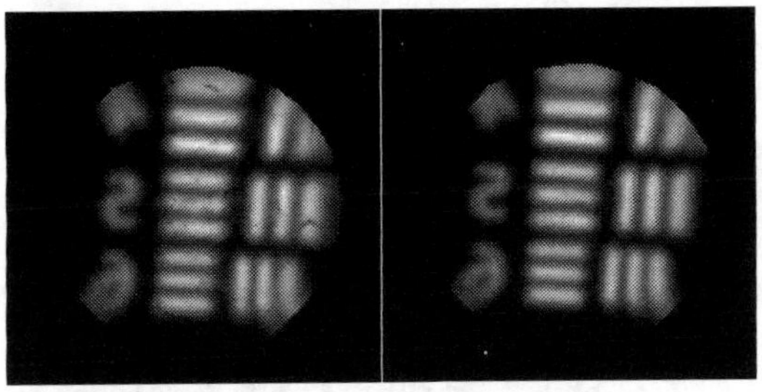

Fig. 3 :Endoscopic imaging of USAF target. System calibrated with 1 (left) and 5 (right) internal references.

imaging, where the fluorescence signal is collected by the probe itself and passed to the detector in the setup.

3 Conclusion

We have demonstrated calibration and imaging with multimode fibre probe. The calibration itself used internal modes of the fibre and removed the necessity to stabilize the external reference path. The price for it is longer time of the procedure, as the technique requires several runs of measurements of the transmission matrix. The resulting calibration is highly uniform, especially when calibrated with 5 or more references, as we have proved by imaging the USAF target in an endoscopic configuration.

References

1. P. McGonigle, Biochem. Pharmacology **87,** 140 (2014)
2. J. Jung, A. D. Mehta, E. Aksay, R. Stepnoski, M. J. Schnitzer, J. Neurophysiology **92,** 3121 (2004)
3. S. P. Mekhail, N. Abudukeyoumu, J. Ward, G. Arbuthnott, S. N. Chormaic, Biomedical Optics Express **9,** 1843 (2018)
4. S. Turtaev, I. T. Leite, T. Altwegg-Boussac, J. M. P. Pakan, N. L. Rochefort, T. Cizmar, Light: Sience & Applications **7** (2018).
5. S. Popoff, G. Lerosey, R. Carminati, M. Fink, A. Boccara, S. Gigan, Phys. Rev. Lett. **104** (2010).
6. T. Cizmar, K. Dholakia, Opt. Express **19**, 18871 (2011).
7. S. Bianchi, V. P. Rajamanickam, L. Ferrara, E. Di Fabrizio, C. Liberale, R. Di Leonardo, Opt. Lett. **38**, 4935 (2013)

Advances in precision freeform manufacturing by plasma jet machining -INVITED

Thomas Arnold[1,2,*], *Georg* Boehm[2], and *Faezeh* Kazemi[2]

[1]Technische Universität Dresden, Institut für Fertigungstechnik, 01062 Dresden, Germany
[2]Leibniz Institute of Surface Engineering (IOM), Permoserstr. 15, 04318 Leipzig, Germany

Abstract. Atmospheric pressure plasma jet machining technology provides a flexible and efficient way to fabricate precise freeform optics. Due to the pure chemical material removal mechanism based on a dry etching process using fluorine containing gas, the choice of materials that can be treated is limited. Fused silica, Si, SiC or ULE® are easy to machine since the etching products formed are solely volatile. Recently, plasma jet machining has been also adopted to treat optical glasses like N-BK7® which contain amongst others alkali metals that form a solid residual layer during etching. In the paper a new approach to apply deterministic plasma jet etching on optical glass coping with complex etch characteristics caused by the residual layer is introduced.

1 Introduction

Current developments in the field of modern optics are aimed at miniaturization, quality improvement and reduction of the assembly effort for optical systems, for example by increasing the complexity of the optical functional surfaces used. Especially optical freeform surfaces have become increasingly important in recent years, since they allow optical assemblies with a reduced number of optical surfaces compared to systems without freeform surfaces, and therefore they are much more compact and lighter. In addition, losses in radiation power can be minimized and, in some cases, completely new optical functionalities can be achieved.

However, the flexible and efficient production of precise optical freeform surfaces is a major challenge. Unlike spherical or aspherical surfaces, freeform surfaces do not have rotational symmetry, but generally have six degrees of freedom, which can be described mathematically with polynomials or splines. Due to the locally changing curvature, locally acting, so-called sub-aperture processes must be used for free-form surfaces. In order to achieve high-precision surfaces correction and finishing techniques are applied. A range of deterministic methods exist, such as bonnet polishing, magnetorheological finishing, or ion beam finishing.

Mechanical abrasive techniques can be applied on nearly every kind of optical material. However, a challenge is to achieve a sufficiently high removal rate to eliminate sub-surface damage (SSD) while maintaining the form after the shaping process, which is usually already within the specification range. Beam based techniques such as ion beam figuring potentially yield ultra-high precision on surfaces but the removal rates are very low and costly vacuum equipment is necessary. As an alternative plasma jet machining was introduced that provides a flexible and efficient way to fabricate precise freeform optics at atmospheric pressure using inexpensive equipment [1].

2 Plasma jet machining

Plasma jet machining (PJM) technology is a sub-aperture surface manufacturing method based on a reactive plasma jet etching process that allows precise material removal controlled by pure chemical interaction between the tool and the surface. The plasma jet tool exerts virtually no forces to the surface and thus induces no SSD. Furthermore, the rotationally symmetric near-Gaussian tool function that can be scaled from 0.3 mm to 4 mm (full width at half maximum) enables to generate a desired freeform shape with nearly no geometric limitations. A microwave-driven plasma nozzle generates an Ar/He plasma jet at atmospheric pressure where reactive gases like CF_4 and O_2 are admixed in order to create free fluorine atoms by means of dissociation. The fluorine atoms react with the substrate material to form volatile compounds, and material removal from the surface is achieved. Figure 1 shows a photograph of the plasma jet interacting with an optical surface and a typical cross section of the arising tool function profile. For easy-to-etch materials such as fused silica, silicon, ULE®, or SiC a simple etch chemistry is utilized leading to exclusively gaseous reaction products in the form of SiF_4 and CO_2. Hence, the material removal rate which can reach up to 20 mm^3/min remains constant as long as a constant substrate surface temperature distribution is maintained during the process. For the surface machining process the plasma nozzle is moved relative to the surface by a CNC motion system. Shaping or surface figure error correction is

* Corresponding author: thomas.arnold@iom-leipzig.de

performed by applying dwell time methods that are usually employed in deterministic surface machining. As in most cases the heat flow from the plasma jet to the surface alters the surface temperature distribution and thus the etching rate in the course of the machining process, it must be taken into account by the dwell-time calculation software based on additional input data obtained from temperature and etch rate measurements.

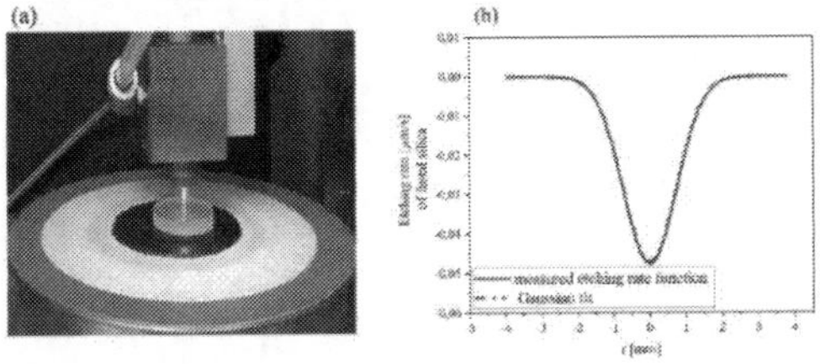

Fig. 1. (a) Plasma jet on substrate, (b) Typical plasma jet etching tool function on fused silica.

3 Plasma jet machining of N-BK7

Different to the above mentioned easy-to-etch materials optical glasses like N-BK7® contain alkali metal oxides in addition to silicon dioxide. Although SiO_2 is also converted into volatile products here, the chemical interaction between the plasma-generated fluorine radicals and the metal components of N-BK7 leads to the formation of non-volatile compounds such as KF and NaF, which remain as residual layers on the surface. Consequently, the fluorine attack to the surface is inhibited or eventually stops completely. Hence, the local etching rate function is not constant in time anymore but rather depends on the local layer thickness that develops depending on the applied etching time. In Figure 2 (a), the radial etch rate distribution is shown depending on the residual layer thickness while the N-BK7 substrate is approx. at room temperature. The data illustrate that the etching rate varies strongly in a nonhomogeneous way over the radial dimension. It is observed that in the center the etch attack stops for layer thickness of approx. 0.3 µm, whereas in the periphery the etching process can proceed. It is clear that such complex and irregular behaviour is not suitable for a deterministic machining process. However, it is shown in Figure 2 (b) that heating the substrate to achieve elevated temperatures of approx. 350 °C results in a much more uniform etching rate distribution maintaining a Gaussian tool function throughout the etching process. The reason for this is a structural change of the residual layer from a densely closed layer occurring at room temperature to a porous layer found at elevated temperature. Thus, the fluorine atoms can penetrate the layer reaching the N-BK7 interface to proceed etching [2, 3]. Although the absolute values of etching rate gradually decrease with increasing layer thickness, such a homogeneous behaviour is more suitable to be taken into account in a dwell-time based machining algorithm. An enhanced approach to dwell time calculation has been tested where the target topography (either a prescribed surface shape to be generated, or a measured surface error topography) is deconvolved by an assumed constant Gaussian tool function in order to obtain a dwell-time

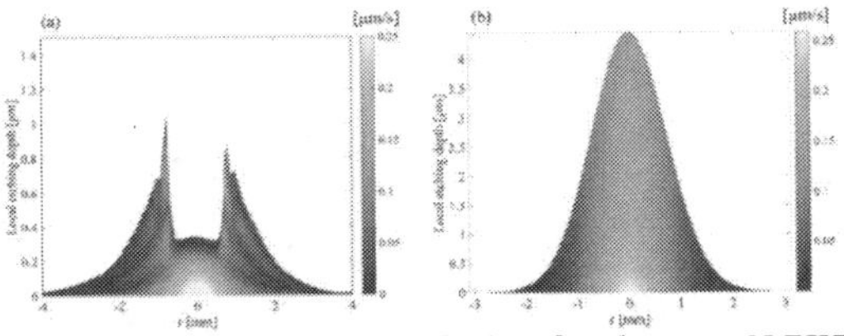

Fig. 2. Radial etch rate distribution functions on N-BK7 obtained from static footprint etching depending on layer thickness at different surface temperatures: (a) T_S=25°C, (b) T_S=350°C.

matrix. In a next step, the calculated local dwell times are convolved with a spatio-temporally varying rate function gained from preliminary measurements (i.e. the rate function in Figure 2 (b) which is additionally corrected for temperature effects). Performing a virtual machining process using a Matlab® model yields then a topography that differs to some extent from the target topography. Taking the difference topography as a correction to the initial target topography and recalculating dwell times by deconvolution yield an improved estimation of machining data. This procedure can be iterated in order to minimize the difference of the prescribed and calculated topography. The procedure resembles the van Cittert algorithm used for numerical deconvolution. As a proof-of-principle, a freeform surface on N-BK7 featuring a sinusoidal test structure with PV of 800 nm has been fabricated by plasma jet machining using the described method. For interferometric surface measurement the residual layer was removed by rinsing in water/ethanol. In Figure 3 the shape and corresponding horizontal cross sections are depicted. Additionally, the target topography is shown. The measured profile has a PV of 817 nm, which is about 2% higher than the prescription profile. Presumably, local surface temperature fluctuations were still not completely corrected in the model.

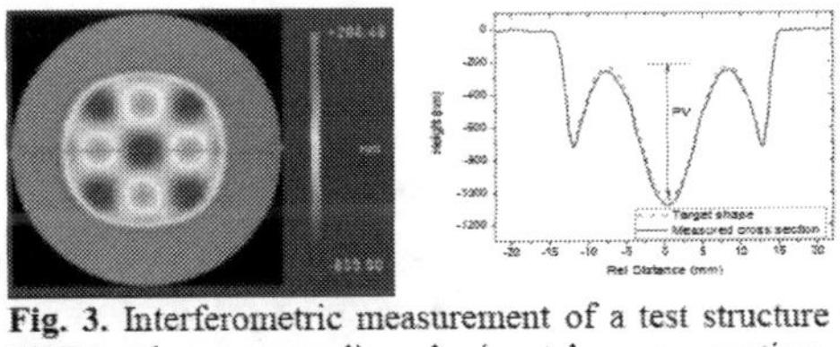

Fig. 3. Interferometric measurement of a test structure on N-BK7 and corresponding horizontal cross section (blue: measured, dotted red: target shape.

The described method and the experimental result show a potential way to cope with residual layers formed in the etching process that locally and temporally influence the etch rate function in plasma jet machining of optical glasses.

References

1. Th. Arnold, G. Boehm, H. Paetzelt, J. Eur. Opt. Soc.-Rapid 11 (2016) 16002
2. F. Kazemi, G. Boehm, T. Arnold, Plasma Process Polym. 16 (2019) 1900119
3. F. Kazemi, G. Boehm, T. Arnold, Proc. SPIE 11385, Optics and Measurement 2019 Int. Conf. (2019) 1138509

Introducing Full-Aperture FJP (faFJP), a Technique for Manufacturing Sub-Millimetre Aspheres

W.A.C.M. Messelink[1,2,*] and *M.* Meeder[2,3]

[1]Edmund Optics Singapore, Singapore, 738100, Singapore
[2]Fisba AG, Sankt Gallen, CH-9016, Switzerland
[3]ETH Zürich, Zürich, CH- 8092, Switzerland

Abstract. We propose a novel approach, "faFJP", to the computer controlled polishing (CCP) of mini aspheres down to 0.2 mm. The generation of locally-tunable, full-aperture footprints is accomplished by applying a Fluid Jet Polishing (FJP) beam larger than the full aperture being polished and varying the incident angle.

1 Introduction

Thanks to their point symmetry, spherical lenses have been produced for centuries in high qualities and quantities, with shape deviations down to dozens of nanometres, by applying chaotic, semi-random tool paths above the rotating workpiece. Unfortunately, optics designers have been insisting on using aspherical lenses, due to their better performance in handling optical rays and because the use of one of these types of lenses replaces up to five spherical lenses within the design of an optical system. The fact that aspherical lenses possess line symmetry, rather than point symmetries, forced optical fabrication engineers to invent numerous dedicated polishing techniques, each featuring the generation and application of sub-aperture footprints (local polishing spots within the clear aperture of the optical surface) such as belts [1], bonnets [2], fluid jets [3, 4], ion beams [5], plasma jets [6], magnetorheological fluids [7], rotating spheres [8] or polyurethane needles. All these techniques distinguish themselves in the type of aspherical shapes producible, quality generable or dimensions manageable. In this paper, we propose an alternative approach, named "faFJP", to the computer controlled polishing (CCP) of mini aspheres featuring clear-apertures down to 0.2 mm. Instead of traversing a static, sub-aperture footprint across the clear-aperture, faFJP uses a locally-tunable, full-aperture footprint by applying a macroscopic Fluid Jet Polishing (FJP) beam, which has a diameter larger than the clear aperture being polished. The variation of the footprints is generated by changing the incident angle of the fluid beam with respect to the asphere axis of symmetry, due to FJP's characteristic wear profile.

2 Theory and simulations

Finnie et al. [9] established the dependency of the volumetric removal on the angle of incidence of an abrasive particle. Particularly it was shown that the highest removal did not occur

*e-mail: PMesselink@edmundoptics.com.sg

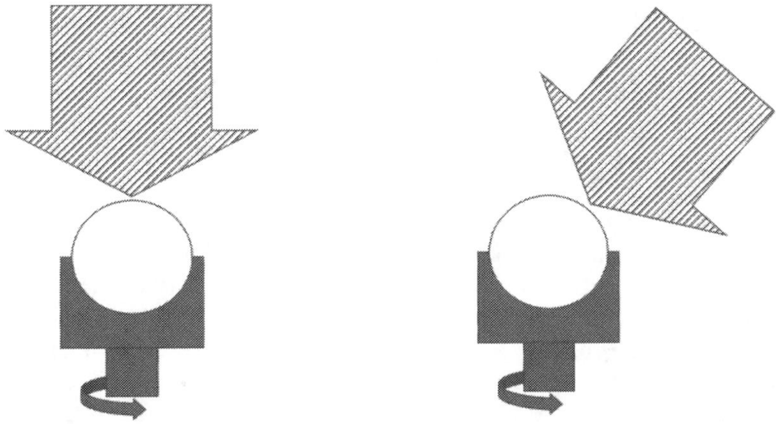

Figure 1: Diagrams of a macroscopic FJP beam, impinging on a spherical substrate with a diameter smaller than the width of the beam, at different angles.

when the particle impacted the surface perpendicularly, but at a specific grazing incidence. For a macroscopic FJP beam, impinging on a spherical substrate with a diameter smaller than the width of the beam (see Figure 1), this results in a rotationally symmetric removal profile, with the highest removal some distance away from the centreline of the beam. Simulations of the removal profiles for different angles of the FJP beam while the workpiece is not rotating are shown in Figure 2. If the workpiece is mounted on a rotary table and spun

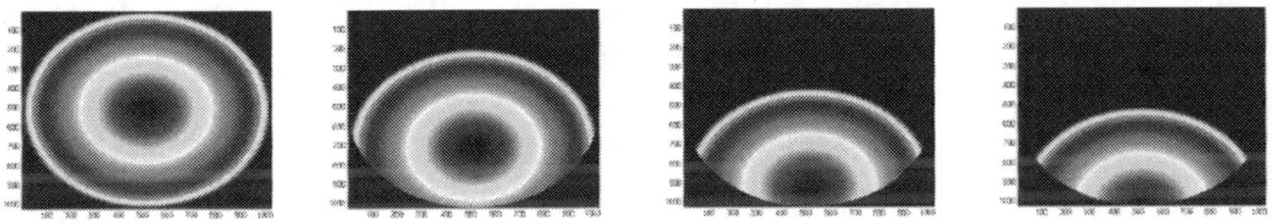

Figure 2: Simulations of the removal profiles for different angles of the FJP beam on a non-rotating workpiece.

underneath the FJP beam, then the removal profile will be distributed tangentially over the surface. A simulation of this for different angles of the FJP beam is shown in Figure 3. Each

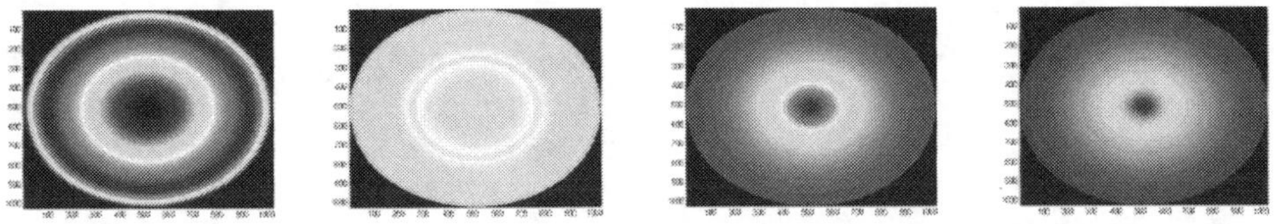

Figure 3: Simulations of the removal profiles for different angles of the FJP beam on a rotating workpiece.

of the removal profiles from Figure 3 produces an asphere. Increasing the time the workpiece spends inside the FJP beam increases the aspheric departure that is generated. And by using a linear combination of (beam) angles and (dwell) times an arbitrary aspheric profile can be generated.

3 Conclusions and future work

We have introduced a novel method of generating and correcting mini aspheres that applies full-aperture, but locally-tunable footprints. As opposed to using a static, sub-aperture footprint that is traversed over the full-aperture, as is the state of the art for macro aspheres. The generation of locally-tunable, full-aperture footprints is accomplished by applying an FJP beam larger than the full aperture being polished and varying the incident angle. The theory of volumetric removal dependency on the angle of incidence has been introduced and the effect on a spherical surface in a macroscopic FJP beam has been derived and simulated.

Arbitrary asphere profiles can be generated by applying a linear combination of FJP beam angles. In practice, this linear combination can be implemented by mounting the FJP beam on a pivot and placing the workpiece in the centre of the pivot. By rotating the pivot, the different FJP beam angles can be obtained sequentially and by varying the speed of rotation the required dwell times for each angle can be achieved.

The next step is to experimentally verify the introduced technique. Future research could examine the orthogonality or joint entropy of the removal profiles for different angles to assess what shapes of aspheres are more straightforward to generate.

4 Acknowledgement

The theoretical work presented in this paper was performed while all the authors were employed by Fisba AG, Sankt Gallen, CH-9016, Switzerland. We would especially like to thank Dr. Oliver Faehnle for his guidance and useful discussions.

References

[1] E. Fess, J. Schoen, M. Bechtold, D. Mohring, *UltraForm finishing*, in *Window and Dome Technologies and Materials IX*, edited by R.W. Tustison (SPIE, 2005), Vol. 5786 of *Proc. SPIE*, pp. 305–309, `http://dx.doi.org/10.1117/12.606468`

[2] D.D. Walker, D. Brooks, R. Freeman, A. King, G. McCavana, R. Morton, D. Riley, J. Simms, *First aspheric form and texture results from a production machine embodying the precession process*, in *Optical Manufacturing and Testing IV*, edited by H.P. Stahl (2001), Vol. 4451 of *Proc. SPIE*, pp. 267–276, `http://dx.doi.org/10.1117/12.453652`

[3] O.W. Faehnle, H. van Brug, H.J. Frankena, Applied Optics **37**, 6771 (1998)

[4] W.A. Messelink, O.W. Faehnle, *Exploiting the Process Stability of Fluid Jet Polishing*, in *Frontiers in Optics 2008/Laser Science XXIV/Plasmonics and Metamaterials/Optical Fabrication and Testing* (Optical Society of America, 2008), p. OThD3, `http://www.opticsinfobase.org/abstract.cfm?URI=OFT-2008-OThD3`

[5] S. Zafran, K.A. Kaufman, M.M. Silver, Optical Engineering **21**, 216002 (1982)

[6] T. Arnold, G. Boehm, I.M. Eichentopf, M. Janietz, J. Meister, A. Schindler, Vakuum in Forschung und Praxis **22**, 10 (2010)

[7] S.D. Jacobs, D. Golini, Y. Hsu, B.E. Puchebner, D. Strafford, I.V. Prokhorov, E.M. Fess, D. Pietrowski, W.I. Kordonski, *Magnetorheological finishing: a deterministic process for optics manufacturing*, in *International Conference on Optical Fabrication and Testing*,

edited by T. Kasai (SPIE, 1995), Vol. 2576 of *Proc. SPIE*, pp. 372–382, `http://dx.doi.org/10.1117/12.215617`

[8] T. Hirata, Y. Takei, H. Mimura, Procedia CIRP **13**, 198 (2014), 2nd CIRP Conference on Surface Integrity (CSI)

[9] I. Finnie, Wear **3**, 87 (1960)

Application of a pressure measuring film for pressure observation in overarm polishing

Michael F. Benisch[1,a], Oliver Faehnle[2], and Rolf Rascher[1]

[1] *Institute for Precision Manufacturing and High-Frequency Technology, Deggendorf Institute of Technology DIT, Dieter-Goerlitz-Platz 1, 94244 Deggendorf, Germany*
[2] *FISBA OPTIK AG, Rorschacherstrasse 268, 9016 St. Gallen, Switzerland*

Abstract. In previous publications the inhomogeneous pressure transfer through the polishing tool onto a glass surface could be shown. This experiment shows polishing trials with different polishing materials and the differences in the homogeneity of the pressure transfer through them. Only the properties of the material will be discussed explicitly, the change of the tool constitution through the process will be part of further publications.

1 Introduction

Previous observations have shown the pressure distribution between the polishing pad and the glass surface. Inhomogeneities in the pressure distribution could also be shown in the plane state. [1]. The pressure during polishing is a relevant process parameter in the Preston equation [2] and directly influences the friction acting in the process [3]. It could also be shown that tilting of the glass surface causes an increase in pressure on the lower side and a decrease in pressure on the upper side. The use of a tool changes the surface of the polishing foil, glass abrasion and cerium oxide are deposited in the pad. For this reason new, non-dressed polishing pads were used and their material properties investigated. The materials used are LP66, LP87 and GR35 (see tab. 1 and fig. 2). A polishing felt was also used for comparison purposes in the case of a flat surface (see fig. 5). It was not used for experiments on the inclined surface.

2 Setup of the experiments

In the experiment, the discoloration of a pressure-sensitive foil (4LW, Fujifilm) under the application of force is recorded using a polishing tool. The polishing tool has a diameter of 30 mm and thus a circular area of about 700 mm^2. A polishing pad or felt (see fig. 1, fig. 2 and 4) is part of this tool. The pressure sensitive foil is applied between tool and glass surface. The glass surface is lying straight or is tilted (approximately 30°). The force is applied via the lever of an overarm polishing machine, which is additionally loaded with a weight of 1.25 kg, see fig. 1. Pressure measuring films of the type 4LW (Fujifilm) are used. The details of the application of these films and their limitations are described in [1].

[a]e-mail: michael.benisch@th-deg.de

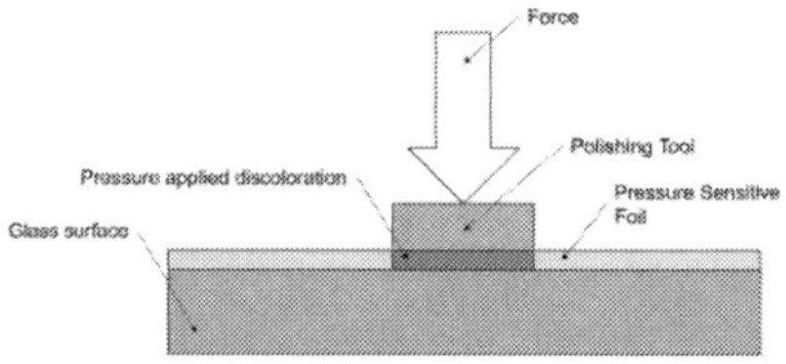

Figure 1. Experimental setup for the case of a plane glass surface: Under the influence of a force, a pressure occurs in the contact area between the polishing pad, pressure measurement film and glass surface. Under this pressure the pressure measuring film changes colour.

3 Basic Material Properties

The properties of the polishing foils are shown in table 1. The films are polyurethane films, which are partly filled with zirconium-oxide (GR35) or cerium-oxide (LP66). The material LP87 has no filler. They are standard material for the production of glasses, crystals and ceramics. The material properties in tab. 1 are manufacturer's (Universalphotonics) data. The approach to determine the modulus of elasticity via tensile tests does not work due to the properties of the polishing pads. Inhomogeneities and porous material areas act as fractures. Therefore the results of tensile tests are poorly comparable.

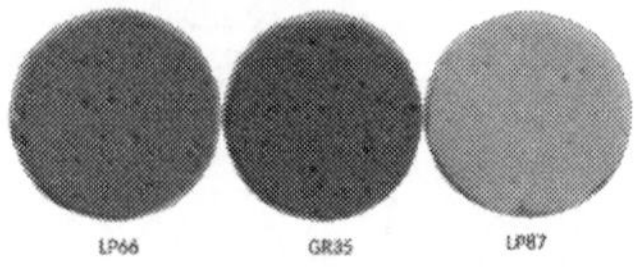

Figure 2. Photography of the polyurethane films used.

 https://doi.org/10.1051/epjconf/202023803003

Table 1. Important technical parameters of the polishing pad materials used. The higher the durometer Shore A value, the more resistant the material is to penetration and the harder it is.

Material	Density [g m^{-3}]	Durometer Shore A
LP66	0.35	66
GR35	0.59	90
LP87	0.43	96

The results of previous investigations, which were carried out with a comparable polishing tool, clearly show that the pressure distribution on the surface is not homogeneous for plane and tilted surfaces. In particular, if the glass lens and thus the polishing pad is tilted, significant deviations from a homogeneous pressure distribution arise [1]. For the following investigations a new polishing pad was used to see the influence of the material.

4 Results

4.1 Flat Surface

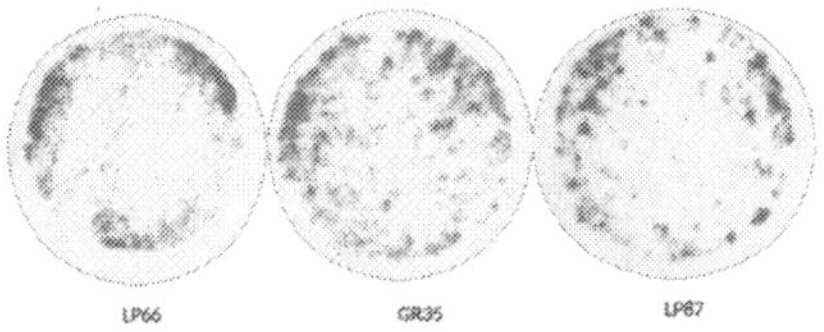

Figure 3. Discoloration of the film after pressure loading on a non-tilted surface.

All three materials show a pressure increase in the edge area. The two harder materials (GR35 and LP87) show additional local pressure increases over the entire contact area between polishing pad and pressure measurement film. The visual impression the pressure distributions is very similar. The somewhat softer LP66 shows a lower tendency to pressure increases (see fig. 3). The cause of these two effects is currently unclear. There may be an influence by the tool shape or the as-new condition of the polishing pad.

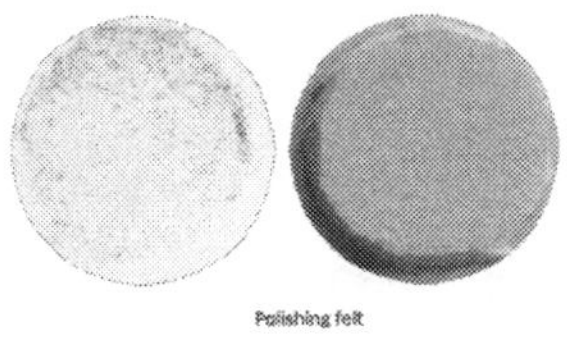

Figure 4. Discoloration of the film after pressure loading on a flat surface with a polishing felt (left) and the polishing felt (right).

If one compares the films with the polishing felt, it is noticeable that the polishing felt has a much more even pressure distribution (see fig. 4). In particular, hardly any pressure increases occur. Pressure marks are present on the entire surface. The discoloration of the pressure measurement film is low (compared to the discoloration caused by the polishing pad). There is a small effect to a one-sided discoloration, which may be caused by the tool shape.

4.2 Tilted Surface

Even with a tilted surface, there is a tendency for the two harder materials to have local pressure increases. With the harder materials there is a more extensive pressure distribution. The LP66 material seems to have the tendency for local pressure increases to a lesser extent (see fig. 5), but also shows traces of excessive pressure. All materials show a higher pressure load on the side of the polishing pad which is downwars lying (see fig. 5). The result from [1] is therefore confirmed.

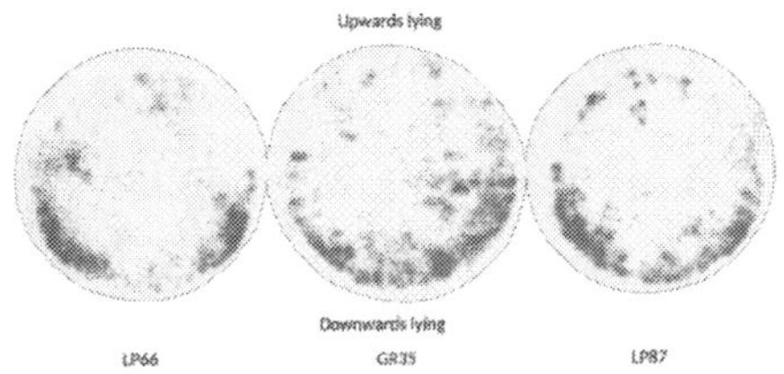

Figure 5. Discoloration of the film after pressure loading on a tilted (approximately 30°) surface.

5 Conclusion

The present tests show the basic behaviour of different polishing pads under pressure. The already mentioned change of the pressure distribution due to a tilting of the glass lens [1] could be confirmed. For further investigations the influence of the dressing of the tool and its wear condition and their influence on the pressure distribution is of iterest. The effect of a pressure increase at the corners could not be observed with a used tool (compare source). When tilted, the materials all show a tendency to local pressure increases. Furthermore, there is a need for a way to monitor the pressure during a process at different points of the tool, especially for the case of a tilted surface to polish (sperhical lens).

References

[1] Benisch, M. F.; Faehnle, O. Rascher R.; Bogner, W.; Force and Pressure Analysis during Overarm Polishing **POM 2020**, (2020)

[2] Preston F. W.; The theory and design of plate glass polishing machines **Journal of the society of glass**, 214-256 (1926)

[3] Belkhir N.; Bouzid D.; Herold V.; Determination of the Friction Coefficient During Glass Polishing **Tribology Letters**, 55-61 (2009)

Octopus – Application test of a simple and effective tool for polishing slurry monitoring

Christian Vogt[1*], *Oliver* Faehnle, and *Patrick* Kochs[2] and *Rolf* Rascher[1]

[1]EUniversity of Applied Sciences, 94469, Deggendorf, Germany
[2]EFisba AG, 9016 ST. Gallen, Switzerland

Abstract. In the digital age it has become a fundamental issue to monitor and control process parameters continuously and automatically. The monitoring of critical process parameters enables better quality at lower production costs and higher yields. In order to investigate the interactions between production quality and polishing parameters, we have developed a solution for automatic and reliable measurement of critical parameters. With this paper we first document the resolution and accuracy of the system compared to manual measuring methods currently in use.

The monitoring of critical process parameters enables better quality at lower production costs and higher yields. In polishing processes, the quality of the polishing agent is crucial. To ensure stable production processes, it is only logical to monitor the most important parameters:
pH value, density and temperature.
In many production lines, measurements are currently still carried out iteratively and manually, partly because automatic density measurement is particularly problematic and prone to errors when data is transferred manually.
In order to investigate the interactions between production quality and parameters, we have developed a solution for automatic and reliable measurements of the three parameters mentioned above as a first step.
The presented method works in situ and in process. The specially developed sensor for determining the density of the polishing agent provides a resolution of 0.001 g/ccm with a maximum deviation of 4%.
The system can be used for continuous monitoring of production processes. Furthermore, it is suitable for the documentation of experiments and the identification of correlations.

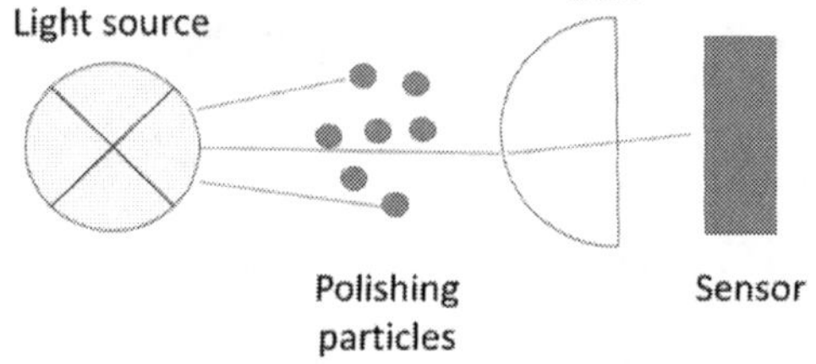

Fig. 1. Operating principle density sensor

The sensor for monitoring density measures the turbidity of the polishing agent and relies on a correlation of density. It is based on an intensity measurement of infrared radiation, whereby the degree of shading of the emitted beam by the polishing agent is measured and then the density is deduced (Fig. 1).

The system is kept very simple, extremely robust and yet has a high resolution. It can be used for several months without further calibration.

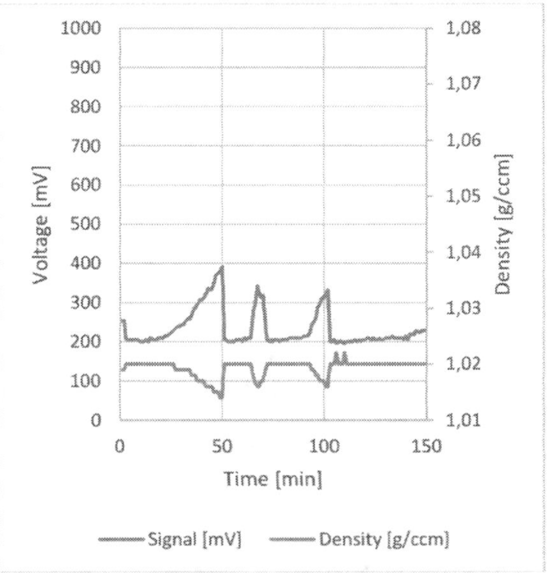

Fig. 2. Sample measurement with signal shape and density derived from it

*Corresponding author: Christian.vogt@th-deg.de

https://doi.org/10.1051/epjconf/202023803004

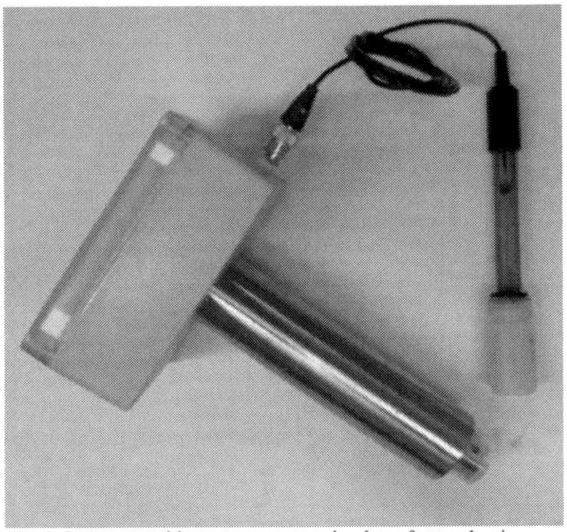

Fig. 3. Prototype with measurement technology for production monitoring, pH-probe

Compact, astigmatism corrected cross-grating spectrometer

Matthias Kraus[1,2,*], *Erik* Förster[1,2], *Verena* Bagusat[1], *Tobias* Hönle[1], *Xavier* Uwurukundo[1], *Patrick* Bohnert[1], *Robert* Brüning[3], *Hartmut* Hillmer[2], and *Robert* Brunner[1,3]

[1]University of Applied Sciences Jena, 07745 Jena, Germany
[2]Institute of Nanostructure Technologies and Analytics (INA), University of Kassel, 34132 Kassel, Germany
[3]Fraunhofer Institute for Applied Optics and Precision Engineering (IOF), 07745 Jena, Germany

Abstract. Compact cross-grating spectrometers are inspired by classical high-end Echelle spectrometers and allow to acquire a large spectrum with a high spectral resolution in a single shot. To decrease the size of an Echelle spectrometer substantially, a cross grating was employed combining both required dispersive functionalities. First, an Echelle grating diffracts light in several higher diffraction orders, while a superposed, perpendicularly oriented cross-disperser is used in first diffraction order. In this contribution, we report on the realization of a cross-grating spectrometer employing a folded reflective beam path. A toric-convex mirror was introduced allowing aberration compensation. We present the basic concept and optical design of the system and discuss the mechanical implementation and the adjustment process. Here, different laser sources and gas emission lamps are used to realize the system and derive the optical performance. In particular, the spectral resolution in the different diffraction orders is determined for the realized device.

1 Introduction

The market for spectroscopy is known as a growing market based on an increasing number of applications, for example in industrial process control, agriculture or environmental monitoring. These new applications are connected to increasing demands on the spectroscopic instruments itself. In particular, compact devices without moving parts are demanded which allow to acquire simultaneously a wide spectral range with high resolution. It is known that these specifications are contradictory and cannot be fulfilled by established compact spectrometers based on a concave grating mount. Classical Echelle spectrometers are able to measure a large spectral bandwidth with an extremely high resolution. However, they require a large volume and are very expensive [1]. Echelle-inspired cross-grating spectrometers are able to bridge the gap between these two instrumental concepts [2-4]. The cross-grating as the key element combines two dispersive functionalities in a single element which are separated in classical Echelle spectrometers. Therefore, it consists of two superposed blazed line gratings: The Echelle grating and the cross-dispersion grating. The Echelle grating is used in moderately higher diffraction orders (here 4th to 11th) and diffracts the incoming light into different overlapping spectral bands with a high spectral resolution. These overlapping diffraction orders are separated by the perpendicularly oriented cross-dispersion grating which is used in 1st diffraction order. The use of the combined cross-grating allows a substantial size reduction compared to classical Echelle spectrometers but requires new optical design approaches.

In this contribution, we report on the realization of a compact cross-grating spectrometer which covers a spectral range from 330 nm to 1100 nm with a resolving power $\lambda/\Delta\lambda$ higher than 500. As a starting point, we present the working principle and the optical design comprising a folded, all-reflective beam-path including a toric-convex mirror used for aberration compensation. Afterwards, we present the mechanical layout and discuss aspects of the system mounting and the adjustment process. Furthermore, optical key parameters like the spectral resolution are determined by measurements.

2 Optical design and implementation

The optical design is based on a folded Czerny-Turner setup with an additional toric correction mirror [3]. The light is fed to the spectrometer by an optical fiber which works as the entrance pinhole. The divergent beam is collimated by a concave mirror and propagates to a toric correction mirror for aberration compensation. In particular, astigmatism has to be corrected which is introduced to the system by the subsequent following cross-grating. It is illuminated at an oblique angle of 22.9 degree and diffracts the light in two orthogonal directions. The light propagates to the focusing mirror and is imaged onto the 2D detector. Corresponding to the separated diffraction orders, linear spectral bands are formed on the detector. Due to the spectral sensitivity of the silicon-based detector, the detectable spectrum ranges from approximately 330 nm to 1100 nm. For the mechanical design of the prototype, standard catalogue elements are

* Corresponding author: Matthias.Kraus@eah-jena.de

https://doi.org/10.1051/epjconf/202023803005

preferentially used. This allows an inexpensive setup at the cost of slightly increasing space demands. Figure 1 shows a photograph of the realized setup. The full system is attached to a stable base plate which is part of the stray light reducing black housing.

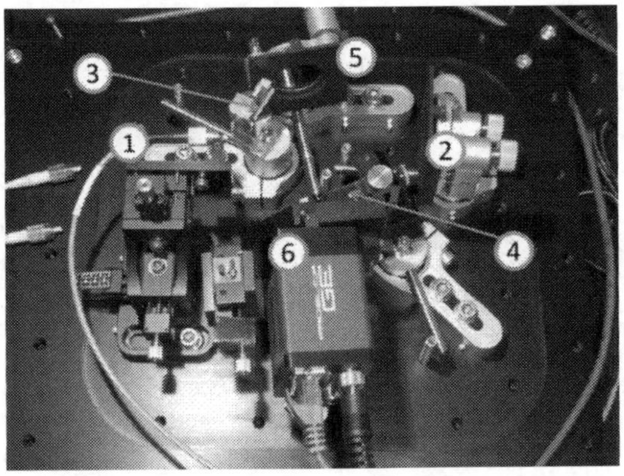

Fig. 1. Photograph of the realized spectrometer. The different mounted components are labelled with the following numbers: fiber (1), collimating mirror (2), toric correction mirror (3), cross-grating (4), focusing mirror (5) and detector (6).

The light is lead into the housing through the optical fiber and is collimated by a spherical mirror and a toric correction mirror. The mirror's radii of curvature are infinity in one direction and approximately 12 m in the other direction. In order to provide a mirror with the required sag height of about 4 µm, we deformed a plane mirror in a precise bending mount. The resulting sag height is controlled with a Fizeau-Interferometer.

In the beam-path follows the cross-grating which is a superposition of two blazed gratings and is manufactured by direct laser beam writing. The Echelle grating has a period of 10 µm and a height of 1.5 µm while the perpendicularly oriented cross dispersion grating has a period of 5.2 µm and a height of 150 nm [4]. The light, which is diffracted by the cross-grating, is imaged onto a monochrome camera (Allied Vision, GE2040) by the focusing mirror. In order to determine the spectral resolution and precise wavelength range for each diffraction order, different gas emission lamps (HgCd, HgAr, Na, Cs) were used. We measured the pixel distance between known spectral lines and the full width half maximum (FWHM) size of each spectral line. Based on the ratio of pixel distance to wavelength distance between these lines, the covered spectral range and spectral resolution for all diffraction orders can be derived. We determined a resolving power $\lambda/\Delta\lambda$ higher than 500 for the full spectral range. The central wavelength range, which is imaged with low aberrations on the detector, an even higher resolving power of more than 800 is obtained. Figure 2 shows an acquired spectrum of a HgAr-lamp. The measured intensity distribution shows distinct spectral lines in different diffraction orders (4th to 6th diffraction order of the Echelle grating) which are labelled with table values [5]. The inset shows a magnified view of specific spectral lines. The wavelength pair 810.37/811.53 nm are clearly separated. This corresponds to a resolving power of ~700.

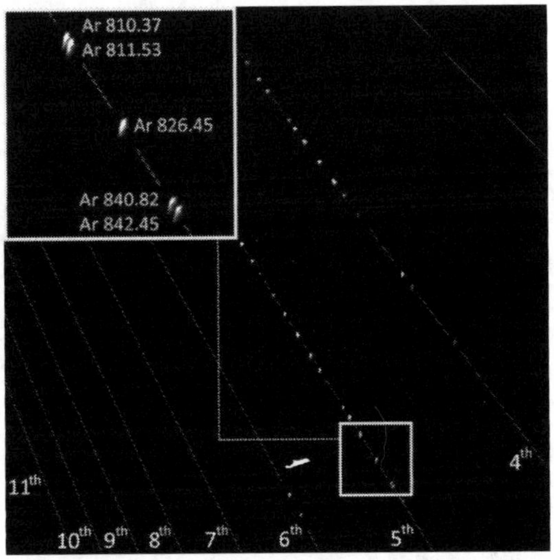

Fig. 2. Acquired line spectra of a HgAr-lamp. Spectral lines appear in the 4th to the 6th diffraction order. The image positions of all usable diffraction orders are marked by dashed lines. The inset shows a magnified view of specific spectral lines.

3 Conclusion

This contribution presents the realization of a compact cross-grating spectrometer with an all-reflective, folded beam path. The utilized cross-grating is a superposition of two blazed line gratings and allows to address a wide spectral range with a high resolution. Starting from the basic concept and the optical design, we report on the mechanical layout and the experimental demonstration of the setup. We proved by measurements that the realized setup covers the spectral range from 330 nm to 1100 nm and allows a resolving power of more than 500.

References

1. T. Eversberg, K. Vollmann, *Spectroscopic Instrumentation*, (Springer, Heidelberg, 2015)
2. D. Thomae, T. Hönle, M. Kraus, V. Bagusat, A. Deparnay, R. Brüning, R. Brunner, Appl. Opt. **57**(25), 7109-7116 (2018)
3. V. Bagusat, M. Kraus, E. Förster, D. Thomae, T. Hönle, R. Brüning, H. Hillmer, R. Brunner, J. Opt. Soc. Am. A **36**(3), 345-352 (2019)
4. M. Kraus, E. Förster, V. Bagusat, T. Hönle, X. Uwurukundo, P. Bohnert, R. Brüning, H. Hillmer, R. Brunner, Appl. Opt. **59**(5), 1338-1346 (2020)
5. Kramida, A., Ralchenko, Yu., Reader, J., and NIST ASD Team (2019). NIST Atomic Spectra Database (ver. 5.7.1), [Online]. Available: https://physics.nist.gov/asd [2020, March 2].

Optimising zero-order suppression in ion-exchanged phase gratings

Carlos Montero-Orille[1,*], *Héctor* González-Núñez[1], *Xesús* Prieto-Blanco[1], *Vicente* Moreno[1], *Dolores* Mouriz[1], *María C.* Nistal[1], and *Jesús* Liñares[1]

[1]Quantum Materials and Photonics Group, Department of Applied Physics, Faculty of Physics / Faculty of Optics and Optometry, University of Santiago de Compostela, Campus Vida s/n, E-15782, Santiago de Compostela, Galicia, Spain

Abstract. Ion-exchange in glass is a well-known technique to fabricate phase optical elements. For elements with reduced dimensions, the side diffusion, intrinsic to ion-exchange processes, can affect the performance of these elements if it is not taken into account. Here we present a procedure to optimise the zero-order suppression of ion-exchanged phase gratings.

1 Introduction

Ion-exchanged phase gratings are very versatile optical elements [1]. They can be used in spectrometry, wavelength division multiplexing, power division or optical pulse compression. On the other hand, ion-exchange is a well established technique to modify the refractive index of a substrate. This modification can be made in different areas by using appropriate masks. Ion-exchange has several advantages, namely, simplicity, low cost and easy control of the fabrication parameters. Moreover, the flat surface of the elements fabricated by ion-exchange permits straightforward cleaning and antireflection coating. The fabrication of high performance elements with reduced dimensions by ion-exchange is a delicate task. In particular, it is essential to take into account the side diffusion inherent to any thermal ion-exchange process which produces a transition region [2]. If this is not done, the phase profile of the grating as well as its performance will be affected.

Here we present the design and fabrication by ion-exchange of a phase diffraction grating with its zero-order suppressed. Removal of the zero-order of diffraction is desirable in most applications, including digital holography, spatial light modulation and grating fabrication [3, 4]. As it is known, a binary phase grating does not have zero-order if its phase profile is antisymmetric and its phase step is π. However, this is not the case for an ion-exchanged grating due to the mentioned side diffusion (see Fig. 1). Therefore, in this work, we show how to compensate the phase introduced in the transition region by a proper design of the ion-exhange process and the mask (lithographical process) used to produce the diffraction grating.

2 Materials and Methods

Soda-lime glasses of 1 mm thickness and refractive index n_s=1.5103, for λ=632.8 nm, have been used as substrates. The grating fabrication was made in these substrates by thermal ion-exchange and using aluminium layers as a mask for the ions. A binary grating pattern was translated from a sheet of heavy paper to the substrate by a two-step photolithographic process. The binary pattern was designed taking into account the side diffusion inherent to the ion-exchange process. The first photolithographic reduction was made by a lens Schneider XENON Sapphire 4.5/95, using a magnification of M=-1/14.1. In the second step a Zeiss S-Planar 1.1/68 lens performing a reduction 1:5 was used. The final result was an aluminium binary grating on the substrate. As for the ion exchanges, we used a salt mixture 5% mole $AgNO_3/NaNO_3$ at 340 °C. The ion-exchange modification produced by this salt in these glasses was analysed by using the in-

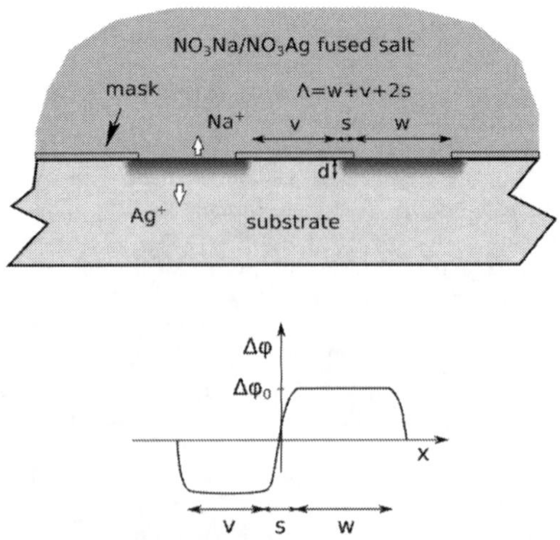

Figure 1. Scheme of an ion-exchanged grating and its phase profile. The side diffusion gives rise to a gradient in the phase profile.

*e-mail: carlos.montero@usc.es

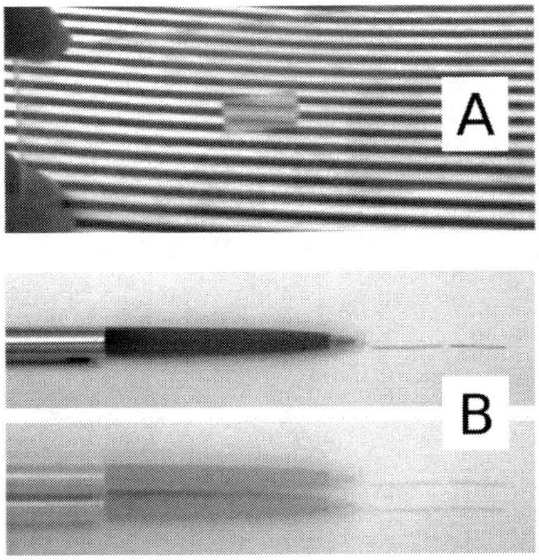

Figure 2. Image through one of the gratings fabricated of: A) a black and white object and B) a long object.

verse Wentzel–Kramers–Brillouin (IWKB) method. As it is known, this method recovers the gradient index profile following an ion-exchange process, which can be represented by a function $n = n_s + \Delta n(x/d)$ for a 1D ion-exchange, with d an effective depth [5]. After determination and integration of this profile, the phase step ($2\Delta\varphi_0$) generated by the ion-exchange can be calculated.

With regard to the design of the grating, we recall that our goal was to minimise its zero-order. As we said, the phase profile must be antisymmetric for this purpose. This means w=v (see Fig. 1) and hence the period of the grating must be Λ=2(w+s). On the other hand, due to the isotropic nature of the ion exchange, the size of the side diffusion (s) is similar to the size of the in-depth diffusion (s≈d). Moreover, both effective depths are proportional to the maximum phase step ($2\Delta\varphi_0$) and this phase step must be set properly. Indeed, for a moderate side diffussion, we must look for a maximum phase step slightly bigger than π to compensate for the minor phase step that the side region gives. Therefore, the design of both the aluminium mask and the ion-exchange process are linked and must be done together. After fabrication of the aluminium mask, we fabricated several gratings by changing the ion-exchange time in order to find the optimal one: the grating with the minimum zero-order. The performance of these gratings was assessed by measuring the efficiency of their zero-order through two methods: 1) directly, by using a He-Ne laser and a power meter, and 2) by measuring their spectral transmittance with a spectrometer.

3 Experimental results

We fabricated two gratings designed to suppress the zero-order at λ=632.8 nm. We chose their periods to be Λ=60 μm and Λ=30 μm. Hence their dimensions were w=v=25 μm and w=v=10 μm, respectively, because we estimated that an in-depth diffusion $d\approx 5\mu$m was needed. In order to look for the properly phase step, a calibration of the ion-exchange process was made based on the diffusion time [5]. We found that an ion-exchange time of t=12.6 min gives rise to $2\Delta\varphi_0=\pi$ in the glasses used. Because a bigger than π maximum phase step should be necessary to minimise the zero-order, we made several gratings with longer times and measured their performance by the two methods commented in the previous section. We found that ion-exchange times of t=15.5 min, for the grating with period Λ=60 μm, and t=28.2 min, for the grating with period Λ=30 μm, gave the best results: gratings with a zero-order efficiency T_0<0.5%. The phase steps of both gratings were $2\Delta\varphi_0=1.11\pi$ and $2\Delta\varphi_0=1.52\pi$, respectively, supporting these results our initial hypothesis.

We show, in Fig. 2, the image of two objects through the grating of Λ=30 μm. In Fig. 2A the grating can be seen in the central rectangular region of the microscope slide used as a substrate, where the zero-order suppression is observed. The white region of the object that is behind becomes blue at the zero-order, because we did the design for red light. On the other hand, Fig. 2B shows a pen without (up) and with (down) the grating in front of the camera lens. The double image proofs that the ±1 orders are the most relevant ones.

4 Conclusion

We have presented the fabrication by thermal ion-exchange of phase gratings with their zero-order suppressed. The design of these gratings has been made by taking into account the side diffusion inherent to any thermal ion-exchange process. In particular, a specific design of the mask needed for the fabrication must be done. The removal of the zero-order that we achieved is excellent: its efficiency falls below 0.5%.

Funding. This research was funded by the 'Ministerio de Economía y Competitividad', Central Government of Spain, grant number AYA2016-78773-C2-2-P, the 'Fondo Europeo de Desenvolvemento Rexional' (FEDER), and by 'Consellería de Educación, Universidade e FP, Xunta de Galicia', under Grant GRC No. ED431C 2018/11.

References

[1] R.P. Salmio, J. Saarinen, Applied Physics Letters **66**, 917 (1995).
[2] J. Liñares, E. López-Lago, C. Montero and X. Prieto, Pure Appl. Opt.A, **5**, 919 (1996)
[3] W.D.G.D. Improso, G.A. Tapang, C.A. Saloma, *Suppression of Zeroth-Order Diffraction in Phase-Only Spatial Light Modulator*, Vol. 222 of *Springer Series in Optical Sciences* (Springer, Cham, 2019).
[4] S.J. Mihailov, C.W. Smelser, P. Lu, R.B. Walker, D. Grobnic, H. Ding, G. Henderson, J. Unruh, Opt. Lett. **28**, 995 (2003).
[5] X. Prieto-Blanco, C. Montero-Orille, V. Moreno, E.F. Mateo, J. Liñares, Applied optics **54**, 3308 (2015).

Scientific-technical investigations for development of optical multifunctional components for deflection, splitting and shaping of light beams

Sebastian Henkel[1,*], *Andrea* Barz[1], *Jens* Bliedtner[1] and *Kurt* Kleinen[2]

[1]Ernst-Abbe University of Applied Sciences, Carl-Zeiss-Promenade 2, 07745 Jena, Germany
[2]MICROS Optics GmbH & Co. KG, Taubenbacher Weg 80, 07318 Saalfeld, Germany

Abstract. Many optical applications require splitting or deflection of light beams. This usually is achieved by usage of standard plane optical elements such as mirror systems or prismatic elements. However, these components are usually limited to one certain function and the manufacturing process for achieving precise functionality is very demanding. The presented research topic shows the development of new optical multifunctional components that introduce certain advantages compared to existing plane optical components. The basic roof-element configuration is based on glass cuboids joined together under a certain angle that can be fine adjusted during the manufacturing process. The components are therefore rather simply built and robust, and can fulfil different optical functions simultaneously by modifications of the glass cuboids. A process chain was developed for efficient and flexible manufacturing of such plane optical elements. Technical investigations were made on novel manufacturing processes such as ultrasonic assisted grinding and ultrafine grinding for fabrication of the cuboid parts, as well as diffusion joining being one possibility to create the basic roof-element shape.

1 Introduction

Optics designers can choose from a broad variety of plane optical elements. Reflective prisms for example exist in many different geometries to achieve all kinds of beam or image splitting, as well as image inversion in optical systems, such as telescopes, cameras or digital projectors. In contrast to spherical optics, plane optics are not used for optical magnification or optical reduction of an image. Instead, these components are used to change the direction of light beams, for example to change the orientation and position of an image or to achieve a parallel offset. A beam path can be advantageously "folded" with these elements within an optical system to reduce its overall length. Beam splitting components, such as splitter plates or splitter cubes, in turn, serve to split a beam into several partial beams. When it comes to functionality all existing parts suffer from the fact, that they can be usually used only for image inversion or beam splitting or light deflection. So, they are not multifunctional and often demand high manufacturing efforts. Due to this a new development of innovative adjustable roof elements, which can fulfil various optical functions simultaneously and are of simple and robust design, was sought. These new multifunctional elements can be used for beam splitting and guidance as well as for beam focusing or expansion of laser and light beams. Their application is also conceivable in measuring systems or laser material processing machines.

2 Basic configuration and possible variations of new multifunctional components

The basic configuration, a so-called roof element, consists of two glass cuboids joined together under an angle of 90° either by an adhesive or by the process of diffusion joining. The cuboids can be coated with arbitrary metal or dielectric coatings chosen for the desired application. What makes the element unique is the possibility of fine adjusting the angle between the cuboids in the range of a few arc seconds during or after the joining process by inserting a small wedge-like element between the cuboids as seen in fig. 1. Therefore, less precision in angle manufacturing is needed, since the element can be fine adjusted afterwards. This is applied for patent [1].

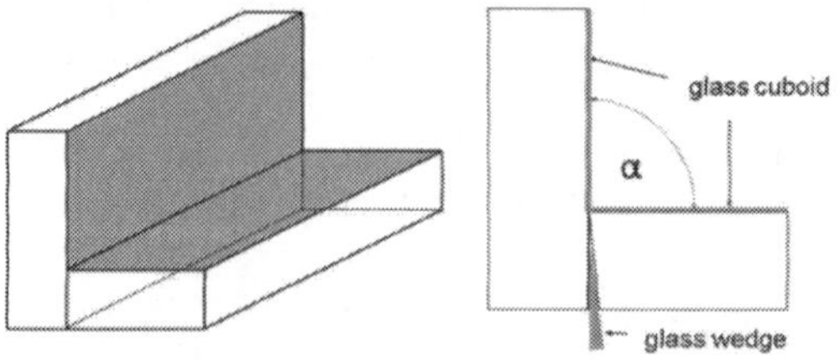

Fig. 1. Schematic view on adjustable roof-element: 3D illustration of basic set-up (left); joining and simultaneous angle adjustment by wedge positioning (right)

* Corresponding author: sebastian.henkel@eah-jena.de

A high modularity and flexibility can be achieved by variations of the basic configuration. A usage as a multiple beam-splitter is possible using partially transmissive coatings (fig. 2 left). Altering the shape of the glass cuboids (e.g. adding spherical or aspherical surfaces) offers the chance of focussing or shaping the incident light beams in addition to a beam split (fig. 2 right).

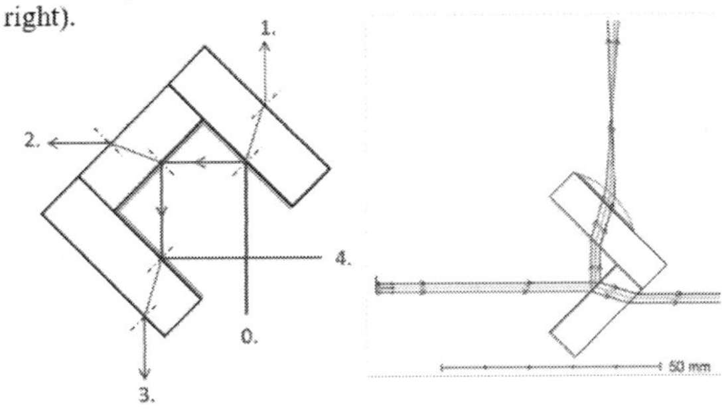

Fig. 2. Examples of component modifications for multiple beam splitting and beam shaping

Further ideas include the addition of a freeform surface or a microstructure on a cuboid surface as well as usage of λ/4 – coatings for changing the polarisation conditions of light rays.

3 Novel process chain for component manufacturing

The presented project also includes investigations on novel manufacturing technologies to offer an improved process chain for component production. It is aimed to replace or reduce lapping or polishing steps, especially for more complex surfaces, which would be normally used in optics manufacturing and are rather time consuming. This can be achieved by implementing the technologies ultrasonic assisted (US) grinding, ultrafine grinding and a more deterministic polishing process using NC-controlled machines and kinematic simulations for uniform material removal. Fig. 3 shows a potential process chain.

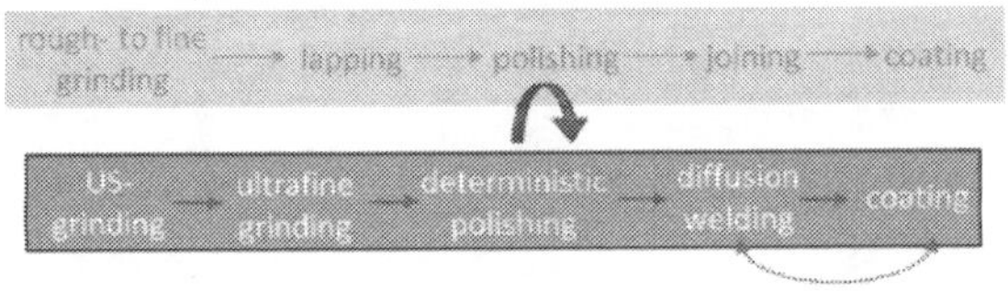

Fig. 3. Conventional process chain (above) and novel process chain (below) for optics manufacturing

US grinding can be done on a 5-axis grinding machine, offering high shape flexibility. This technology is characterized by a high frequency, longitudinal oscillation of the tool in addition to its rotation. This kinematic superimposition results in higher cutting speed [2] and reduction of process forces up to over 40 %. Own investigations and literature have shown the tendency to decrease sub surface damages. The application of an ultrafine grinding process afterwards on the same 5-axis machine can strongly reduce the surface roughness of the glass components. Special resin bond tools from company GÜNTER EFFGEN GMBH are used to achieve roughness values down to Rq = 10 nm on fused silica. This shortens the time for an afterwards polishing process over 50 % compared to polishing a fine lapped surface, regarding a polishing roughness goal of Ra = 5.5 nm (tactile measurement). A further innovative technology of the process chain is diffusion joining. Applying pressure and heat to the joined elements enables a material bond with a strength comparable to the bulk material of the glass.

4 Results and summary

Next to the preparation of the glass cuboids and the process investigations, a set-up for mounting the elements and carrying out the angle fine-adjustment was constructed and fabricated. The mounting device with the clamped glass cuboids was inserted into the beam path of a goniometer. Using suitable fine adjustment screws, the internal angle of the roof-element can be successively changed via a freely movable aluminium leg and measured inline until the desired angle is reached. The wedge is then inserted to keep the position stable and the joint is fixated by UV-curing an adhesive film. This set-up enables a high precision angle adjustment with deviations to about 3 arc seconds. During curing of the adhesive, the deviations will increase but still stay below 20 arc seconds. This corresponds to the specified requirements of 90° prisms for higher demands. [3][4]

The investigations have shown, that it is possible to build novel multifunctional optical elements with fine adjustable angle precision based on a simple and cost-effective basic roof-element configuration. Different functions of light deflection, beam splitting, beam shaping and more can be achieved, which was also proofed with some simple demonstrators. Furthermore, technological investigations on a novel process chain for manufacturing such elements were carried out. Based on the basic roof-element design it is possible to create inexpensive, lightweight and monolithic optical elements flexibly adaptable to many different required applications.

The authors gratefully acknowledge financial support by the German Federal Ministry of Economy and Energy in the funding program ZIM (funding reference: ZF4039608DF7).

References

1. EP 3 528 027 A1: Optical assembly for changing the direction of light beams with at least two mirror parts. (2019)
2. C. Schopf, R. Rascher, P. Sperber: Vor- und Nachteile der Ultraschalltechnologie beim Schleifen von optischen Flächen. *Future Glass – from the Tradition to the High-Tech-Product, Focus: Optical Glass*, Zwiesel. (2010)
3. https://www.thorlabs.de/newgrouppage9.cfm?objectgroup_id=142; viewed on 24.03.2020.
4. https://www.thorlabs.de/newgrouppage9.cfm?objectgroup_id=142; viewed on 24.03.2020.

Investigation of extended depth-of-field f/8 camera with optimized cubic phase mask and digital restoration

Po-Sheng Chiu[1,2*], *Kurt* Vonmetz[2], *Federico* Canini[2], and *H. Paul* Urbach[1]

[1]Optics Research Group, Dept. of Imaging Physics, Delft University of Technology, Lorentzweg 1, 2628 CJ, Delft, the Netherlands
[2]Datalogic IP Tech S.r.l., Via San Vitalino 13, 40012, Calderara di Reno (BO), Italy

Abstract. An investigation of extended depth-of-field camera with optimized phase mask and digital restoration is presented. The goal of this paper is to implement the wavefront coding technique without affecting much of the original design, and the design has taken the complexity of imaging system into consideration. The optimized strength of cubic phase mask (CPM) is based on the analytical optimal solution for the task-based imaging system [J. Opt. Soc. Am. A 25, 1064 (2008)]. The noisy intermediate images of CPM system with highest spatial frequency of interest can be effectively restored by vector-based Richardson-Lucy algorithm. Restoration from the system with generalized CPM produces precise image position than the system with CPM does. In general, the CPM system procures modulation transfer function higher than 0.195 in the whole depth-of-field, and the mean squared error of the restored images are less than 5 %.

1 Introduction

Wavefront coding involves modification of the wavefront such that the imaging system is less sensitive to defocus, the cubic phase mask (CPM) is initially chosen [1]. The intermediate images of the wavefront coded system are blurred and it can be restored with a single kernel [2]. Moreover, the analytical expressions for the modulation transfer function (MTF) of system with CPM are given in the literature [3]. CPMs for extending depth-of-field (DOF) has been used in applications, such as, iris recognition [4], and microscopy [5]. In this paper, we implement CPM in a f/8 camera without affecting much of the original design. The challenge of the application is to capture a sufficiently good image in the range of DOF.

2 Image defocus, formation, restoration and evaluation

We choose to place CPM in the exit pupil plane has not only the design purpose that the total amount of light capture in the camera will pass through the exit pupil, but also the practical reason that the exit pupil is right in front of the compound lens of the system. The object distance is defined as the the distance from object plane to first principal plane (FPP), and the image distance is defined as the distance between second principal plane (SPP) and image plane. The focal plane of the camera is designed at 170 mm in front of FPP, hence by the thick-lens model we can calculate the location of image plane. The path length error W in the normalized exit pupil plane (x, y) can be determined by subtracting the ideal phase distribution from the out of focus phase distribution. The location of the image plane in the camera is fixed, and the object distance is varied. Therefore, W is given by

$$W(x,y) = \frac{1}{2}\left(\frac{1}{\left(\frac{1}{f}-\frac{1}{\widetilde{d_o}}\right)^{-1}-d_{exit}} - \frac{1}{\left(\frac{1}{f}-\frac{1}{d_o}\right)^{-1}-d_{exit}}\right)(x^2+y^2), \quad (1)$$

where f is the effective focal length, $\widetilde{d_o}$(170 mm) is the focal plane of the object, d_o is the arbitrary object plane between 50 mm (1.203 λ) and 300 mm (-0.231 λ), and d_{exit} is the distance from SPP to exit pupil. Image formation (I) by a spatially invariant system under incoherent illumination can be written as the convolution of its PSF with the object (O) by the following

$$h * O = I \leftrightarrow H \cdot G_o = G_i, \quad (2)$$

where h is the system's PSF, $*$ is the convolution operator, and noise usually appears in detected images; the capital letters correspond to the spatial frequency by the Fourier transformation, and the H is the system optical transfer function (OTF), which equals to the Fourier transform of the PSF. The PSF of an incoherent system is given as $h(x,y,W) = \left|\iint p(x,y)\, e^{-i2\pi(xu+yv)}\, dxdy\right|^2$, where u, v are the normalized spatial coordinates. The pupil function in our system $p(x, y)$ is written as

$$p(x,y) = \exp\{ik[\phi(x,y) + W(x^2+y^2)]\}\,\mathrm{circ}(x,y), \quad (3)$$

where ϕ is the modulation introduced by the phase mask, and W is the defocus parameter as given in equ (1), and the circ function is defined as 1 inside the circular pupil and 0 outside of it. The modulation introduced by a CPM is expressed as $\phi_{CPM}(x,y) = \alpha(x^3+y^3)$, where α is the CPM parameter. Another commonly used mask is generalized cubic phase mask (GCPM), and given by $\phi_{GCPM}(x,y) = \alpha(x^3+y^3) + \beta(x^2y+xy^2)$, where $\beta = -3\alpha$

* Corresponding author: p.chiu@tudelft.nl

is usually assumed [6]. The favor of choosing GCPM is because it can produce a PSF which is approximately the same in each direction. The restored image will be the deconvolution of the intermediate image and the system PSF. Due to the ill-posed problem for inverse calculation, we investigate two methods for solving this issue in the restoration. The first one is the Wiener filter restoration,

$$I_W = \mathcal{F}^{-1}\{(H_{dec}{}^T H_{dec} + \lambda^2 I)^{-1} H_{dec}{}^T G_i\}, \quad (4)$$

where $\mathcal{F}^{-1}$ stands for inverse Fourier transform, H_{dec} is the OTF restoration kernel (assume zero defocus OTF of the wavefront coded system), T is the transpose of the matrix, and λ is regularization parameter determined by L-curve [7]. The second method is the vector-based Richardson-Lucy algorithm [8], it calculates the direction of next point as the different between the current iteration and the previous iteration. If x_k is the iterated point after k iterations, y_k the predicted point, h_k the direction vector, and α_k the acceleration parameter then:

$$y_k = x_k + \alpha_k h_k, \quad (5)$$

where $h_k = x_k - x_{k-1}$, $x_{k+1} = y_k + g_k$, $g_k = \psi(y_k) - y_k$. ψ is called the Richardson-Lucy function which expresses as $y_{k+1} = y_k\left(h \otimes \frac{I}{h * y_k}\right) \equiv \psi(y_k)$, where $\otimes$ is the correlation operator, and $h * y_k$ is referred to as the reblurred image after k iterations. We use MTF as the quality assessment in optical design. Furthermore, we choose mean squared error (MSE) and structural similarity (SSIM) for the image evaluation metrics between the restoration image and the reference object.

3 Simulation

USAF 1951 test chart is our reference object, and we focus on group 1, element 4 which has the resolution of 2.83 line-pairs/mm. The following simulation is based on camera with pixel number 1024 x1024, pixel size 3 μm and 35 dB signal-to-noise ratio. Figure 1. displays restoration of the near field object with defocus of 1.203 λ by using different phase masks and different algorithms. Vector-RL algorithm generate visually good images compare with the result from Wiener filter, besides, images form the conventional system cannot be restored with a single kernel. The effect of image shift of the CPM can be observed in row profile picture in Figure. 2. These artifacts result from the phase mismatch between convolution kernel and deconvolution kernel. From Table 1, iterative method gives better result than Wiener filter does in near field, besides, system with GCPM provides higher SSIM and smaller MSE in same condition. Further, MTF of GCPM system is lower than CPM system because GCPM introduces more aberrations in the pupil function.

Table 1. Overview of MTF, MSE and SSIM of restored USAF 1951 images. The values in MSE and SSIM are in the order of Wiener filter/vector-RL 10 iterations/vector-RL 50 iterations

Phase mask	Defocus (λ)	MTF	MSE	SSIM
CPM	1.203	0.425	0.040398/0.031871/0.035072	0.3011/0.5110/0.44207
	0	0.284	0.009372/0.002035/0.002933	0.4997/0.8270/0.7433
	-0.2316	0.195	0.009635/0.002046/0.003124	0.4854/0.8105/0.71439
GCPM	1.203	0.208	0.039017/0.018688/0.021098	0.2103/0.5701/0.5441
	0	0.061	0.018683/0.012639/0.004558	0.3369/0.6819/0.7251
	-0.2316	0.035	0.018593/0.012655/0.004643	0.3297/0.6955/0.7322

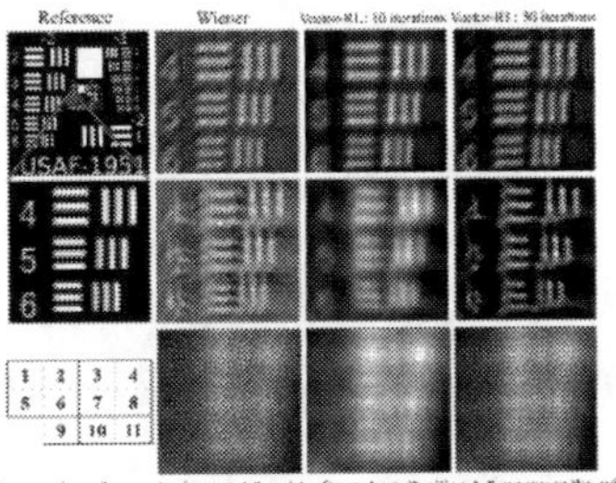

Fig. 1. Restoration of group 1, element 4,5 and 6 of test chart. Position 1,5 represent the reference object. Position 2,3,4 use CPM, 6,7,8 utilize GCPM, and 9,10,11 without wavefront coding.

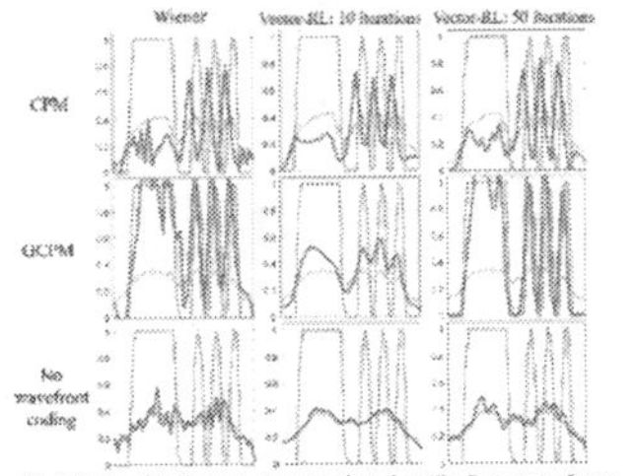

Fig. 2. Row profile of group 1, element 4 of test chart. Blue lines mean reference object, green lines mean intermediate image, and red lines shows restoration.

4 Conclusion & Outlook

In conclusion, we investigate wavefront coding technique in a f/8 camera for extending DOF. The design has taken lens thickness, pupil planes, and principal planes into account. The image restoration shows that the noisy images with highest spatial frequency of interest can be effectively restored. Furthermore, due to the antisymmetric characteristic PSF of the CPM system, we further examine the possibility of using GCPM in the imaging system. The outcome provides us a solution to mitigate the image shift, however, system with CPM provide higher MTF values in the same range of DOF. The fabricated CPM and GCPM will be implemented in prove of concept experiment. Moreover, how to quantify the image artifacts and restore without image shift from the CPM system is the future direction.

This research has received funding from the European Union's Horizon 2020 research and innovation program under the Marie Skłodowska-Curie Grant Agreement No. 675745.

References

1. E. R. Dowski, Jr., W. T. Cathey, Appl. Opt. **34**(11), 1859-1866 (1995)
2. W. T. Cathey and E. R. Dowski, Appl. Opt. **41**(29), 6080-6092 (2002)
3. S. Bagheri et al., J. Opt. Soc. Am. A **25**(5), 1051-1063 (2008)
4. S. Bagheri et al., J. Opt. Soc. Am. A **25**(5), 1064-1074 (2008)
5. Y. Zhou et al., Opt. Express **26**(6), 7563-7577 (2018)
6. S. Prasad et al., Proc. SPIE **5108**, 1-12 (2003)
7. P. C. Hansen, *Discrete inverse problem: insight and algorithms* (SIAM, Philadelphia, 2010), chap.4
8. D. S. C. Biggs, M. Andrews, Appl. Opt. **36**(8), 1766-1775 (1997)

Efficiency evaluation of a specialist in the field of optical design during education and work

Irina Livshits[1,*] and *Aleksandr Petukhov*[1]

[1]ITMO University, 49 Kronverksky, Saint-Petersburg, Russia, 197101

Abstract. Our goal is to evaluate the efficiency of a specialist in the field of optical design. The classification of specialists is given depending on their level of education and work experience in the field of optical design. The analysis of factors that determine the efficiency of the specialist is performed, on the basis of which the equation for calculating efficiency is derived. As an example, we present empirical data for calculating the efficiency of some specialists.

1 Introduction

The creation of optical devices is closely related to the development of technologies. In our century, the pace of their improvement has accelerated so much that specialists of one generation become participants in several technical revolutions. Such events include, for example, the introduction of computer technology in optical calculations in the 60s of the 20th century [1], the improvement of calculation technologies (Forbes aspherics [2] - the beginning of the 21st century, etc.) and the manufacture of aspherical surfaces, the use of new materials [3], the creation of methods for calculating and manufacturing free-form surfaces [4], the use of artificial intelligence in optical design [5], etc.

The desire of specialists to remain updated is the main motivation for continuing their education in the field of optical design and related fields. Therefore, these specialists are constantly trained, improved, conduct research, analyze the results obtained, exchange them at conferences and in various publications.

2 Specialist classification

The community of specialists is very heterogeneous in its composition. The article provides a classification of specialists depending on their level of education and work experience in the field of optical design, please, see Table 1.

Table 1. Classification of optical design specialists.

N	Level of specialists	Time of study/work, years	Indicator of professional level
1	beginner	3-5	0.3
2	advanced	6-20	0.7
3	expert	>20	1

A beginner is a student or employee who has recently graduated from a University and started working. He/she already has the theoretical knowledge and set of necessary skills to be an optical system design engineer.

Advanced specialist is an employee who feels confident in the profession. Such an employee applies technical skills freely. Usually he/she specializes in one of the types of optical systems.

Expert is a specialist who already has a lot of experience behind him. Such an employee can easily calculate any type of system. He/she is the generator of new solutions and schemes.

3 Efficiency equation

We live in the age of digital technologies, when most phenomena and events, organizations, etc. are classified, and then their ratings are determined.

We propose to represent the efficiency of specialists working as optical designers as a number that can be compared. This can be useful, for example, for employment or when choosing an invited speaker for a conference, choosing a research supervisor and opponent, and so on. There are known attempts to classify specialists in such systems as ORCID [6], Research Gate [7], ResearcherID [8], and some others.

The information contained in these systems was partially used by us, but we set the task of a deeper analysis of the efficiency of specialists in one subject area-optical design. As a result of this assessment, we have obtained expressions that take into account the main factors of efficiency of a University graduate throughout his/her career - from a student to an expert.

The number of publications, patents, and monographs is taken into account. We also consider it important to evaluate aspects that have not been considered before – "loyalty to the profession" and the indicator of creativity.

* Corresponding author: ecenter-optical0@yandex.ru

It seems necessary to take into account the factors defined by equations (1-3):

$$ODS=L*\sum_{i=1}^{3} P_i*T_i+C \quad (1)$$

$$L=\frac{N_p}{N_a} \quad (2)$$

$$C=K_p*M_p+K_a*M_a+K_m*M_m \quad (3)$$

where:
ODS - efficiency of a specialist in the field of optical design;
L - loyalty to the profession;
Np - number of years of work by profession;
Na - total work experience;
Pi – indicator of professional level;
Ti - number of years at level I;
C - creativity indicator;
Mp - number of patents;
Kp - the weight ratio of patents;
Ma - number of articles;
Ka - the weight ratio of articles;
Mm - number of monographs;
Km - the weight ratio of monographs.

We analyzed the effectiveness of two-generation opticians in the Soviet Union and Russia and determined the values of the coefficients of equation (1), which are used to calculate the creativity indicator (*Kp*=0.2; *Ka*=0.1; *Km*=1).

Table 2 shows an example of calculating the effectiveness of existing specialists, but the names and surnames have been changed. Ivanov is a second-year student with one patent and one article. Petrov is a young specialist, studied for 5 years (two years as an electrical engineer and 3 years as an optical design student), works for two years by profession, he has two patents and four articles. Ivanova is a specialist, she has studied optical design for 5 years, works for 5 years in her specialty, she has 5 articles and 5 patents. Petrova is an expert, she has studied optical design for 5 years, she has been working for 20 years in her specialty, she has 50 patents, 100 articles and 5 monographs.

Table 2. ODS calculation examples.

N	Name(sex)	L	$\sum_{i=1}^{3} P_i*T_i$	C	ODS
1	Ivanov(m)	1	0.3	0.3	0.6
2	Petrov(m)	0.6	2.1	0.8	2.0
3	Ivanova(f)	1	5	1.5	6.5
4	Petrova(f)	1	15	25	40

Thus, from table 2 it can be seen that ODS for a beginner (Ivanov, Petrov), advanced specialist (Ivanova) and expert (Petrova) differs seriously. The idea of including the creativity into equation (1) looks important as creativity has sometimes more value than even professional skills.

In terms of developing the presented idea, it is desirable to develop a number of tests that are appropriate for this subject area. To evaluate the effectiveness of a student, graduate, and post-graduate student, you can use the results of certification exams.

In conclusion, it should be noted that equations (1-3) can be used in other professional subject areas, provided that coefficients are identified in a specific subject area.

References

1. I.L. Livshits, D.C. Dilworth, Advanced Optical Technologies, **7**, 335 (2018)
2. http://www.opticsinfobase.org/abstract.cfm?URI=OFT-2010-OTuA6
3. https://doi.org/10.1364/OME.393986
4. J. Miñano, P. Benitez, et al. *Proc. SPIE*. **10590**, *International Optical Design Conference* (2017)
5. Р.В. Анитропов, И.Л. Лившиц, Д.В. Свищук, Известия высших учебных заведений. Приборостроение, **53**, 77 (2010)
6. https://orcid.org/
7. https://researchgate.net/
8. https//researcherid.com/

Efficient and innovative polishing processes for smoothing complex surface geometries and internal contours on brittle-hard components

Marcel Binder[1*], *Sebastian* Henkel[1], *Anne-Marie* Schwager[1], *Christoph* Letsch[1], *Jens* Bliedtner[1], , *Vladislav* Matusevich[2] and *Andrew* Matusevich[2]

[1]Ernst-Abbe University of Applied Sciences Jena, Carl-Zeiss-Promenade 2, 07745 Jena, Germany
[2]VM-TIM GmbH, Hans-Knöll-Str. 6, 07745 Jena, Germany

Abstract. The material fused silica, as well as other brittle-hard materials such as glass ceramics, have great potential for use in a wide range of applications due to their special material properties. The technical advantages of these materials require sophisticated processing technologies, including polishing steps, in order to be able to use these interesting materials advantageously. In addition, a current trend in modern optical manufacturing is the use of free-form surfaces and monolithic components that combine several optical and mechanical functions in one part. Novel or improved processes are needed in order to meet future requirements for resource-saving and effective production methods at the same time.

1 Methods – A Novel Process Chain

The production of complex glass components involves a great deal of effort and requires the use of advanced CNC technology. Especially the final surface treatment to create transparent, optically usable surfaces is a time-consuming and costly task. Because of this a new process chain was developed which is described below

Ultrasonic-assisted grinding is applied as a first step, to produce workpieces of any shape as well as free-form surfaces. The special kinematic principle, which is accompanied by high-frequency longitudinal tool oscillation, enables efficient manufacturing processes. By the determined reduction of process forces using ultrasound, either higher removal rates or a more gentle processing for tool and material can be achieved. [1] The surfaces produced in this way enable the application of new types of smoothing processes, which have considerable advantages over conventional polishing. The application of modern ultra-fine grinding processes with novel resin-bond tools enables average surface roughnesses which are already comparable to conventional pre-polishing. These tools offer numerous advantages in the CNC-controlled glass processing compared to conventional diamond tools. The diamond grains (D16) are embedded in a special resin bond, which has a certain elasticity comparable to polishing pads. This results in a sort of ductile grinding mode and a lower resulting roughness with already semi-transparent glass surfaces. The roughness values can be further reduced in a very short time by subsequent polishing with a CO_2 laser. With this process, very short polishing times of 4.8 s/cm^2 are possible, while at the same time a high surface quality can be achieved. [2] Polishing with CO_2 laser radiation is a thermal process in which the energy of the laser radiation is converted into thermal energy. Due to the resulting material heating and reduction of the viscosity at the sample surface, the material starts to flow. The surface roughness is smoothed by the surface tension, roughness peaks are removed and roughness grooves are filled in. [2] Another polishing process that is considered is the wet-chemical polishing of the described optical components. In contrast to laser beam polishing, there is no smoothing of the surface, but a removal of the upper defective material layer. This is achieved by etching the glass in liquid solutions using hydrofluoric acid (HF) and sulphuric acid (H2SO4). [3] As a special feature, ultrasonic support will be used for the first time in the planned experiments to increase process efficiency.

The presented development of a novel process chain consisting of ultrasonic grinding, ultra-fine grinding and the polishing processes laser beam polishing and wet-chemical polishing thus offers the possibility of a flexible and fast production of complex glass components, while still ensuring a high optical quality of the surfaces produced.

In addition to the complex outer contour requirements, there are also geometrical requirements for inner contours of complex glass elements, that require a very high aspect ratio (e.g. component length to component diameter). The special challenge here is the high quality of the drill hole wall, which must correspond to a nearly polished surface. The methods of ultrasonic-assisted grinding, ultra-fine grinding, laser beam polishing (LBP) and wet-chemical

* Corresponding author: Marcel.Binder@eah-jena.de

polishing described above should also be applied here. Figure 1 shows a schematic representation of the desired process chain.

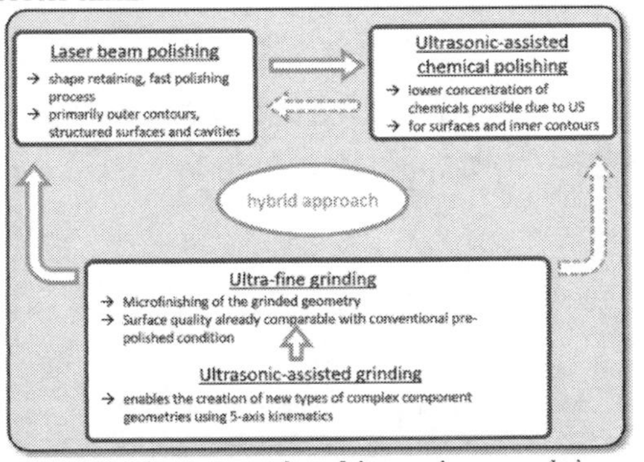

Fig. 1. schematic representation of the novel process chain

2 Examinations & Results

First results of the novel process chain are presented below. The methods were initially applied to external contours in order to gain a basic understanding of the process and to investigate the influence of LBP. Subsequently, first inner contours were realized on which the described polishing processes are soon to be tested.

The novel process chain was investigated using the example of a free-form made of fused silica. Grinding and ultra-fine grinding were performed with a spherical tool (ø 20 mm). A meander pattern was chosen with a line distance of 0.1 mm and a quasi point tool contact. This kinematic approach resulted in narrow grooves, which are visible in the direction of grinding. These can be smoothed by laser beam polishing carried out afterwards. Low roughness values of max. R_q = 72 nm can be already achieved after ultra-fine grinding. With the help of LBP and smoothing of the resulting grooves, the roughness can be reduced even further, to max. R_q = 12 nm. All roughness values were obtained by tactile measurement.

In order to investigate the feasibility of introducing inner contours with a large aspect ratio, drill holes with different diameters (2 mm, 3 mm, 5 mm, 7.5 mm & 10 mm) and a length of 50 mm were drilled into samples made of fused silica (*Corning HPFS 79805F*), ZERODUR® and CLEARCERAM®. In addition, diamond drills of different diameters (2 mm, 5 mm & 10 mm) were used for this purpose, whereby the matching bore diameters were realized by using a deep drilling process. The remaining diameters (3 mm & 7.5 mm) were drilled by using helical machining kinematics. Of particular interest were the resulting roughnesses of the borehole walls after the grinding / drilling process, which are compared in figure 2. The measurements were taken at the tool entry point as well as at the tool exit point to determine any roughness differences in the course of the hole.

Although the cutting speeds used were calculated to be approximately constant for each bore diameter, it is apparent that there are differences in roughness depending on the material, bore diameter and measuring point.

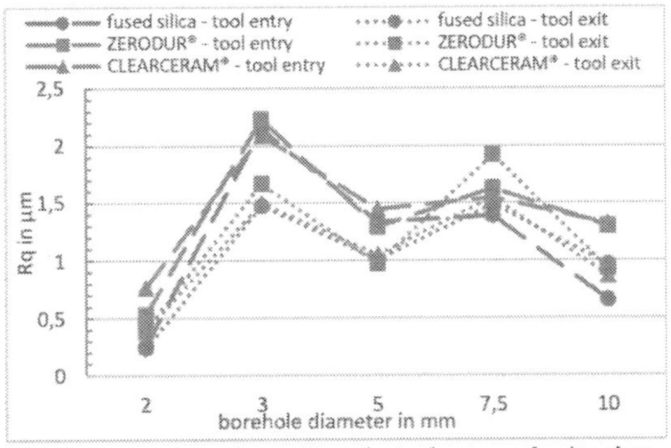

Fig. 2. measured roughness *Rq* at the tool entry and exit points

It can be seen, that for all materials, the 2, 3 & 5 mm diameter holes have higher roughness at the tool entry than at the tool exit. The same applies to the 10 mm diameter, with the exception of fused silica. Possibly the material has an influence on the process result here, which is currently still being investigated. In addition, the roughness at the tool entry tends to be lower or the same for all three materials for the 7.5 mm diameter borehole. The roughness differences between the two machining techniques are most noticeable. The roughness values for helical machining are generally higher (3 mm & 7.5 mm). The optimal parameter combinations for both kinematics are also currently being investigated.

3 Conclusion and Outlook

It could be demonstrated that the new process chain enables a fast and drastic reduction of roughness on complex outer geometries of brittle-hard materials. In the future it will be necessary to investigate whether this process chain can be used for the machining of internal contours with a high aspect ratio. The technical implementation of ultrafine grinding and laser beam polishing will be tested on the bores shown. Furthermore, the wet-chemical polishing for external and internal contours is to be investigated and compared with the described process chain in terms of economic efficiency and achievable surface qualities.

The authors gratefully acknowledge financial support by the German Federal Ministry of Economy and Energy in the funding program ZIM (funding reference: ZF4039613TV8).

References

1. M. Rost, J. Bliedtner, S. Henkel: Einflüsse des ultraschallunterstützten Schleifens auf silikatische Werkstoffe. In: Eltmann: *Diamond Business* **1/2016**, pp. 68-77. (2016)
2. S. Henkel, A.-M. Schwager, J. Bliedtner, et al.: New surface smoothing technologies for manufacturing of complex shaped glass components. In: *Proceedings of SPIE* Vol. **10448**. (2017)
3. H. Junjiang, Y. Jungxin, C. Wie, Z. Changhe: Experimental investigation of en-chancing the subsurface damage threshold of Nd-doped phosphate glass, In: *Chinese Optics Letters*, Vol. **6**, Issue 9, pp. 681-684. (2008)

 https://doi.org/10.1051/epjconf/202023803011

Ultra-fine grinding of inorganic non-metallic materials using various types of bonds and processing strategies

Christian Schulze[1,*], *Sebastian* Henkel[1], and *Jens* Bliedtner[1]

[1]Ernst-Abbe University of Applied Sciences Jena, Carl-Zeiss-Promenade 2, 07745 Jena, Germany

Abstract. Ultra-fine grinding is a technology, which can be used to process inorganic non-metallic materials. High quality surfaces with a low roughness, and high flatness are achievable in a low processing time. The experimental setup which is described shall compare face grinding with lateral grinding in a single processing step using various materials, bonds and processing parameters. The results are supposed to deepen the knowledge about this grinding technology to further improve the properties of fabricated components.

1 Ultra-fine grinding

Due to the developments of mankind in regard to machining materials, such as using high accurate CNC machines in collaboration with advancing tool technologies, high quality surfaces can be achieved. Ultra-fine grinding can be considered as part of precision grinding with low removal rates using tools with a small diamond grain size as shown in Fig. 1. [1]

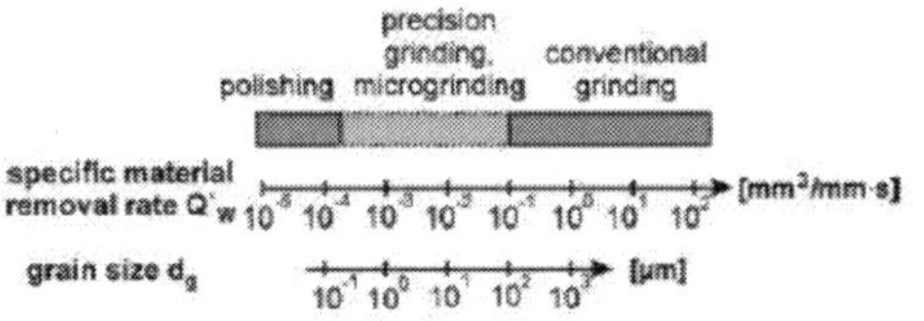

Fig. 1. Properties of grinding [1]

Especially when processing glass, high quality surfaces with a low roughness, high flatness and low processing times are demanded. Two strategies of grinding to fulfil those demands are face grinding and lateral grinding. [2] The operating principles are displayed in Fig. 2.

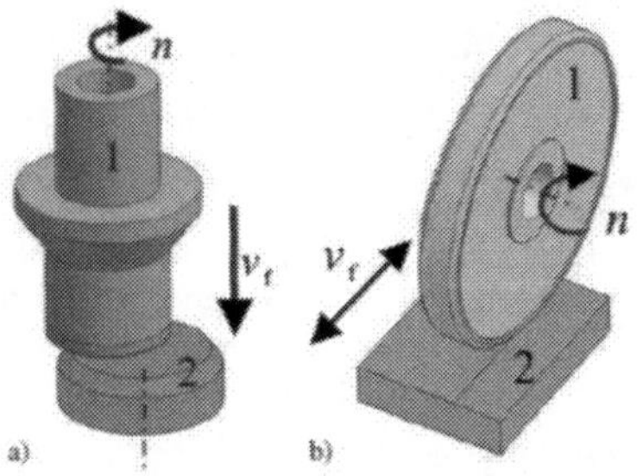

Fig. 2. Grinding strategies: a) face grinding b) lateral grinding 1: tool 2: sample [2]

2 Experimental setup

2.1. Description

The conducted experiments are based on former experiments and shall reveal the differences in lateral grinding in comparison with face grinding as well as a combination of lateral and face grinding in a single processing step, in regards to achievable surface roughness, occurring process forces as well as subsurface damages. In order to realise this experiment, a special sample geometry was designed and prepared, as shown in Fig. 3.

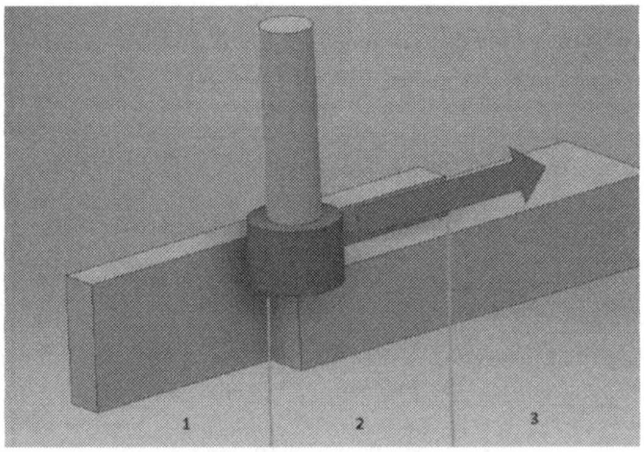

Fig. 3. Sample geometry and tool kinematic.

The samples used are 20x20x100 mm^3 in size. The tool will move in a straight line, processing 3 different areas. The first area will be lateral grinding only (1), the second area will be a combination of face and lateral grinding (2) and the third area will be face grinding only (3) as marked in Fig. 3. Due to the geometry of the sample as well as the tool, the cutting speed will have the same value for face grinding as well as lateral grinding.

* Corresponding author: christian.schulze@eah-jena.de

2.2 Parameters

A 5-axis CNC machine from company SAUER (DMG) will be used to conduct the experiments as pictured in Fig. 4.

Fig. 4. CNC grinding machine: Ultrasonic 20 linear.

Besides enabling a rotational speed up to 42000 rpm, an ultrasonic assistance is integrated in this system. It can induce a high-frequency oscillation (about 20-30 kHz) of a few µm alongside the z-axis of the machine in addition to the tool rotation. This results in numerous benefits: possible reduction of roughness and subsurface damages, wear reduction of the tool as well as force reduction. [3]

The process influence of the following parameters will be investigated:

- 2 materials: fused silica, BK7
- 2 diamond tools, ø14mm D16: metal bond, resin bond
- values of cutting depth
- values of cutting width
- values of cutting speed
- values of feed rate
- state of ultrasonic assistance (on, off)

The following parameters will be examined:

- roughness
- waviness
- subsurface damages
- process forces

3 Perspective

Ultra-fine grinding processes combined with modern tool technologies have proved themselves to significantly improve the quality of glass surfaces in already conducted experiments. As a result, transparent, reflective surfaces are achievable with a roughness improvement up to $Ra = 8$ nm and $Rq = 10$ nm by face grinding fused silica with a resin bonded tool as shown in Fig. 5. [4]

This results in a significant reduction of polishing time needed in following processes, reducing the necessary total processing time about 40 % (Fig. 6). [4]

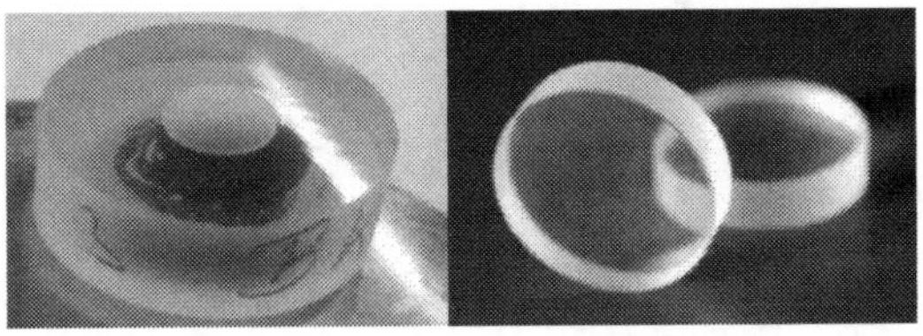

Fig. 5. Achieved surface qualities by ultra-fine face grinding

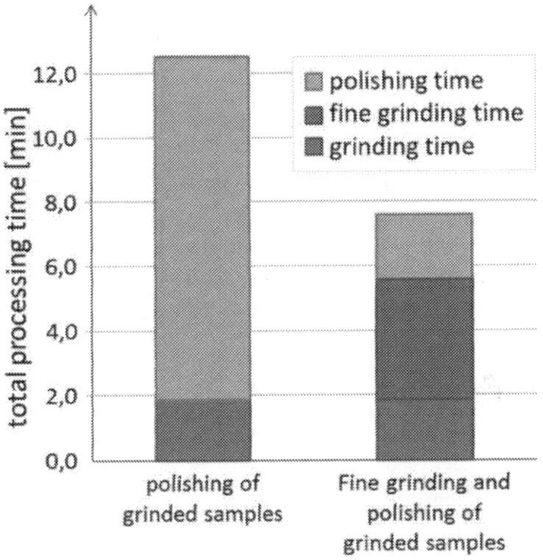

Fig. 6. Comparison of polishing time of fused silica samples by fine grinding using a resin bonded tool and samples grinded with a metallic bond D35 Tool. [4]

Therefore, more investigations are necessary in order to further improve the technology of ultra-fine grinding in terms of achievable surface quality as well as its economic aspects. The experiments focusing lateral and face ultra-fine grinding are currently in preparation. First results are soon to be expected.

References

1. E. Brinkmeier, Y. Mutlugünes, F. Klocke, J.C. Aurich, P. Shore, H. Ohmori : Ultra-precision grinding. *CIRP Annals, Manufacturing Technology*. Vol. **59**, Page 652-671. (2010)
2. J. Bliedtner, G. Gräfe, R. Hector : *Optical technology*. McGraw-Hill Education. (2011)
3. J. Bliedtner, S. Henkel, A.-S. Schwager, E. Rädlein, M. Gerhardt: Ultrafine grinding and laser beam polishing processes for the manufacturing of freeform optics. Delft (NL): *EOSAM 2018*, Presentation. (2018)
4. S. Henkel, J. Bliedtner, E. Rädlein, C. Schulze, M. Rost, M. Gerhardt, M. Fuhr : Ultra-fine grinding of silicate materials under the use of new resin bond diamond tools. *Proceedings of SPIE* Vol. **10326**, Teisnach: *4th European Seminar on Precision Optics Manufacturing*, 10326 05. (2017)

https://doi.org/10.1051/epjconf/202023803012

Rotation-free industrial alignment of high performance optics

Dennis Leenman[1,*], *Frederic* Berndt[1], and *Stefan* Beyer[1]

[1]Berliner Glas KGaA Herbert Kubatz GmbH & Co., Waldkraiburger Str. 5, 12347 Berlin, Germany

Abstract. Conventional production methods for high precision lens alignment typically rely on lens rotation. In the case of rotationally non-symmetric optics and mounts, this is problematic. Here, we report a new concept for alignment bonding without lens rotation, based on a high precision linear bearing as position reference and a hexapod actuator for lens manipulation. For the optical axis of two test lenses after bonding, <1 arcmin element tilt and <10 µm decentre was achieved. This is confirmed by an independent measurement. Our alignment device and process can be applied for any lens and mount geometry. This will be especially useful for high-end products with small size and for rotationally non-symmetric systems.

1 Conventional lens alignment

The assembly of high performance imaging lens systems requires extreme accuracy. Lens tilt below 2' and decentre below 10µm cannot be achieved by mechanical positioning, due to machining tolerances. For this reason, alignment turning processes are widely used for industrial production of high performance lenses [1]. This involves the mounting of each lens in a subcell and the subsequent adaptation of the mechanical interfaces to the optical axis of the lens by use of a turning machine with integrated optical measurement instruments. Finally, each subcell is integrated in the optical system.

1.1 Limitations

Inherent disadvantages of alignment turning include:

- It is not suitable for rotationally non-symmetric lenses and mounts
- Alignment precision is determined by the used turning machines for the subcells and also by the mechanical tolerances of the main mount
- The double mounting is space consuming

In medical imaging, for example, these limitations are highly relevant. Endoscopic devices must be small and do not allow for subcell mounting. Large optics with flat sides (for stereoscopic imaging), or prisms cannot be processed. Alignment of lens systems requiring separate mounts (e.g. zoomsystems), remains complicated.

2 New alignment device

In optical terms, the lens position is defined by the centres of curvature of both optical surfaces. The line through these centres is called the lens's optical axis. The position of optical surfaces can be measured by an autocollimator [2]. For alignment turning, the optical axis is made to coincide with the well-defined rotation axis, using the lens's rotational invariance.

In contrast, we use the axis of an autocollimator's linear bearing as reference. Furthermore, we use a hexapod actuator to tilt the lens about the centre of curvature, which is the point of rotational invariance in all directions, of the first lens surface. The lens is aligned and bonded directly in the optical system, avoiding any further position tolerances.

2.1. Process description

The optical surface positions are measured by an autocollimator on a high precision linear bearing. For curved surfaces, a head lens is attached, see **Fig. 1**.

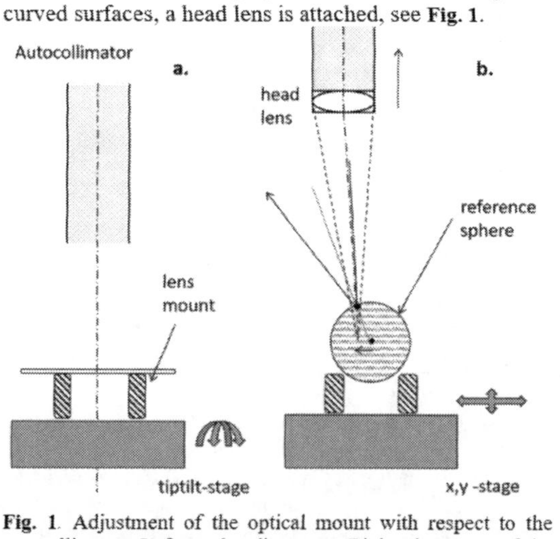

Fig. 1. Adjustment of the optical mount with respect to the autocollimator. Left: angle adjustment. Right: the centre of the reference sphere is adjusted to coincide with the focal point of the autocollimator head lens.

To measure different optical surfaces, the auto-collimator is translated along its axis (z). The focal point

[*] Corresponding author: dennis.leenman@berlinerglas.de

of the head lens thereby describes a line parallel to the autocollimator's translation axis. This line is the reference axis for the adjustment of the lens surfaces. When the head lens focal point and a centre of curvature lie in the same horizontal plane (x,y), the autocollimator measures the lateral distance between these points.

First, using reference mirrors and manual stages, the lens mount is adjusted in tilt (**Fig. 1**a) and translation (**Fig. 1**b), to reflect the autocollimator signal to the sensor's optical centre. Second, the lens is connected to a hexapod actuator and inserted into the mount. By lateral adjustment of the lens, the first centre of curvature C1 is moved to the reference axis (**Fig. 2**a). Then the hexapod's pivot point is placed (by software control) in C1. Tilting the lens about this point will not influence the first lens surface, while it allows the alignment of the second surface (**Fig. 2**b) until C2 and the reference axis overlap.

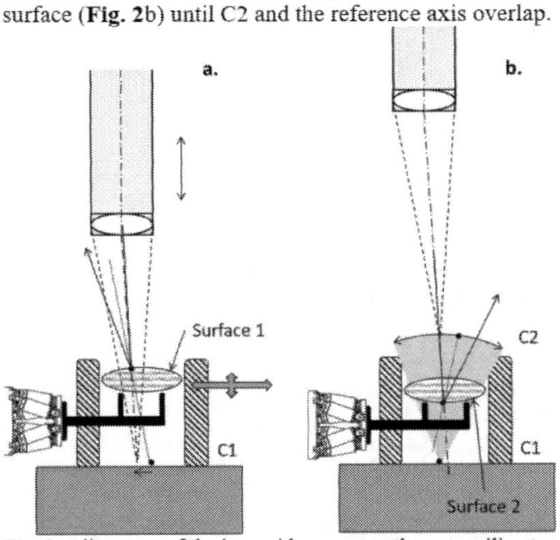

Fig. 2. Alignment of the lens with respect to the autocollimator. Left: lateral adjustment of the lens to move the first center of curvature C1 to the reference axis. Right: tilt of the lens to adjust surface 2, such that C1 is unchanged. Note that the mount position is not changed.

As a result of these process steps the mount front surface is orthogonal to the autocollimator's optical axis, and the lens optical axis is centred to the mount and parallel to the autocollimator's translation axis.

The process can be repeated for further lenses in the same mount. All lenses will be aligned to exactly the same reference axis. Since the centering errors are independent from mechanical interfaces of lens and mount, the mechanics can be kept relatively simple.

3 Test setup and results

Our alignment device contains an autocollimator AKG 300/40 (Möller-Wedel), a hexapod H-811 (PI), and a linear bearing LIMES 170 (OWIS), and is based on an existing alignment platform at Berliner Glas [3].

Two identical lenses (**Table 1**) were bonded in one mount, with 0.4mm adhesive gap. UV curing adhesive was used for lens fixation; an additional epoxy was applied for mechanical and temperature stability.

Table 1. Test system lens data.

Radius /mm	Index of refraction	Thickness /mm	Diameter /mm
R80.220	1.881	2.5	22
R39.625	1.497	4.6	22
inf	air	15	22
R80.220	1.881	2.5	22
R39.625	1.497	4.6	22
inf	air	-	22

Reproducibility measurements were conducted to characterize the setup, yielding <1µm measurement and alignment precision, and +-2µm stability of the reference axis over 500mm. Minor drift of the lens position was observed during UV curing and removal of the lens gripper. The largest error was caused by the cure of the additional epoxy.

The test assembly was measured independently on a centering measurement instrument OptiCentric (Trioptics) using the multilens software. The results of both measurements (**Table 2**) confirm that the displacement of the second lens centres of curvature with respect to the first lens optical axis is below 10µm. These results correspond to a tilt between the optical axes below 1 arcmin.

Table 2. Decentre measurement of the second lens with respect to the optical axis of the first lens.

		new alignment device	Opticentric
Surface decentre	1	6.1 µm	7.0 µm
Surface decentre	2	9.9 µm	3.5 µm
Tilt of the optical axis		0.3'	0.9'

The feasibility of high precision alignment by this new device is thus shown by two independent measurements. Further investigation of the measurement errors and improvement of the second adhesive bonding will establish the process limits. Based on the results for measurement and alignment reproducibility, we expect to reduce the centering error to 0.5 arcmin and 5 µm.

Rotation-free alignment bonding avoids the typical disadvantages of alignment turning, and can be considered a viable alternative for space sensitive products and rotationally non-symmetric optical systems.

Competing interests

Patent application 10 2020 107 298.8

References

1. P. Langehanenberg, J. Heinisch, C. Buß, C. Wilde *Optik&Photonik* (Wiley-VCH-Verlag, 2/2014)
2. Möller-Wedel Optical, *Beispiele zur Anwendung von Kollimatoren.* (Stand 18 März 2013)
3. S. Beyer, *Microalignment in der optischen Industrie* (tibb news 01/19)

Optical coatings – precision engineering on non-flat surfaces

Tammo Böntgen, Marc Neufert, Lars Jensen

Laser Zentrum Hannover e.V. – Hollerithallee 8, 30419 Hannover, Germany

Abstract. Manufacturing of optical coatings is about the precise deposition of material and its control on the nanometer scale. This has to be achieved for a stack of multiple layers with a total thickness that can reach several micrometer. In addition, this has to be maintained for the complete functional area. In this contribution we demonstrate a process capable of delivering these accuracies not only to flat but also to curved surfaces.

1 Introduction

Functional coatings are a fundamental part of most optical instruments. They enhance performance by supressing unwanted reflections, provide high reflective mirrors or deliver complex functionalities like radiation filtering or polarization dependent beam splitting. The fabrication of these coatings requires the deposition of alternating layer materials. While anti-reflective coatings are often comprised of only two or four layers the number of layers quickly rises as the desired functionality becomes more complex. When considering a bandpass or edge filter the number of layers increases and the total thickness of these coatings can exceed tenth of micrometers. This puts a high demand on the precision of the manufacturing process as the final spectral performance of a coating is brought about by the combined contributions from every layer. These are in turn defined by the exact thickness of the individual layers. With a process in place that can deliver the needed accuracy it is now also necessary to deliver it to the desired surface requiring also a lateral uniformity. On flat surfaces state of the art coating processes can deliver the needed performance and lateral uniformity. The coating of lenses and other curved substrates presents an additional challenge. Due to the curvature a coating applied without additional means will be inhomogeneous. This will in turn lead to a varying optical performance over the lens surface. To overcome this, two things need to be added. First an additional rotation of the curved substrate on the (rotating) calotte is necessary. This will ensure a rotational symmetry of the coating and eliminate any shadowing effects that the surface might cause. The second is a shaper mask that is specifically tailored to the lens surface that is coated. Production and verification of the mask can be time consuming and difficult. We will show that a system of specialized jigs can be used to overcome these difficulties and allow an easy verification of the mask and process. Moreover, this can be done using simple witness samples without the need for an actual lens to test the process. The jig also allows for having several witness samples in a coating run that unravel the optical performance of the coating at different positions on the lens. The ease of analysis of these witness samples adds control and reliability to the process.

2 Coating system setup

All coatings were fabricated using custom build ion beam sputtering (IBS) plants. The IBS coating technology is known to have a high stability and repeatability. With that it is well suited for the deposition of complex coatings. Non the less a highly accurate process control system is needed to harness these capabilities. A key component here is a broadband optical monitor (BBM). The BBM will continuously measure the transmission of the optic being coated in a wide spectral range during the deposition process. The measured spectrum is then compared to the expected theoretical curve. This allows the calculation of the actual layer thickness and growth rate at any time during the process. This delivers the needed precision and is used to control and terminate the deposition process. The achieved accuracies for an individual layer are in the range of ~1nm. These accuracies are routinely established for flat surfaces. More complex geometries however require additional steps. First the rotational symmetry of a substrate needs to be exploited. This is done by using a planetary rotation system. In this each substrate is placed in a separate fixture that rotates around its own axis. All fixtures than rotate around a common central axis. The complex movement in the coating chamber is necessary because in sputtering systems the (sputtered) coating material is not distributed evenly in the chamber. A single rotation ensures an azimuthal uniforms distribution for flat surfaces. For curved surfaces another rotation is necessary to compensate for the substrate geometry. While the coating is now distributed evenly along an azimuthal trajectory it is still non-uniform along the radius. In a last step, a shadow mask is now added to the system. This mask blocks a portion of the coating material from reaching defined radii on the substrate surface. While the setup of the plant does in general not depend on the exact geometry of the substrate surface the shadow mask has to be tailored carefully to its geometry. Keeping in mind the

* Corresponding author: t.boentgen@lzh.de

necessary accuracy for complex coatings the development of the masks is essential. To allow this, a systematic approach was developed to iteratively find masks for lens shaped surfaces. For each individual surface a dedicated jig is manufactured that can fit several flat witness samples. The samples are placed in the jig such that each represents a defined spot on the surface of the substrate that is to be coated. This is true for the position the surface and the inclination angle of the surface (Fig. 1). When measuring these samples a map of the uniformity of a coating on the surface of the lens can be created. The data collected from these measurements is then used to create a shadow mask that ensures uniformity for the specific surface. Initially the uniformity of a surface with no mask is used as input to calculate an initial mask. In an iterative process this mask is then optimized to fulfill the performance needs.

Fig. 1. Schematic view of witness sample positions. A cut of a lens is depicted in blue and two possible positions for witness samples in red. Any position along the lens surface can be chosen.

3 Experimental

Using the described procedure a total of 32 different surfaces (16 lenses) were processed and masks for uniformity needs of at least 1% were fabricated. The coated lenses have diameters from Ø50mm to Ø80mm and feature radii of curvature from 45mm to 3300mm. The vast number of surfaces demonstrates the versatility of our approach. This allows the deposition of coatings which are not sensitive to the distribution of the angle of incidence (e.g. angular tolerant anti-reflective coatings) across the surface.

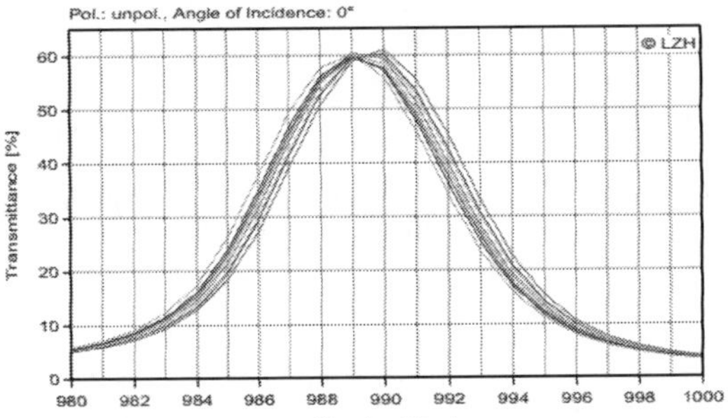

Fig. 2. Filter centered at 990nm. All curves are from measurements taken on positions from center to the edge representing a distance of 40mm.

For more complex coatings a better uniformity is needed. To further asses the capabilities of this approach a designated surface was chosen for which the uniformity was increased beyond the 1% mark. Ultimately on an area of Ø80mm a uniformity of 0,1% was achieved (Fig.2). For coatings with steep edges in the spectral performance (e.g. Bandpass filters) another effect has to be considered. When the coating is to be applied on a non flat substrate the angle of incidence (AOI) is usually not constant over the substrate. Depending on the curvature it can vary substantially. A coating will, to a first degree, be shifted spectrally if viewed at different angles of incidence. A larger angle of incidence will shift the coating to shorter wavelength. Using a dedicated shadow mask this issue can be addressed because the spectral performance of the coating will also shift if the coating thickness is spatially inhomogeneous. It is assumed, that the uniformity will affect each layer in the coating in the same way. Thus areas that deviate from the target (+-0%) will be red shifted (thicker / positive uniformity) or blue shifted (thinner / negative uniformity) with respect to the target. Since the AOI and its spread are defined by the application they can usually not be altered. However, the uniformity of the coating can be used to compensate the shift caused by the AOI spared. This is done by computing the non-uniformity necessary to compensate the shift caused by the AOI and subsequent fabrication of a shadow mask. This process is essentially the same as the one used to find masks that result in homogenous coatings. The coating deposited with the use of the fabricated shadow mask is shown in Figure3. Measurements were done on the witness samples. Each sample is viewed at its specific AOI that the corresponding part of the lens will be subjected to. The spectral shift of the coating is significantly reduced resulting in a fixed edge for any position and AOI on the lens.

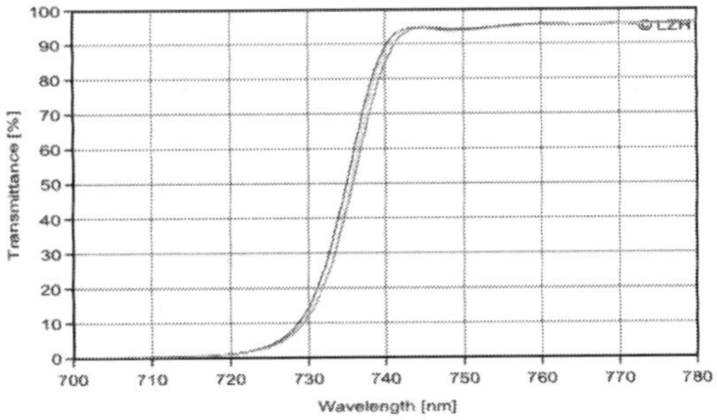

Fig. 3. Measured performance of the coating on the lens. Measurement was done on witness samples. Each sample is viewed at its specific AOI.

3 Summary

We have demonstrated a versatile approach for the coating of non-flat substrates that can increase the performance of simpler AR coatings but more importantly opens the way for the application of complex coatings with high demands on the coating process ultimately increasing freedom and allowing more compact optical designs.

EPJ Web of Conferences **238**, 03014 (2020) https://doi.org/10.1051/epjconf/202023803014
EOSAM 2020

PanDao fabrication cost impact analysis software tool for optical designers

Marco Tinner[a], Irina Livshits[b] and Oliver Faehnle[a]

[a] PanDao GmbH, St. Gallen, Switzerland,
[b] CAD for Opto-Information & energy saving systems", ITMO University, St. Petersburg, Russia

Abstract. A new software tool, called PanDao, for optical system designers is presented enabling during the design stage a simulation of the optimum fabrication chain and technologies needed as well as a fabrication cost impact analysis of design parameters and tolerances.

During the generation of *optical systems*, three different entities are involved subsequently:
(a) initially, *optical system designers* are translating performance parameters into optical system parameters such as types of glass used, lens geometries, surface shape accuracies, roughness and mid-spatial frequency errors as well as types of coatings applied.
(b) subsequently, *optics fabrication designers* translate optical systems' parameters into a well-designed fabrication chain employing machinery with optimized sets of fabrication parameters such as abrasives being used, machining kinematics, resonance frequencies, or sputtering rates being applied.
(c) finally, *production managers* are using the installed optical fabrication chain to manufacture the optical system applying optimized batch sizes and well-trained operators at high automatization levels.

Traditionally, it's the optical designers who are negotiating with customers about the product "optical system" including specifications, prices and issues concerning fabrication, coating, mounting and delivery of product.

While optical designers are well supported by software tools helping them to design the optimum set of optical elements, there are no such tools existing to design the optimum fabrication chain needed for production. As far as optical fabrication is concerned, optical designer's decisions are restricted to their personal experiences from earlier negotiations with their company's or their suppliers optics workshop. This is because optical fabrication technologies are not part of optical designer's training; patricularly chemical engineering, material science, machine tool metrology, mechanical engineering, abrasive machining, fabrication process' parameters control and a profound understanding on the science behind the "golden hands of the opticians".

Recently, within the PanDao software project[1], such a tool has been developed. Based on the know how gained from decades of academic and industrial State-of-the-Art manufacturing and experiences, the PanDao expert system has been developed. The PanDao software has been exposed to stress testing within both industrial production environments and within optical workshops located in research institutes for the generation of high end optical systems for space, medical an defense applications.

Reading in lens parameters and tolerances as generated by commercially available optical design software tools and, additionally, considering commercial order's informations such as "total number of lenses to be produced" and "batch sizes to be delivered", PanDao takes all known optical fabrication techniques (such as SPDT[2], bonnet polishing[3], IBF[4], FJP[5]) into account to generate informations about fabrication during the design stage. That way, optical designers can test their optics designs for producibility and the essential optical fabrication technologies needed for production. Furthermore, fabrication cost-impact analyses of optical design parameters are enabled based on which design parameters can be optimized for minimum fabrication cost. Thereforth, PanDao describes during the optical design process the whole fabrication chain including fabrication cost, coating cost, testing cost and centering cost.

In the following, PanDao's fabrication cost and technology analysis of a ***plano-aspherical lens*** (see Fig.1) is presented, featuring 220 mm diameter to be produced in 500 pieces quantity.
PanDao (Fig.2) identifies sub-aperture cnc grinding with a final ccp-bonnet polishing to be the optimum fabrication chain to be applied. Total fabrication cost per lens is about 185 € (taking an average sized central-European company into account: without companys overhead cost and material cost); total orders fabrication cost is 92'500 €. In the following, three examples are given, demonstrating how optical design adjustments can impact on fabrication cost.

- Changing lens diameter from 220 mm to 210 mm (and filling in the space by e.g. a metal ring), brings cost per lens down to 176.8 € resulting in 4.4 % (or 4'100 €) fabrication cost reduction overall (Fig.3).
- Changing lens' asphericity (maximum distance from best removal spherical surface to aspherical surface to be generated) from 4 µm down to 2 µm reduces fabrication cost per lens down to 155.5 € per lens resulting in 15.9 % (or 14'750 €) fabrication cost reduction overall (Fig.4). In this case, it is cheaper to grind and polish the removal spherical surface first and to use aspherics polishing techniques subsequently, rather than to grind and polish the aspherical surface directly.
- Finally, introducing more stringent mid-spatial tolerancing, e.g. reducing the smallest allowed MSW (mid-spatial wavelength) from 5 mm down to 0,2 mm leads to an increase in fabrication cost of 16.6 % (or 15'355 €) because a final IBF shaping step is needed (Fig.5).

In conclusion, a new optical design software tool, called PanDao, has been developed enabling the design of the optimum fabrication chain depending on optical design parameters and tolerances. That way, fabrication cost can be minimized already in the design stage of optical systems.

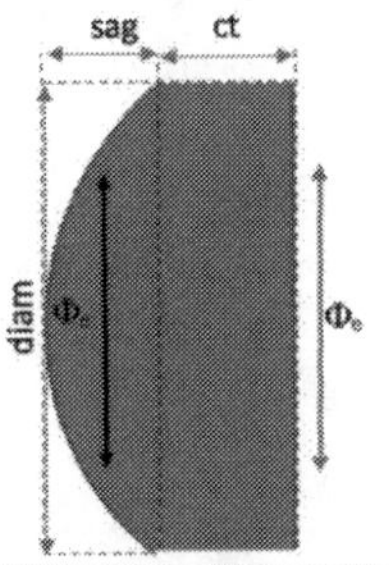

SIDE 1	PANDAO	SIDE 1
clearAperture Φe 200 mm	glass: BK7	clearAperture Φe 206 mm
AR coating	**DIMENSIONS**	AR coating
3/3(2)	center thickness ct 60 mm	3/1(0.2)
4/ 1'	ct tolerance 0.05 mm	4/ -
5/ 3*0.064	diameter 220 mm	5/ 3*0.063
Sq roughness: 1 nm rms	diam tolerance 0.1 mm	Sq roughness: 1.5 nm rms
sag 40 mm	**PRODUCTION ORDER**	sag 0 mm
smallest mitSpatialWavelength 5 mm	totalNumber pcs 500	
radius of removal sphere 660 mm	batchSize pcs: 50	
asphericity 4 µm		
smalles radius of curvature 673 mm		
contains concave parts? no		

Fig.1. Plano-aspherical BK7 lens with 500 pieces to be produced: to be analized by PanDao software tool

	Stage 1:	Stage 2:	Stage 3:
Side 1:	cnc sub aperture rough grinding fabrication price 11.79€	cnc sub aperture grinding fabrication price 29.77€	ccp bonnet fabrication price 99.19€
Side 2:	curve generator rough grinding fabrication price 9.29€	cnc full aperture flats grinding fabrication price 17.62€	cnc full aperture flats polishing fabrication price 17.41€
Total fabrication cost for this lens: 185.08€.			
Investbudget for installing machinery for this fab-chain: 770000€			

Fig.2. PanDao analysis of the plano-aspherical BK7 lens (Fig.1) with per lens 185 € fabrication cost excluding 8.7 € center grinding cost, 18.5 € coating cost and 18.2 € testing cost.

	Stage 1:	Stage 2:	Stage 3:
Side 1:	cnc sub aperture rough grinding fabrication price 11.28€	cnc sub aperture grinding fabrication price 28.43€	ccp bonnet fabrication price 94.71€
Side 2:	curve generator rough grinding fabrication price 8.89€	cnc full aperture flats grinding fabrication price 16.84€	cnc full aperture flats polishing fabrication price 16.63€
Total fabrication cost for this lens: 176.79€.			
Investbudget for installing machinery for this fab-chain: 770000€			

Fig.3. PanDao analysis of the plano-aspherical BK7 lens (Fig.1): reducing lens diameter from 220 mm down to 210 mm gains 4.4 % total fabrication cost reduction.

	Stage 1:	Stage 2:	Stage 3:	Stage 4:
Side 1:	curve generator rough grinding fabrication price 9.29€	curve generator grinding fabrication price 19.89€	cnc full aperture polishing fabrication price 19.89€	ccp bonnet fabrication price 62.10€
Side 2:	curve generator rough grinding fabrication price 9.29€	cnc full aperture flats grinding fabrication price 17.62€	cnc full aperture flats polishing fabrication price 17.41€	
Total fabrication cost for this lens: 155.50€.				
Investbudget for installing machinery for this fab-chain: 560000€				
Bestfit for side 1 is used				
Fabrication chain applying two succeeding steps: a) generation of best starting spherical surface (bsss) and b) asphərizing bsss				

Fig.4. PanDao analysis of the plano-aspherical BK7 lens (Fig.1): reducing asphericity from 4 μm down to 2 μm gains 15.9 % total fabrication cost reduction.

	Stage 1:	Stage 2:	Stage 3:	Stage 4:
Side 1:	cnc sub aperture rough grinding fabrication price 11.79€	cnc sub aperture grinding fabrication price 29.77€	ccp bonnet fabrication price 99.19€	ccp ion beam figuring fabrication price 30.72€
Side 2:	curve generator rough grinding fabrication price 9.29€	cnc full aperture flats grinding fabrication price 17.62€	cnc full aperture flats polishing fabrication price 17.41€	
Total fabrication cost for this lens: 215.80€.				
Investbudget for installing machinery for this fab-chain: 1270000€				

Fig.5. PanDao analysis of the plano-aspherical BK7 lens (Fig.1): reducing mid-spatial wavelength from 5 mm down to 0,2 mm generates 16.6 % more fabrication cost.

References

1. Webside: https://www.PanDao.ch
2. Jiwang Yan, J. Tamaki, K. Syoji, T. Kuriyagawa, "Single-point diamond turning of CaF2 for nanometric surface", The International Journal of Advanced Manufacturing Technology, Volume 24, Issue 9, pp 640-646, 2004
3. David Walker ; A. T. Beaucamp ; David Brooks ; Richard Freeman ; Andrew King ; Gerry McCavana ; Roger Morton ; David Riley ; John Simms; Novel CNC polishing process for control of form and texture on aspheric surfaces. Proc. SPIE 4767, Current Developments in Lens Design and Optical Engineering III, 2002;
4. Franz Thomas, and Thomas Hänsel, "Ion beam figuring (IBF) solutions for the correction of surface errors of small high performance optics." Optical Fabrication and Testing. Optical Society of America, 2008;
5. Oliver Faehnle, H. van Brug and H. Frankena, "Fluid Jet Polishing of optical surfaces", Applied Optics 37(28), 6771-6773, 1998;

EPJ Web of Conferences **238**, 03015 (2020) https://doi.org/10.1051/epjconf/202023803015
EOSAM 2020

Polishing characteristics of sensitive glasses - experimental investigation

Christian Vogt and *Rolf* Rascher

University of Applied Sciences,94469 Deggendorf, Germany

Abstract. Polishing is according to the current state of the art, one of the key processes in precision optics. Many specialists deal with the most diverse applications of these technologies and develop stable, reproducible techniques. In this article we present results for the investigation of different effects in the initial polishing phase in combination with different materials.

Computer Controlled Polishing (CCP) [1] is state of the art when it comes to fine and ultrafine processing of optical glasses. These sub-aperture methods are based on the principle of feed-controlled removal, i.e. in areas of the sample with a high excess of material, polishing takes place for longer, i.e. it lasts longer, then in other areas.
The accuracy of the processes therefore depends on how stable the removal rate is over time. Methods with a high removal rate usually show a higher absolute variance than those with a low removal rate. Accordingly, they are used sequentially and specified with different, achievable accuracies.
For example, polishing with ball tools with polishing foils is usually found in the pre-processing stage, where several micrometers of height of arrow are removed within a few minutes (depending on material and dimension). The achievable form deviations are usually in the range of several 100 nm.
In order to achieve even higher accuracies, processes follow, e.g. with bonnet or fluid jet tools, which enable more precise work due to lower relative speeds and pressures.
This is followed by processes such as MRF or IBF for the high-precision sector.
To make the entire production chain more efficient, it is worth taking a closer look at the foil-based pre-machining processes. While standard glasses, such as quartz or BK7 generally do not cause any or only minor problems, the situation is quite different for very soft glasses with hardness below 400 Knoop.

Particularly when polishing new tools, there are large fluctuations in the removal rate, which we are currently investigating more closely with the aid of a specially developed laboratory test rig.
In a first step the polishing behavior of new tools was investigated. Windows with 12x16mm size were polished on different glasses (table 1).

Figures 1-2 clearly show the fluctuations that occur during the initial polishing phase. Figures 2-4 show a more stable performance after initial wear has occurred. All glasses were polished under identical conditions. SF57 showed the largest fluctuations of all glasses during initial polishing. A separate polishing before the actual processing is definitely reasonable here, as it is also practiced by experienced fine opticians.
The standard in the industry is often to bring newly installed polishing processes directly into production. In general, it is unknown how long a certain constellation of polishing processes needs until stable results are delivered. In the presented project, we have experimentally determined the volume needed for polishing, depending on the glass. Based on these results, scrap can be reduced in industrial production, which is currently under investigation.
We are very grateful to Dr. Oliver Faehnle for valuable discussions and essential thoughts on this topic.

* Corresponding author: christian.vogt@th-deg.de

https://doi.org/10.1051/epjconf/202023803015

Polishing pad material:	Polyurethane, Zirconia filled Density (lb/ft3): 34-42 Hardness: 33-43
Tool main body	3D-printed, PLA, diameter 57mm
Elastic layer	3D-printed polyurethane, shore-hardness ca. 75, 70% filling
Parameters	6.4N tool load, tool speed 2000 rpm, feed speed 20 mm/min
Spotdiameter	5mm
Glas materials	N-LAF21, N-SF57, SF57

Table 1. Process information

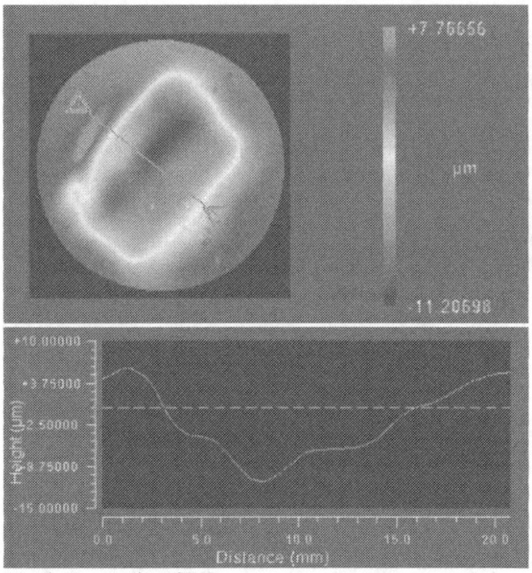

Fig. 1. New tool on SF57

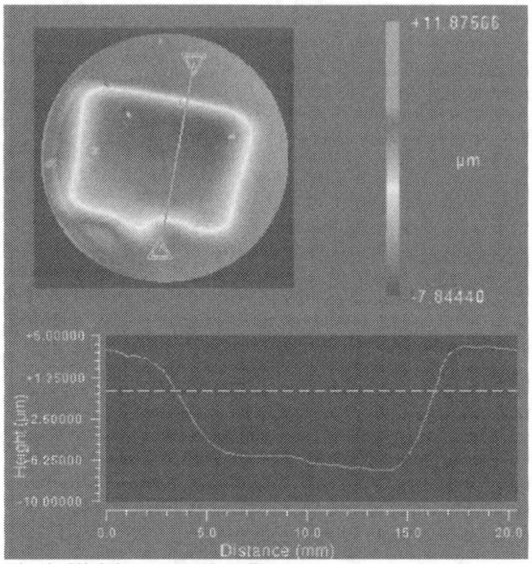

Fig. 2. Slightly worn tool on SF57

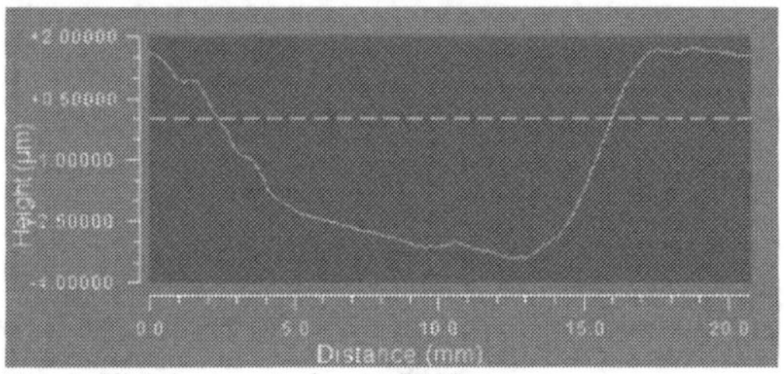

Fig. 3. Slightly worn tool on N-SF57

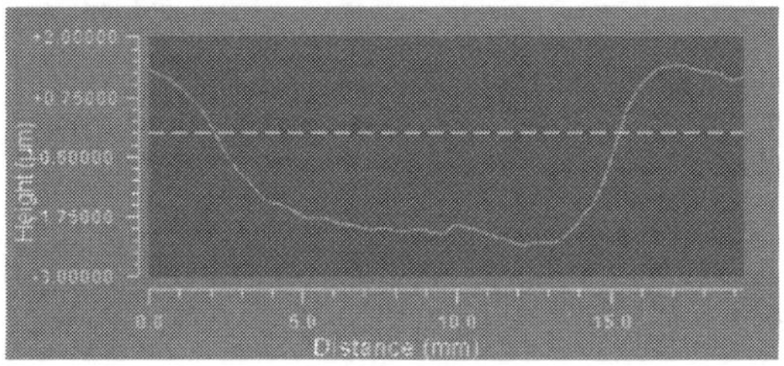

Fig. 4. Slightly worn tool on N-LAF21

References

[1] M. Schinhaerl a. o., Precision Engineering, VOL. 32, pages 47-54, [2008],

* Corresponding author: christian.vogt@th-deg.de

Towards SERS-based multiplexed monitoring of protease activity using non-natural aromatic amino acids

Nina Turk[1,2,*], *Hans* Demol[3,4], *Andre* Skirtach[2,5], *Roel* Baets[1,2] and *Kris* Gevaert[3,4]

[1]Ghent University - IMEC, Photonics Research Group, Technologiepark 126, 9052 Ghent, Belgium
[2]Center for Nano- and Biophotonics, Ghent, Belgium
[3]VIB-UGent Center for Medical Biotechnology, Ghent, Belgium
[4]Ghent University, Department of Biomolecular Medicine, Belgium
[5]Ghent University, Department of Biotechnology, Ghent, Belgium

Abstract. Surface Enhanced Raman Spectroscopy (SERS) allows sensitive and selective detection of protease activity by monitoring the cleavage of specific peptide substrates. Furthermore, it offers the possibility for multiplexing, during which the activity of two (or more) proteases with different specificities is detected simultaneously. To distinguish between the contributions of different proteases, different aromatic amino acids with non-overlapping SERS peaks need to be used as Raman reporters. As the three natural aromatic amino acids only offer limited possibilities for multiplexing, we examined several non-natural aromatic amino acids with the aim of expanding multiplexing possibilities. We recorded their SERS spectra for the Raman shifts of 300-1700 cm^{-1} and identified their characteristic SERS peaks. Of the examined non-natural aromatic amino acids, 3-nitro-tyrosine and two phenylalanines containing stable heavy isotopes seem particularly promising for multiplexing applications. Besides exhibiting characteristic SERS peaks in the spectral region of interest, these non-natural aromatics provide strong SERS peaks compared to natural aromatic amino acids, consequently improving detection sensitivity.

Raman spectroscopy allows for highly specific detection of molecules based on their vibrational modes. Raman signals are however inherently weak, but Surface-Enhanced Raman Spectroscopy (SERS) can be employed to enhance such weak Raman signals near the surface of plasmonic nanostructures. Among others, SERS can be used for sensitive and selective detection of protease activity using gold nanodomes to achieve the SERS enhancement [1]. Proteases are enzymes that catalyse the hydrolysis of peptide bonds and, amongst others, these enzymes play important roles in various human diseases [2]. Many proteases cleave their peptide substrates specifically and by incorporating two aromatic amino acids as Raman reporters in the peptide substrate with the protease cleavage site separating these aromatic amino acids, SERS can be used to monitor protease activity as schematically shown in Figure 1.

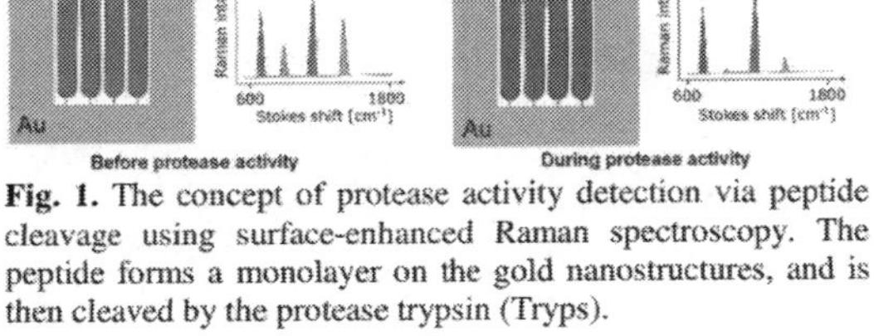

Fig. 1. The concept of protease activity detection via peptide cleavage using surface-enhanced Raman spectroscopy. The peptide forms a monolayer on the gold nanostructures, and is then cleaved by the protease trypsin (Tryps).

SERS also holds great potential for multiplexing, this to simultaneously detect the activity of two (or more) proteases with different specificities. To achieve this, it is crucial that one can distinguish the contributions of different proteases by using aromatic amino acids with non-overlapping SERS peaks. Since the three natural aromatic amino acids, namely phenylalanine, tyrosine and tryptophan, offer only limited opportunities for multiplexing measurements, we turned to non-natural aromatic amino acids. We investigated selected non-natural amino acids for non-overlapping SERS peaks to expand the possibilities of multiplexing. Such non-natural aromatics can additionally offer potentially higher Raman cross-sections, which provide higher SERS signal strengths than those of natural aromatic, thus improving the signal-to-noise ratio. In this paper, we report on the SERS spectra and SERS peaks of different aromatic amino acids and assess their suitability for multiplexing applications.

Experiments and results

We investigated all three natural and eight selected non-natural aromatic amino acids (Table 1). We have chosen to investigate several derivatives of tyrosine [3] and phenylalanine [4], as well as phenylalanine molecules containing stable heavy isotopes of hydrogen, carbon and nitrogen [5]. We incorporated the aromatic acids in a

* Corresponding author: nina.turk@ugent.be

peptide chain NH_2-CALNNXG-COOH, where X denotes the individual aromatic amino acid. In this peptide sequence, cysteine (C) allows the peptide to bind to the gold nanodomes, which provide the SERS enhancement [1]. The SERS spectra were obtained using a laser power of 1 mW and an integration time of 3 s for the Raman shifts of 300-1700 cm^{-1}. Background-subtracted SERS spectra of the aromatic amino acids are shown in Figure 2 for the phenylalanine derivatives and in Figure 3 for the tyrosine derivatives.

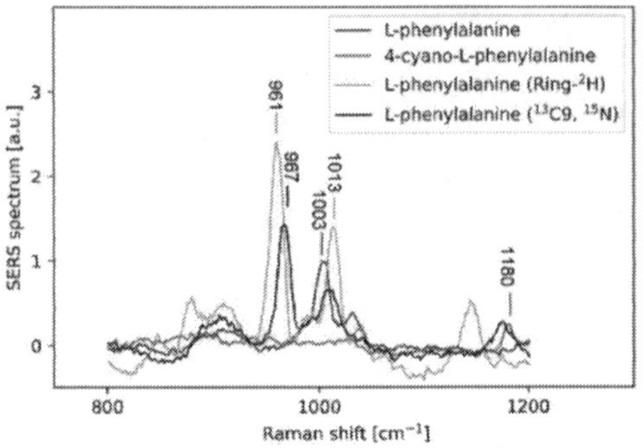

Fig. 2. SERS spectra of phenylalanine derivatives in the spectral region 800-1300 cm^{-1}. The spectra are normalized on the 1003 cm^{-1} phenylalanine peak and most prominent peaks denoted.

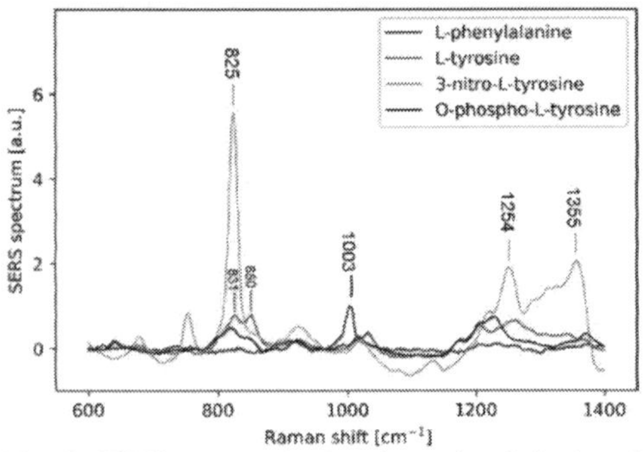

Fig. 3. SERS spectra of selected tyrosine derivatives in the spectral region 600-1400 cm^{-1}. The SERS spectrum of phenylalanine is included for comparison with Figure 2, and the spectra are normalized on the 1003 cm^{-1} phenylalanine peak.

After subtracting the SERS background from the spectra, we identified SERS peaks pertaining to the investigated aromatics (Table 1). For our analysis we assumed that aromatic amino acid residues provide the dominant features in the SERS spectra of peptides [6].

Most of the examined aromatic amino acids provided characteristic SERS peaks in the examined spectral region, more specifically in the 600-1400 cm^{-1} range. We noticed that the SERS peaks of some aromatics overlap, so special care has to be taken to choose compatible aromatics as Raman reporters during multiplexing experiments. We additionally noticed that several non-natural aromatics provide strong SERS peaks compared to their corresponding natural aromatic amino acids, thus improving the signal-to-noise ratio. Of the examined non-natural aromatics, 3-nitro-tyrosine and the two phenylalanine molecules containing stable heavy isotopes seem particularly promising for multiplexing applications.

Table 1. Measured prominent SERS peak positions of different aromatic amino acids with relative SERS peak strengths. Abbreviations: vs: very strong, s: strong, m: medium, w: weak.

Aromatic amino acid	SERS peak position [cm^{-1}] with relative SERS peak strength
L-phenylalanine	1003 (s), 1031 (m)
L-tyrosine	831 (m), 850 (m), 1260 (w)
L-tryptophan	754 (m), 1118 (w), 1356 (m)
4-cyano-L-phenylalanine	1180 (m)
4-benzoyl-L-phenylalanine	1003 (m), 1031 (m), 1152 (m)
3,4-dichloro-L-phenylalanine	-
L-phenylalanine (Ring-^2H)	961 (vs), 1013 (s), 1145 (s)
L-phenylalanine ($^{13}C9$, ^{15}N)	967 (s), 1009 (m), 1173 (m)
3-nitro-L-tyrosine	752 (m), 825 (vs), 1254 (s), 1355 (s)
O-phospho-L-tyrosine	820 (w), 843 (w), 1229 (w)
3-amino-L-tyrosine	Strong, but unstable SERS signal

We have already successfully incorporated cyano-phenylalanine into a peptide substrate and showed peptide cleavage by the trypsin protease [7]. In the future, we intend to incorporate the more promising non-natural aromatic amino acids into peptide substrates to demonstrate multiplexing capabilities of SERS detection of the protease activity.

Nina Turk acknowledges Research Foundation Flanders (FWO) for the predoctoral fellowship (1179319N).

References

1. P.C. Wuytens, H. Demol, N. Turk, K. Gevaert, A.G. Skirtach, M. Lamkanfi, R. Baets, Faraday Discuss, **205** (2017)
2. M. Drag, G.S. Salvesen, Nat Rev Drug Discov, **9** (2010)
3. J. Moger, P. Gribbon, A. Sewing, C.P. Winlove, Biochim Biophys Acta, **1770** (2007)
4. C.L. Weeks, A. Polishchuk, Z. Getahun, W.F. DeGrado, T.G. Spiro, J Raman Spectrosc, **39** (2008)
5. H.J. van Manen, A. Lenferink, C. Otto, Anal Chem, **80** (2008)
6. F. Wei, D. Zhang, N.J. Halas, J.D. Hartgerink, J Phys Chem B, **112** (2008)
7. N. Turk, A. Raza, P.C. Wuytens, H. Demol, A. Skirtach, K. Gevaert, R. Baets, 23rd Annual Symposium of the IEEE Photonics Benelux (2018)

EPJ Web of Conferences **238**, 04002 (2020) https://doi.org/10.1051/epjconf/202023804002
EOSAM 2020

In vivo endoscopic multifunctional optical coherence tomography imaging of lungs periphery before and after bronchial thermoplasty

Margherita Vaselli[1], *Fabio* Feroldi[1], *Joy* Willemse[1], *Maximilian G.O.* Graefe[1], *Dirck* van Iperen[1], *Annika W.M.* Goorsenberg[1], *Peter I.* Bonta[1], *Jouke T.* Annema[1], and *Johannes F.* de Boer[1]

[1]Department of Physics and Astronomy, LaserLab Amsterdam, VU University Amsterdam, the Netherlands

Abstract. We demonstrate the use of a motorized distal scanning endoscope to acquire *in vivo* data in lungs of severe-asthma patients before and after an asthma treatment procedure called bronchial thermoplasty. Conventional optical coherence tomography (OCT) intensity images and polarisation sensitive OCT images were extracted from the acquired data. PS-OCT endoscopy allowing the visualization and segmentation of airway smooth muscle layer - which plays a key role in bronchoconstriction during asthma attacks - showed its potential as means to evaluate the effectiveness of bronchial thermoplasty.

1 Introduction

Over the past 20 years OCT has been widely adopted in ophthalmology, cardiology and gastroenterology. Its adoption in interventional pulmonology has been slower, this can be attributed to the lack of commercially available OCT endoscopes suitable for lungs imaging. In this study we present *in vivo* images of airways at the periphery of the lungs, performed in a severe asthma patient before and after undergoing a recently-developed asthma treatment procedure called bronchial thermoplasty (BT) [1]. Imaging of lung tissue has been performed with a distally scanning catheter connected to a multifunctional OCT system. OCT intensity images, attenuation coefficient (AC) images, and polarisation sensitive OCT (PS-OCT) images showing both phase retardation, optic axis uniformity (OAxU) and optic axis (OA) orientation were extracted from the data. In particular optic axis orientation was used to segment the airway smooth muscle (ASM) layer, which is remodelled by BT therapy.

2 Methods

2.1 Imaging system

PS-OCT is an extension of conventional OCT that besides performing structural imaging of tissue allows to image sample birefringence, from which specific information about the tissue can be extracted [2]. The schematic of our endoscopic PS-OCT system is shown in fig. 1.a. Light emitted from a swept source laser is split into the two arms of a Mach-Zehnder interferometer. In the sample arm a polarisation delay unit (PDU) creates two depth-encoded polarisation states. A circulator (C) redirects the light to the endoscope; the backscattered light in the sample arm is then interfering with the reference arm light in a polarisation diversity detection module (PDDM), which separates the lights into orthogonal polarisation components on separate balanced detectors.

2.2 Endoscope

Imaging the airways periphery was possible through an in-house built endoscope, inserted in the lungs through the working channel of a standard bronchoscope. The distal scanning catheter we developed (fig. 1.b) is a motorized endoscope with a 1.35 mm diameter that allows to scan circumferentially the airways at 52 fps B-scan rate. The reliability of the micromotor placed at the tip of the catheter was improved by doubling the windings of the coil around the micromagnet to two. In order to provide clinicians with a cost effective imaging tool for research studies, the endoscope was designed and constructed to be reused several times.

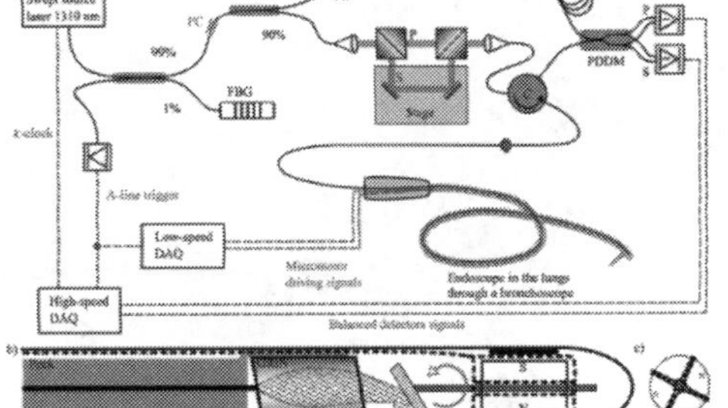

Fig. 1. a) Schematic of the PS-OCT system. b) Schematic of the OCT endoscope

* Corresponding author: m.vaselli@vu.nl

2.3 Intensity and PS-OCT images

On top of conventional OCT intensity images, the data acquired *in-vivo* in human lungs of a severe asthma patient were used to extract images based on the depth-resolved attenuation coefficient (AC) of tissue. AC images have been realized exploiting the algorithm developed by Vermeer *et al.* [3], which makes use of chromatic dispersion corrected and roll-off compensated intensity OCT images in linear scale. The attenuation coefficient of each depth location was estimated by dividing the intensity of each depth pixel by the sum of the intensities of the pixels underneath. Intensity based OCT images have a resolution and a depth of penetration sufficient to image the ASM, however backscattered light from ASM does not provide sufficient contrast to distinguish it from surroundings layers. Since ASM is arranged in fiber-like structure, it exhibits form birefringence, that can be highlighted by PS-OCT. In this study, birefringence properties have been extracted, adopting the Mueller-Stokes formalism, by using an algorithm developed by Villiger in 2016 [4]. Output of this method is a 3D vector: the birefringence vector γ, whose magnitude ρ corresponds to the amount of retardation induced by the sample and its direction $[\eta, v, \mu]$ represents the orientation of the sample optic axis (OA) in a Poincare sphere representation. OA shows consistent orientation throughout the sample if the fibers are oriented along the same direction. We also introduced a simple metric that evaluates the spatial uniformity of the sample OA: the optic axis uniformity OAxU [5].

$$OAxU = \sqrt{\langle\eta\rangle^2 + \langle v\rangle^2 + \langle\mu\rangle^2} \quad (1)$$

OAxU is particularly useful to highlight areas of a sample with similar birefringence characteristics.

3 Results

Intensity based OCT images acquired at the periphery of the lungs in a patient affected by severe asthma are shown in fig. 2.

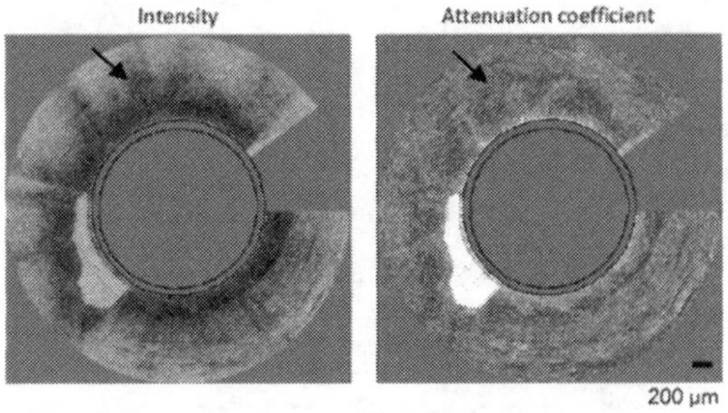

Fig. 2. Cross section frame from a distal location along the pullback. Intensity OCT B-scan and corresponding AC image.

AC images results to provide superior contrast of the lung airways compared to intensity images. In particular, epithelial folds (indicated with a black arrow in fig. 2) are better visualized in the AC images.

PS-OCT images highlighted the presence of several layers in the airway wall; in particular OA orientation and OAxU were used to delineate in the airway lumen the smooth muscle layer. Tissue layers characterized by high optic axis uniformity show high intensity signal in OAxU images (fig. 3); different orientations of these layers are then represented in different colours in OA images.

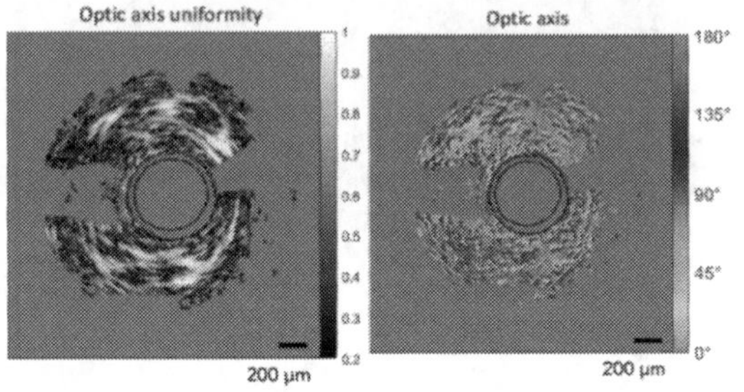

Fig. 3. Cross section frame from a proximal location along the pullback. OAxU B-scan and corresponding OA image.

Knowing ASM is made of consistently oriented fibers located within 100µm from the lumen surface, it was possible to discern this layer from other birefringent structures and segment it. Segmentation of the ASM layer allowed us to compare its thickness before and after BT treatment. Fig. 4 shows a 3D image of the segmented ASM acquired pulling back the catheter from a distal to a proximal location in the same lung segment before and after BT. As can be seen from fig. 4, the thickness of ASM after BT is reduced in the proximal area of the pullback, which corresponds with the area mainly treated by BT.

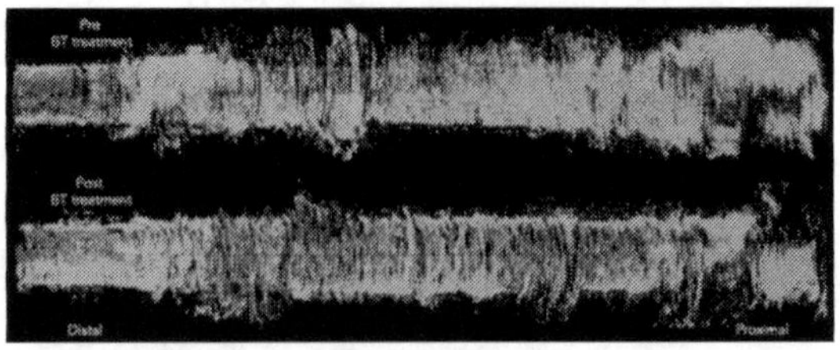

Fig. 4. 3D image of segmented ASM acquired during pullback.

References

1. G. Cox, et al., AJRCCM, **173**, p. 965-969, (2006)
2. J. Li, et al., Optics Express, **23**, p. 3390-3402, (2015)
3. K.A. Vermeer, et al., BOE, **5**, p. 322-337, (2014)
4. M. Villiger, et al., Scientific Reports, **6**, (2016).
5. J. Willemse, et al., Optics Letters, **44**, p. 3893-3896, (2019)

EPJ Web of Conferences **238**, 04003 (2020) https://doi.org/10.1051/epjconf/202023804003
EOSAM 2020

Polarization-sensitive optical coherence tomography as a tool to visualize the fiber direction of retinal nerves and peripapillary sclera

Joy Willemse[1,*], *Frank* Verbraak[2], and *Johannes* de Boer[1,2]

[1]LaserLab, Department of Physics and Astronomy, Vrije Universiteit Amsterdam, de Boelelaan 1081, 1081 HV Amsterdam, the Netherlands
[2]Amsterdam UMC, Vrije Universiteit Amsterdam, Ophthalmology Dept., de Boelelaan 1117, Amsterdam, the Netherlands

Abstract. Polarization sensitive optical coherence tomography (PS-OCT) has been used to visualize the orientation of the nerves in the retinal nerve fiber layer (RNFL) and to visualize the orientation of the collagen fibers in peripapillary sclera in retinas of healthy volunteers. Optic axis orientation images clearly visualize the nerve fibers leaving the optic nerve head (ONH) in all radial directions. Sclera orientation images show that the sclera consist of two layers, an inner layer with an orientation parallel to the RNFL orientation, and a deeper layer where the collagen is circularly oriented.

1 Purpose

Optical coherence tomography (OCT) is an imaging technique which is used in ophthalmology to create high-resolution retinal images. Additional information on the tissue can be obtained with polarization sensitive OCT (PS-OCT), where polarization changing effects of the tissue can be captured and included in the analysis of the backscattered light. In the retina, this includes birefringence of the retinal nerve fiber layer and sclera, and depolarization of the retinal pigment epithelium (RPE). Furthermore, the optic axis orientation can be extracted, which shows the (relative) direction of fibers. It is hypothesized that changes in birefringence of the retinal nerve fiber layer (RNFL) may precede the thinning of the RNFL in the early symptoms of glaucoma [1]. Therefore, the possibilities to image RNFL phase retardation and fiber orientation have been explored previously [2, 3]. In addition, deviations in peripapillary sclera structure might precede damage to the optic nerve head (ONH) [4]. In this experimental proof-of-concept study polarization-sensitive optical coherence tomography (PS-OCT) was used in healthy volunteers to evaluate the local birefringence and direction of the fibers of the RNFL and sclera.

2 Methods

Seven healthy volunteers were scanned with our in-house built PS-OCT system. PS-OCT imaging included intensity, local phase retardation, degree of polarization uniformity (DOPU), and relative optic axis images. The data was corrected for polarization mode dispersion (PMD) by an algorithm by Braaf et al. [5]. Local birefringence images and optic axis orientation were extracted using the differential Mueller matrix method developed by Villiger et al. [6]. This algorithm was modified to correct for the birefringence of preceding layers. To visualize the birefringence with higher signal-to-noise ratio, optic axis uniformity (OAxU) images were used [7]. The setup contained a swept-source laser centered at 1060 nm with a repetition rate of 100 kHz. Segmentation on the RNFL was performed by using a combination of DOPU, OAxU and intensity images. Segmentation on the sclera was performed using OAxU images only. Sclera orientation en face images were created by averaging the optic axes in the first 150 µm of sclera pixels.

3 Results

Figure 1a shows an en face intensity map of the retina of a healthy volunteer The RNFL was segmented automatically, and an en face RNFL thickness map was made (figure 1b). Local birefringence of the RNFL is shown in figure 1c. The relative orientation of the RNFL was found by averaging the measured local optic axis of the RNFL in depth. Figure 1d shows the en face intensity map with an overlay of red lines which delineate the orientation of the RNFL. It can be seen that the nerve fibers are leaving the optic nerve head (ONH) in all radial directions. Minimum RNFL thickness to extract the relative optic axis reliably was about 60 µm.

* Corresponding author: j.willemse@vu.nl

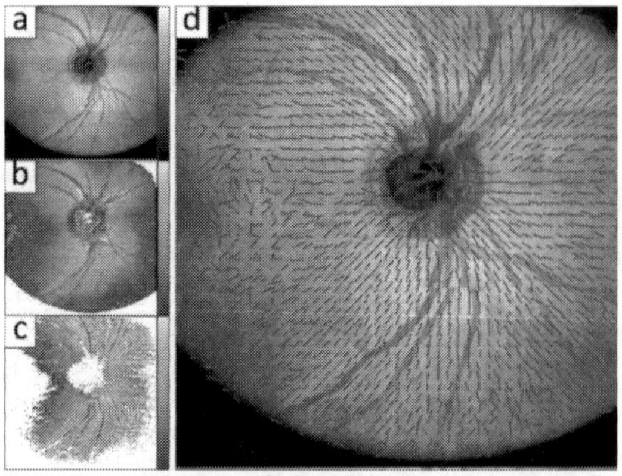

Fig. 1. PSOCT images of a healthy volunteer (a) Intensity en face image, (b) RNFL thickness image, (c) local birefringence image and (d) Intensity en face image overlaid with RNFL orientation as red lines. Color range in (b) indicates 0 – 300 μm. Color range in (c) indicates 0.5-0.15 °/μm. Images sizes: 5.7 x 6.2 mm (depth x height).

Figure 2 shows the orientations measured in the sclera of another healthy volunteer. Figure 2b shows the orientations measured in the sclera, where the white dotted lines correspond to the B-scans shown on the right of the image. These B-scans show the intensity with an overlay of the optic axis orientation for pixels with high OAxU. Two different orientations where measured for the sclera. The inner layer of sclera consists of collagen oriented parallel to the RNFL, i.e. radial from the ONH. A deeper layer was found where the orientation is circular around the ONH. Close to the ONH, the second layer displaces the first layer, as indicated with the red arrows shown in figure 2d. This results in a ring of different orientation as can be seen in the sclera orientation en face image (Figure 2b).

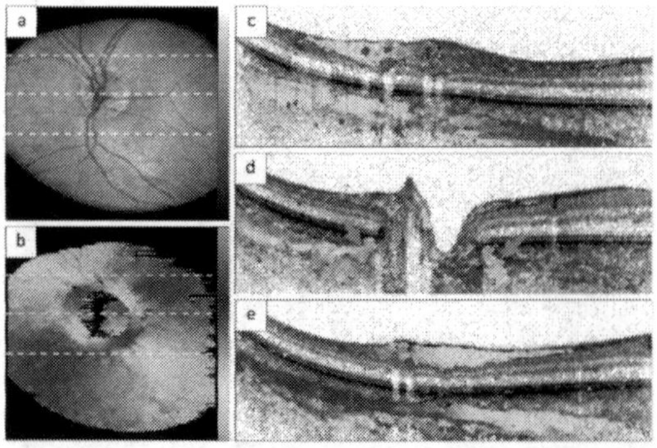

Fig. 2. Sclera orientation images of a healthy volunteer (a) Intensity en face image, (b) sclera orientation en face image, (c-e) cross sections of frames indicated with white in the en face images. Intensity is overlaid with the orientation of pixels with high OAxU. Color range in (b) indicates 0 – 180°. Images sizes: (a - b) 5.7 x 6.2 mm (depth x height), (c- e) 1.2 x 6.2 mm (depth x height)

Eyes of five out of seven volunteers showed these two orientations for the sclera, with the second layer coming closer to the surface in a ring around the ONH. In two volunteers, the retina absorbed the light before reaching the sclera, leaving too little signal to perform the analysis.

Conclusions

PS-OCT has been used to successfully extract the birefringence and orientation of the retinal nerves and sclera. To our knowledge, PSOCT is the only technique which is able extract 3D orientation of fibrous structures of the human retina *in vivo*. The relative direction of the fibers around the ONH have been visualized, demonstrating a radial pattern for RNFL and the first layer of sclera. A second sclera layer with circumferential orientation was found. In RNFL, the fiber direction and its changes could possibly be correlated with areas of visual field loss, and might be used to identify areas of (expected) RNFL loss in glaucoma. Deviations in peripapillary sclera structure might precede damage to the optic nerve head (ONH) in glaucoma. In future research, we will scan glaucoma patients with the PS-OCT system to test these hypotheses.

References

[1] B. Cense, T. C. Chen, B. H. Park, M. C. Pierce, and J. F. de Boer, Invest. Ophth. Vis. Sci. J. **45**, 2606-2612 (2004).

[2] S. Zotter, M. Pircher, E. Gotzinger, T. Torzicky, H. Yoshida, F. Hirose, S. Holzer, J. Kroisamer, C. Vass, U. Schmidt-Erfurth, and C. K. Hitzenberger, Invest Ophth Vis Sci **54**, 72-84 (2013).

[3] M. Sugita, M. Pircher, S. Zotter, B. Baumann, P. Roberts, T. Makihira & C. K. Hitzenberger, Biomed Opt Express, **6**(3), 1030-1054 (2015).

[4] C. F. Burgoyne, J. C. Downs, A. J. Bellezza, and R. T. Hart, Invest. Ophth. Vis. Sci. J. **45**(12), 4388-4399 (2004).

[5] B. Braaf, K. A. Vermeer, M. de Groot, K. V. Vienola, and J. F. de Boer, Biomed Opt Express **5**, 2736-2758 (2014).

[6] M. Villiger, D. Lorenser, R. A. McLaughlin, B. C. Quirk, R. W. Kirk, B. E. Bouma, and D. D. Sampson, Sci Rep-Uk **6** (2016).

[7] J. Willemse, M.G.O. Gräfe, J. A. van de Kreeke, F. Feroldi, F.D. Verbraak, and J. F. de Boer,). Optics letters, **44**(15), 3893-3896 (2019).

EPJ Web of Conferences **238**, 04004 (2020)
EOSAM 2020
https://doi.org/10.1051/epjconf/202023804004

Attenuation coefficient as a quantitative parameter for analyzing cataracts with optical coherence tomography

Pablo Eugui[1,*], *Danielle J. Harper*[1], *Johanna Gesperger*[1,2], *Antonia Lichtenegger*[1], *Conrad W. Merkle*[1], *Martin Glösmann*[3] and *Bernhard Baumann*[1]

[1]Center for Medical Physics and Biomedical Engineering, Medical University of Vienna, Austria
[2]Division of Neuropathology and Neurochemistry, Department of Neurology, Medical University of Vienna, Austria
[3]Core Facility for Research and Technology, University of Veterinary Medicine, Austria

Abstract. Crystalline lenses of mice were imaged in vivo with a custom-made swept-source optical coherence tomography system. The use of the attenuation coefficient as a quantitative parameter for investigating the lens opacities magnitude is proposed, demonstrating a significant difference between the values retrieved from cataractous and normal mouse lenses.

1 Introduction

Cataracts are the major cause of blindness worldwide, they are characterized by a progressive loss of transparency of the ocular lens which ultimately leads to poor vision [1]. Optical coherence tomography (OCT) has been proposed as a non-invasive imaging technique to visualize and quantify the morphology of these opacifications in human patients and animal models [2,3]. In biomedical applications, the attenuation coefficient of the tissue recovered by optical methods has been used as a parameter for classification and diagnosis [4]. Previously we presented OCT as a tool to visualize different opacifications in vivo in the crystalline lens of mice [3]. Here we want to extend our previous results with the use of the optical attenuation coefficient as a further measure to analyze and characterize cataract formations.

2 Methods

In this paper we used the custom-made OCT system illustrated in Fig. 1(A), which operates at a central wavelength of 1310 nm and provides an axial resolution of ~6.5 μm in air with a sensitivity of 100 dB. The OCT device was described in detail elsewhere along with the experimental procedure of the data used for this work [3]. Data sets with 400 × 400 × 768 pixels (3 × 3 × 5 mm) were acquired in the mouse anterior segment, repeating 4 scans per position to increase the signal-to-noise ratio. A post-processing pipeline was implemented in MATLAB (MATLAB, R2015b, MathWorks). Every acquired spectrum was aligned with a unique spectrum using cross-correlation to reduce shifts induced by A-line trigger jitter. Background removal and spectral shaping were then performed before averaging the resulting repeated scans after Fourier transformation. The lens surface was segmented and the equation

$$y = A \cdot \exp(-\mu z) \tag{1}$$

was fitted to the intensity profiles in all A-scans for a range spanning 0.5 mm in depth underneath the surface. In the fit, the scaling factor A and the attenuation coefficient μ were the free running parameters to be optimized by the algorithm. En-face attenuation maps were then calculated. A similar region of interest consisting of 0.16 mm^2 was analyzed for cataract and non-cataract eyes and compared as shown in Figure 1(B). A Student's t-test was used to analyze the difference between the average attenuation coefficients of the two groups.

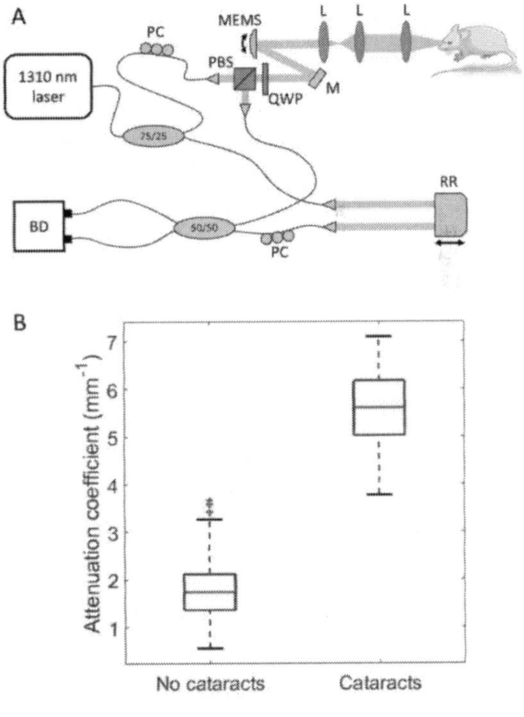

Fig. 1. A) Home-built OCT device used for mouse lens imaging. PBS: beam splitter, BD: balanced detector, L: lens, M: mirror, MEMS: microelectromechanical scanner, PC: polarization controller, QWP: quarter wave plate, RR: retroreflector. B) Average attenuation coefficient compared between normal and cataract eyes.

* Corresponding author: pablo.euguiarrizabalaga@meduniwien.ac.at

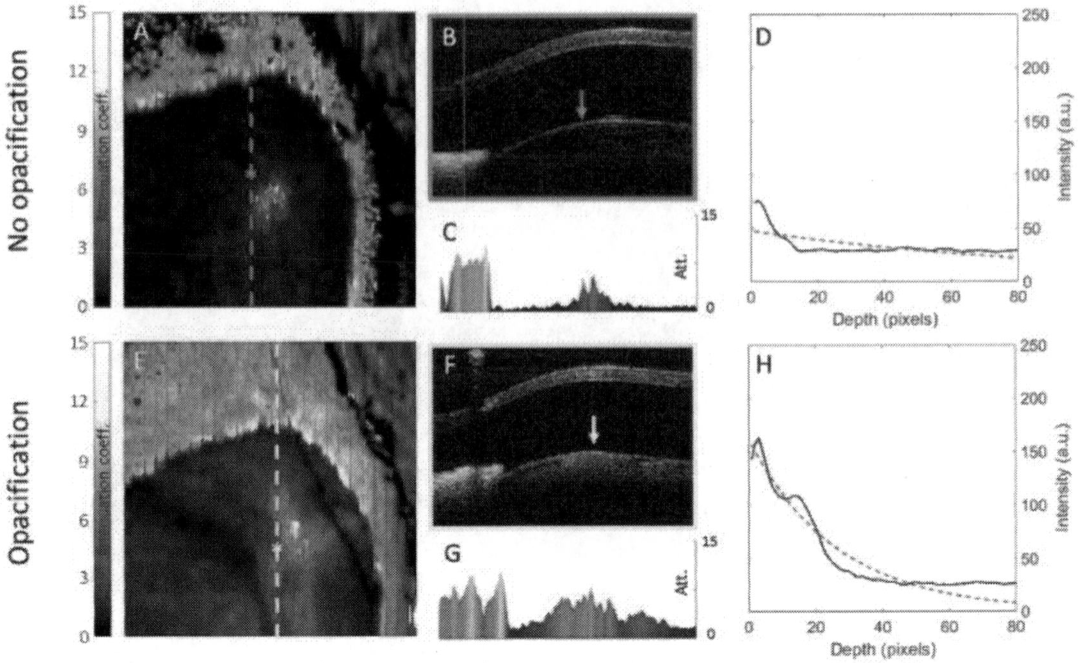

Fig. 2. A) Attenuation map of a lens without opacification. Aside from the high values caused by a strong surface reflection in the center at the anterior lens pole, only low attenuation is observed in the lenticular area. B) When looking at an individual B-scan marked with a yellow line in A), only the lens surface is visible without showing any scattering underneath. C) Attenuation coefficient profile of the lens in the B-scan shown in B). Low values can be seen in the lens except the central reflection area and those values corresponding to the iris on the left. D) A-scan profile and exponential fit of the pixels selected underneath the lens surface in the location indicated by the green arrow in B). E) Attenuation map of a lens with an opacity present in the lens cortex. Higher attenuation coefficients can be observed in the lens. F) Cross-section of B-scan located at the position indicated by the yellow line in E). A cloudy opacity can be seen underneath the lens surface. G) Attenuation profile of the lens in the B-scan shown in F) where higher values can be observed in the lens. H) Depth profile and exponential fit of the selection underneath the lens indicated by the yellow arrow in F).

3 Results

An attenuation map of a mouse eye without cataract is shown in Figure 1(A). In the eyes where no opacification was observed, only a very subtle scattering signal was detected in the lens due to its transparency, which led to low attenuation coefficients as shown in the Figure 1(B) and 2(ACD). An intensity profile of a lens without opacification is shown in Figure 2(D). When an opacity was present in the mouse eye, the attenuation within the lens was stronger as shown in Figure 1(E). This was due to a cloudy hyperscattering signal caused by the cataract formed in the lens cortex as shown in Figure 2(F). In the lateral attenuation profile of a B-scan with opacification, higher values can be observed in addition to the ones corresponding to the iris on the left as shown in Figure 2(G). This is also visible in the axial intensity profile shown in Figure 2(H) where higher values, rising as well as a steeper slope of the fitted exponential can be observed. When comparing similar regions between cataractous and non cataractous eyes, a higher attenuation (5.5 ± 0.7 mm^{-1}) was found in cataractous areas compared to the non-cataracts regions (1.8 ± 0.5 mm^{-1}), showing a significant difference.

4 Conclusion

A custom-made OCT system was used for volumetric in vivo visualization of the crystalline lens in mice. A post-processing pipeline was implemented to obtain the attenuation coefficient, which revealed a significant difference between cataracts and normal eyes.

This research was funded by the European Research Council (ERC StG 640396 OPTIMALZ).

References

1. Thylefors, B. Int. Ophthalmol. 14(3) 211-219 (1990)
2. de Castro, A, et al. Invest. Ophthalmol. Vis. Sci. 59(2) 897-903 (2018)
3. Eugui, P. et al. Biomed. Opt. Express. 11(4) 2085-2097 (2020)
4. Chan, S. et al. J. Biomed. Opt. 24(9), 090901 (2019)

* Corresponding author: pablo.euguiarrizabalaga@meduniwien.ac.at

EPJ Web of Conferences **238**, 04005 (2020)
EOSAM 2020
https://doi.org/10.1051/epjconf/202023804005

Light sheet fluorescence microscope for microfluidic chip

Caroline Bissardon[1], *Xavier* Mermet[1], *Sophie* Morales[1], *Frédéric* Bottausci[1], *Marie* Carriere[2], *Florence* Rivera[1], and *Pierre* Blandin[1,*]

[1]Univ. Grenoble Alpes, CEA, LETI, DTBS, F-38000 Grenoble, France
[2]Univ. Grenoble-Alpes, CEA, CNRS , IRIG, SyMMES, F-38000 Grenoble, France

Abstract. We present a light sheet fluorescence microscope dedicated to image "Organ-on-chip"-like biostructures in microfluidic chip. Based on a simple design, the setup is built around the chip and its environment to allow 3D imaging inside the chip in a microfluidic laboratory. The experimental setup, its optical characterization and first volumetric images are reported.

1 Introduction

Nowadays, the organ-on-chip technology presents a large interest for biomedical researches. Important issues, such as the vascularization of organoids or tissues, remain and must be solved to enhance the potential of this application technique.

In the present study, the optical system has been developed to response to vascularization problems encountered in Organ-on-chip imaging. The visualization and understanding of the capillarization process occurring within the organoid-like samples is crucial to characterize it. To follow up the vascularization process, a microfluidic chip associated with an adequate microfluidic injection system have been developed at the CEA LETI.

Obtaining information of the central core of the sample is not satisfying with classical fluorescence microscopy techniques such as inverted fluorescence microscope, spinning-disk microscopy. Light Sheet Fluorescence Microscopy (LSFM) seems to be a promising solution to analyze the inner parts of samples. The main interest of this microscopy is to allow a 3D morphological and functional characterization of organoids and other tissues cultured in a microfluidic chamber. The samples will not be transformed or clarified before analyses in order to be physico-chemically and photo-preserved.

LSFM is an optical imaging technique that has been increasingly used in biological applications ranging from molecular biology to whole mount tissue analysis. This technique presents a lot of advantages in microdevice field[1–5]. The basic idea of this technique is described in the reviews[6–8]. This confinement of laser light confers optical sectioning capability to the microscope. The main advantages of this technique are the high-speed volumetric acquisition rate and the low photo-toxicity[6–8]. These two particularities are essential for rapid 3D but also for 4D imaging.

2 Optical microscope system

2.1 Positioning

To limit the disturbance in the sample development and avoid any microfluidic injection problem, the image acquisition need to be done on a horizontal microfluidic chip. Inspired by the inverted selective plane illumination microscopy developed by Wu et al. 2011[9], we conceived a new microscope for specific "Organ-on-chip" application. Long distance objective is required to be able to collect the fluorescence coming from the microfluidic disposal system and having enough space to insert, position and move it. The detection and excitation paths are placed at 90 degrees from each other on a vertical breadboard in order to observe horizontally the sample positioned at 45 degrees from both optical arms. The light sheet (LS) is controlled with a cylindrical lens in order to form a fine and homogeneous LS on the field of view of the camera. 3D imaging is then possible thanks to the translation of this LS at different depth of the sample. Contrary to other studies focusing on super-resolution or large field of view, the main purpose of this instrument is to obtain the best compromise between large field of view and resolution that can be achieved to morphologically analyze samples.

2.2 Experiment set-up

The beam from a 488 nm Ar laser is enlarged with an afocal system. The expanded Gaussian laser beam is shaped into a light sheet with a cylindrical lens. This laser plane illuminates the sample contained in the microfluidic chip with an angle of 45 degrees, as shown in Fig. 1. The microfluidic chip is mounted on translation stages in order to be able to scan the whole sample. The fluorescence from the sample plane illuminated with the light sheet is collected by a long working distance 20x objective,

* Corresponding author: pierre.blandin@cea.fr

associated with a long-pass filter and a tube lens. The wide-field image from this orthogonal viewing direction is acquired on a sCMOS camera.

Fig. 1. Scheme of the experimental set-up of the light-sheet fluorescent microscope for microfluidic analysis on chip realized with Solidworks. Disposal centered on the sample holder system with an inset presenting the support of the microfluidic chip and the microfluidic injection system location.

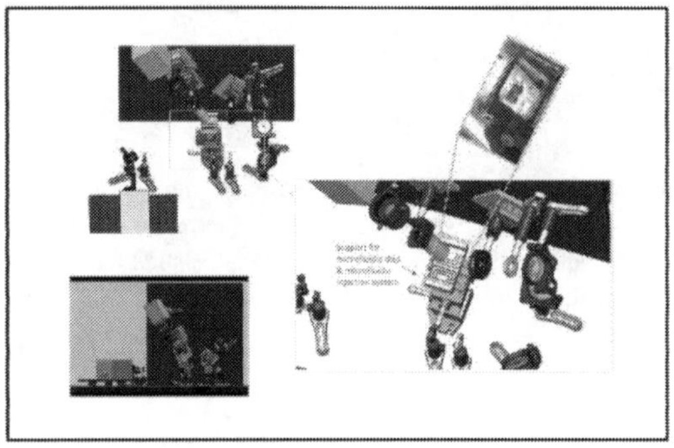

2.3 Characterisation of the optical system

The system was classically optically characterized using green-yellow fluorescent microspheres with a size lower than the expected resolution. These beads were disposed on a classical microscope slide to determine the spatial resolution of this microscope system. We imaged a xy plane parallel to the camera which corresponds to a plane at 45 degrees from the microfluidic chip. Fig. 2 shows the xy plane of the selected ROI and the corresponding intensity profiles of the PSF along x- and y-directions. The fitted Gaussian curve gives a value of $\sim 3 \pm 0.3\mu m$ for the lateral resolution measured at the FWHM. We also investigated the resolution that can be achieved within the microfluidic chamber with these same fluorescent microbeads. We proceeded with different buffer solutions in order to estimate their potential influences. Furthermore, we imaged in 3D diverse volumetric samples (pollen, distribution of several substances, and characterization of the chamber structures…). 3D rending can then be obtained with data treatments and analysis. Further experiments on biological samples are currently on-going and could be very soon presented.

3 Conclusion

The development of this LSFM leads to an innovative microscope dedicated to imaging in microfluidic chip. The system being well characterized, we are currently implementing the next step, which consists in realizing a real time acquisition on a 3D volume of a continuously perfused biological sample. This simple set-up presented will probably be optimized for the biological sample of interest that will be studied. The set-up must stay easy to handle with a minimal data analysis.

Fig. 2. xy plane of the selected ROI and the corresponding intensity profiles of the PSF along x-direction and y-direction with the fitted Gaussian curve associated. White scale bar: 25 µm

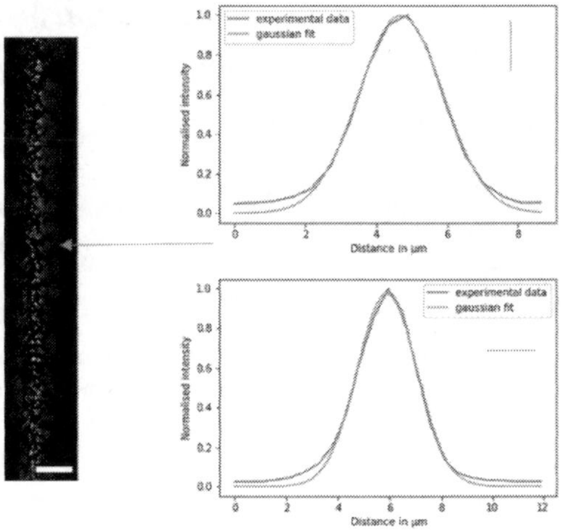

We acknowledge the financial support of the Cross-Disciplinary Program on Instrumentation and Detection of CEA, the French Alternative Energies and Atomic Energy Commission.

References

1. I. Albert-Smet, A. Marcos-Vidal, JJ. Vaquero, M. Desco, A. Muñoz-Barrutia, J. Ripoll. *Front Neuroanat.***13**:1 (2019)
2. E. De Santis, E. Shardlow, F. Stellato, et al. *Condens Matter.*;**4**(1):13 (2019)
3. R. Galland, G. Grenci, A. Aravind, V. Viasnoff, V. Studer, J-B. Sibarita. 3 *Nat Methods.*;**12**:641 (2015)
4. A-K. Gustavsson, PN. Petrov, WE. Moerner.. *Opt Express.*;**26**(10):13122-13147. (2018)
5. JW. Krieger, AP. Singh, CS. Garbe, T. Wohland, J. Langowski. *Opt Express.***22**(3):2358-2375 (2014)
6. J. Huisken,J. Swoger, F. Del Bene, J. Wittbrodt, EHK Stelzer. *Science (80-).*;**305**(5686):1007(2004)
7. OE. Olarte, J. Andilla, EJ. Gualda,P. Loza-Alvarez P. *Adv Opt Photonics.***10**(1):111 (2018)
8. LA. Royer, WC . Lemon, RK. Chhetri, et al. *Nat Biotechnol.***34**(12):1267-1278 (2016)
9. Y. Wu, A. Ghitani, R. Christensen, et al. *Proc Natl Acad Sci U S A.*;**108**(43):17708-17713 (2011)

Non-linear label-free imaging through a multimode graded index optical fibre

Angel Cifuentes[1,*], *Johanna* Trägårdh[1], *Tomáš* Pikálek[1], *Mojmír* Šerý[1], *Denis* Akimov[2], *Tobias* Meyer[2,3], *Jürgen* Popp[2,3], *Rodrigo Amezcua-Correa*[4] *and Tomáš* Čižmár[1,2,5]

[1]Institute of Scientific Instruments of the CAS, v. v. i., Královopolská 147, 612 64 Brno, Czech Republic
[2]Leibniz Institute of Photonic Technology, Albert-Einstein-Straße 9, 07745 Jena, Germany
[3]Institute of Physical Chemistry, Friedrich Schiller University of Jena and Abbe Center of Photonics, Helmholtzweg 4, 07743 Jena, Germany
[4]CREOL, The College of Optics and Photonics, University of Central Florida, 4304 Scorpius St., 32816-2700 Orlando, Florida, United States of America
[5]Institute of Applied Optics, Friedrich Schiller University Jena, Fröbelstieg 1, 07743 Jena, Germany

Abstract. In recent years, great advances have been made in developing minimal footprint micro-endoscopes using multimode optical fibres (MMF) [1]. By employing wavefront shaping methods the seemingly random speckle pattern resulting from the guiding of coherent light through an MMF can be formed into a diffraction limited spot. This enables the implementation of multiple laser scanning techniques. In this work we show that this approach can be employed to realize label-free non-linear microscopy techniques such as coherent anti-Stokes Raman scattering (CARS) and second harmonic generation (SHG). Backscattered light makes epi-detection possible even though these processes preferably emit light in the direction of beam propagation [2].

1 Introduction

The potential for MMF to be utilized as micro-endoscopes has come into light in recent years. By measuring the transmission matrix (TM) of a MMF it is possible to obtain a tightly focused spot on the far side of said fibre. The spot can be swept across the facet allowing for imaging techniques to be implemented. Label-free imaging offers the opportunity to observe and analyse samples without the use of exogenous markers, making it particularly alluring for endoscopic applications. Furthermore, graded index (GRIN) MMFs push the potential even further due to their resilience to bending induced distortions [3] and larger bandwidth (allowing the guiding of ultrashort pulses).

In this work we demonstrate the use of GRIN MMFs for imaging using two non-linear label-free techniques, namely, coherent anti-stokes Raman scattering (CARS) and polarization resolved second harmonic generation (P-SHG).

2 Methods

2.1. Calibration

We refer to calibration as the procedure used to measure the TM. For this, a set of input and output modes is defined. In our case, the proximal facet of the MMF was placed in the Fourier plane of a SLM and, by displaying the appropriate gratings, a grid of individual input spots, or modes, can be projected onto the proximal facet. Each input mode is addressed one at a time and the resulting speckle pattern at the distal end of the fibre is imaged on to a digital camera. Here, phase shifting interferometry allows for the determination of the complex amplitude of the field. For convenience, the set of output modes is defined by the camera pixels themselves. The process is carried out for each input mode resulting in the TM of the fibre. We may then select a desired output mode and calculate the light pattern required at the fibre proximal end to obtain constructive interference at a single well defined point in the distal end, effectively creating a focus.

The calibration procedure must be repeated for each wavelength used, tuning it without recalibrating is only possible for a small range (tens of nanometres depending on the fibre used). Moreover, an independent calibration must be effectuated for two orthogonal polarization states at the proximal end to improve the efficiency with which light may be focused into a single spot. Additionally, if full control over the polarization state of the resulting spot is required, a set of output polarization orthogonal TMs must be obtained, this can be achieved by rotating the polarization of the reference beam by 90°.

2.2 Experimental setup

To perform the calibration and imaging, the setup seen in figure 1 was utilized. A liquid crystal phase-only SLM is used to create the required input patterns on the fibre

* Corresponding author: cifuentes@isibrno.cz

proximal facet during calibration and imaging. The orthogonal polarization states of the input light are independently controlled, shaped, and overlapped (both spatially and temporally) before projection onto the fibre facet through a 20x/0.45NA microscope objective.

On the distal side, a 20x/0.45NA microscope objective and corresponding tube lens image the sample plane onto a digital camera where the speckle pattern is interfered with a reference beam. A delay line is used to obtain the temporal overlap required.

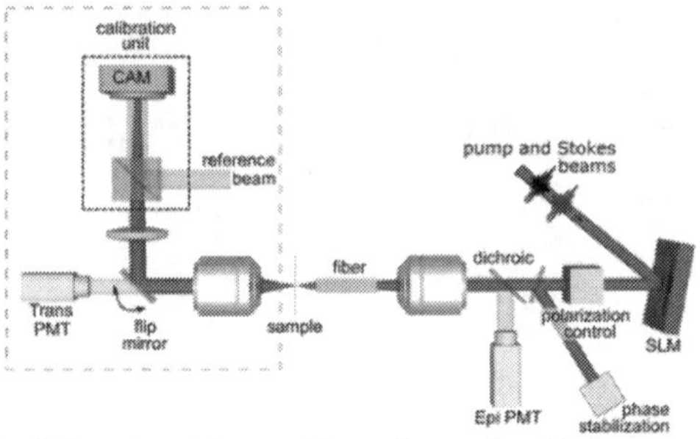

Fig. 1. Experimental setup. The section enclosed in the grey dashed line is only necessary when calibrating or imaging in transmission. Peripheral optical elements are not shown.

3 Results

3.1 CARS

CARS imaging and its inherent chemical selectivity was demonstrated by imaging 2μm polystyrene beads (PS) and 2.5μm PMMA beads (Figure 2), with a pixel integration time as low as 1 ms for epi-detection. A 125 μm cladding/50μm core 0.29NA GRIN fibre was used. The beams used were 1-2 ps long. The pump beam used was tuneable from 660-675 nm while the Stokes was fixed at 831 nm. [4]

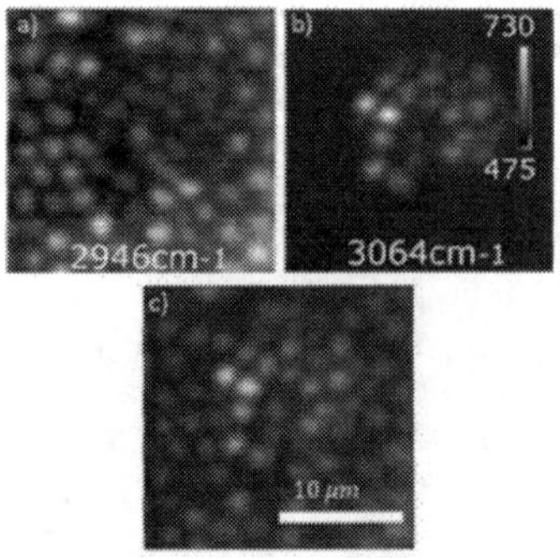

Fig. 2. Epi CARS imaging of polymethyl methacrylate (PMMA) and polystyrene (PS) beads through a GRIN MMF using 1-2 ps laser pulses showing chemical selectivity. a) PMMA beads, Raman shift=2946 cm-1 b) PS beads, Raman shift=3064 cm-1, c) Composite image

3.2 P-SHG

In the case of P-SHG a 1040 nm, ~100 fs long pulse was used to image a starch grain through a 0.29 NA 62.5μm core 100μm cladding GRIN MMF. The polarization of the exciting spot was varied from 30° to 180°. An additional image using right-handed circular polarization was also taken. The response of the starch grain to the change in polarization is clearly visible in Fig. 3.

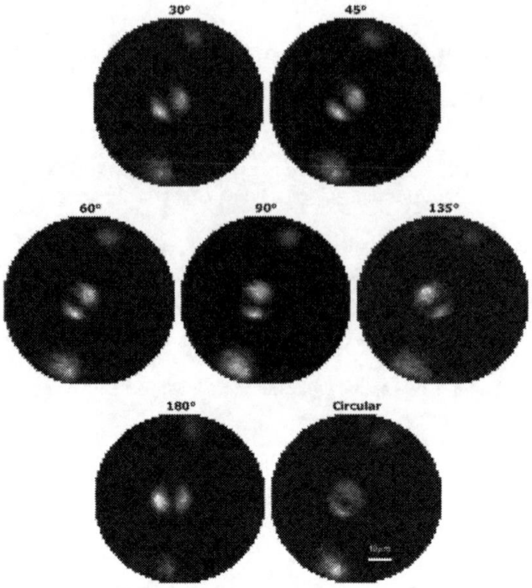

Fig. 3. SHG images of a potato starch grain in epi detection. The polarization angle of the linearly polarized excitation was varied between 30° and 180°. The resulting image for circular polarization (RH) excitation spot is also shown.

A potential application of P-SHG is in determining collagen fibril morphology in tissue. Other biological markers responding to SHG are myosin and microtubules. [5]

4 Conclusion

We have shown the capability of GRIN MMF to be used in label-free non-linear applications using the TM based fibre imaging approach. This opens a broad range of possible applications where hard to access samples may be observed in a relatively undisturbed state.

5 References

1. S. Turtaev, I. T. Leite, T. Altwegg-Boussac, J. M. Pakan, N. L. Rochefort, T. Čižmár, Light: Science & Applications **7**, 1 (2018).
2. F. Légaré, C. Pfeffer, B. R. Olsen, Biophys. J **93**.4, 1312 (2007).
3. D. E. B. Flaes, J. Stopka, S. Turtaev, J. F. De Boer, T. Tyc, T. Čižmár, Physical review letters, **120**.23, 233901 (2018).
4. J. Trägårdh, T. Pikálek, M. Šerý, T. Meyer, J. Popp, T. Čižmár, Optics express, **27**(21), 30055-30066 (2019).
5. R. Cicchi, N. Vogler, D. Kapsokalyvas, B. Dietzek, J. Popp, F.S. Pavone J. Biophotonics **6**.2, 129 (2013).

EPJ Web of Conferences **238**, 04007 (2020) https://doi.org/10.1051/epjconf/202023804007
EOSAM 2020

Combined Brillouin and Raman system for biomedical applications

Jan Rix[1,*], *Edmund* Koch[1], *Matthias* Kirsch[2, 3, 4], *Gabriele* Schackert[3], *Ortrud* Uckermann[3], and *Roberta* Galli[1, 4]

[1]Clinical Sensoring and Monitoring, Department of Anesthesiology and Intensive Care Medicine, Faculty of Medicine Carl Gustav Carus, TU Dresden, Fetscherstrasse 74, D-01307 Dresden, Germany
[2]Asklepios Kliniken Schildautal, Karl-Herold-Strasse 1, D-38723 Seesen, Germany
[3]Neurosurgery, Faculty of Medicine Carl Gustav Carus, TU Dresden, Fetscherstrasse 74, D-01307 Dresden, Germany
[4]National Center for Tumor Diseases (NCT), Partner Site Dresden, Fetscherstrasse 74, D-01307 Dresden, Germany

Abstract. A VIPA based Brillouin spectrometer consisting of commercially available components has been combined with a standard Raman spectrometer and a confocal microscope allowing simultaneous and co-localized measurements of biological samples and tissues.

1 Introduction

In the biomedical field, information about the stiffness of a tissue is of great importance as several cellular processes and mechanisms are depending on biomechanics. Especially in oncology, tumor tissue possesses a different stiffness compared to normal tissue [1], which might have an effect on cancer spreading and metastasis. Brillouin spectroscopy enables to analyze the mechanics of biological tissues in a non-destructive, label and contact-free manner. Furthermore, it can be combined with Raman spectroscopy, which enables addressing the compositional fingerprint of cells and tissue. The combined investigation of the biomechanical information obtained by Brillouin spectroscopy and the biochemical information by Raman spectroscopy provides a unique possibility of understanding complex biological mechanisms, as the molecular origin of changes in tissue's stiffness can be examined.

Therefore, a combined system for simultaneous acquisition of Raman and Brillouin spectra was built inside the research laboratory at the University Hospital of Dresden. First simultaneously acquired Brillouin and Raman spectra of different biomedical tissues were acquired in order to demonstrate the general practicability of combined measurements. Nearly transparent samples as well as opaque tissue samples were successfully measured.

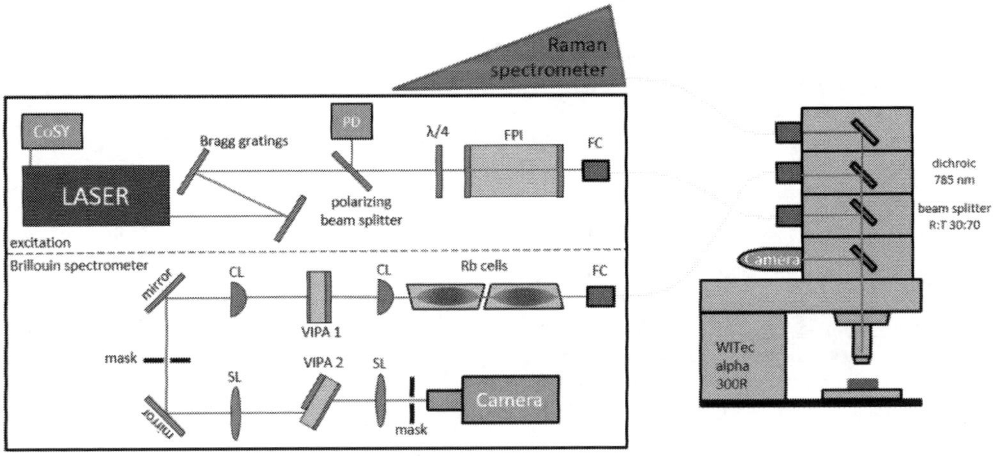

Fig. 1. Scheme of the experimental setup showing the excitation, the confocal microscope and both spectrometers. CoSY: Compact Saturation Spectroscopy, PD: Photodiode, FPI: Fabry-Pérot, FC: fiber coupler, CL/SL: cylindrical spherical lens

* Corresponding author: jan.rix@mailbox.tu-dresden.de

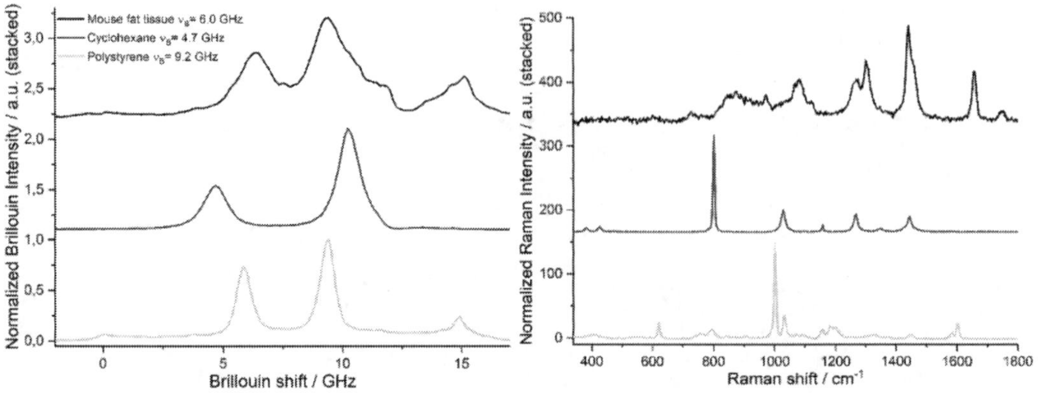

Fig. 2. Exemplary Brillouin and Raman spectra of polystyrene, cyclohexane and fat tissue of a mouse. Water has been used for axis calibration of the Brillouin spectra. As the free spectral range (FSR) of the VIPAs amounts 15 GHz, the Rayleigh bands appear at 0 GHz and 15 GHz, because of an incomplete filtering out of the elastically scattered light.

2 Results

A schematic representation of the system is given in Fig. 1. Scattering was excited using a laser source emitting at a wavelength of 780.24 nm that is best for opaque biological samples. The scattering radiation is acquired by a commercial confocal microscope (WITec Alpha 300R, WITec GmbH, Ulm, Germany). The Rayleigh and the Brillouin scattered light is separated from the Raman scattered light by a dichroic beam splitter (edge at 785 nm) and transferred to the respective spectrometer by single mode fibers. For Raman spectroscopy, a standard Raman spectrometer (WITec UHTS 400, WITec GmbH, Ulm, Germany) has been used. A Brillouin spectrometer based on a two-stage virtually imaged phased array (VIPA) configuration, as first described by Scarcelli et al. [2], has been designed and assembled. The system consists entirely of commercial components and is adapted from a set-up by Schlüßler et al. [3]. Three combined spectra, which have been acquired simultaneously, are exemplary presented in Fig. 2. Whereas the Brillouin spectra indicate a substantial difference in the samples' biomechanics, the Raman spectra show their characteristic chemical fingerprints.

3 Conclusion

It was shown that first combined measurements on biological samples are possible with the described setup. In the near future the system will be dedicated to investigations of biomechanical properties of brain tumor cells, with the aim of better understanding progression, spreading and metastasis of aggressive tumors within the central nervous system.

We thank the NCT Dresden for financing the experimental system. Furthermore, Jan Rix gratefully acknowledges the financial support of the Free State of Saxony. Special thanks go to Dr.-Ing. Raimund Schlüßler for the precious help and the fruitful discussions.

References

1. M. Troyanova, Z. Meng, V.V. Yakovlev, Biomed. Opt. Express **10**, pp. 1774–1781 (2019)
2. G. Scarcelli, P. Kim, S. H. Yun, Opt. Lett. **33**, pp. 2979–2981 (2008)
3. R. Schlüßler, S. Möllmert, S. Abuhattum, G. Cojoc, P. Müller, K. Kim, C. Möckel, C. Zimmermann, J. Czarske, J. Guck, Biophys. J. **115**, pp. 911–923 (2018)

EPJ Web of Conferences **238**, 04008 (2020)
EOSAM 2020
https://doi.org/10.1051/epjconf/202023804008

Imaging birefringent tissue in the human tympanic membrane by polarization-sensitive optical coherence tomography

Svea Steuer[1,*], *Jonas* Golde[1], *Steffen* Ossmann[2], *Lars* Kirsten[1], *Joseph* Morgenstern[2], *Matthias* Bornitz[2], *Marcus* Neudert[2], and *Edmund* Koch[1]

[1]Clinical Sensoring and Monitoring, Anesthesiology and Intensive Care Medicine, Technische Universität Dresden, Fetscherstrasse 74, 01307 Dresden, Germany
[2]Otorhinolaryngology, Carl Gustav Carus Faculty of Medicine, Technische Universität Dresden, Fetscherstrasse 74, 01307 Dresden, Germany

Abstract. Acousto-mechanical properties of the human tympanic membrane mainly depend on the connective tissue in its layered structure. Using polarization-sensitive optical coherence tomography, a depth-resolved imaging technique which provides additional tissue specific contrast, polarization changes of the birefringent layers in the human tympanic membrane were detected. By depicting estimated local retardances, distinguishing different tissue types was possible. This suggests the ability to image pathological alterations of the connective tissue with PSOCT, which extends the conventional diagnostic methods in middle ear surgery.

1 Introduction

Normal hearing requires the acousto-mechanical conduction of sound waves and impedance matching in the middle ear, consisting of the tympanic membrane (TM) and the ossicles. Stability and movement properties of the TM are mainly defined by a connective tissue layer [1], which is embedded in between an epidermal layer of the ear canal and mucosal tissue of the middle ear. If pathological changes of the TM are present, e.g. in succession of a chronic otitis media, the oscillation behavior of the eardrum and thus the sound conduction of the middle ear is impaired which has a direct effect on hearing. In clinical practice, otoscopy provides a general overview of the semi-transparent eardrum and adjacent structures, which is supported by tympanometry and audiometry for the assessment of middle ear function. However, alterations of the layered TM structure are not accessible with these methods. Optical coherence tomography (OCT) is a non-invasive and three-dimensional imaging technique which has been demonstrated as a promising tool for middle ear diagnostics [2]. Conventional OCT measures the backscattered signal of a sample, but is limited in providing tissue-specific contrast. Polarization-sensitive OCT (PSOCT) additionally detects polarization-changing properties of a sample, i.e. birefringence and depolarization [3]. Here, it is shown that the connective tissue layer of the TM can be assessed by PSOCT due to its birefringent collagen fiber structure. Utilizing an algorithm for the visualization of local retardance, the layered TM structure of a temporal bone specimen is examined.

*e-mail: svea.steuer@mailbox.tu-dresden.de

2 Method and measurements

To perform *ex vivo* PSOCT imaging of the TM, a right human cadaveric temporal bone, which was stored frozen after extraction, was prepared for measurements with a bench-top scanner by removing the auditory canal. A commercial spectrometer-based PSOCT system (Thorlabs TEL220PS) with a broad-bandwidth light source centered at 1300 nm and a telecentric scan lens (Thorlabs OCT-LK3) was utilized. While the measurement and the acquisition of the raw spectra was done using the provided commercial software (Thorlabs ThorImage OCT), further data handling was implemented in Matlab, including OCT processing and polarization analysis. After spectral shaping, k-linearization and Fourier transforming of the raw spectra to obtain complex depth profiles for both polarization channels, Stokes vectors were calculated and spatially averaged using a 3D Gaussian kernel with a full width at half maximum of 30 µm in axial direction and 60 µm in both lateral directions. As the utilized PSOCT system measures the backreflected signal for a single circularly polarized illumination state with polarization-diverse detection, the acquired cumulative polarization contrast is highly depending on the arrangement and orientation of the subsequent layers in a complex sample. To derive the local sample retardance, we calculated the angle between the averaged Stokes vectors of subsequent pixels over depth. As the result is only a valid retardance for certain constellations including circularly illuminated layers, the presented measure is only an estimate of the local retardance. For qualitative evaluation of birefringent and non-birefringent TM layers, both intensity images and visualizations of estimated local retardance with overlayed intensity are presented (see Fig. 1).

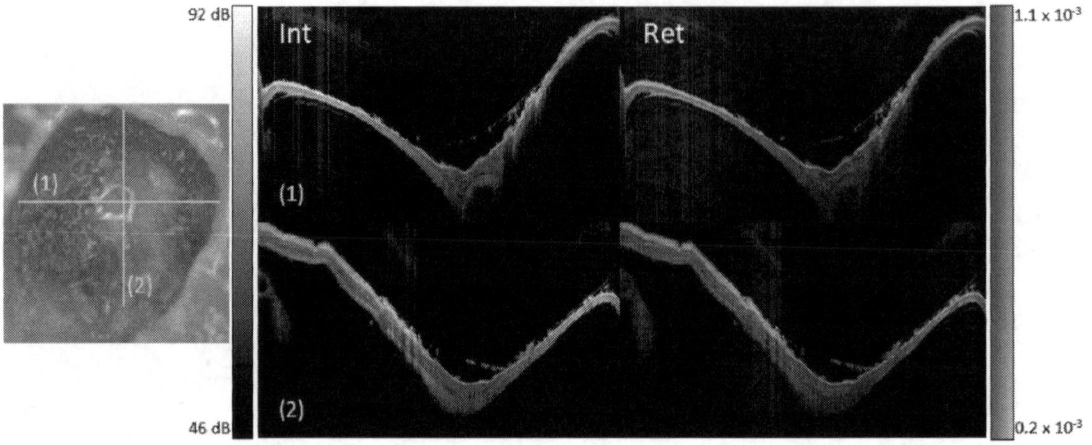

Figure 1. Photographic overview of the tympanic membrane (left) of a right temporal bone specimen with marked positions, (1) and (2), of the exemplary PSOCT B-scans. Intensity (middle column) and estimated local retardance (right column) images are 7.6 mm in lateral direction and 3.15 mm in axial direction.

3 Results

The connective tissue of the TM is commonly described as a two-layer arrangement of radially and circularly aligned collagen fibers surrounding the malleus. Although limited in the field of view, non-linear microscopic recordings of TMs, based on second harmonic generation, indicate that the fibrous structure of the connective tissue is interwoven in a more complex way. By using PSOCT and the estimated local retardance within the sample, the layered structure of the TM was assessed. The birefringent connective tissue causes local polarization changes which were detected and visualized by the colored contrast depicted in Fig. 1. Differentiation of the connective tissue with high retardance (blue) from the less or non-birefringent epidermal and mucosal tissue (yellow) was possible thereby. This was particularly true for the outer regions of the membrane in both scanning directions. In contrast to the intensity images, not only structures but also different tissue types could be distinguished. Thus, the results indicate the direct coupling of the connective tissue to the handle of malleus. However, resolving finer structures, e.g. individual collagen fibers, was impossible due to the limited axial and lateral resolution as well as applied spatial Stokes averaging.

4 Conclusion

Visualizing the connective tissue of the human tympanic membrane was achieved by PSOCT and an algorithm for the qualitative evaluation of the local retardance. As different diseases affecting the human tympanic membrane and its acousto-mechanical properties are related to alterations of the fibrous structure, e.g. tympanosclerosis, distinguishing the collagen fiber layer from the embedding tissue might provide additional diagnostic value in comparison to conventional OCT.

Acknowledgments

The author Svea Steuer is supported by the European Union/European Social Fund (ESF) and the Free State of Saxony with a doctoral scholarship (Project No. 100380876). Temporal bone experiments were approved by the Institutional Review Board (IRB00001473) at TU Dresden (EK108032015).

References

[1] K.N. O'Connor, M. Tam, N.H. Blevins, S. Puria, The Laryngoscope **118**, 483 (2008)

[2] L. Kirsten, M. Schindler, J. Morgenstern, M.T. Erkkilä, J. Golde, J. Walther, P. Rottmann, M. Kemper, M. Bornitz, M. Neudert et al., Journal of Biomedical Optics **24**, 1 (2018)

[3] J.F. de Boer, C.K. Hitzenberger, Y. Yasuno, Biomed. Opt. Express **8**, 1838 (2017)

Towards quantitative demineralization imaging for the assessment of carious lesions based on PS-OCT

Jonas Golde[1,*], *Florian* Tetschke[1,2], *Robin* Vosahlo[2], *Lars* Kirsten[1], *Julia* Walther[1,3], *Christian* Hannig[2], and *Edmund* Koch[1]

[1]Clinical Sensoring and Monitoring, Carl Gustav Carus Faculty of Medicine, TU Dresden, Fetscherstrasse 74, 01307 Dresden, Germany
[2]Policlinic of Operative and Pediatric Dentistry, Carl Gustav Carus Faculty of Medicine, TU Dresden, Fetscherstrasse 74, 01307 Dresden, Germany
[3]Department of Medical Physics and Biomedical Engineering, Carl Gustav Carus Faculty of Medicine, TU Dresden, Fetscherstrasse 74, 01307 Dresden, Germany

Abstract. Assessing the stage and progression of enamel demineralization non-invasively is of high interest in conservative dentistry. By examining tooth samples with suspected occlusal lesions, we show the potential of depolarization imaging based on polarization-sensitive optical coherence tomography for the assessment of carious lesions and validate the results by co-registered X-ray micro-computed tomography volumes.

1 Introduction

Assessing the stage and progression of caries non-invasively and preferably without utilizing ionizing radiation is of high interest in dentistry. Besides the early detection of carious lesions, conservative therapies, aiming for the remineralization of the demineralized enamel, depend on reliable measures of the enamel state to evaluate the success of the treatment [1]. Multiple techniques for qualitative and quantitative enamel analyses in vivo have been developed, which can be applied on a regular basis as they are based on optical technologies, e.g. trans-illumination and fluorescence. However, both heterogeneity and geometry of the intraorally examined tooth structures complicate the quantitative and repeated assessment for a long-term progression analysis.

Optical coherence tomography (OCT) is a non-invasive, interferometric imaging technique based on near-infrared light, that provides depth-resolved images and volumes of semi-transparent samples. Conventional OCT images represent the backscattering properties and are thus susceptible to various characteristics and alterations of the enamel, e.g. the alignment of the enamel rods, surface stains and demineralization. Additionally exploiting the polarization changes in the sample, i.e. utilizing polarization-sensitive OCT (PS-OCT), provides a more specific measure of those characteristics. We have recently shown that depolarization imaging based on PS-OCT is able to detect early stages of enamel demineralization [2] and facilitates the differentiation of surface staining and discoloration from carious lesions [3].

To assess the validity of this contrast mechanism, we acquired data sets of tooth samples with suspected occusal lesions using PS-OCT and X-ray micro-computed tomography (μCT), an in dental research established radiographic imaging technique for demineralization imaging. By software-based registration of the volumes, we present here the spatially aligned comparison of PS-OCT-based depolarization images and μCT sections of a human molar tooth.

*e-mail: jonas.golde@tu-dresden.de

2 Methods and materials

The 1300 nm swept-source PS-OCT system [2] as well as the processing [3] of the depolarization images, based on calculating the noise-immune degree of polarization, have been described previously. After PS-OCT imaging of the extracted molar tooth samples, which were collected during clinical treatment and visually inspected by an experienced dentist, μCT volumes were acquired and pre-processed using a commercial high-resolution x-ray tomograph (Bruker Skyscan 1172, 85 kV source voltage). For all subsequent processing steps including registration and visualization, custom MATLAB implementations were used. After rescaling both volumes to uniform voxels sizes of 8x8x8 μm³, a 3D canny edge detection algorithm was used to segement the tooth surface. The surface segmentation data sets were transformed into point clouds and downsampled for speed and registration improvements. Subsequent registration of the point clouds was performed by rigid transformation based on the "iterative closest point" algorithm as implemented in the MATLAB Computer Vision Toolbox. Fig. 1 shows three representative slices of the spatially aligned volumes and a photograph of the occlusal surface.

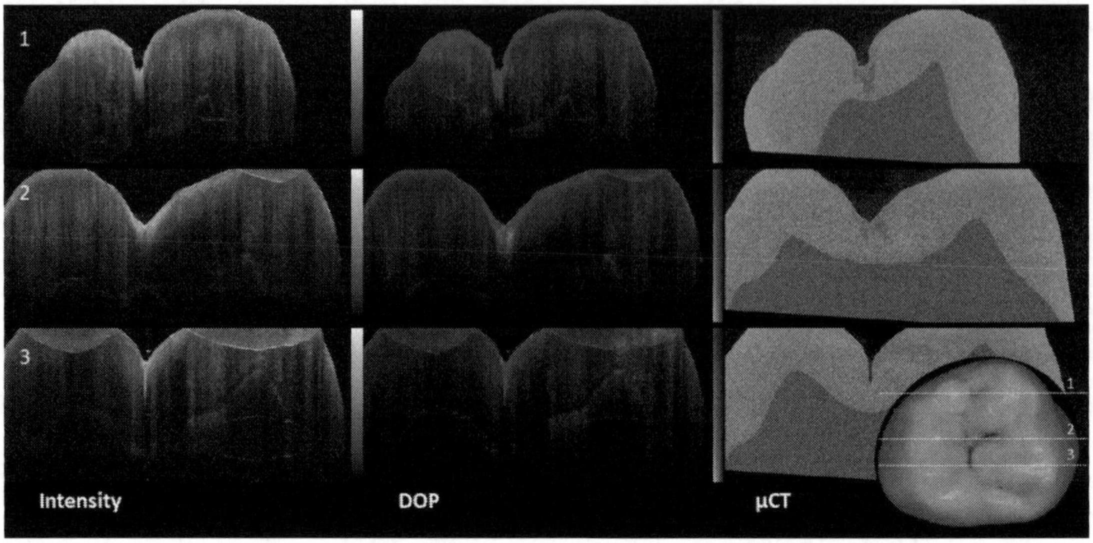

Figure 1. Comparison of PS-OCT and µCT slices from co-registered volumes: Intensity (-2...41 dB), degree of polarization (DOP) with intensity overlay (DOP: 0.25...1, Int: -2...28 dB) and spatially aligned µCT images, spanning 9.6 mm in lateral and 4.3 mm in axial direction. Photograph (bottom right) of the occlusal surface with slice assignments.

3 Results and conclusion

The presented molar tooth shows considerable stains in the occlusal fissures and indications of demineralization. Depolarization images and µCT slices of selected regions (Fig. 1) confirm the suspected demineralization in the enamel in both the outer (slice 1) and the central pit (slice 2), as reduced DOP values for PS-OCT and reduced absorption, i.e. intensity, for µCT indicate. Furthermore, both modalities are in good agreement assessing only stains but no considerable demineralization in the dark-colored fissure (slice 3). In comparison of the PS-OCT depolarization and intensity images, the more specific contrast of the DOP is highlighted as both staining and demineralization cause an altered scattering contrast in the intensity images, but DOP facilitates the differentiation.
Although the applied registration provides corresponding depth sections for both modalities, the light propagation, i.e. surface refraction and optical vs. geometric thickness of the material, was not taken into account for the alignment of µCT and PS-OCT images. Thus, the approach is currently limited to a qualitative comparison of enamel demineralization in corresponding regions. Both more sophisticated alignment approaches for the analysis of tooth samples and the examination of artificial, less complex samples with demineralization might provide the basis for the quantitative assessment of demineralization by depolarization imaging.

Acknowledgments

This project is co-financed with tax funds based on the budget decided by the members of the Saxon State Parliament. This project was supported by the European Union/ European Social Fund (ESF) and the Free State of Saxony within the ESF junior research group "Optical Technologies in Medicine" (project number 100270108). The author Jonas Golde was supported by the European Union/ European Social Fund (ESF) and the Free State of Saxony within a doctoral scholarship (project number 100284305).

References

[1] H. Meyer-Lueckel, S. Paris, Operative dentistry **41**, S35 (2016)
[2] J. Golde, F. Tetschke, J. Walther, T. Rosenauer, F. Hempel, C. Hannig, E. Koch, L. Kirsten, Journal of biomedical optics **23**, 071203 (2018)
[3] J. Golde, F. Tetschke, R. Vosahlo, J. Walther, C. Hannig, E. Koch, L. Kirsten, *SPIE 11078, Optical Coherence Imaging Techniques and Imaging in Scattering Media III* (2019), p. 24

Second harmonic generation enhancement by polarization-matched nanostructures -INVITED

Christoph Dreser[1,2], *Emre* Gürdal[1,2], *Godofredo* Bautista[3], *Anke* Horneber[2,4], *Xiaorun* Zang[3], *Dominik A.* Gollmer[1,2], *Alfred J.* Meixner[2,4], *Dieter P.* Kern[1,2], *Dai* Zhang[2,4], *Martti* Kauranen[3], and *Monika* Fleischer[1,2,*]

[1]Institute for Applied Physics, Eberhard Karls University of Tübingen, Germany
[2]Center for Light-Matter Interaction, Sensors and Analytics LISA+, Eberhard Karls University of Tübingen, Germany
[3]Laboratory of Photonics, Tampere University of Technology, Finland
[4]Institute for Physical and Theoretical Chemistry, Eberhard Karls University of Tübingen, Germany

Abstract. Frequency conversion plays an important role in both fundamental and applied nano-optics. Doubling the frequency of light by second harmonic generation (SHG) is a vital process e.g. in laser optics or high-resolution microscopy. SHG can be created through symmetry breaking at plasmonic nanostructures, or the local high electric near-fields of plasmonic nanoantennas can be utilized to further enhance the SHG e.g. from nonlinear crystals. Examples of SHG microscopy using cylindrical vector beams in combination with tilted nanocones and radially symmetric oligomers are shown as well as enhancement studies of the SHG from nonlinear crystals decorated with polarization-matched nanostructures.

1 Introduction

In the interaction of light with nanostructures many intricate effects can be observed. For light with high local intensities also nonlinear effects, such as second harmonic generation (SHG), can be observed [1,2]. In plasmonic nanostructures, which are most often created from gold, silver or aluminium, collective oscillations of the free electron density (so-called localized surface plasmon polaritons) are excited, leading to high scattering and absorption cross-sections, a high local concentration of optical energy near the nanostructures, strong electric near-fields, and resonances that depend on the geometry, material, arrangement, and dielectric environment [3,4]. Advanced microscopy and spectroscopy techniques are being developed to further probe these interactions.

One possibility for investigating optical nanoantennas is the use of cylindrical vector beams (CVBs) [5]. These CVBs can exhibit either azimuthal or radial polarization with rotationally-symmetric electric or magnetic field components. In a tight focus they contain either exclusively in-plane or predominantly out-of-plane components. In the present work SHG microscopy is performed by scanning plasmonic nanostructures through the focus of a CVB and recording the resulting SHG intensity for each relative position of the focus and the nanostructure [6,7]. The resulting scanning images give clear indication of the polarization-dependent SHG induced by the nanoantennas, where maximum intensity can be found for polarization-matched nanostructure arrangements.

For non-linear applications that require larger volumes for SHG, non-linear crystals such as lithium niobate ($LiNbO_3$) are commercially available. For applications, typically high signal intensities are required, whereas the SHG efficiency is typically limited. To further enhance the overall SHG intensity, the non-linear crystals can be decorated with plasmonic nanoparticles. The polarization direction of maximum near-field enhancement should be matched to the extraordinary axis of the crystal. In this work, the interaction of individual gold nanodiscs with y-cut crystals is investigated under excitation with in-plane polarized light [8] as well as the SHG enhancement by self-assembled gold nanospheres on a z-cut crystal in the focus of a radially polarized CVB [9,10].

2 Results and Discussion

For the first sample type (see Fig. 1, left), a strategy for the fabrication of gold nanocones with a pre-defined tilt of the cone axis was developed [6]. The asymmetric geometry supports the transformation of a transversal electric far-field (oriented parallel to the substrate) to a longitudinal plasmonic excitation (perpendicular to the substrate). Nanocones with a range of different tilting angles were examined by angle-dependent extinction spectroscopy. This way the ratio of the electric field components oriented parallel vs. perpendicular to the respective cone axes was systematically changed, as could be observed in the respective experimental and simulated spectra. The cones were imaged by SHG microscopy using tightly focused either azimuthally or radially polarized CVBs, and the data compared to simulations. It could be demonstrated in both the linear and the non-linear optical measurements that an excitation of the cone tip is achieved even with purely in-plane polarized excitation fields.

* Corresponding author: monika.fleischer@uni-tuebingen.de

For the second sample type, oligomers consisting of gold nanorods that were either radially (Fig. 1, center) or azimuthally (Fig. 1, right) arranged were lithographically fabricated. The longitudinal plasmon resonance of the nanorods was designed to coincide with the fundamental laser wavelength (1060 nm laser with 140 fs pulses and a repetition rate of 80 MHz), and the oligomer geometry was carefully designed to match the electric field distributions of the focused CVBs to maximize interaction and analyse the ensuing collective effects [7]. Maximum SHG was observed for radial oligomers excited by radial, and for azimuthal oligomers excited by azimuthal CVBs. Whereas for radial oligomers, the SHG intensity increased with the number of nanorods, the SHG intensity for azimuthal oligomers went through a maximum, with increasing silencing effects for increasingly small gaps.

Fig. 1. Tilted gold nanocone (left), radial oligomer (center) and azimuthal oligomer (right) that are imaged by SHG imaging microscopy with cylindrical vector beams.

For the third non-linear study, SHG enhancement was pursued by nanofabricating plasmonic nanostructures on top of nonlinear $LiNbO_3$ crystals. Depending on the cut of the crystal, the extraordinary optical axis can be oriented in different crystal directions. The gold nanoparticles were designed to be resonant at the fundamental harmonic of the exciting pulsed laser (774 nm laser with 110 fs pulses and a repetition rate of 89 MHz), and the cut was chosen for the extraordinary axis to be oriented parallel to the respective dipolar localized plasmon excitation of the nanoparticle. The two systems outlined in Fig. 2 were investigated numerically, as well as with linear dark-field and non-linear SHG spectroscopy. Individual nanodiscs were lithographically prepared on a y-cut crystal and examined under linearly polarized excitation [10], and dense arrays of self-assembled gold nanospheres [8] were created on a z-cut crystal and analyzed in the focus of a radially polarized CVB [9]. While the SHG of the nanostructures themselves proved negligible compared to the SHG intensity from the crystal, 3-fold SHG enhancement could be demonstrated for a single disc on $LiNbO_3$, and 60-fold SHG enhancement for the dense arrays on $LiNbO_3$, due to the interaction of the hybrid systems.

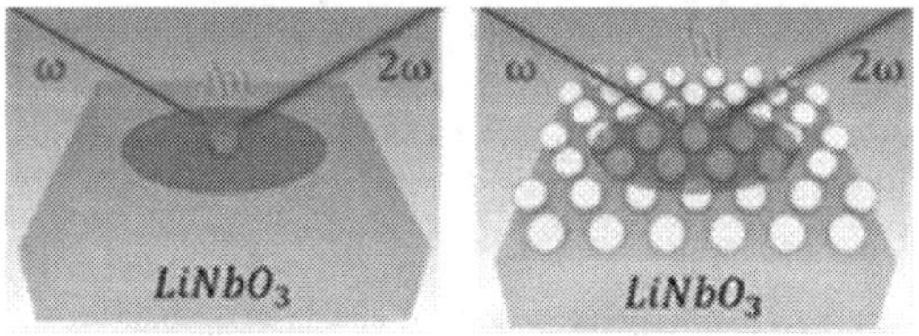

Fig. 2. SHG enhancement by a single gold nanodisc (left) and an array of gold nanospheres from diblock-copolymer micellar lithography (right) on non-linear $LiNbO_3$ crystals (from [11]).

3 Conclusions

SHG from nanostructures that are designed to exhibit localized surface plasmon resonances at the fundamental harmonic of the pulsed excitation laser and are matched to the polarization direction of the electric field components in the laser focus is investigated. In a first example, SHG scanning images are recorded of tilted nanocones, which prove that the hotspots at the tip apex can be excited even with purely in-plane polarized illumination. From SHG scanning images of radially and azimuthally oriented oligomers of gold nanorods it can be shown that collective modes are excited when the orientation of the nanorods in the oligomer matches the orientation of the electric field components in the focus. In a third example, strong local near-field enhancement is created by matching the resonance wavelength of gold nanodiscs or nanospheres that are positioned on the surface of a non-linear crystal to the fundamental of the incident laser beam, leading to up to 60-fold SHG enhancement factors within the laser focus.

Financial support by the European Cooperation in Science and Technology through COST Action MP1302 Nanospectroscopy, the Landesgraduiertenförderung Baden-Württemberg, the Academy of Finland, Tampere University of Technology, and the German Excellence Strategy is gratefully acknowledged.

References

[1] M. Kauranen, A. V. Zayats, Nature Photonics **6**, 737 (2012)

[2] P. Reichenbach, A. Horneber, D. A. Gollmer, A. Hille, J. Mihaljevic, C. Schäfer, D. P. Kern, A. J. Meixner, D. Zhang, M. Fleischer, L. M. Eng, Opt. Express **22**(13), 15484 (2014)

[3] M. Fleischer, D. Zhang, K. Braun, S. Jäger, R. Ehlich, M. Häffner, C. Stanciu, J. Hörber, A. Meixner, D. Kern, Nanotechnol. **21**, 065301 (2010)

[4] C. Schäfer, D. A. Gollmer, A. Horrer, J. Fulmes, A. Weber-Bargioni, S. Cabrini, P. J. Schuck, D. P. Kern, M. Fleischer, Nanoscale **5**, 7861 (2013)

[5] T. Züchner, A. V. Failla, A. J. Meixner, Angew. Chem. Int. Ed. **50**, 5274 (2011)

[6] C. Dreser, D. A. Gollmer, G. Bautista, X. Zang, D. P. Kern, M. Kauranen, M. Fleischer, Nanoscale **11**, 5429 (2019)

[7] G. Bautista, C. Dreser, X. Zang, D. P. Kern, M. Kauranen, M. Fleischer, Nano Lett. **18**, 2571 (2018)

[8] E. Gürdal, A. Horneber, N. Shaqqura, A. J. Meixner, D. P. Kern, D. Zhang, M. Fleischer, J. Chem. Phys. **152**, 104711 (2020)

[9] E. Gürdal, S. Dickreuter, F. Noureddine, P. Bieschke, D. P. Kern, M. Fleischer, Beilstein J. Nanotechnol. **9**(1), 1977 (2018)

[10] E. Gürdal, A. Horneber, A. J. Meixner, D. P. Kern, D. Zhang, M. Fleischer, under revision, (2020)

[11] E. Gürdal, PhD Thesis, Eberhard Karls University Tübingen (2020)

Resonant diffraction gratings with polarization-dependent efficiencies

Julian Wüster[1,*], *Patrick* Feßer[1], *Arne* Behrens[1], and *Stefan* Sinzinger[1]

[1]Technische Universität Ilmenau, Center for Micro- and Nanotechnologies, FG Technische Optik, Am Helmholtzring 1, 98693 Ilmenau

Abstract. Subwavelength-structures with different fill factors in the lateral dimensions result in unique phase shifts for the different polarization states of transmitted light.
By using this additional degree of freedom for diffractive optical elements, we yield additional functionalities for compact optical systems with DOEs.
As a fully operable example we present a binary subwavelength-grating which acts as a polarizing beamsplitter for TE- and TM-polarization over a wide range of incidence angles. We show our design approach, the manufacturing process with Soft-UV-Nanoimprint-Technology, as well as experimental results. We will also lay out possibilities for the design and application of multilevel polarization-functionalized gratings.

1 Introduction

Polarizing beam splitters are classically realized with dielectric coatings or birefringent materials. Alternatively, beam splitters can also consist of high-frequency gratings (single-order gratings) under Bragg (or Littrow) incidence [1] or subwavelength-structures, e.g. under an incidence of 45° [2]. For a compact in-line measurement system it is necessary to achieve a beamsplitting functionality while maintaining one optical axis over the whole system. This is possible in systems using a diffraction grating as shown in [3] and [4] which uses the 0th order for illumination and as the light reflected from the surface-to-be-measured passes the grating again, the higher orders are used for the detection. For this it is advantageous to design a grating, which has low diffraction efficency in the direction of illumination and high efficiency for the reflected light. The distinction is possible with the help of a quarterwaveplate switching between TE- or TM-polarization, respectively.

2 Design approach

The basic idea is to combine a diffraction grating with a period greater than the wavelength with subwavelength-structures. Those can, in a first step, be considered as an effective medium with different effective indices for TE- and TM-polarization, which is the so-called form birefringence [5, 6]. The final structure can be seen in Fig. 1. It realizes a phase shift of $\phi_{TM} = \pi$ and $\phi_{TE} = 2\pi$. The initial design has been laid out with the Effective Medium Approach with correction terms for larger structures [5] and then refined with RCWA-simulations. With the parameters height, fill factor and material, the desired phase difference $\Delta\psi$ is tunable.

*e-mail: julian.wuester@tu-ilmenau.de

Simulations show that the high phase difference $\Delta\psi = \pi$ within the effective medium mainly relies on the refractive index n_g. Due to manufacturing restraints in terms of the aspect ratio (see Section 3) we choose the highly refractive material silicon and a design wavelength of $\lambda = 1400$ nm. The structure sizes are $p_1 = 500$ nm, $p_2 = 10$ µm, the fill factor $f = 0.5$ and the etching depth $h = 700$ nm.

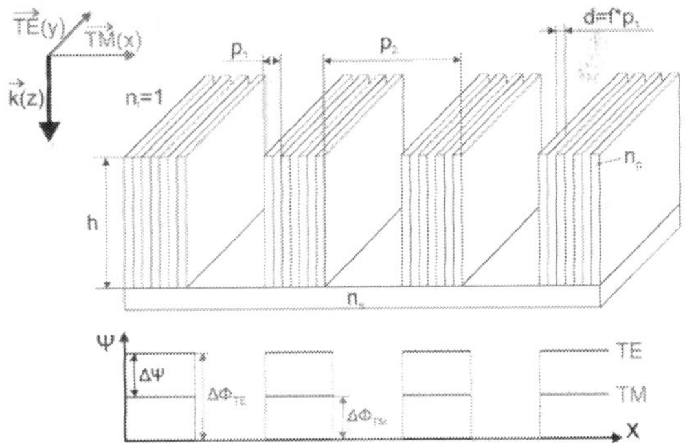

Figure 1. Design layout

3 Manufacturing

Due to the potential cost-efficiency (reproducability of large structured areas) in comparison to direct-writing techniques, Soft-UV-Nanoimprint-Lithography was chosen for the manufacturing process. The main limitation within the design is the limited aspect ratio one can expect from the lithographic process, which also limits our choice of the fill factor. Therefore various steps are needed to manufacture the grating in the high-refractive silicon.

https://doi.org/10.1051/epjconf/202023805002

We started from an electron-beam-structured chromium mask on fused silica, which was fabricated externally. All the other steps were carried out in-house.

The first dry-etching pattern transfer of the chromium layer to the fused silica wafer (RIE CHF_3/SF_6) results in our master for the following NIL-steps. The master with structure heights of about 250 nm is coated with FDTS for anti-sticking. The diluted PDMS is poured onto the master and after hardening we yield the PDMS stamp. We chose a SiO_2 hard mask. For that we executed a dry thermal oxidation process with a silicon wafer up to an oxide layer thickness of 200 nm. On top of that we spincoat the NIL-resist NIL 210. First we execute the imprinting process with the again anti-sticking-treated stamp to reproduce the structures in the resist layer. The master can be reused. Two more etching steps are needed to come to the final structure in silicon, which can be seen in Fig 2. The transfer from the resist pattern to the oxide layer is carried out with a RIE process (CHF_3/AR PlasmaPro 100) and the transfer from the oxide mask to the silicon wafer with ICP-RIE (SF_6/O_2 PlasmaPro 100 Cobra). The most critical step due to the different scales (nm/μm/cm) within the whole process is the molding of the PDMS stamp.

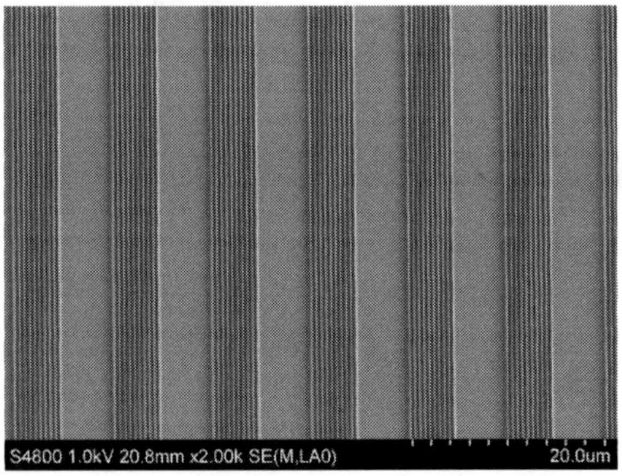

Figure 2. SEM image of the grating in silicon

4 Experimental results

We used an SC fiber laser, set to a wavelength of λ = 1400 nm. The fiber is not polarization-maintaining, so we let the light pass through a linear polarizer before interacting with the grating from the back side of the polished waver. The intensities are measured with Thorlabs photodiodes S122C. For obtaining the different polarization states just the polarizer is rotated. The results can be seen in Fig. 3. Note that due to the refractive index n = 3.49 the Fresnel losses are very high. About 50% of the light is lost in transmittance. The results show the diffraction efficiencies relative to the intensity of the incident light under an angle of 0°, including the Fresnel losses. This means that the efficiencies could be significantly increased with a lower refractive material or anti-reflection coating. Also experiments were carried out with varying wavelengths and varying incidence angles. Within a range of 1400 nm ± 60 nm there is no significant change in efficiency.

Measuring the intensities in relation to the incidence angle shows, that the grating is operable within a range of 0° ± 30°, as the efficiency contrast between TE and TM is still very high.

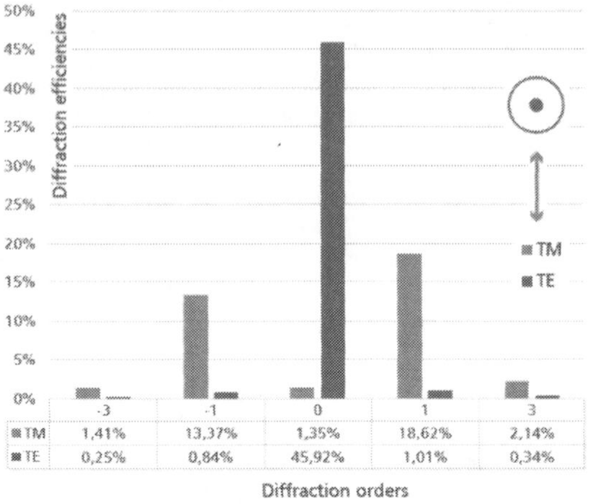

Figure 3. Diffraction efficiencies

5 Summary and Outlook

We presented an approach for the design of diffractive optics with polarization-dependent diffraction efficiencies. As an example we chose a binary grating with subwavelenghth-structures which was fabricated with Soft-UV-Nanoimprint-Lithography. It acts as a polarizing beamsplitter over a broad range of incidence angles and with tolerance in wavelength or etching depth, respectively. The design approach to combine those subwavelength gratings with diffractive optics with conventional dimensions is also applicable for multilevel elements and is therefore a promising way to add new functionalities to DOEs.

References

[1] H. Bartelt et al., Optik **112**, 283-288 (2001)
[2] P. Lalanne et al., J. of Optics A **215**, 214-219 (1999)
[3] M. Bichra et al., Appl. Optics **56**, A13-A17 (2017)
[4] M. Bichra et al., Appl. Optics **57**, 3808-3816 (2018)
[5] H. Haidner, *Dissertation* (FAU, Erlangen, 1994)
[6] P. Lalanne et al., J. Modern Opt. **43**, 2063-2085 (1996)
[7] J. Wüster et al., EOS Diffractive Optics, Jena (2019)

Towards active electro-optic lithium niobate metasurfaces

Flavia Timpu[1], *Helena* Weigand[1], *Fabian* Kaufmann[1], *Felix U.* Richter[1], *Viola-Valentina* Vogler-Neuling[1], *Artemios* Karvounis[1], and *Rachel* Grange[1*]

[1]ETH Zurich, Institute for Quantum Electronics, Department of Physics, Auguste-Piccard-Hof 1, 8093 Zurich, Switzerland

Abstract. We present the design and fabrication advances on active lithium niobate metasurfaces. We determine by numerical calculations a metasurface design with electro-magnetic resonances in the visible and near-infrared, by taking into account the constraints for fabrication on thin films of lithium niobate. We suggest that the optical properties of the metasurface can be switched using the electro-optical properties of lithium niobate.

1 Metasurface design for electro-optic switching

We design metasurfaces of lithium niobate ($LiNbO_3$, LNO) starting from a monocrystalline 200 nm layer on 500 μm quartz. LNO is a non-centrosymmetric metal-oxides with a high refractive index, second-order susceptibility tensor and electro-optic coefficients [1]. By nanostructuring LNO at the sub-wavelength scale, it is possible to control the optical properties by tuning the confinement of light in the material.

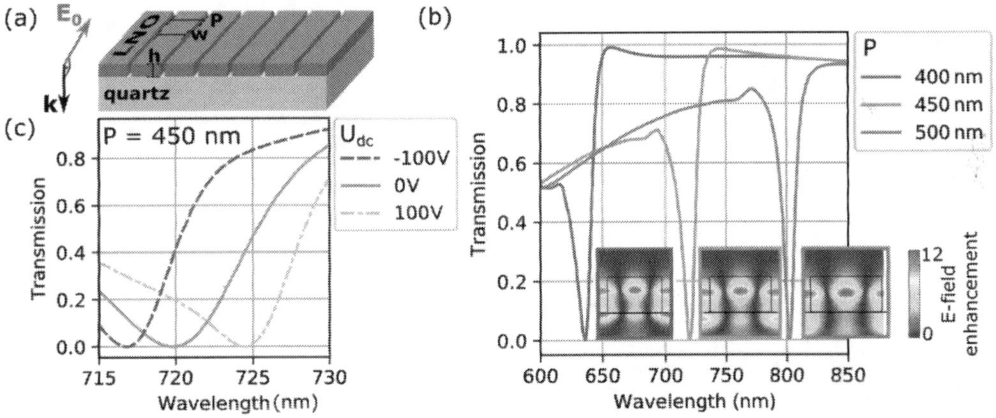

Figure 1. (a) Schematic of LNO gratings on a quartz substrate: period P, width $w = P - 100$ nm, height $h = 200$ nm. (b) Calculated transmittance of a plane wave polarized along the gratings at normal incidence. The insets show the electric field enhancement at the transmission resonance in a plane perpendicular to the grating. (c) Calculated transmission under actuation with a DC-bias voltage for the LNO grating with a period of 450 nm.

In Figure 1(a), we present a basic design of LNO gratings on a quartz substrate. For a well-defined wavelength that increases with the period, we observe the total suppression of the transmitted light (Fig. 1(b)). At these wavelengths, the electric field is enhanced up to 12 times compared to an unpatterned layer of LNO.

We further show that the application of an electrical voltage U can actively tune the optical transmittance of the LNO gratings. We assume that the extraordinary optical axis of LNO and

*e-mail: grange@phys.ethz.ch

the external DC-bias field are oriented along the gratings. Thus, the extraordinary refractive index n_e changes due to U and due to the field enhancement factor, defined as the average field enhancement in the volume of LNO [2]. We estimate a 5 nm shift of the transmission resonance for U = 100 V (Fig. 1(c)).

2 Fabrication approach and preliminary measurements

We verify our numerical results by fabricating LNO gratings on quartz using focused ion beam (FIB) milling. We measure their optical transmission in an optical setup described elsewhere [3]. For all three LNO gratings with periods from 400 nm to 500 nm, we observe a transmission resonance at wavelengths that increase with their period. In comparison to the simulation, the wavelength of the transmission resonance is red shifted due to geometry differences between simulated and fabricated gratings. Furthermore, the reduced suppression of the measured transmission is a consequence of the measurement setup, which includes not only the normal-incident wave, but several angles of incidence [3].

In addition to the fabrication by FIB, we are currently developing a top-down fabrication approach using e-beam and inductively coupled plasma (ICP/RIE) etching (Fig. 2 (c)) to improve the fabrication yield. Preliminary fabrication results of a LNO metasurface consisting of holes (Fig. 2 (d)) are promising.

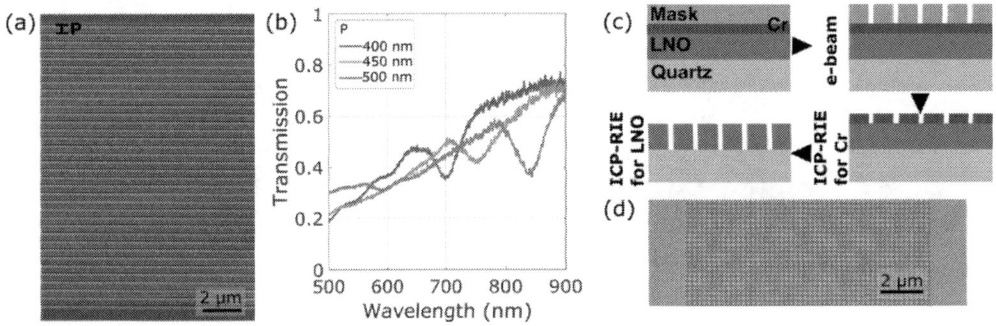

Figure 2. (a) SEM image of LNO gratings fabricated by FIB. (b) Measured transmission of the LNO gratings. (c) Schematic flow of the ICP fabrication protocol of LNO metasurfaces. (d) SEM image of a metasurface after the ICP-RIE for chromium (Cr).

3 Conclusions

We show initial designs of LNO periodic arrays with efficient electro-optical switching. We demonstrate the fabrication of such gratings by FIB and we propose more efficient fabrication by ICP dry etching. Preliminary optical measurements show good agreement with the simulation. In a next step, more complex designs will be produced by ICP. Furthermore, the electro-optical switching will be measured.

References

[1] M.J. Weber, *Handbook of optical materials* (CRC Press LLC, 2003)

[2] Z. Huang, H. Lu, H. Xiong, Y. Li, H. Chen, W. Qiu, H. Guan, J. Dong, W. Zhu, J. Yu et al., Nanomaterials **9**, 69 (2019)

[3] F. Timpu, M. Reig Escalé, M. Timofeeva, N. Strkalj, M. Trassin, M. Fiebig, R. Grange, Advanced Optical Materials **7**, 1900936 (2019)

Strong light-matter coupling in dielectric metasurfaces

Gabriel W. Castellanos[1,*], *Shunsuke* Murai[2], *T.V.* Raziman[1], *Shaojun* Wang[1,3], *Mohammad* Ramezani[1], *Alberto G.* Curto[1], and *Jaime* Gómez Rivas[1]

[1]Department of Applied Physics and Institute for Photonic Integration, Eindhoven University of Technology, P.O. Box 513, 5600 MB Eindhoven, The Netherlands
[2]Department of Material Chemistry, Graduate School of Engineering, Kyoto University, Nishikyo-ku, Kyoto 615-8510, Japan
[3]MOE Key Laboratory of Modern Optical Technologies and Jiangsu Key Laboratory of Advanced Optical Manufacturing Technologies, School of Optoelectronic Science and Engineering, Soochow University, Suzhou 215006, China

Abstract. We demonstrate the strong coupling between excitons in organic molecules and all-dielectric metasurfaces formed by arrays of silicon nanoparticles supporting Mie surface lattice resonances (MSLRs). Compared to Mie resonances in individual nanoparticles, MSLRs have extended mode volumes and much larger quality factors, which enables to achieve collective strong coupling with very large coupling strengths and Rabi energies. Moreover, due to the electric and magnetic character of the MSLR given by the Mie resonance, we show that the hybridization of the exciton with the MSLR results in exciton-polaritons that inherit this character as well. Our results demonstrate the potential of all-dielectric metasurfaces as novel platform to investigate and manipulate exciton-polaritons in low-loss polaritonic devices.

1 Introduction

The use of dielectric metasurfaces supporting Mie resonances as optical cavities to control light-matter interactions has triggered many interesting works in the last years. Due to the low absorption losses and the presence of resonances with both electric and magnetic character, dielectric metasurfaces have an exciting prospect for the development of novel applications.

However, such work has been limited to the control of emission, a weak light-matter coupling phenomenon. Achieving the strong coupling regime with Mie resonances has remained challenging due to their moderate field enhancements and large radiative losses.[1, 2] Strong coupling results in the hybridization of light and matter into quasiparticles known as exciton-polaritons, with remarkable properties such as enhanced transport, long-range energy transfer, condensation, and non-linear response.

Here, we demonstrate experimentally the strong coupling between excitons in organic molecules and all-dielectric metasurfaces formed by arrays of poly-crystalline silicon nanoparticles supporting Mie surface lattice resonances (MSLRs) with electric and magnetic character.[3] We first discuss the bare MSLRs in the array without molecules and explain the origin of their electric and magnetic character. Next, we deposit an organic dye on the array and show the formation of polaritonic states. Using a coupled oscillator model, we fit these states and extract the values of the Rabi energy. Finally, using simulations we discuss the character of the exciton-polaritons formed in this system.

2 Mie surface lattice resonances (MSLRs)

MSLRs emerge from the enhanced radiative coupling between the Mie resonances of nanoparticles through in-plane diffracted orders known as Rayleigh anomalies (RAs). Because of the low-loss characteristic of diffracted orders, MSLRs have large quality factors. Furthermore, due to the collective nature of the resonance, MSLRs have extended mode volumes that allow interaction with many emitters.

Our array has a periodicity a= 400 nm and the particles have a diameter d= 120 nm with height h= 90 nm. To characterize the optical response of the MSLRs we measure the extinction as function of the incident in-plane moment $\mathbf{k}_\parallel$ ($k^2 = k_\parallel^2 + k_z^2$) of the metasurface with a deposited layer of poly(methyl methacrylate) (PMMA) with a thickness of t = 290±10 nm (see Figure 1a and 1b). The extinction is defined as 1-T_0, where T_0 is the zeroth-order transmission. The measurements for TE (Figure 1a) and TM (Figure 1b) polarizations reveal several dispersive modes with Fano lineshapes, which correspond to the bare MSLRs arising from the RAs (indicated by the grey lines).

MSLRs can have an electric or magnetic character that is given by the presence of both electric and magnetic dipole resonances in the individual silicon nanoparticles. The character of the MSLR can be understood with the help of the bottom diagram that shows the preferential coupling between the excited dipoles for a given polarization of the incident light field.

* Corresponding author: g.w.castellanos.gonzalez@tue.nl

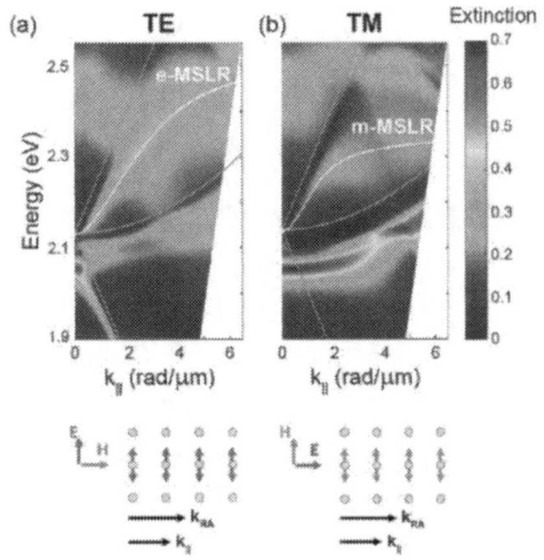

Fig. 1. Extinction of the MSLRs as function of the incident in-plane moment $\mathbf{k}_{\parallel}$ for TE (a) and TM (b) polarizations. The RAs are identified by the grey lines. The fit to the MSLRs of interest corresponds with the white solid curves. This corresponds with an electric MSLR (e-MSLR) and a magnetic MSLR (m-MSLR) for TE and TM polarizations, respectively. The character of the MSLR can be understood with the help of the bottom diagram that shows the preferential coupling between the excited dipoles for a given polarization of the incident light field.

3 Strong light-matter coupling with MSLRs

The interaction between the dye molecules and the MSLR corresponds to the collective coupling between multiple emitters with a common cavity mode. This collective interaction results in a coupling strength proportional to the number of emitters within the mode volume of the cavity.

Consequently, to achieve strong coupling we deposit a layer of PMMA doped with organic molecules in a concentration of 35 wt%. This molecule is characterized by a main electronic transition, that we label as X_1, and a vibronic replica of the main electronic transition, that we label as X_2. Due to the different dispersion of the MSLRs, only the fitted MSRLs in Figure 1a and 1b interact with the molecule. This corresponds with an electric MSLR (e-MSLR) and a magnetic MSLR (m-MSLR) for the TE and TM polarizations, respectively. The extinction as function of the incident in-plane moment $\mathbf{k}_{\parallel}$ in Figure 2 shows the formation of exciton-polariton states which have a different dispersion compared with that of the bare cavity and bare excitons. We analyse the new eigenstates using a coupled oscillator model that defines the interaction between the MSLRs (the cavity mode) and the electronic transitions as function of the incident in-plane momentum. The fitted lower, middle and upper polariton states (LP, MP and UP, respectively) are identified in Figure 2a and 2b by the white dashed lines. Extracting the Rabi energy from the fitting we obtain the value of $\hbar\Omega$~200 meV for the splitting between the LP and MP for TE, and $\hbar\Omega$~200 meV for the splitting between the LP and UP in TM. To claim strong coupling from an experimental optical spectrum, as is the case with the extinction, it is common to use the condition of clear visibility of the exciton-polariton states in the spectrum, which requires that $(2g)^2 > (\gamma_{MSLR}^2 + \gamma_{exciton}^2)/2$. Comparing the obtained values of the Rabi energy with the losses of the system, we conclude that the strong coupling has been achieved in both TE and TM polarizations.

Finally, the electric or magnetic character of the MSLR not only influences the interaction strength with the dye, but it also determines the character of the hybridization. Our simulations (Figures 2c and 2d) of the spatial distribution of the electric field show that due to the extended mode volume of the MSLR, this electric and magnetic character is inherited as well in the exciton-polariton formed at the strong coupling regime.

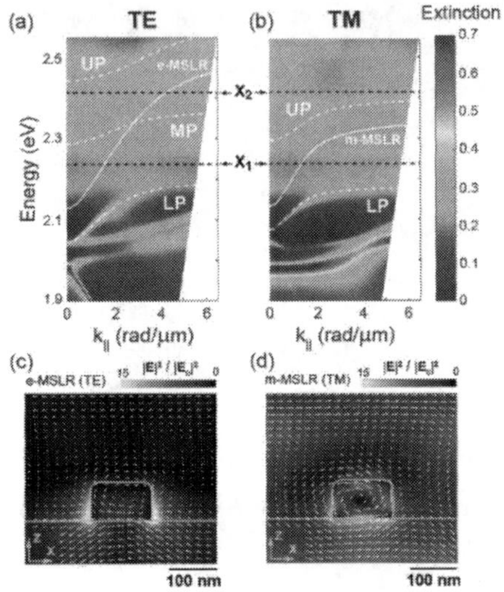

Fig. 2. Extinction in presence of organic dye as function of the incident in-plane moment $\mathbf{k}_{\parallel}$ for TE (a) and TM (b) polarizations. The formation of lower, middle and upper polaritons (LP, MP and UP respectively) and their fit (white dashed lines) indicates the strong coupling between the excitons and the MSRLs. In (c) and (d) the spatial distribution of the electric fields shows the electric and magnetic character of the MSLR is shown for each polarization. The color scale represents the intensity of the electric field ($|\mathbf{E}|^2/|\mathbf{E}_0|^2$) and the white arrows represent the vectorial projection of the electric field in the xz-plane.

These results demonstrate the potential of all-dielectric metasurfaces as a novel platform to investigate and manipulate exciton-polaritons in low-loss polaritonic devices.

References

1. F. Todisco, R. Malureanu, C. Wolff, P. Gonçalves, A. Roberts, N. Mortensen, C. Tserkezis, Nanophotonics **9** (4), 803-814 (2020).
2. R. Heilmann, AI. Väkeväinen, JP. Martikainen, P. Törmä, Nanophotonics **9** (2), 267-276 (2020).
3. G.W. Castellanos, S. Murai, T.V. Raziman, S. Wang, M. Ramezani, A.G. Curto, and J. Gómez Rivas, ACS Photonics (accepted).

https://doi.org/10.1051/epjconf/202023805005

Versatile architecture of ultra-narrow band absorbing photonic nanostructure

Clément Verlhac[1,*], *Hasnaa* El-Ouazzani[1], *Mathilde* Makhsiyan[1], *Riad* Haidar[1,2], *Jérôme* Primot[1], and *Patrick* Bouchon[1,**]

[1]DOTA, ONERA, Université Paris-Saclay, F-91123 Palaiseau, France
[2]École polytechnique, Département de Physique, 91128 Palaiseau, France

Abstract. Ultra narrow-band absorbing surfaces answer a growing demand of precise control of absorption spectra. Starting from a design of guided mode resonator (GMR), we present here a versatile architecture presenting resonances of very high quality factors (up to 10 000) and spectrally adjustable. This design uses multiple slices of dielectrics to focus resonating field in the middle of a metallic waveguide, diminishing its extension in the metal and thus the loss rate of the propagating mode.

1 Introduction

Controlling the absorption spectrum of a surface - and its thermal emission as a consequence of Kirchhoff's law - is crucial for a wide panel of applications ranging from filtering elements to light sources and thermal management. Metamaterials and more generally nanostructures offer the possibility of controlling the electromagnetic response of an effective medium by taking advantage of photonic and plasmonic responses of a sub-wavelength structuring of matter [1–3]. Since it has been shown such architectures were able to tune the impedance of a surface and adapt it to free space [4], using metamaterials to tailor spectral and angular absorption met a growing interest and numerous designs were proposed depending on the desired application [5]. Broadband absorbers are adapted to passive radiative cooling [6, 7], thermophotovoltaics [8] and domestic light sources [9]. On the contrary narrow-band absorbers are promising for gas sensing applications [10, 11] or as field enhancing structures since subwavelength narrow-band absorption is often synchronous with very high field concentration [12]. As we come to ultra-narrow band absorbing structures, guided-mode resonators (GMR) are good candidates for they are known to present resonances with quality factors up to several thousands under certain circumstances [13]. GMR are based on a superficial diffraction grating that couples incident waves to guided modes propagating along the interface. Considering architectures with metal-dielectric waveguides, the narrowness, and thus the quality factor of the resonance, depends a lot on the propagation range of the guided mode and thus on its losses which are mainly subordinated to mode's penetration in the metallic components. Starting from this observation, this paper presents a multi-dielectric architecture of a GMR in which it is possible to control the bandwidth of the resonance by limiting the penetration in metal of the propagating mode.

*e-mail: clement.verlhac@onera.fr
**e-mail: patrick.bouchon@onera.fr

2 Description of the structure

An usual architecture of a guided-mode resonator consists in the association of a metallic-dielectric waveguide and a metallic grating. The response of an usual one-dielectric structure is given by the red curve on fig.1(a) and its magnetic field map at resonance is depicted on fig.1(c). Under right condition of incidence, the grating diffracts incoming radiation along the surface which thus fuels a mode of the waveguide. If impedance matching is achieved, all the incoming energy is transferred into the guided-mode and then absorbed as the mode propagates. Under TM normal illumination, the structure resonates not far from $\lambda_r = p * n$ where p is the period of the grating and n the optical index of the dielectric. We propose to modify this design by subdividing the core of the waveguide into three slices of dielectrics where higher optical index is placed at the center and lower indices on the sides. As presented on fig.1, this architecture resonates under the same conditions as the previous one. However, as depicted on the magnetic field map of fig.1(b), the resonating field is much more focused at the center of the waveguide. The loss rate of the guided mode is thus reduced, resulting in a 20-fold reduction of the bandwidth of the peak. Both architectures are very close except for the width of gold's ribbon which is tuned to achieve perfect impedance matching.

This design presents a very good versatility. A modification of the geometrical parameters makes possible the tuning of the peak's wavelength throughout the whole 3-5 μm atmospheric transparency window and the adjustment of its quality factor from 100 up to 10 000.

To provide for an experimental proof of principle, a sample was designed and its angular and spectral reflectivity was measured using an adapted Fourier transform spectrometer. Even if, because of losses in the high index

https://doi.org/10.1051/epjconf/202023805005

dielectric part, the quality factor of the resonance revealed itself to be not as high as expected ($Q \sim 80$), numerical simulations and experimental curves are in very good agreement, proving the applicability of this concept.

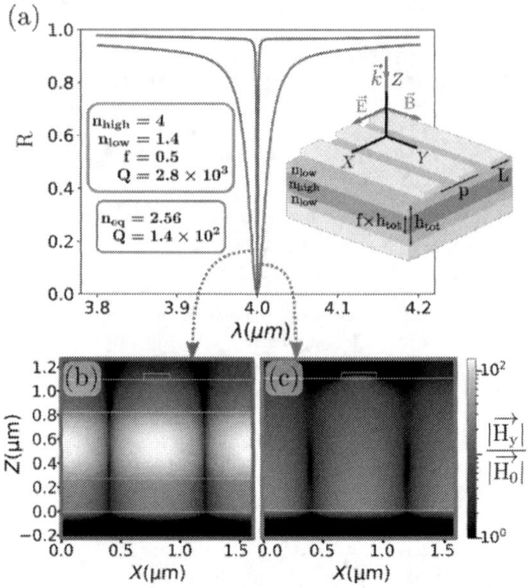

Figure 1. **(a)** General design of the structure in shown on insert. For both curves period p is set to 1.6 μm and dielectric height h_{tot} to 1.1 μm. Blue and red curves respectively correspond to the responses of three- and one-dielectric design under TM normal illumination. For theses two architectures, the width of gold ribbon is respectively set to 220 and 290 nm. **(b)** and **(c)** present the enhancement of magnetic field at the resonance (4 μm) for both structures. The maximal values are respectively 140 and 27.

3 Conclusion

This architecture seems quite promising in terms of versatility and potential applications. Quality factor of several thousand had already been observed with guided mode resonators but few nanophotonic concepts present as high values as depicted here. Losses in the metal becomes so low that, in practice, losses in the dielectric itself can limit the bandwidth. This apparent limitation opens the perspective of simultaneously concentrating losses and high field enhancement in the dielectric part of the waveguide which would make this design adapted to active optics such as exaltation of non-linear effects. Besides, the possibility of precisely choosing the spectral position of the resonance and its bandwidth makes this structure a promising candidate as thermal source for molecular sensing.

References

[1] V.G. Veselago, Soviet physics uspekhi **10**, 509 (1968)

[2] D.R. Smith, W.J. Padilla, D. Vier, S.C. Nemat-Nasser, S. Schultz, Physical review letters **84**, 4184 (2000)

[3] R.A. Shelby, D.R. Smith, S. Schultz, science **292**, 77 (2001)

[4] N.I. Landy, S. Sajuyigbe, J.J. Mock, D.R. Smith, W.J. Padilla, Physical review letters **100**, 207402 (2008)

[5] Y. Cui, Y. He, Y. Jin, F. Ding, L. Yang, Y. Ye, S. Zhong, Y. Lin, S. He, Laser & Photonics Reviews **8**, 495 (2014)

[6] E. Rephaeli, A. Raman, S. Fan, Nano letters **13**, 1457 (2013)

[7] P.C. Hsu, C. Liu, A.Y. Song, Z. Zhang, Y. Peng, J. Xie, K. Liu, C.L. Wu, P.B. Catrysse, L. Cai et al., Science Advances **3**, e1700895 (2017)

[8] A. Lenert, D.M. Bierman, Y. Nam, W.R. Chan, I. Celanović, M. Soljačić, E.N. Wang, Nature nanotechnology **9**, 126 (2014)

[9] O. Ilic, P. Bermel, G. Chen, J.D. Joannopoulos, I. Celanovic, M. Soljačić, Nature nanotechnology **11**, 320 (2016)

[10] A. Lefebvre, D. Costantini, G. Brucoli, S. Boutami, J.J. Greffet, H. Benisty, Sensors and Actuators B: Chemical **213**, 53 (2015)

[11] H. Miyazaki, T. Kasaya, M. Iwanaga, B. Choi, Y. Sugimoto, K. Sakoda, Applied Physics Letters **105**, 121107 (2014)

[12] P. Chevalier, P. Bouchon, J.J. Greffet, J.L. Pelouard, R. Haïdar, F. Pardo, Physical Review B **90**, 195412 (2014)

[13] A. Sharon, S. Glasberg, D. Rosenblatt, A.A. Friesem, Journal of the Optical Society of America A **14**, 588 (1997)

Fabrication and characterization of plasmonic band-stop filter using Ag grating

Atsushi Motogaito[1,*], *Ryoga* Tanaka[1], and *Kazumasa* Hiramatsu[1]

[1]Mie University, Graduate School of Engineering, 1577 Kurima-machiya-cho Tsu, Japan

Abstract. This study proposes a plasmonic band-stop filter with surface plasmon resonance in a double-layer wire grid structure targeting short-wavelength visible and near-ultraviolet regions for applications in ultraviolet photography. Using Ag and Al, the rigorous coupling wave of analysis method revealed that the maximum absorption was approximately 90% at 450 nm and 375 nm. The experiments using Ag produced similar results in a simulation. These results demonstrate that plasmonic band-stop filters in the visible and near-UV region can be realized at 450 nm and 375 nm using Ag or Al.

1 Introduction

Ultraviolet (UV) photography is used in various fields, including medicine, science, engineering, criminal investigation, and visual arts. The most popular method of photographing is to expose UV rays to an imaging film or device. The most convenient and simple UV light source is direct sunlight. However, the intensity of the UV rays is highly dependent on the atmospheric conditions. For example, in sunny and dry highlands, the UV intensity of sunlight is high. Electronic flashes are additional examples of UV light sources that can be used in everyday life. Surface reflection by an aluminum reflector can be used for more effective irradiation of UV rays by an electric flash.

In using non-monochromatic light for UV photography, it is necessary to prepare appropriate filters to transmit or stop transmitting a specific wavelength. To solve this problem, we propose a plasmonic filter using the propagating surface plasmon polariton in a wire grid (WG) structure. In a previous study, experiments demonstrated by designing the structure of the double-layer WG structure, the extraordinary transmission can be achieved at a specific incident angle and light wavelength [1, 2]. In addition, results of transmittance mapping indicate that transmission phenomena can be explained by the propagating surface plasmon polariton in the double-layer WG structure.

If the conditions of the structure for absorbing light can be determined, the WG structure can be used as a band-stop filter. In a previous study, Hu simulated extraordinary transmission by a nanohole array in the UV region [3]. In addition, Jaksic fabricated a band-pass filter for the UV region by surface-plasmon-polariton-assisted metal-dielectric multilayers [4]. Mazulquim fabricated a band-pass color filter enabled by resonant modes and plasmons near the Rayleigh anomaly [5]. Furthermore, Liang fabricated plasmonic structural color thin films through the aluminum oxide process in an anodic aluminum oxide thin film [6].

In this study, we propose band-stop filters that stop transmitting light at 450 nm and 375 nm using extraordinary absorption by propagating surface plasmon resonance of the Ag double-layer WG structure.

2 Design and fabrication method

As illustrated in Fig. 1, the double-layer WG structures were designed and fabricated on a glass substrate. The light wavelength was 450 and 375 nm. The thickness of the Ag and Al layer and the electron beam (EB) resist are presented in Table 1. Reflectance mapping was obtained by the rigorous coupling wave of analysis method using MATLAB-based commercial software (rcda-1d [SourceForage, http://rcwa-1d.sourceforge.net/]) by arranging the calculation results of the transmittance with respect to the incident angle obtained for each period. The double-layer WG structure was fabricated using EB lithography (EBL) and direct current (DC) magnetron sputtering. It should be noted that the transmittance, reflectance, and absorption of p-polarized light depends on the incidence angle.

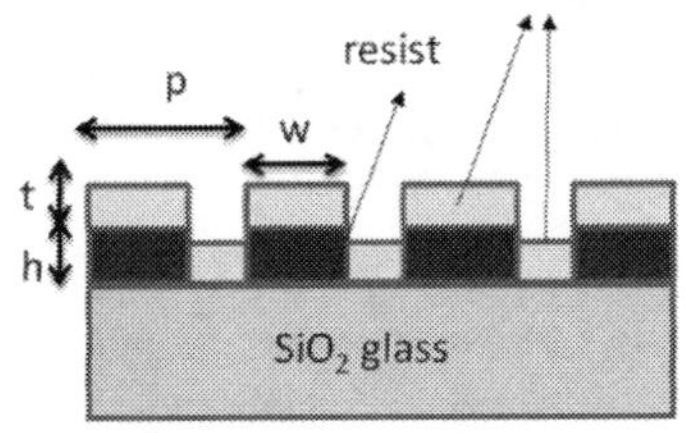

Fig. 1. Schematic drawing of double-layer wire grid structure.

* Corresponding author: motogaito@elec.mie-u.ac.jp

Table 1. Parameter of wire grid structures.

Wavelength λ (nm)	450		375	
Metal	Ag	Al	Ag	Al
Period p (nm)	350	350	350	350
Width of resist w (nm)	262.5	122.	182.0	287.0
duty ratio w/p	0.75	0.35	0.52	0.82
Thickness of resist h (nm)	100	100	100	100

3 Results and discussion

Fig. 2 presents the angle spectra of the absorption at 450 nm. The results demonstrate that the absorption peak occurred together with the reflectance dip at 16.4° and that the maximum absorption was 91.2%. In addition, in the experiment, a peak in the absorption rate was observed at the almost same angle. In the case of Al, the absorption peak occurred at almost the same angle. The structure using Ag was sharper; however, the average absorption was larger in the structure using Al. These results demonstrate a structure capable of absorbing light with a wavelength of 450 nm.

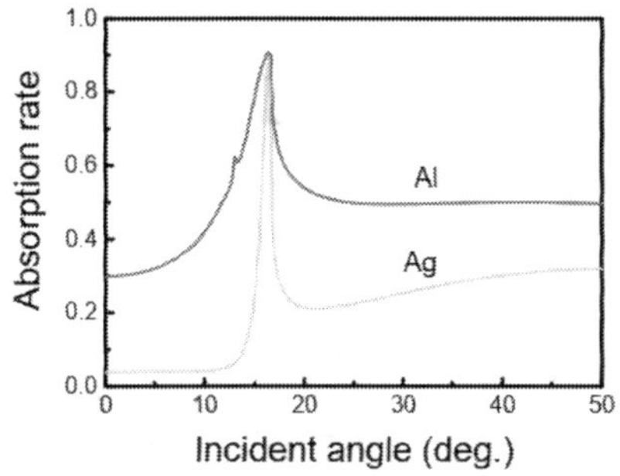

Fig. 2. Angle spectra of absorption rate at 450 nm.

To consider the peak of the absorptance, Fig. 3 presents the reflectance color mapping, in which the abscissa axis represents the structural period and the ordinate axis represents the incident angle. Here the two-layer Ag WG structure is used as an example. This figure reveals the dispersion relation, and the curve formed by the low reflectance corresponds to the excitation curve of the surface plasmon. The incident angle of 16.4° at which the absorption peak (reflectance dip) occurred in the structure using Ag (structure period 350 nm) is on the excitation curve. Therefore, the generated abnormal absorption phenomenon can be considered an effect of surface plasmons. This phenomenon was also confirmed in the structure using Al. The absorption characteristics at 375 nm were also simulated. At 4.0°, a sharp absorption peak was obtained using Al and Ag. In the experiment using Ag, the absorption peak at almost the same angle was obtained. These results can be used to obtain plasmonic band-stop filters in the visible and near-UV region.

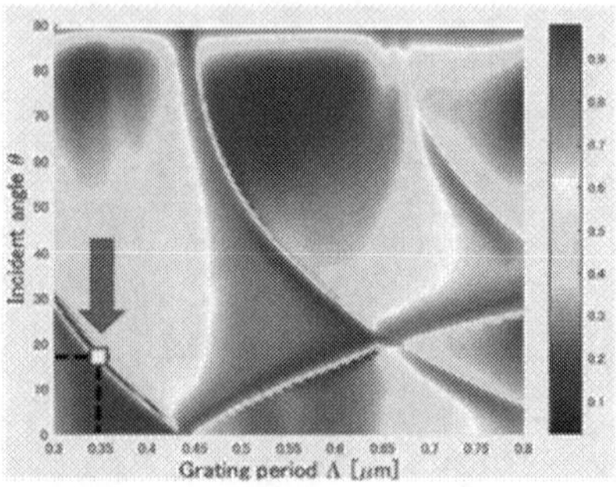

Fig. 3. Reflectance color mapping.

4 Conclusion

In this study, we fabricated plasmonic band-stop filters for visible and near-UV light using a WG structure. The results demonstrate that plasmonic band-stop filters in the visible and near-UV region can be realized at 450 nm and 375 nm using Ag or Al.

Acknowledgements

We would like to thank Prof. Takeshi Kato and Dr. Daiki Ohshima for using EBL system. We also thank Prof. Yoshimasa Kawata and Prof. Wataru Inami for using UV laser. This work was conducted at Nagoya University and supported by the Nanotechnology Platform Program of the Ministry of Education, Culture, Sports, Science and Technology (MEXT), Japan, Grant Number JPMXP09F19NU0018. This work was partially supported by JSPS KAKENHI Grant Number 20K05359. We also thank Enago (www.enago.jp) for English language editing.

References

1. A. Motogaito, Y. Morishita, H. Miyake, K. Hiramatsu, Plasmonics **10**, 1657-1662 (2015)
2. A. Motogaito, T. Nakajima, H. Miyake and K. Hiramatsu, Appl. Phys. A **123**, 729 (2017)
3. J. Hu, M. Shen, Z. Li, X. Li, G. Liu, X. Wang, C. Kan, Y. Li, Nanotechnology **28**, 215205 (2017)
4. Z. Jaksic, M. Maksimovic, M. Sarajlic, D. Tanaskovic, Acta Phys. Pol. A **112**, 953-958 (2007)
5. D. B. Mazulquim, K. J. Lee, J. W. Yoon, L. V. Muniz, B.-H. V. Borges, L. G. Neto, R. Magnusson, Opt. Exp. **22**, 30843-30851 (2014)
6. C.J. Liang, K.Y. Huang, L.T. Hung, C.Y. Su, Surface & Coatings Technology **319**, 170-181 (2017)

 https://doi.org/10.1051/epjconf/202023805007

Modal expansion of optical far-field quantities using quasinormal modes

Felix Binkowski[1], *Fridtjof* Betz[1], *Rémi* Colom[1], *Martin* Hammerschmidt[2], *Lin* Zschiedrich[2], and *Sven* Burger[1,2]

[1]Zuse Institute Berlin, Takustraße 7, 14195 Berlin, Germany
[2]JCMwave GmbH, Bolivarallee 22, 14050 Berlin, Germany

Abstract. We discuss an approach for modal expansion of optical far-field quantities based on quasinormal modes (QNMs). The issue of the exponential divergence of QNMs is circumvented by contour integration of the far-field quantities involving resonance poles with negative and positive imaginary parts. A numerical realization of the approach is demonstrated by convergence studies for a nanophotonic system.

Introduction

For the study of physical phenomena in nano-optical systems, a modal description is the most instructive approach. Modal expansion techniques using QNMs [1–3] have been proposed to analyze light-matter interaction in nanoresonators [4–7]. As the QNMs are the solutions to open systems, they decay in time and are characterized by complex eigenfrequencies. State-of-the-art approaches use the electromagnetic fields of the QNMs in the near-field region of the resonant systems to expand near-field quantities of interest. However, far-field properties of optical systems are important for many applications because typical experiments perform measurements in the far-field region. QNMs diverge exponentially in the far-field region [2, 3], which is a key issue for modal expansion techniques. Approaches using model approximations with real-valued frequencies have been proposed to overcome the divergence problem [8–10].

Here, we discuss an approach for modal expansion of optical far-field quantities [11]. The approach is based on the complex eigenfrequencies of QNMs. The divergence issue in the far-field region is circumvented by introducing contour-integral-based expressions of the far-field quantities involving resonance poles with negative and positive imaginary parts. In this way, one can derive nondiverging expansions of the far-field quantities while the model with complex-valued frequencies of the resonant systems can be retained. We demonstrate the approach by convergence studies for a nanophotonic system.

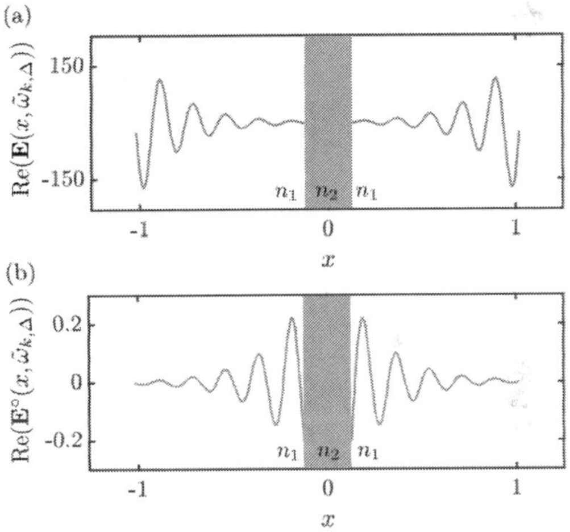

Figure 1. One-dimensional resonator defined by different refractive indices, where $n_2 > n_1$. Solving the Helmholtz equation with a source term corresponding to incoming plane waves yields solutions for the electric field, $\mathbf{E}(x,\omega)$ and $\mathbf{E}^{\circ}(x,\omega)$. For simplification, only the real parts of the scattered fields (a.u.) outside the resonator are shown. (a) Diverging field $\mathbf{E}(x,\tilde{\omega}_{k,\Delta}) = Ae^{i(n_1\tilde{\omega}_{k,\Delta}/c)|x|}$, where $\tilde{\omega}_{k,\Delta} = \tilde{\omega}_k + \Delta\tilde{\omega}_k$ is a frequency close to $\tilde{\omega}_k$. The frequency $\tilde{\omega}_k$ is a resonance pole of $\mathbf{E}(x,\omega)$. (b) Nondiverging field $\mathbf{E}^{\circ}(x,\tilde{\omega}_{k,\Delta}) = Be^{-i(n_1\tilde{\omega}_{k,\Delta}/c)|x|}$.

Modal expansion of far-field quantities

In nano-optics, QNMs, $\tilde{\mathbf{E}}(\omega_0) \in \mathbb{C}^3$, are solutions to the time-harmonic Maxwell's equations in second-order form,

$$\nabla \times \mu_0^{-1} \nabla \times \tilde{\mathbf{E}}(\omega_0) - \omega_0^2 \epsilon(\omega_0) \tilde{\mathbf{E}}(\omega_0) = 0, \qquad (1)$$

where $\omega_0 \in \mathbb{R}$ is the angular frequency, μ_0 is the vacuum permeability, and $\epsilon(\omega_0)$ is the permittivity tensor. For simplification of the notation, we omit the spatial dependence of the quantities. The eigenfrequencies $\tilde{\omega}_k \in \mathbb{C}$ corresponding to the QNMs have negative imaginary parts as the QNMs have to satisfy outgoing radiation conditions.

The approach proposed in [11] is demonstrated by decomposing the energy flux density,

$$s(\mathbf{E}(\omega_0), \mathbf{E}^*(\omega_0)) = \frac{1}{2}\mathrm{Re}\left(\mathbf{E}^*(\omega_0) \times \frac{1}{i\omega_0\mu_0}\nabla \times \mathbf{E}(\omega_0)\right) \cdot \mathbf{n},$$

where the field $\mathbf{E}^*(\omega_0)$ is the complex conjugate of the electric field $\mathbf{E}(\omega_0)$ and $\mathbf{n}$ is the normal vector on the associated far-field sphere. The Riesz projection expansion (RPE) is used to expand $s(\mathbf{E}(\omega_0), \mathbf{E}^*(\omega_0))$ into modal contributions [7, 12]. The RPE is based on complex contour integration, which means that $s(\mathbf{E}(\omega_0), \mathbf{E}^*(\omega_0))$ has to be evaluated for complex frequencies. This is not straightforward as $s(\mathbf{E}(\omega_0), \mathbf{E}^*(\omega_0))$ is nonholomorphic. This challenge has been addressed by exploiting the relation $\mathbf{E}^*(\omega_0) = \mathbf{E}(-\omega_0)$ for $\omega_0 \in \mathbb{R}$, which is also a solution to Eq. (1). The field $\mathbf{E}(-\omega_0)$ has an analytical continuation into the complex plane $\omega \in \mathbb{C}$, denoted by $\mathbf{E}^\circ(\omega)$. This field yields the required analytical continuation given by $s(\mathbf{E}(\omega), \mathbf{E}^\circ(\omega))$. With this, Cauchy's integral formula,

$$s(\mathbf{E}(\omega_0), \mathbf{E}^\circ(\omega_0)) = \frac{1}{2\pi i} \oint_{C_0} \frac{s(\mathbf{E}(\omega), \mathbf{E}^\circ(\omega))}{\omega - \omega_0} \mathrm{d}\omega,$$

is exploited for the closed integration path C_0 around ω_0, where $s(\mathbf{E}(\omega), \mathbf{E}^\circ(\omega))$ is holomorphic inside of C_0. Cauchy's residue theorem leads to

$$\begin{aligned} s(\mathbf{E}(\omega_0), \mathbf{E}^\circ(\omega_0)) = &- \sum_{k=1}^{K} \frac{1}{2\pi i} \oint_{\tilde{C}_k} \frac{s(\mathbf{E}(\omega), \mathbf{E}^\circ(\omega))}{\omega - \omega_0} \mathrm{d}\omega \\ &- \sum_{k=1}^{K} \frac{1}{2\pi i} \oint_{\tilde{C}_k^*} \frac{s(\mathbf{E}(\omega), \mathbf{E}^\circ(\omega))}{\omega - \omega_0} \mathrm{d}\omega \\ &+ \frac{1}{2\pi i} \oint_{C_r} \frac{s(\mathbf{E}(\omega), \mathbf{E}^\circ(\omega))}{\omega - \omega_0} \mathrm{d}\omega, \end{aligned} \quad (2)$$

where $\tilde{C}_1, \ldots, \tilde{C}_K$ are contours around the resonance poles of $\mathbf{E}(\omega)$, given by $\tilde{\omega}_1, \ldots, \tilde{\omega}_K$, and $\tilde{C}_1^*, \ldots, \tilde{C}_K^*$ are contours around the resonance poles of $\mathbf{E}^\circ(\omega)$, given by $\tilde{\omega}_1^*, \ldots, \tilde{\omega}_K^*$. The contour C_r comprises ω_0, the resonance poles $\tilde{\omega}_1, \ldots, \tilde{\omega}_K$ and $\tilde{\omega}_1^*, \ldots, \tilde{\omega}_K^*$, and no further poles. The Riesz projections

$$\begin{aligned} \tilde{s}_k(\mathbf{E}(\omega_0), \mathbf{E}^\circ(\omega_0)) = &- \frac{1}{2\pi i} \oint_{\tilde{C}_k} \frac{s(\mathbf{E}(\omega), \mathbf{E}^\circ(\omega))}{\omega - \omega_0} \mathrm{d}\omega \\ &- \frac{1}{2\pi i} \oint_{\tilde{C}_k^*} \frac{s(\mathbf{E}(\omega), \mathbf{E}^\circ(\omega))}{\omega - \omega_0} \mathrm{d}\omega \end{aligned}$$

are modal contributions for the energy flux density. The contribution

$$s_r(\mathbf{E}(\omega_0), \mathbf{E}^\circ(\omega_0)) = \frac{1}{2\pi i} \oint_{C_r} \frac{s(\mathbf{E}(\omega), \mathbf{E}^\circ(\omega))}{\omega - \omega_0} \mathrm{d}\omega$$

is the remainder containing nonresonant components as well as contributions corresponding to eigenfrequencies outside of the contour C_r.

The presented approach is based on computing the quantity $s(\mathbf{E}(\omega), \mathbf{E}^\circ(\omega))$ by solving Eq. (1) for ω and for $-\omega$. Due to the compensation of the factors $e^{i(n\omega/c)r}$ and $e^{-i(n\omega/c)r}$ of the fields in the far-field region, this yields a nondiverging quadratic form $s(\mathbf{E}(\omega), \mathbf{E}^\circ(\omega))$, where a product of $\mathbf{E}(\omega)$ and $\mathbf{E}^\circ(\omega)$ is involved. In this way, modal expansions of far-field quantities can be computed.

To illustrate this, a one-dimensional resonator with the fields $\mathbf{E}(x, \omega)$ and $\mathbf{E}^\circ(x, \omega)$ fulfilling the corresponding Helmholtz equation is considered. Figure 1(a) sketches the diverging field $\mathbf{E}(x, \tilde{\omega}_{k,\Delta})$, which relates to a QNM of the problem as $\tilde{\omega}_{k,\Delta} = \tilde{\omega}_k + \Delta\tilde{\omega}_k$ is a frequency close to the eigenfrequency $\tilde{\omega}_k$. Figure 1(b) shows the nondiverging field $\mathbf{E}^\circ(x, \tilde{\omega}_{k,\Delta})$ outside of the resonator. Note that the frequency $\tilde{\omega}_{k,\Delta}$ represents a point on an integration contour $\tilde{C}_k$ from Eq. (2). The product $\mathbf{E}(x, \tilde{\omega}_{k,\Delta}) \cdot \mathbf{E}^\circ(x, \tilde{\omega}_{k,\Delta})$ shows a nondiverging behavior and relates to the energy flux density. The approach also applies to arbitrary three-dimensional problems, where, in the far-field region, $\mathbf{E}(\mathbf{r}, \omega) \sim e^{i(n\omega/c)r}(1/r)$ and $\mathbf{E}^\circ(\mathbf{r}, \omega) \sim e^{-i(n\omega/c)r}(1/r)$.

We acknowledge funding by the Deutsche Forschungsgemeinschaft (DFG, German Research Foundation) under Germany's Excellence Strategy - The Berlin Mathematics Research Center MATH+ (EXC-2046/1, project ID: 390685689, AA4-6). We acknowledge the Helmholtz Association for funding within the Helmholtz Excellence Network SOLARMATH, a strategic collaboration of the DFG Excellence Cluster MATH+ and Helmholtz-Zentrum Berlin (grant no. ExNet-0042-Phase-2-3).

References

[1] M. Zworski, Notices Amer. Math. Soc. **46**, 319 (1999)

[2] P. Lalanne, W. Yan, K. Vynck, C. Sauvan, J.P. Hugonin, Laser Photonics Rev. **12**, 1700113 (2018)

[3] P.T. Kristensen, K. Hermann, F. Intravaia, K. Busch, arXiv:1910.05412 (2019)

[4] E.A. Muljarov, W. Langbein, R. Zimmermann, EPL **92**, 50010 (2010)

[5] F. Zolla, A. Nicolet, G. Demésy, Opt. Lett. **43**, 5813 (2018)

[6] W. Yan, R. Faggiani, P. Lalanne, Phys. Rev. B **97**, 205422 (2018)

[7] L. Zschiedrich, F. Binkowski, N. Nikolay, O. Benson, G. Kewes, S. Burger, Phys. Rev. A **98**, 043806 (2018)

[8] P.Y. Chen, D.J. Bergman, Y. Sivan, Phys. Rev. Appl. **11**, 044018 (2019)

[9] R.C. Ge, P.T. Kristensen, J.F. Young, S. Hughes, New J. Phys. **16**, 113048 (2014)

[10] J. Ren, S. Franke, A. Knorr, M. Richter, S. Hughes, Phys. Rev. B **101**, 205402 (2020)

[11] F. Binkowski, F. Betz, R. Colom, M. Hammerschmidt, L. Zschiedrich, S. Burger, arXiv:2003.11305 (2020)

[12] F. Binkowski, L. Zschiedrich, M. Hammerschmidt, S. Burger, Phys. Rev. B **100**, 155406 (2019)

EPJ Web of Conferences **238**, 05008 (2020) https://doi.org/10.1051/epjconf/202023805008
EOSAM 2020

Quasi-normal mode expansion as a tool for the design of nanophotonic devices

Rémi Colom[1,*], *Felix* Binkowski[1], *Fridtjof* Betz[1], *Martin* Hammerschmidt[2], *Lin* Zschiedrich[2], and *Sven* Burger[1,2,**]

[1]Zuse Institute Berlin, Takustraße 7, 14195 Berlin, Germany
[2]JCMwave GmbH, Bolivarallee 22, 14050 Berlin, Germany

Abstract. Many nanophotonic devices rely on the excitation of photonic resonances to enhance light-matter interaction. The understanding of the resonances is therefore of a key importance to facilitate the design of such devices. These resonances may be analyzed by use of the quasi-normal mode (QNM) theory. Here, we illustrate how QNM analysis may help study and design resonant nanophotonic devices. We will in particular use the QNM expansion of far-field quantities based on Riesz projection to design optical antennas.

1 Introduction

Nanophotonic devices allow to control and tailor light-matter interactions at the nanoscale. Many applications such as quantum infomation technology, bio-sensing or the use of metasurfaces as miniature optical components may benefit from the use of nanophotonic devices. This control of light-matter interactions is often achieved through the excitation of the photonic resonances of these nanophotonic devices. The resonant interaction between light and nanometric structures may readily be described by using QNM theory [1, 2]. The design of nanophotonic devices could consequently be facilitated by using QNM analysis. QNM are however diverging with the distance from the resonator. This complicates the analysis of far-field quantities using QNM [1, 2]. Here, we will use QNM expansion of far-field quantities based on Riesz projection [4] to understand and design nanophotonic devices. We will in particular focus on the design of optical antennas.

2 Design of optical antennas using quasi-normal mode expansions

When designing nanophotonic devices, it is usually necessary to shape and control both the far-field and near-field properties of photonic structures. This may necessitate the careful tuning of many parameters or degrees of freedom of the structure under consideration and consequently result in time-consuming and costly parameter scanning. Even if this design may be assisted by using optimization approaches [5, 6], it would still be valuable to fully characterize the optical properties of a photonic structure in terms of a few number of modes. These modes are solution of the source-free Maxwell's equations:

$$\nabla \times \mu_0^{-1} \nabla \times \mathbf{E}_k(\mathbf{r}) - \omega_k^2 \varepsilon(\mathbf{r}, \omega_k)\mathbf{E}_k(\mathbf{r}) = 0 \quad (1)$$

with outgoing boundary conditions. The eigenfrequencies ω_k are in general complex valued.

The design of optical antennas is an example where one needs to control both the near-field and far-field optical behavior of photonic structures. Tuning the near-field properties of the photonic stucture allows one to have a control over the Purcell factor enhancement $\Gamma(\omega)$:

$$\Gamma(\omega) = -\frac{\mathrm{Re}\,(\mathbf{E}(\mathbf{r}_d, \omega) \cdot \mathbf{j}^*)}{2\Gamma_b} \quad (2)$$

$\mathbf{j} = -i\omega\mathbf{p}$, $\mathbf{p}$ being the dipole moment while Γ_b is the dipole emission in the homogeneous background material and $\mathbf{r}_d$ is the position of the dipole. This quantity may be expanded on the QNM basis by means of the Riesz projection $\Gamma(\omega) = \sum_{k=1}^{k_{\max}} \tilde{\Gamma}_k(\omega) + \Gamma_r(\omega)$ where $\tilde{\Gamma}_k(\omega)$ are the resonant contribution and $\Gamma_r(\omega)$ is the non resonant background contribution which are defined as follows [3]:

$$\begin{aligned} \tilde{\Gamma}_k(\omega) &= \frac{1}{2\Gamma_b}\mathrm{Re}\left[-\frac{1}{2i\pi}\left(\oint_{C_k} \frac{\mathbf{E}(\mathbf{r}_d, z)}{z-\omega}dz\right)\cdot \mathbf{j}^*\right] \\ \Gamma_r(\omega) &= \frac{1}{2\Gamma_b}\mathrm{Re}\left[\frac{1}{2i\pi}\left(\oint_{C_r} \frac{\mathbf{E}(\mathbf{r}_d, z)}{z-\omega}dz\right)\cdot \mathbf{j}^*\right] \end{aligned} \quad (3)$$

where the contours C_k are closed contours around the eigenfrequencies ω_k while the contour C_r is a large contour permitting to take into account the non-resonant background. One sees that the Purcell factor can thus be described in terms of a small number of resonant and non-resonant terms. One can then vary the opto-geometrical parameter of a photonic sctructure to tune the peak position and width of resonant contributions $\tilde{\Gamma}_k(\omega)$ in order to optimize the Purcell factor for a given application.
Besides the near-field behavior, the far-field properties of optical antennas are also of particular interest. Optical antennas can indeed allow to control the radiation pattern of a quantum emitter. This is of particular importance for certain applications where the light emitted has then to be

*e-mail: colom@zib.de
**e-mail: burger@zib.de

collected by a lens or an optical fiber with a given numerical aperture [6, 7]. It would then be very useful to describe the far-field response of these photonic structures in terms of their modes. In particular, a modal expansion of the Poynting vector would be of a special interest. This is however quite challenging since quasi-normal modes are exponentially diverging with the distance from the resonator [1, 2, 4]. To circumvent this divergence problem, we developed a method for expanding quadratic quantities and in particular the Poynting vector $\mathbf{S}(\mathbf{r},\omega) = s(\mathbf{E}(\mathbf{r},\omega),\mathbf{E}^*(\mathbf{r},\omega)) = \frac{1}{2}\mathrm{Re}\left(\mathbf{E}^*(\mathbf{r},\omega)\times\frac{1}{i\omega\mu_0}\nabla\times\mathbf{E}(\mathbf{r},\omega)\right)$. We found that the Poynting vector admits the following modal expansion [4]:

$$\begin{aligned}\mathbf{S}(\mathbf{r},\omega) = &-\sum_{k=1}^{k_{\max}}\frac{1}{2i\pi}\oint_{C_k}\frac{s(\mathbf{E}(\mathbf{r},z),\mathbf{E}^{\circ}(\mathbf{r},z))}{z-\omega}dz\\ &-\sum_{k=1}^{k_{\max}}\frac{1}{2i\pi}\oint_{C_k^*}\frac{s(\mathbf{E}(\mathbf{r},z),\mathbf{E}^{\circ}(\mathbf{r},z))}{z-\omega}dz\\ &+\frac{1}{2i\pi}\oint_{C_r}\frac{s(\mathbf{E}(\mathbf{r},z),\mathbf{E}^{\circ}(\mathbf{r},z))}{z-\omega}dz\end{aligned} \tag{4}$$

where $\mathbf{E}^{\circ}(\mathbf{r},z)$ is defined as the analytic continuation of $\mathbf{E}(\mathbf{r},-\omega)$ to the entire complex plane. The contour C_k and C_r are analogous to the contours used for the expansion of the Purcell factor in Eq. (3) while C_k^* are the contours around the resonance poles of $\mathbf{E}^{\circ}(\mathbf{r},z)$ equal to ω_k^*. The important point is that the expansion provided in Eq. (4) is not diverging as $|\mathbf{r}|$ increases. This expansion can thus be used to study the far-field properties of photonic resonators. One can in particular describe the emission pattern in terms of resonant and non-resonant contributions. This can help shape the emission pattern by tuning the shape of the photonic structure under consideration to selectively excite the modes allowing to obtain the desirable emission pattern.

We will illustrate the usefulness of QNM expansion to design several types of optical antenna. We will start by presenting results on the circular Bragg gratings [4, 7]. We will also show how QNM expansion can used to understand the structures supporting quasi-bound states in the continuum originating from the avoided crossing of two resonances [8, 9].

Acknowledgement

We acknowledge funding by the Deutsche Forschungsgemeinschaft (DFG, German Research Foundation) under Germany's Excellence Strategy - The Berlin Mathematics Research Center MATH+(EXC-2046/1, project ID: 390685689, AA4-6). We acknowledge the Helmholtz Association for funding within the Helmholtz Excellence Network SOLARMATH, a strategic collaboration of the DFG Excellence Cluster MATH+and Helmholtz-Zentrum Berlin (grant no. ExNet-0042-Phase-2-3)

References

[1] P. Lalanne et al., Laser and Photonics Reviews, **12**, 1700113 (2018).
[2] P. Kristensen et al., arXiv:1910.05412, (2019).
[3] L. Zschiedrich et al., Physical Review A, **98**, 043806 (2018).
[4] F. Binkowski et al., arXiv:2003.11305, (2020).
[5] P. Schneider et al., ACS Photonics, **6**, 2726-2733 (2019).
[6] P. Schneider et al., Optics Express, **26**, 8479-8492 (2018).
[7] L. Rickert et al., Optics Express, **27**, 36824-36837 (2019).
[8] M. Rybin et al., Physical Review Letters, **119**, 243901 (2017).
[9] K. Koshelev et al., Science, **367**, 288-292 (2020).

Disruptive space telescope concepts, designs, and developments: OASIS and Nautilus -INVITED

Dae Wook Kim[1,2,*], *Christopher K.* Walker[2], *Dániel* Apai[2,3], *Tom D.* Milster[1], *Yuzuru* Takashima[1], *Glenn* Schneider[2], *Rongguang* Liang[1], *Young-sik* Kim[1], *Chuck* Fellows[2], *Yingying* Zhang[1], *Andras* Gasper[2], *I. Steve* Smith[4], *Devon* Crowe[5], *Alex* Bixel[2], *Stefan* O'Dougherty[2], *Siddhartha* Sirsi[2], *Aman* Chandra[6], *Andy* Phan[1], *Zichan* Wang[1], *Kira* Purvin[1], *Henry* Quach[1], *Marcos* Esparza[1], *Joel* Berkson[1], *Oliver* Spires[1], *Heejoo* Choi[1], *Geon Hee* Kim[7], and *Jonathan* Arenberg[8]

[1]James C. Wyant College of Optical Sciences, The University of Arizona, Tucson, AZ 85721, USA
[2]Steward Observatory, The University of Arizona, Tucson, AZ 85721, USA
[3]Lunar and Planetary Laboratory, The University of Arizona, Tucson, AZ 85721, USA
[4]Southwest Research Institute, 6220 Culebra Road, San Antonio, TX 78238, USA
[5]Raytheon Technologies, 10618 E. Rose Hill St., Tucson, AZ 85747, USA
[6]Aerospace and Mechanical Engineering, The University of Arizona, Tucson, AZ 85721, USA
[7]Korea Basic Science Institute, 169-148, Daejeon 34133, Republic of Korea
[8]Northrop Grumman Aerospace Systems, Redondo Beach, CA 90278, USA

Abstract. Two disruptive space telescope concepts are being designed and developed at the University of Arizona; these are the 20-meter OASIS (Orbiting Astronomical Satellite for Investigating Stellar Systems) and 8.5-meter Nautilus. OASIS combines break-through inflatable aperture and adaptive optics techniques to realize the dream of a 20+ meter class spaceborne terahertz/far-infrared telescope. In the Nautilus visible/near-infrared telescope concept, conventional primary mirrors are replaced by an ~8.5-meter MODE (Multi-order diffractive engineered) lens with 10 times lower areal density and up to 100 times lower mis-alignment sensitivity over traditional systems, enabling large-diameter optical space telescopes. The OASIS and Nautilus concepts have the potential to greatly reduce mission costs and risks compared to the current state of the art.

1 Introduction

The realization of a very large, space-based telescope for far-infrared/terahertz studies has long been a goal to study the origins of stars, planets, molecular clouds, and galaxies by providing an essential means of following-up on tantalizing results from recent successful missions such as Spitzer and SOFIA. The OASIS concept combines break-through technologies utilizing inflatable spherical reflectors and adaptive optics to realize a 20m terahertz space telescope [1,2].

In visible and near-infrared astronomy, an outstanding goal is to study the diversity of potentially Earth-like planets. Nautilus is a space telescope concept that builds on an engineered material diffractive-transmissive optical element [3, 4]. The primary mirrors typical of current space telescopes are replaced by an ~8.5-meter MODE (multi-order diffractive engineered) lens with 10 times lower areal density, thereby enabling a lightweight structure. MODE lens telescopes also have up to 100 times relaxed alignment tolerances, compared to mirrors.

The OASIS and Nautilus concepts have the potential to greatly reduce mission costs and risks compared to current space telescope paradigms through light-weighted optical design and technology.

2 Disruptive telescope concepts

Two paradigm-breaking space telescopes, OASIS and Nautilus, are summarized in Table 1.

Table 1. Summary of OASIS and Nautilus [1-4]

	OASIS	Nautilus
Science wavelength	Terahertz / Far-infrared	Visible / Near-infrared
Primary aperture diameter	~20-meter	~8.5-meter (with array option)
Primary optic technology	Inflatable Hencky mirror	Replicable MODE lens
Main instrument	Heterodyne receivers	Grism-based spectrometer

2.1. 20-meter class inflatable OASIS

The science objectives of OASIS are met by utilizing a 20-meter inflatable aperture with heritage from the Inflatable Aperture Experiment (IAE) mission, which demonstrated in-orbit deployment of a 14-meter aperture system shown in Figure 1. OASIS will utilize a Hencky reflector geometry together with proven adaptive optics techniques to yield a wide-field-of-view inflatable Mylar mirror operating at submillimeter wavelengths (i.e., terahertz frequencies). Mylar is a polyester film made

* Corresponding author: letter2dwk@hotmail.com

from stretched polyethylene terephthalate called BoPET (Biaxially-oriented polyethylene terephthalate).

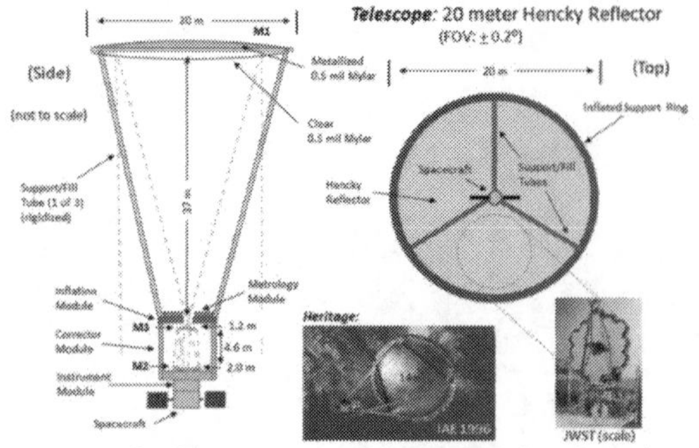

Fig. 1. Inflatable 20-meter Mylar balloon-based terahertz OASIS telescope concept design and configuration. [2]

2.2 8.5-meter class replicable Nautilus

The baseline concept for the Nautilus unit telescope in compact launch and deployed configuration is presented in Figure 2. A cross sectional view of a deployed Nautilus unit telescope is shown in Figure 2(left). The telescopes use an 8.5 m diameter MODE lens as the light-collecting element for exoplanet transit spectroscopy observations. Each Nautilus unit will be a stand-alone telescope equipped with two visual/near-infrared band detectors and a low-resolution spectrograph (e.g., R ~ 200) optimized for the ~0.45–1.6 μm wavelength range.

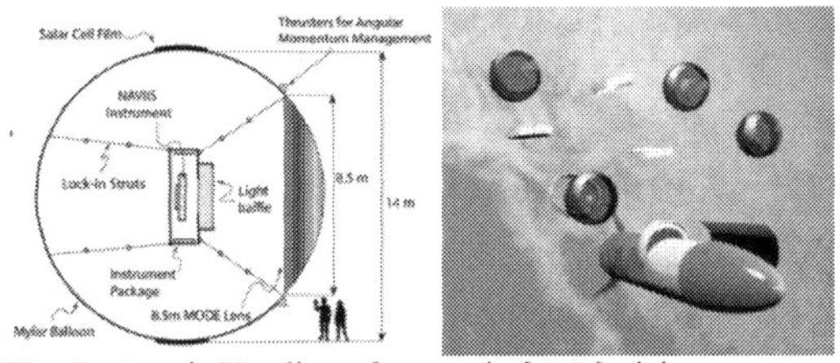

Fig. 2. A unit Nautilus telescope is launched in a compact configuration and inflated in orbit (left). Due to the compact launch configuration, multiple Nautilus units could be launched simultaneously in next-generation rocket fairings (right). [3, 4]

3 Technology developments

3.1. 1-meter Mylar prototype for OASIS

The 1-meter class OASIS Mylar reflector demonstration setup has been successfully manufactured, inflated, and tested as shown in Figure 3. The Hencky surface shape and its stability as a function of time and balloon pressure has been measured using a non-null deflectometry setup.

3.2 0.25-meter MODE lens camera for Nautilus

The MODE lens has been designed considering both the diffractive and refractive nature of light and fabricated as small-scale prototypes.

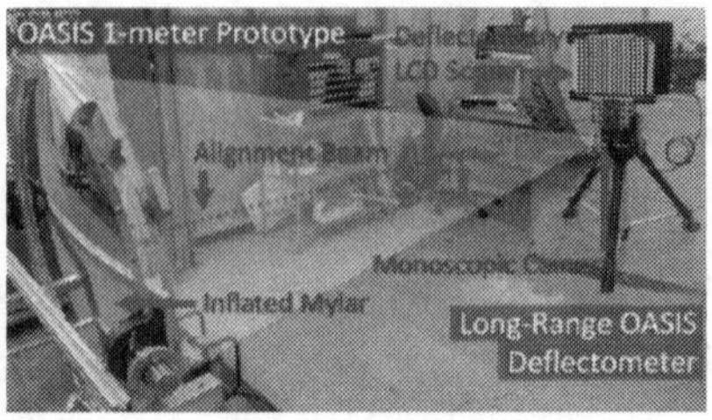

Fig. 3. 1-meter class Mylar OASIS prototype tested using the long-range deflectometry measurement setup (top-right inset).

The prototype lenses are small (~20 – 250 mm in diameter in Figure 4) and ideal for understanding the details of the fabrication trade-offs and evaluating as-manufactured optical performance. An advanced optical metrology environment has been assembled to demonstrate quantitative imaging performance.

Fig. 4. MODE lens prototypes designed and fabricated with two diffractive surfaces of a front MODE surface and a single-order diffractive lens back-surface (left) [4] and a direct diamond turned 240 mm diameter MODE lens prototype (right).

4 Concluding remarks

OASIS utilizes a space-based inflatable reflector and has the potential of making the dream of space-based 20-meter class telescopes a reality. Nautilus will transform astrophysics through its paradigm-breaking, scalable, and ultralight MODE lens technology.

The >10 fold increase in sensitivity OASIS provides will enable unprecedented access to the physical conditions of planetary bodies within our own solar system, as well as those surrounding distant stars. Large aperture visible/near-infrared wavelength spectroscopy with Nautilus will address one of the most fundamental questions in modern astrophysics by providing the first comprehensive, high-quality spectral atlas of small exoplanet atmospheres. Disruptive optics technology enables new discoveries.

Acknowledgement

This research was made possible in part by the Gordon and Betty Moore Foundation and the II-VI Foundation.

References

1. C. K. Walker, et al., IEEE MTT-S (IMS), pp. 1884-1887 (2017).
2. C. Walker, et al., White-Paper (2020).
3. D. Apai, et al., The Astronomical Journal, 158:83 (21pp) (2019).
4. D. Apai, et al., Proc. SPIE 11116, 1111608 (2019).

From 2D to 3D super-resolution imaging through glass microspheres -INVITED

Stephane Perrin*, *Sylvain* Lecler, and *Paul* Montgomery

ICube Laboratory, CNRS - INSA de Strasbourg - University of Strasbourg, 67412 Illkirch, France

Abstract. Microsphere-assisted microscopy is a new imaging technique which allows the diffraction limit to be overcome using transparent microspheres. It makes it possible to reach a resolution of up to 100 nm in air while being label-free and full-field. An overview of the imaging technique is presented showing the influence of the photonic jet on the image nature and the unconventional behaviour of the magnification factor. Moreover, interferometry through microspheres is demonstrated for the 3D reconstruction of nano-elements.

1 Introduction

In optical imaging, observation of sub-micrometre features is a real challenge. According to theory of diffraction, the smallest discernible size through a classical microscope is of $\lambda/2$ (where λ is the wavelength of the broadband light source). In experiments, the resolving power appears to be reduced due to the optical aberrations, misalignments and numerical aperture of the system (NA $\leq$ 1.0 in air). Since the appearance of the notion of ultramicroscopy in 1930's, several approaches have been suggested making it possible to overcome the diffraction barrier [1] such as confocal microscopy, scanning near field optical microscopy, negative-index-superlenses microscopy, structured illumination microscopy, and stimulated emission depletion microscopy super resolved fluorescence microscopy. However, these sub-diffraction-limit techniques are often photo-toxic for the sample, are complex to implement, need markers or require long-acquisition times.

2 Direct imaging through microspheres

More recently, Wang *et al.* experimentally demonstrated the principle of full-field super-resolution microscopy by only introducing a transparent microsphere in a classical white-light microscope [2]. This easy-to-implement and non-invasive nanoscopy technique allows the diffraction limit to be overcome by a factor of 3 ($\lambda/6$ in air) [3]. Through a probable conversion of the near-field information to propagating waves [4,5], microsphere-assisted microscopy is thus able to achieve a lateral resolution that is superior to that attainable by confocal microscopy and by the solid immersion lens [6] and similar to that achieved from structured illumination microscopy and meta-material-based superlenses. Numerical studies and experiments have made it possible to highlight the influence of the photonic jet on the image nature (despite the performance of the photonic jet not being able to be used to predict the phenomenon behind the super resolution) [7,8] and the unconventional behaviour of the magnification factor [9], respectively. Moreover, microsphere-assisted microscopy has recently been adapted for biological imaging using fluorescence [11] and dark-field imaging for label-free visualization of transparent objects [12].

3 Interferometry through microspheres

Furthermore, microsphere-assisted microscopy has successfully been combined to interference microscopy in

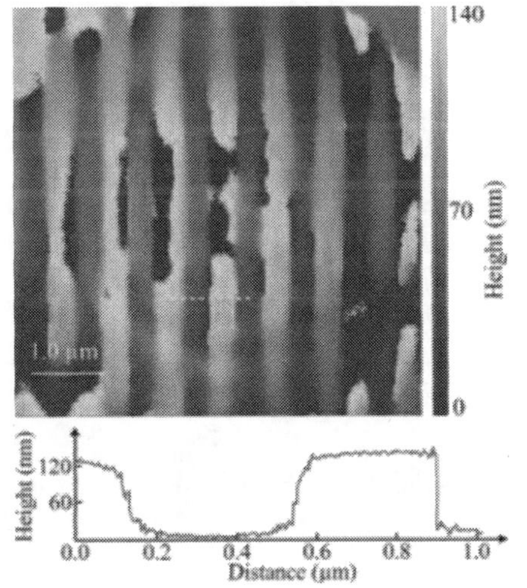

Fig. 1. Topography reconstruction of an 800-nm-period standard grating using microsphere-assisted interference microscopy compensated by two 50-µm-diameter microspheres. The height profile is along the dashed-white line of the high-resolution surface-topography reconstruction.

* Corresponding author: stephane.perrin@unistra.fr, ORCID: 0000-0003-2470-7636

order to reconstruct surface topographies of sub-wavelength features [12,13]. Indeed, the lateral resolution in interference microscopy is improved while keeping a high axial accuracy (around 10 nm). We can notice that in interference microscopy, the lateral resolution is not only limited by diffraction of light, but also by the ability of the system to retrieve the height distribution. This leads to a micrometre-scale resolution (instead of $\lambda/2$). Thus, in order to further increase the 3D spatial resolution, an experimental method has recently been proposed. It consists in introducing a second microsphere in the reference arm (similar to the one in the object arm) [14]. In this way, the wave-front distortion due to the optical aberrations is avoided and the optical path difference in the interferometer arms is compensated. Topographies of 300-nm-width grating lines and 250-nm-diameter transparent nanopillars have been retrieved. In order to provide a better understanding of the phase behaviour through microspheres, a numerical model of microsphere-assisted interference microscopy is being implemented.

Acknowledgements: This work received funding from SATT Conectus Alsace and was also supported by the University of Strasbourg.

References

1. Editorial, Nat. Photon. **3**, 361 (2009)
2. Z. Wang, W. Guo, L. Li, B. Luk'yanchuk, A. Khan, Z. Liu, Z. Chen, and M. Hong, Nat. Commun. **2**, 218 (2011)
3. V.N. Astratov, Label free super resolution microscopy (Springer, Cham, 2019)
4. Y. Ben-Aryeh, J. Opt. Soc. Am. A **33**, 2284 (2016)
5. F. Wang, Y. Cao, S. Yang, X. Wang, J. Wang, T. Chen, and Y.-H. Ye, Proc. SPIE **10964**, 109643H (2018)
6. A. Darafsheh, N.I. Limberopoulos, J.S. Derov, D.E. Walker, and V.N. Astratov, Appl. Phys. Lett. **104**, 061117 (2014)
7. S. Lecler, S. Perrin A. Leong-Hoi, and P. Montgomery, Sci. Rep. **9**, 4725 (2019)
8. A.V. Maslov and V.N. Astratov, Phys. Rev. Appl. **11**, 064004 (2019)
9. S. Perrin, H. Li, S. Lecler, and P. Montgomery, Opt. Laser Technol. **114**, 40 (2019)
10. H. Yang, N. Moullan, J. Auwerx, and M.A.M. Gijs, Small **10**, 1712 ((2014)
11. S. Perrin, H. Li, K. Badu, T. Comparon, G. Quaranta, N. Messaddeq, N. Lemercier, P. Montgomery, J.-L. Vonesch, and S. Lecler, Phys. Status Solidi RRL **13**, 1800445 (2019)
12. F. Wang, L. Liu, P. Yu, Z. Liu, H. Yu, Y. Wang, and W.J. Li, Sci. Rep. **6**, 24703 (2016)
13. Lucie Hüser and Peter Lehmann, Appl. Opt. **59**, 1695 (2020)
14. S. Perrin, Y.J. Donie, P. Montgomery, G. Gomard, and S. Lecler, Phys. Rev. Appl. **13**, 014068 (2020)

Picometer mechanical displacement measurement using heterodyne interferometer with phase-locked loop

Masato Aketagawa[1,*], and *Thanh Dong* Nguyen[1]

[1]Department of Mechanical Engineering, Nagaoka University of Technology, 940-2188 Kamitomioka, Nagaoka, Niigata, Japan

Abstract. In this paper, we show picometer-order mechnical displacmment measurements using a heterodyne interferometer with a phase-locked loop (PLL). A heterodyne light source for the interferometer is implemented with a frequency stabilized HeNe laser and two acousto-optic modulators. A real time phase measurement is performed by the PLL, whose software is programmed in a field-programmable gate array (FPGA). A stiff parallel spring stage combined with a high-voltage piezoelectric actuator is used to generate picometer-order mechanical motion. With the above implementations, mechanical displacement of 10 picometer or less can be measured.

1 Introduction

With the rapid progress of ultraprecision engineering and semiconductor manufacturing, precise metrology systems based on laser interferometers with resolution of 10 picometer order have become necessary for displacement/distance measurements [1]. In recent years, displacement-measuring heterodyne interferometers have become suitable for industrial applications [2], because they have traceability to the standard definition of the meter if frequencies of light sources are stable. In this paper, we show picometer-order mechnical displacmment measurements using a heterodyne interferometer with a phase-locked loop (PLL). A heterodyne light source for the interferometer is implemented with a frequency stabilized HeNe laser and two acousto-optic modulators (AOMs). A real time phase measurement is performed by the PLL, whose software is programmed in a field-programmable gate array (FPGA). A stiff parallel spring stage combined with a high-voltage piezoelectric actuator is used to generate picometer-order mechanical motion. With the above implementations, mechanical displacement of 10 picometer or less can be measured.

2 Implementation and Experiment

Figure 1 shows a heterodyne interferometer with a heterodyne light source (= a HeNe laser + two AOMs). The heterodyne light source emits two bemas with difference frequencies f_1 and f_2 ($f_2 > f_1$). The interferometer consists of two beam splitters (BSs) and two retro reflectors (target & refrence reflectors: TR & RR). TR is moved by the parallel spring stage, whose resonant frequency is higher than 1kHz, combined with the high-voltage piezoelectric actuator. The two beams enter to two BSs, TR, and RR. Then the mixed beams go to photo detectors (PD1 and PD2). At PD1, the reference beat signal ($\Delta f = f_2 - f_1$) is measured, at PD2, the measurement beat signal ($f_2 - f_1 + \Delta f_D = \Delta f + \Delta f_D$, where Δf_D is Doppler frequency due to the movement of TR) is measured. The reference and measurement beat signals go to the phase

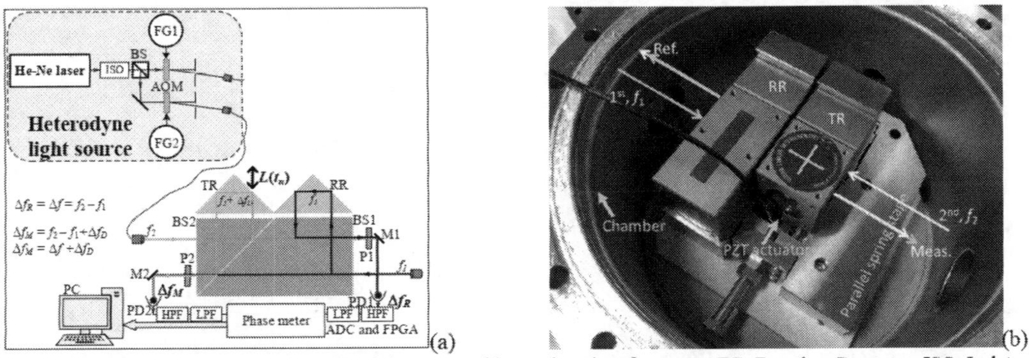

Figure 1. Schematic diagram (a) and photo (b) of the proposed heterodyne interferometer. FG: Function Generator, ISO: Isolator, BS: Beam splitter, TR&RR: Target and reference retro reflectors, PD: Photo detector, P: Polarizer, LPF&HPF: Low-pass & high-pass filters, ADC: Analogue to digital converter, M: Mirror, PZT: Piezo electric transducer.

* Corresponding author: masatoaa@vos.nagaokaut.ac.jp

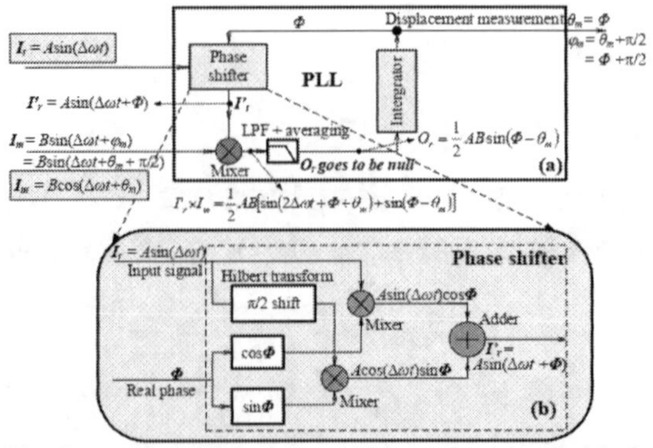

Figure 2. Schematic diagram of the phase meter using two sinusoidal heterodyne signals: (a) whole system and (b) phase shifter. PLL, phase-locked loop, LPF + averaging, low-pass filter combined with an averaging operation.

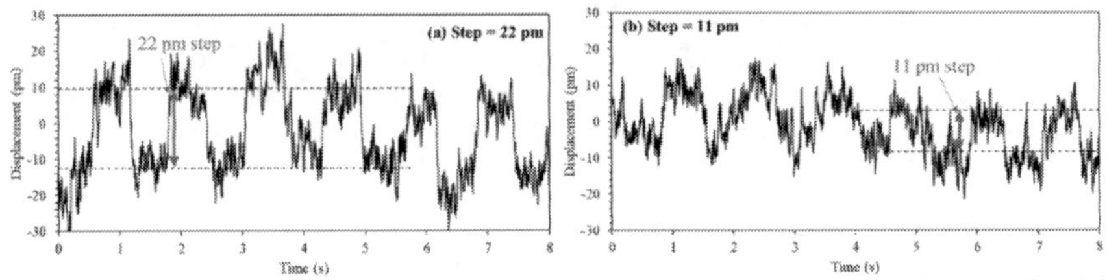

Figure 3. Step displacement measurement results obtained by the proposed system. (a) 22 pm step (0.44 mrad) and (b) 11 pm step (0.22 mrad).

meter to determine the displacement of TR. Figure 2 shows the phase meter with PLL, whose softoware is programmed in the FPGA. The PLL consists of an active phase shifter with Hilbert transform, a mixer, a low-pass-filter (LPF) with an averaging operation, and an integrator. The 120MHz/16bit analogue-to-digital converters (ADC) are used to digitize the reference and measurement beat signals. In the experiment, the beat frequency ($\Delta f = f_2 - f_1$) is 2.4MHz. The input signals of the phase meter are the reference I_r and measurement I_m signals from PD1 and PD2 as shown in Figure 1. These signals I_r and I_m are

$$I_r = A \sin 2\pi \Delta f t = A \sin \Delta\omega t \quad (1),$$

$$I_m = B \sin(\Delta\omega t + \varphi_m(t)) = B \cos(\Delta\omega t + \theta_m(t)) \quad (2),$$

where $A(B)$ $\Delta\omega$, and $\varphi_m(t)$ ($\varphi_m(t)=\theta_m(t)+\pi/2$) are amplitudes of I_r and I_m, the angular frequency difference of the heterodyne light source and a phase to be measured. The active phase shifter adjusts the phase of I_r, then the phase-shifted refrence signal I'_r is,

$$I'_r = A \sin(\Delta\omega t + \Phi) \quad (3),$$

where Φ is an adjustable phase in the PLL. At the mixer, the product of I_m and I'_r is obtained as follows;

$$I'_r I_m = \frac{1}{2} AB \sin(2\Delta\omega t + \Phi + \theta_m(t)) + \frac{1}{2} AB \sin(\Phi - \theta_m(t)) \quad (4)$$

To remove a high-frequency component (the double-beat frequency appearing in Eq. (4)), the product is passed through the LPF, and the output signal after the LPF (= the error signal (O_r)) is calculated as

$$O_r = \frac{1}{2} AB(\Phi - \theta_m(t)) \approx \frac{1}{2} AB(\Phi - \theta_m(t)) \quad (5),$$

where we assume $\Phi - \theta_m << 1$. The error signal O_r should be zero by a negative feedback loop, which utilizes the integrator to create the null condition so that Φ becomes θ_m ($\Phi = \theta_m$). When the PLL is locked, the measured Φ is adjusted to cancel out θ_m and is accumulatively identified by the integrator (a detailed description is given in Section 2.3). Finally, the displacement information can be obtained from the phase shift between the two input heterodyne signals. The detail of operation principle of the phase meter can be seen in [3].

Figures 3(a) and (b) show the measurement results of the displacement measurements of 22 pm and 11 pm mechanical steps, respectively. At the moment, we only confirm 11 pm step using the above implementation.

We would like to express our sincere thanks to Chuo Precision Industrial Co., Ltd for the use of the equipment.

References

[1] R. Saathof, G. Schutten, J. Spronck, R. Munnig Schmidt, *Precis. Eng.*, **41**, (2015) 102-110, https://dx.doi.org/10.1016/j.precisioneng.2015.03.004
[2] F. C. Demarest, *Meas. Sci. Technol.* **9** (1998) 1024–1030, https://doi.org/10.1088/0957-0233/9/7/003
[3] T. D. Nguyen, Q. A. Duong, M. Higuchi, T. T. Vu, D. Wei, M. Aketagawa, *Sens Actuator A*, **304**, (2020), 111880. https://doi.org/10.1016/j.sna.2020.111880

Intensity dependent deflection spectroscopy for absorption measurements

Walter Dickmann*[1,2], *Tom* Götze[2], *Mark* Bieler[2], and *Stefanie* Kroker[1,2]

[1] Technische Universität Braunschweig, LENA, Langer Kamp 6, 38106 Braunschweig, Germany

[2] Physikalisch-Technische Bundesanstalt (PTB), Bundesallee 100, 38116 Braunschweig, Germany

Abstract. We report on a method for the characterization of optical absorption in semiconductors at photon energies below the bandgap energy. We use intensity dependent deflection spectroscopy to measure spatially resolved the optical absorption and to separate the occurring absorption mechanisms. To this end, we take advantage of the different intensity scaling of these mechanisms and extract the material parameters by fitting the intensity dependent absorption to a physical model. Our method enables a simple but sufficient determination of crucial optical loss properties (e.g. impurity related absorption and two-photon absorption) in various semiconductor systems, e.g. substrates for optical components or solar cells.

1 Introduction

The performance of a variety of optical applications, e.g. in photovoltaics and telecommunication, is strongly influenced by optical absorption processes [1], [2]. The determination of the strength of each of these processes is desirable in order to physically understand the optical loss behaviour of the respective application and potentially enable ways for their tuning. The widely used method to distinguish between different absorption processes is time-resolved absorption spectroscopy: in this approach, the different time scaling of various loss processes (e.g. free carrier creation, phonon-assisted absorption and pure field effects) is utilized [3]. However, this method requires synchronized short pulse excitation and detection which makes it experimentally challenging and expensive concerning measurement infrastructure. We developed a different approach, combining well established spatially resolved deflection spectroscopy [4] and intensity dependent measurements. Fitting the intensity dependent deflection signal to a physical model that describes the different power scaling of optical loss effects enables the separation of these effects by simple cw detection. The extraction of actual material parameters such as the linear attenuation coefficient and the two-photon absorption coefficient is performed by calibrating the setup and utilizing affiliated finite element (FEM) computations. We reproduced well known absorption parameters for silicon and gallium arsenide in order to verify our measurement method. Furthermore, we determined the strength of the Franz-Keldysh effect in cadmium telluride for the first time as $C_{\mathrm{FK}} = [8.7 \ldots 16.9] \times 10^{13}\ \mathrm{m^{-1}s^{-1/2}}$.

2 Experimental setup

The experimental setup is sketched in figure 1. We utilize deflection spectroscopy [4] with a cw laser diode at a wavelength of 1310 nm as probe source. Its deflection is probed by a quadrant detector. The pump source is an Er-doped fs fibre laser with centre wavelengths of about 1600 nm (used for Si and GaAs) or 800 nm (used for CdTe). The ultra-short pulses lead to peak intensities of above $10^{13}\ \mathrm{W/m^2}$ at which nonlinear absorption effects are significant loss contributions. The pump intensity is scanned over more than two orders of magnitude in order achieve an intensity dependent deflection signal. Spatial resolution is enabled by shifting the readout point where pump and probe beam cross each other.

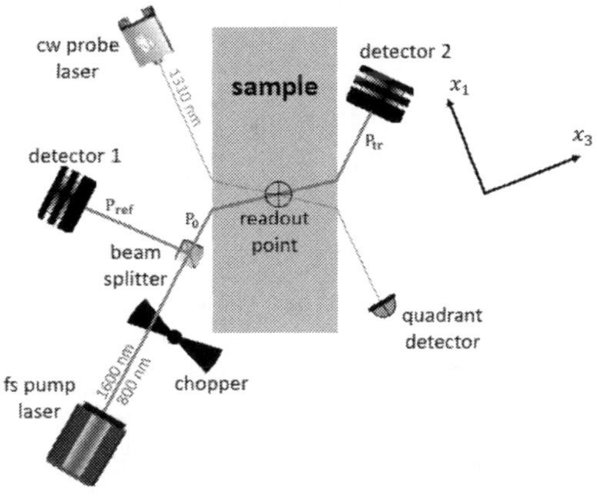

Fig. 1. Deflection setup for absorption measurements.

* Corresponding author: w.dickmann@tu-bs.de

3 Modelling of absorption mechanisms

The modelling of the absorption mechanisms and their influence on the deflection signal is necessary to determine quantitative material parameters. The modelling is divided into three parts: first, free charge carriers are created due to linear absorption from defect states (the corresponding material parameters is the linear attenuation coefficient α) and due to quadratic two-photon absorption [5] from the valence band (two-photon absorption coefficient β). This directly leads to a refractive index change and thus to a deflection signal. The dynamics of this process is described by the carrier diffusion equation that is numerically solved by FEM.

Second, the free charge carriers recombine after a certain lifetime and are generating heat. This leads to a local temperature rise in the sample and consequently to another signal contribution. As this process is much slower than the free carrier recombination (few ms half-life compared to several ns charge carrier lifetime), the temperature related deflection signal is much stronger than the charge carrier related signal. An exemplary FEM computed temperature field in the GaAs sample is depicted in figure 2.

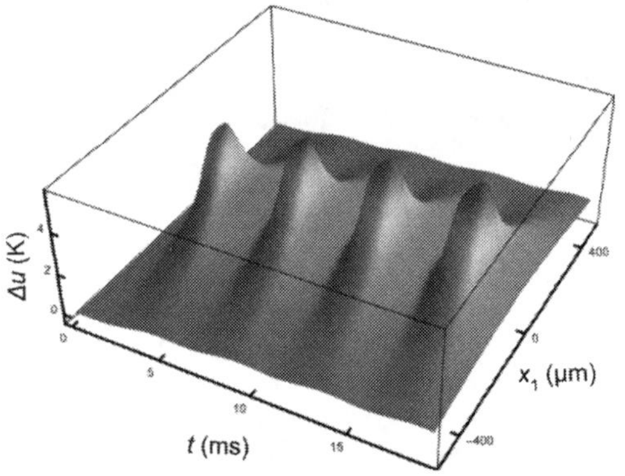

Fig. 2. Computed spatially and time-resolved temperature rise in the GaAs sample for a peak intensity of 10^{13} $\mathrm{W/m^2}$.

Third, the electric field of the pump laser itself leads to a change of the dielectric function of the sample material. The dominating pure field effect in the investigated wavelength and intensity regime is the Franz-Keldysh effect (FKE) [6]. It changes both the refractive index (primary effect) and the absorption coefficient, which consequently influences the charge carrier and heat creation (secondary effect). Its strength is described by the material parameter C_{FK}.

We finally formulate a model function based on FEM computations that describes the measured deflection signal X as a function of the pump beam intensity I_0 with the material parameters α, β and C_{FK} as free coefficients.

4 Experimental results

The intensity dependent deflection signal $\{X|I_0\}$ is measured for each sample material at several readout positions. The measurements are fitted by the model function $X(I_0)$ and the three absorption related material parameters are extracted. One measured data set for the GaAs sample and the corresponding fit function are shown exemplarily in figure 3.

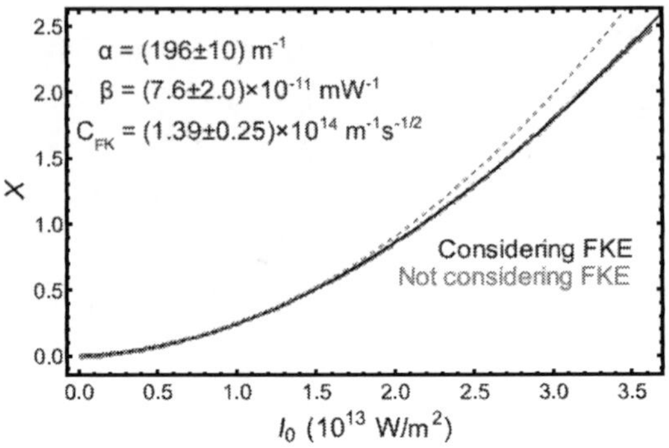

Fig. 3. Measured intensity dependent deflection signal (blue dots) in the GaAs sample, corresponding fit functions with (black line) and without (red line) considering the Franz-Keldysh effect and extracted material parameters.

For very low intensities ($I_0 \leq 10^{12}\ \mathrm{W/m^2}$) the absorption behaviour is predominately linear. For higher intensities, two-photon absorption is the dominant effect. However, for very high intensities ($I_0 \geq 1.5 \times 10^{13}\ \mathrm{W/m^2}$) the FKE is an additional contribution that must be considered. To conclude, we established a simple experimental method for the spatially resolved characterization of absorption mechanisms in semiconductors by combining deflection spectroscopy with intensity dependent measurements. Utilizing this method, we reproduced absorption related literature values for GaAs and Si. Furthermore, we determined the strength of the FKE in CdTe as $C_{\mathrm{FK}} = [8.7 \ldots 16.9] \times 10^{13}\ \mathrm{m^{-1}s^{-1/2}}$.

References

[1] W. Winkler, K. Danzmann, A. Rüdiger, R. Schilling, Physical Review A, **44**(11), 7022 (1991)

[2] T. Siefke et al., Advanced Optical Materials, **4**(11), 1780 (2016)

[3] M. Topp, P. M. Rentzepis, R. P. Jones, Journal Of Applied Physics, **42**(9), 3415 (1971)

[4] A. C. Boccara, D. Fournier, W. B. Jackson, N. M. Amer, Optics Letters, **5**(9), 377 (1980)

[5] M. Bass, E. W. Van Stryland, D. R. Williams, W. L. Wolfe, *Handbook of Optics volume I,* McGraw-Hill (1995)

[6] H. I. Ralph, Journal of Physics C: Solid State Physics, **1**(2), 378 (1968)

https://doi.org/10.1051/epjconf/202023806005

Spatial spectroscopy for high resolution imaging

Arturo Villegas[1,2,*] and *Juan* P. Torres[1,3]

[1]ICFO-Institut de Ciències Fotòniques, Mediterranean Technology Park, 08860 Castelldefels, Barcelona, Spain
[2]The Barcelona Institute of Science and Technology, 08036 Barcelona, Spain
[3]Department of Signal Theory and Communications, Universitat Politècnica de Catalunya, 08034 Barcelona, Spain

Abstract. Quantum estimation theory provides bounds for the precision in the estimation of a set of parameters that characterize a system. Two questions naturally arise: *Is any of these bounds tight?* And if this is the case, *what type of measurements can attain such a limit?* In this work we show that for phase objects, it is possible to find a tight resolution bound. Moreover one can find a set of spatial modes whose detection provides an optimal estimation of the complete set of parameters for which we propose a homodyne detection scheme. We call this method *spatial spectroscopy* since it mimics in the spatial domain what conventional spectroscopy methods do in the frequency domain employing many frequencies (hyperspectral imaging).

1 Resolution limits for multi-parameter estimation of phase objects

The quantum Cràmer-Rao inequality sets a bound for the resolution of any ensemble of unbiased estimators $\hat{\theta} = \{\hat{\theta}_i\}$ ($i = 1...N$) corresponding to a set of parameters $\theta = \{\theta_i\}$ that characterize relevant features of an object [1]. The variance of the set of unbiased estimators fulfills the matrix inequality

$$V(\hat{\theta}) \geq F_Q^{-1}, \quad (1)$$

i.e. $V(\hat{\theta}) - F_Q^{-1}$ is a positive semi-definite matrix. $V(\hat{\theta})$ is the covariance matrix with elements $[V(\hat{\theta})]_{ij} = E\left[(\hat{\theta}_i - \theta_i)(\hat{\theta}_j - \theta_j)\right]$ and F_Q is the Quantum Fisher Information Matrix (QFIM). For a pure state $|\Psi(\theta)\rangle \equiv |\Psi\rangle$ [1], the QFIM is

$$[F_Q]_{ij} = 4\,\mathrm{Re}\left\{\langle \partial_i \Psi | \partial_j \Psi\rangle - \langle \partial_i \Psi | \Psi\rangle\langle \Psi | \partial_k \Psi\rangle\right\}, \quad (2)$$

where ∂_i corresponds to the partial derivative with respect to the i–th parameter. In general, Eq. (1) is not a tight precision bound, so F_Q^{-1} is not always the most informative bound. However if

$$\mathrm{Im}\left\{\langle \partial_i \Psi | \partial_j \Psi\rangle\right\} = 0, \quad \forall\ i,j \in [1,N] \quad (3)$$

the inequality in Eq. (1) is tight and it is attainable when using the appropriate (optimal) measurement [2, 3].

Quantum estimation theory is of great relevance for optical probing and imaging. It has raised great interest in the last few years as a tool to determine the *ultimate* resolution limits. For instance, contrary to common belief, it has been demonstrated that one can distinguish the location of two incoherent point sources no matter how close they are to each other [4–7].

*e-mail: Arturo.Villegas@icfo.eu

More general problems can be analyzed. Consider the probing of a phase object. The illumination beam can be modelled as M independent photons whose quantum state is pure. The state of each illumination photon is

$$|\Psi\rangle = \int \mathrm{d}\mathbf{x}\ \Psi_0(\mathbf{x})\, a^\dagger(\mathbf{x})\, |\mathrm{vac}\rangle, \quad (4)$$

where $\mathbf{x}$ is the spatial coordinate, $a^\dagger(\mathbf{x})$ is the electromagnetic creation operator, $\Psi_0(\mathbf{x})$ is the spatial profile of the illumination beam and $|\mathrm{vac}\rangle$ is the electromagnetic vacuum state. After the interaction with the phase object, the quantum state of a reflected (or transmitted) photon is

$$|\Psi(\theta)\rangle = \int \mathrm{d}\mathbf{x}\ \Psi(\mathbf{x},\theta)\, a^\dagger(\mathbf{x})\, |\mathrm{vac}\rangle, \quad (5)$$

where $\Psi(\mathbf{x},\theta) = \Psi_0(\mathbf{x}) \exp\left[i\varphi(\mathbf{x},\theta)\right]$. Here $\varphi(\mathbf{x},\theta)$ is the phase imprinted on the field by the object under study.

The quantum state in Eq. (5) fulfills the condition (3) that allows to determine the best resolution that can be attained when estimating the parameters of the object. This is of great interest since one can *a priori* determine the limits of what can be measured in different scenarios. Thus, the corresponding resolution bound depends on the type of interaction between light and object and on the quantum nature of the illumination.

2 Spatial mode projection as imaging method

If the condition given by Eq. (3) is fulfilled, it has been theoretically demonstrated that projecting light after interacting with the object onto a set quantum states $\{|\gamma_i\rangle\}$ formed by a particular set of spatial modes should allow to achieve the fundamental resolution limit given by Eq. (1)

[3]. The quantum states corresponding to these modes are combinations of the quantum states $|\partial_i\Psi\rangle$.

How to measure such projections? In a series of pioneering experiments, Delaubert et al. [8, 9] considered a homodyne detection scheme to measure with great resolution two parameters of a Gaussian beam: 1) the displacement and 2) the tilt; both with respect to a reference axis. For the spatial shape of the mode in the local oscillator they chose the one resulting from deriving a Gaussian mode, the TEM_{10}. In their experiments there are two particular aspects that are not obvious to hold in a general imaging scheme for phase objects: i) the spatial mode carrying information about the displacement and tilt is the same, i.e., $|\partial_1\Psi\rangle \propto |\partial_2\Psi\rangle$ and ii) it is possible to discriminate the values of the displacement and tilt by changing the phase of the local oscillator.

In this work we show how a homodyne detection scheme allows the estimation of N parameters that characterize an arbitrary phase object when using a particular set of spatial modes in the local oscillator. We thus generalize the method put forward by Delaubert et al. and show that this method is generally useful for measuring phase objects.

For the sake of clarity we present the case of the estimation of two parameters, θ_1 and θ_2. We choose the reference state $|\Psi_0\rangle \equiv |\Psi(\theta_0)\rangle$ for fixed values of the parameters $\theta_0 = (\theta_{1_0}, \theta_{2_0})$. Using $\Delta_1 = \theta_1 - \theta_{1_0}$ and $\Delta_2 = \theta_2 - \theta_{2_0}$, the state that describes photons after reflection (transmission) can be written as:

$$|\Psi(\theta_1, \theta_2)\rangle = \Gamma_0 |\Psi_0\rangle + \Gamma_1 |\gamma_1\rangle + \Gamma_2 |\gamma_2\rangle . \quad (6)$$

$|\Psi_0\rangle$, $|\gamma_1\rangle$ and $|\gamma_2\rangle$ form an orthonormal basis of vectors. The coefficients in expression (6) are explicitly: $\Gamma_0 = 1 + \Delta_1 A + \Delta_2 B$, $\Gamma_1 = \Delta_1 N_1 + \Delta_2(C - A^* B)/N_1$ and $\Gamma_2 = \Delta_2 N_2$. A, B and C are complex numbers corresponding to the mode overlaps $A = \langle\Psi_0|\partial_1\Psi\rangle$, $B = \langle\Psi_0|\partial_2\Psi\rangle$ and $C = \langle\partial_1\Psi|\partial_2\Psi\rangle$. N_1 and N_2 are normalization constants ensuring $\langle\gamma_i|\gamma_j\rangle = \delta_{ij}$. All constants are determined by the object and parameters under consideration.

Writting explicitly the coefficients Γ_i in terms of the real and imaginary parts of A, B and C, equation (6) shows that by projecting onto the spatial mode $|\gamma_2\rangle$ one can determine Δ_2 and equivalently, by projecting onto the spatial mode $|\gamma_1\rangle$ and setting the phase difference between probe and the local oscillator to 0 ($\mathrm{Re}\{\Gamma_1\}$) one can determine Δ_1. Alternatively, it is sufficient to project onto the spatial mode $|\gamma_1\rangle$. When the phase difference is $\pi/2$ ($\mathrm{Im}\{\Gamma_1\}$) one can determine Δ_2; then setting the phase difference to 0 allows to determine Δ_1.

This procedure is general and applies to any phase object characterized by two parameters; an example is hereby presented. Consider the estimation of the sidewall angle $\beta \equiv \theta_1$ and height $h \equiv \theta_2$ of a cliff-like nanostructure. The transverse section is modelled as $S(x) = \frac{h}{2}(1 + \tanh \alpha x)$, with $\alpha = 2\tan\beta/h$. The input beam is assumed to have a Gaussian profile $\Psi_0(x)$ of width $\sigma = \lambda/\pi NA$, λ is the wavelength of the probe beam and NA the numerical aperture of the imaging system. The phase added to the field after interaction with the cliff-like structure is $\varphi(x, \theta) = kh(1 - \tanh \alpha x)$. Using a homodyne detection scheme with the mode $|\gamma_1\rangle$ in the local oscillator one can measure the coefficient Γ_1 in Eq. (6) and determine the parameters β and h by setting the phase difference to 0 and $\pi/2$, respectively.

3 Spatial spectroscopy as an imaging technique

Spectroscopy is the study of the interaction between matter and electromagnetic radiation. In the most usual case, *spectroscopic schemes* choose one or more wavelengths to interrogate a sample and obtain information about particular aspects of interest, such as the presence or concentration of a chemical compound. Here we choose spatial modes to obtain information about the spatial properties of the object. This is why we term the technique *spatial spectroscopy*, the use of one or more spatial modes to obtain information of particular spatial features of the object.

The proposed scheme also allows to determine the modal decomposition (amplitude and phase) of the light reflected (or transmitted) from the object in a convenient basis. Although this does not always correspond to an optimal measurement (the conditions for such measurement to be optimal will be shown), it might be an useful / alternative imaging technique as one can select spatial modes that are more sensitive to the interaction with the object. When the basis are modes embedded with orbital angular momentum, this is referred as *spiral imaging* [10].

References

[1] C.W. Helstrom, Journal of Statistical Physics **1**, 231 (1969)

[2] K. Matsumoto, in *Asymptotic Theory Of Quantum Statistical Inference: Selected Papers* (2005), pp. 305–350

[3] L. Pezzè, M.A. Ciampini, N. Spagnolo, P.C. Humphreys, A. Datta, I.A. Walmsley, M. Barbieri, F. Sciarrino, A. Smerzi, Physical review letters **119**, 130504 (2017)

[4] M. Tsang, Optica **2**, 646 (2015)

[5] M. Tsang, R. Nair, X.M. Lu, Physical Review X **6**, 031033 (2016)

[6] M. Paúr, B. Stoklasa, Z. Hradil, L.L. Sánchez-Soto, J. Rehacek, Optica **3**, 1144 (2016)

[7] J. Řehaček, Z. Hradil, B. Stoklasa, M. Paúr, J. Grover, A. Krzic, L. Sánchez-Soto, Physical Review A **96**, 062107 (2017)

[8] V. Delaubert, N. Treps, M. Lassen, C.C. Harb, C. Fabre, P.K. Lam, H.A. Bachor, Physical Review A **74**, 053823 (2006)

[9] V. Delaubert, N. Treps, C. Fabre, H.A. Bachor, P. Réfrégier, EPL (Europhysics Letters) **81**, 44001 (2008)

[10] L. Torner, J.P. Torres, S. Carrasco, Optics express **13**, 873 (2005)

Imaging Mueller matrix ellipsometry setup for optical nanoform metrology

Tim Käseberg[1,*], *Jana* Grundmann[1], *Johannes* Dickmann[1,2], *Stefanie* Kroker[1,2], and *Bernd* Bodermann[1]

[1]Physikalisch-Technische Bundesanstalt, Bundesallee 100, 38116 Braunschweig, Germany
[2]Laboratory for Emerging Nanometrology, Technische Universität Braunschweig, Langer Kamp 6a/b, 38106 Braunschweig, Germany

Abstract. We designed, realized, and characterised an imaging Mueller matrix ellipsometry setup for the pixelwise measurement of the Mueller matrices in microscope images. Our setup is capable of performing measurements in reflection as well as in transmission in a broad range of angles of incidence for wavelengths between 400 nm and 700 nm. We compared measurements of specially designed nanostructured samples with AFM and SEM measurements as well as with numerical simulations using the finite element method.

1 Introduction

The reliable production of structures at the nanometre scale remains a crucial issue in modern nanotechnology and it relies on the precise characterisation of the structures using fast and non-invasive measurement techniques. One popular technique for this task is Mueller matrix ellipsometry, which is a non-imaging optical technique that allows for the determination of the polarizing characteristics and by this to derive information about structural features of a sample [1, 2]. The polarization state of light can be fully described using Stokes vectors. The 4x4 elements of the Mueller matrix provide the mathematical transform for these Stokes vectors of incident and outgoing light. As an integral measurement method, ellipsometry is capable of examining a rather large integration area, while restricting its measurement results to mean values of the whole illumination area only. Given usually large beam sizes in common ellipsometric setups, these areas rarely fall below several 10 µm in diameter, whereas applications such as in semiconductor industries deal with much smaller measurement targets.

This limitation can be overcome by upgrading the classic ellipsometric setup with an imaging system. By this, the Mueller matrix of the sample can be determined for each pixel in an image, allowing for evaluations of sample features within much smaller interaction areas than the illumination area. This can be complemented by the investigation of off-diagonal elements of the Mueller matrix which have been shown to provide additional information about subwavelength sized features [3].

Our investigations include the construction of such an imaging ellipsometry setup and its application for the characterisation of different metastructure samples [4, 5]. Our setup provides a flexible composition that allows for

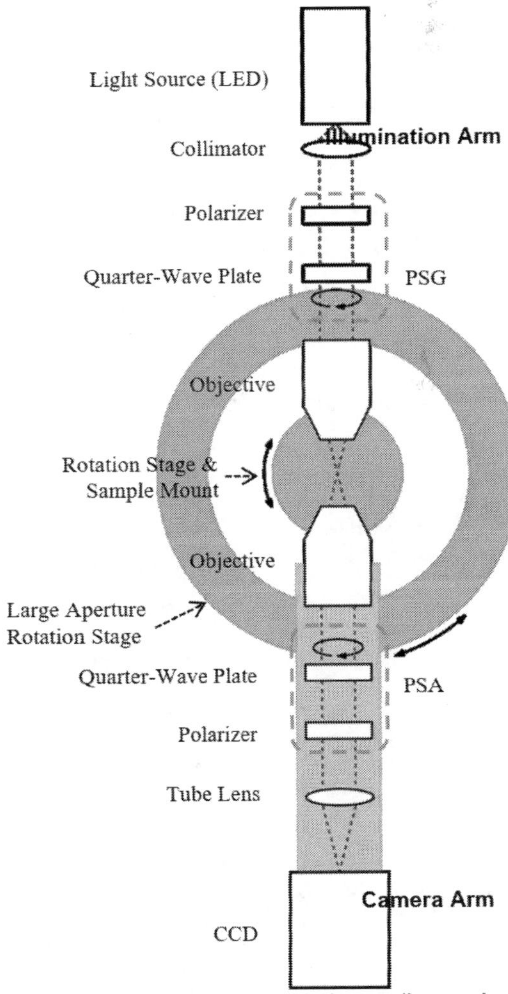

Fig. 1. Schematic representation of the Mueller matrix microscope setup.

* Corresponding author: tim.kaeseberg@ptb.de

measurements in both reflection and transmission for a wide range of angles.

2 Experimental Setup

The setup of our Mueller matrix microscope is sketched out in figure 1. It is composed of two separate arms: a fixed illumination arm and a rotating camera arm. The illumination arm consists of a white light LED, limited to single wavelengths using bandpass filters, as well as a collimating lens and a polarisation state generator (PSG) system. The PSG contains a linear Glan-Thompson polarizer and a rotating quarter-wave plate. An objective focuses the light onto the sample which is mounted on top of a couple of linear and rotation stages for the precise adjustment of the sample in all degrees of freedom. The camera arm is attached to a rotation stage with a large aperture of 30 cm diameter with the same axis of rotation as the smaller sample rotation stage at its centre. An infinity corrected objective with a long working distance of 25.5 mm, a magnification of 50x and a numerical aperture of NA = 0.42 combined with a tube lens creates an image of the sample that is collected by a cooled monochromatic CCD camera at the end of the camera arm. In between the objective and the CCD, a polarization state analyser (PSA) identical to the PSG but in reverse order is positioned. The setup forms a conventional dual-rotating retarder Mueller matrix ellipsometer where the two retarders are rotated at different angular frequencies at a certain ratio. By this, the whole Mueller matrix can be determined for each pixel in the obtained image. To avoid collisions of the arms, limit switches restrict their movement. The setup as well as the measurement procedures are controlled by a LabVIEW interface.

3 First Measurements

The mechanical and optical components of the measurement system have been characterized. The system can be used to investigate in the polarizing properties of arbitrary samples. Figure 2 shows the very first preliminary Mueller matrix image recorded as a test of our setup for 11 μm wide lines in transmission mode at normal incidence. As a first step, we examine plasmonic lenses that can be used as resolution enhancement devices. Our goal is to include the use of these structures in advanced Mueller matrix ellipsometry setups for an enhanced sensitivity to structures at the nanometre scale. Therefore, we produced a set of plasmonic lenses for applications at several visible wavelengths using electron beam lithography methods. First tests with these lenses were performed at our setup. Additionally, we developed concepts on the combination of ellipsometric setups and plasmonic lenses. To further examine the connection between subwavelength sized structural features and off-diagonal Mueller matrix elements, we designed and fabricated a set of geometrical samples with different shapes, ranging between squares and circles [4]. The measurements of all these distinct structures are compared to numerical simulations, performed with the finite element tool JCMsuite [6]. Apart from these structures, measurements of samples with well-known properties at our setup were compared to measurements with scanning electron microscopy (SEM) and atomic force microscopy (AFM) setups.

Fig. 2. First preliminary Mueller matrix image of 11 μm wide lines. Each image contains the pixelwise values of one element of the 4x4 Mueller matrix.

This work is funded through the project 17FUN01 "BeCOMe" within the Programme EMPIR. The EMPIR initiative is co-founded by the European Union's Horizon 2020 research and innovation program and the EMPIR Participating Countries

References

1. M. Losurdo, K. Hingerl, *Ellipsometry at the Nanoscale* (Springer, Berlin, 2013)
2. A. Diebold, A. Antonelli, N. Keller, APL Materials **6**, 058201 (2018)
3. N. Kerwien, *On the influence of polarization effects in microscopic image formation*, Berichte aus dem Insitut für Technische Optik **58** (2007)
4. T. Käseberg, J. Dickmann, T. Siefke, M. Wurm, S. Kroker, B. Bodermann, Proc. SPIE **11057**, 110570R (2019)
5. T. Käseberg, T. Siefke, S. Kroker, B. Bodermann, Meas. Sci. Technol. (manuscript accepted), https://doi.org/10.1088/1361-6501/ab7e6b
6. JCMsuite, JCMwave GmbH, Berlin. https://www.jcmwave.com/

Non-destructive testing of a rotating glass-fibre-reinforced polymer disc by swept source optical coherence tomography

Jonas Golde[1], *Christian* Schnabel[1], *Angelos* Filippatos[2], *Tino* Wollmann[2], *Maik* Gude[2], *Edmund* Koch[1]
[1]Clinical Sensoring and Monitoring, Anesthesiology and Intensive Care Medicine, TU Dresden, 01307 Dresden, Germany
[2]Institute of Lightweight Engineering and Polymer Technology (ILK), TU Dresden, 01307 Dresden, Germany

Abstract. Composite materials are used for high-performance rotating blades, e.g. in turbines and wind power plants. Here, optical coherence tomography was used to visualize large areas of fibre-reinforced-polymer discs at rotation speeds of up to 1200 rpm. These measurements allowed to visualize the fibre structure over large areas of the disc. By recording the front and back reflex, the wobble of the disc was measured precisely. Additionally, the recorded structure was used to detect even small deviations from a uniform rotation.

1 Introduction

Composite materials are of increasing interest due to a rising demand in efficiency and reliability of fast-rotating blades as used in turbines, power plants and pumps. The tunable successive damage behaviour of multi-layer materials is especially advantageous for the design of rotor blades with a successive and failure-tolerant characteristic. A detailed knowledge about damage formation and structural disorders is necessary to optimize component's life cycle and to achieve cost-efficient and secure turbine operation. Because vibration characteristics change with different and increasing structural damage, the beginning of those changes needs to be visualized.

Optical Coherence tomography (OCT) is a high-resolution three-dimensional imaging modality, suitable to investigate the structure of glass-fibre-reinforced polymer (GFRP) composites as shown previously [1, 2]. Stifter et al. [3] were able to measure the evolving internal stress distribution with polarization-sensitive OCT (PS-OCT). In this and future studies, we want to visualize the structure and structural damage of a GFRP disc under dynamic strain caused by fast rotation.

2 Experimental setup

A disc of GFRP with a diameter of 200 mm and nominal 4 mm thickness was rotated by a stepper motor with velocities ranging from 120 to 1200 rpm. The motor controller consisted of an Arduino, generating a stable frequency signal and the ramps for acceleration and deceleration, and a TMC 2208 stepper controller. The beam of a self-developed swept source OCT system (SS-OCT), similar to [4], with a centre wavelength of 1300 nm and an A-scan rate of 200 kHz was focused on the outer part of the rotating disc. The OCT system has an axial measurement range of 8 mm and an axial resolution of 16 µm in air. This corresponds to a range of 5.17 mm and a resolution of approximately 10 µm in GFRP assuming a refractive index of 1.547. The optical head, consisting of a fibre collimator (F260APC-C, f = 15.52 mm), a deflecting mirror and a focusing lens (AC127-025-C-ML, f =25 mm), was placed on a linear stage and could be positioned to image the outer 27.5 mm of the disc. The resulting lateral resolution was approximately 15 µm; thus, the linear stage was driven with a step size of 10 µm. The OCT system enabled data acquisition at full speed for extended periods of time. A photo sensor in conjunction with a wire glued to the disc gave a regular signal for every rotation, which was used to trigger the OCT system (B-scan trigger) or recorded as a marker during continuous acquisition. For 3D datasets, the position of the linear stage was changed after each rotation. Due to the stable trigger, the data sets could be sliced in radial direction to form radial cross-sections or combined to enface images.

3 Results

High quality OCT images of the fibre structure inside the GFRP could be acquired while the disc was rotating at speeds from 120 to 1200 rpm (Figure 1). Higher velocities were not tested due to safety reasons and will be performed in further studies. From these images, the front and back surface signals were extracted (Figure 2), allowing to precisely measure the wobble of the disc. Moreover, it was found that the thickness of the disc varied between 4.0 and 4.3 mm (Figure 2 bottom), which was later confirmed by a mechanical measurement. The frequency spectrum (not shown) revealed strongly decreasing amplitudes of the harmonics of the rotational frequency. The highest peak in the frequency spectrum not being a harmonic was at 78 Hz with an amplitude of ~0.5 µm, which could be a resonance frequency of the disc. The noise level was well below 10 nm.

* Corresponding author: Edmund.Koch@tu-Dresden.de

Furthermore, the fingerprint-like unique pattern of the fibres inside the GFRP allowed detecting even faint deviations from a uniform revolution. While the pattern could be matched to a single A-scan (5 µs), the noise level of a periodic deviation was less than 0.1 µs, corresponding to 10^{-6} of a revolution (Figure 3).

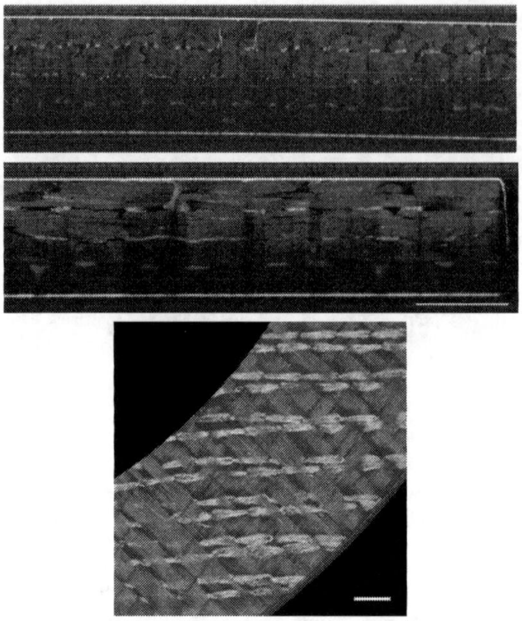

Fig. 1. OCT images of the rotating disc captured at 1200 rpm. Top: Azimuthal B-scan, the width corresponds to a length of 115 mm while the height is only 8 mm in air. Middle: Cross section in radial direction. Bottom: enface image at approx. 1.3 mm depth within the disc. Scale bars are 5 mm in air.

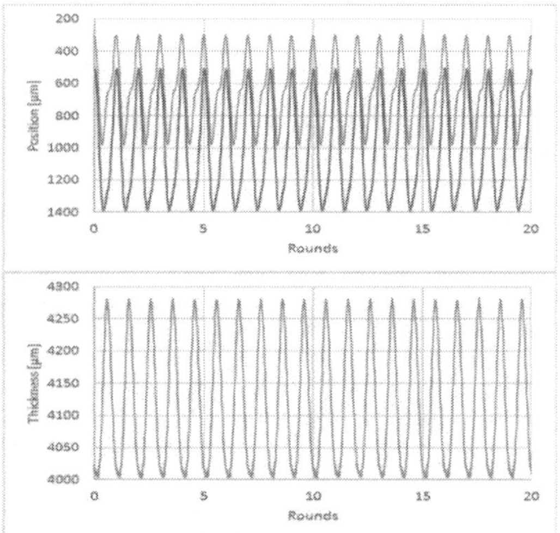

Fig. 2. Top: Signal of the front (blue) and back (red) surface of the disc over 20 rotations, measured from one side of the disc. The signal of the back surface (lower trace) was shifted to be visible on the same scale. Bottom: Thickness of the disc calculated from the signals of both surfaces.

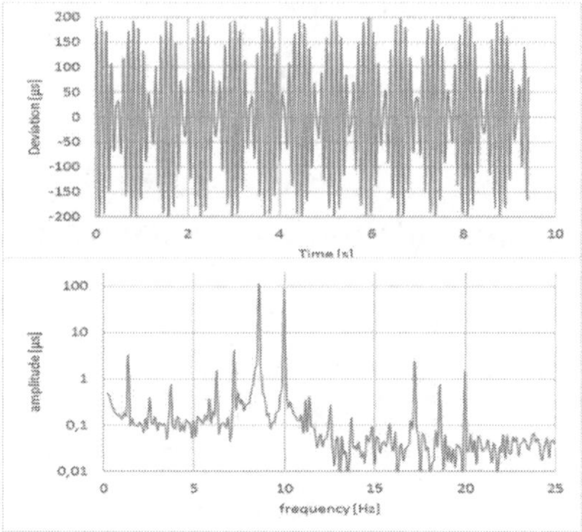

Fig. 3. Top: Deviations of the angular position of the disc from a uniform rotation as indicated by the unique pattern of the fibres inside the GFRP measured while rotating at 600 rpm. Bottom: Corresponding frequency spectrum showing peaks at 10 Hz (rotation frequency) and 8.6 Hz with similar amplitude. Other frequencies (harmonics and mixing frequencies) have amplitudes, which are at least one order of magnitude smaller. The noise level, visible at frequencies above 12 Hz, is below 0.1 µs. The frequency of 8.6 Hz seems to be a consequence of the magnetic forces given by the motor and the moment of inertia of the disc.

4 Conclusions

Swept source OCT can be used to measure the structure of GFRP while rotating at fast speed. The surface position as well as the (optical) thickness of the disc could be measured with high accuracy. The structure of the GFRP allowed identifying the position with a resolution of 5 µs, corresponding to 0.018° at 600 rpm. While radial acceleration at this velocity is more than 150 g, this does not endanger the structure of the disc. Measurements at discs rotating 5 to 10 times faster, possible with faster OCT systems, could show the degradation of the material and lead to a better understanding of the dynamics during material failure.

References

1. P. Liu, Roger M. Groves, R. Benedictus. Ndt & E International **64**: 52-58 (2014)
2. Joy P Dunkers, et al., Composites Part A: Applied Science and Manufacturing 30.2: 139-145 (1999).
3. D. Stifter, E. Leiss-Holzinger, Z. Major, B. Baumann, M. Pircher, E. Götzinger, C. K. Hitzenberger, B. Heise, Opt. Express 18, 25712-25725 (2010).
4. L. Kirsten, M. Schindler; J. Morgenstern; M. T. Erkkilä; J. Golde; J. Walther; P. Rottmann; M. Kemper; M. Bornitz; M. Neudert; T. Zahnert; E. Koch, Journal of biomedical optics **24**.3: 031017 (2018).

Fast 3D Shape Measurement of Transparent Glasses by Sequential Thermal Fringe Projection

Martin Landmann[1,2*], *Henri* Speck[1], *Patrick* Dietrich[1,2,3], *Stefan* Heist[1], *Peter* Kühmstedt[1], and *Gunther* Notni[1,3]

[1]Fraunhofer Institute for Applied Optics and Precision Engineering IOF, Albert-Einstein-Str. 7, 07745 Jena, Germany
[2]Friedrich Schiller University, Institute of Applied Physics, Abbe Center of Photonics, Albert-Einstein-Str. 6, 07745 Jena, Germany
[3]Ilmenau University of Technology, Department of Mechanical Engineering, Gustav-Kirchhoff-Platz 2, 98693 Ilmenau, Germany

Abstract. We present an approach for measuring the shape of transparent glasses, which enables us to significantly reduce the comparatively long measurement time while increasing the measurement accuracy. Instead of using area-like patterns, we irradiate the object to be measured successively in a locally strongly restricted area with considerably higher irradiance.

1 Introduction

Pattern projection-based 3D measurement systems are widely used to measure objects quickly, contactlessly, and non-destructively. However, the necessary condition of diffuse reflection of the projected patterns is strongly limited or not fulfilled for transparent, translucent, shiny, or absorbing materials. In recent years, we have developed a two-step optical method [1–5] for the measurement of such uncooperative objects based on the scanning from heating approach [6]: 1. projection of area-like thermal patterns onto the measurement object and 2. stereo recording of the thermal radiation that is absorbed and re-emitted by the object surface.

In this contribution, we present an approach that allows the significant reduction of the comparatively very long measurement time (> 20 s) while increasing the measurement accuracy. Instead of using area-like patterns, we irradiate the object to be measured successively in a locally strongly restricted area with considerably higher irradiance. This allows significantly shorter irradiation times and reduces contrast decrease due to thermal diffusion. Despite a higher number of individual projections, this results in a substantially lower total measurement time.

2 Sequential Fringe Projection Principle

The measurement principle is similar to the one presented by Brahm et al. [3] but with an improved projection method. The temperature contrast between dark and bright fringes is the key quantity for the thermal 3D measurement quality. It is induced by the irradiance of the projection pattern. However, while a longer irradiation period can lead to a higher temperature contrast, thermal diffusion usually counteracts a temperature contrast build-up. Hence, in order to improve the contrast, we increase the irradiance while we reduce the time until recording the camera images. Higher irradiance can be easily achieved by shaping the laser beam into locally strongly restricted areas, e.g., few fringes or even a single fringe per projection, instead of using a diverging laser beam in combination with a mask. Theoretical considerations show that in our setup, the irradiance can be increased by a factor of about 30. In order to achieve a temperature contrast, which is similar to that of the reference method, the irradiation period can be reduced drastically. In this way, thermal diffusion has very low influence on the temperature contrast. At the end of the projection of a single fringe, we record a pair of camera images. After that, the projection unit directs the laser sheet onto the next position and the irradiation and recording process is repeated. The high irradiance allows us to reduce the irradiation period so far that we can operate the cameras at their highest frame rates. This leads to a substantially lower total measurement time despite a higher number of individual projections.

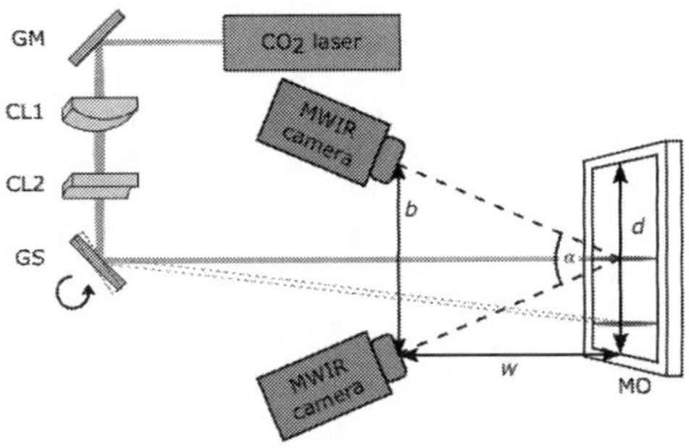

Figure 1. Schematic setup of our MWIR 3D system based on sequential thermal fringe projection consisting of projection unit (CO_2 laser, gold mirror GM, ZnSe lenses CL1 and CL2, galvanometer scanner equipped with gold mirror GS), measurement object MO, and MWIR cameras.

*e-mail: martin.landmann@iof.fraunhofer.de

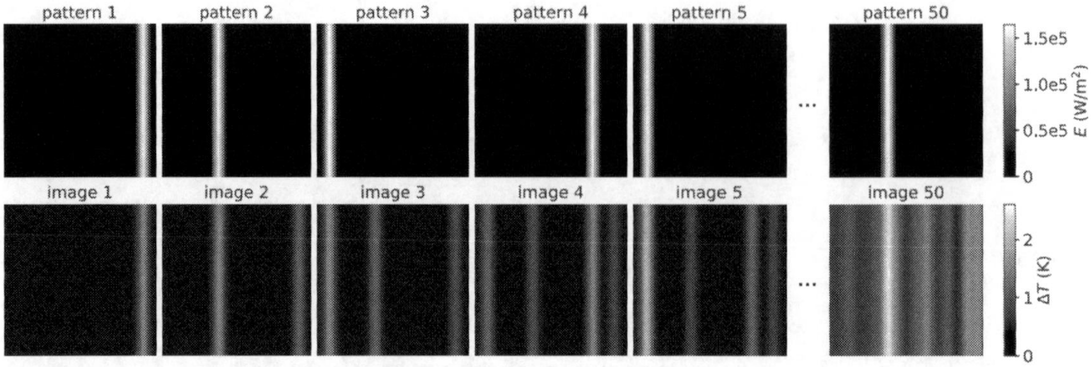

Figure 2. Example projection sequence of single fringe patterns for a sequence length of $N = 50$ and random positioning in the top row and corresponding MWIR camera images in the bottom row.

3 Measurement Setup

The measurement system comprises a projection unit, a measurement object, and two mid-wave infrared (MWIR) cameras. Figure 1 shows a schematic setup. The laser beam is deflected by a gold mirror GM. To obtain a thin laser beam in the horizontal direction on the measurement object MO, a convex cylinder lens CL1 slightly focuses the laser beam. A further concave cylinder lens CL2 provides a stronger diverging beam in the vertical direction. Behind CL2, a galvanometer scanner GS equipped with a gold mirror enables the projection unit to quickly change the fringe position in the measurement field. The central fringe position and a second arbitrary position are depicted. The projected fringe is partly absorbed by the measurement object and converted into heat. Local temperature increase leads to thermal diffusion. The measurement object emits the heat distribution which is recorded by two MWIR cameras.

4 Measurement Process

As we successively project single fringes instead of many simultaneous fringes covering the whole measurement field, we have to perform more projections and camera recordings compared to Brahm et al. [3]. Some example projection patterns for a sequence length of $N = 50$ and a random fringe positioning algorithm are shown in the upper row of Fig. 2. The corresponding MWIR camera images are illustrated in the bottom row. It can be observed that fringes of previous projections are still seen by the cameras. This is not a drawback but leads to additional information in the correspondence search between the two camera views. The entire measurement object warms up slightly during the measurement.

For the 3D reconstruction, we search for corresponding pixels by detecting the maximum of the normalized cross-correlation function [7–9]. Therefore, previous knowledge of the emitted pattern is not necessary. Once we have found the corresponding pixel pairs, we calculate the object point coordinates via triangulation.

This research project (2018 FGR 0098, project acronym "3DWÄRME") is supported by the Free State of Thuringia, the European Social Fund (ESF) of the European Union, and the Thüringer Aufbaubank (TAB).

References

[1] A. Brahm, C. Rößler, P. Dietrich, S. Heist, P. Kühmstedt, G. Notni, Proc. SPIE **9868**, 98680C (2016)

[2] A. Brahm, E. Reetz, S. Schindwolf, M. Correns, P. Kühmstedt, G. Notni, Adv. Opt. Technol. **5**, 405 (2017)

[3] A. Brahm, S. Schindwolf, M. Landmann, S. Heist, P. Kühmstedt, G. Notni, Proc. SPIE **10667**, 106670D (2018)

[4] M. Landmann, S. Heist, A. Brahm, S. Schindwolf, P. Kühmstedt, G. Notni, Proc. SPIE **10667**, 1066704 (2018)

[5] M. Landmann, S. Heist, P. Kühmstedt, G. Notni, Proc. SPIE **11056**, 1105615 (2019)

[6] G. Eren, O. Aubreton, F. Meriaudeau, L.S. Secades, D. Fofi, A.T. Naskali, F. Truchetet, A. Ercil, Opt. Express **17**, 11457 (2009)

[7] S. Heist, A. Mann, P. Kühmstedt, P. Schreiber, G. Notni, Opt. Eng. **53**, 112208 (2014)

[8] P. Albrecht, B. Michaelis, IEEE Transactions on Pattern Recognition **14**, 845 (1998)

[9] A. Wiegmann, H. Wagner, R. Kowarschik, Opt. Express **14**, 7692 (2006)

Fringe Pattern Denoising using U-Net based neural network

J. M. Crespo[(2,4,1)*], *V. Moreno*[(1)] *Juan Ramón Rabuñal*[(2,4)], *Alejandro Pazos*[(2,3)], *Monica Canabal Carbia*[(1)]

[1] Departamento de Fisica Aplicada. Universidade de Santiago de Compostela. Spain
[2] RNASA-IMEDIR, Computer Science Faculty, University of A Coruna, 15071 A Coruña, Spain
[3] CITIC, Computer Science Faculty, University of A Coruna, 15071 A Coruña, Spain
[4] CITEEC, Computer Science Faculty, University of A Coruna, 15071 A Coruña, Spain

Abstract. Fringe visibility and noise removal, are key success factors in interferometric techniques, where novel deep learning techniques can be applied. We test the use U-Net deep convolutional network applied to the obtained interference images, trained with an ad-hoc generated image dataset with complex fringe patterns, computed using high order Zernike polynomials.

1 Introduction

Interferometric techniques have been used along years as one of the most important methods for non-contact measurements and quality assessment of wavefronts coming from objects surfaces, but one of the most relevant caveats for accurate measurements is the noise level presented in the obtained interferograms. To deal with this caveat image filtering techniques have been applied to remove the noise (specially speckle or salt & pepper noise) from the interferograms, using more or less complex filters based on Fourier transforms[1], or other filters for high frequency noise in the spatial domain [2]. More specifically for the speckle pattern interferometry, the issue coming from the background noise and the need to increase fringe visibility are key factors to the success of the techniques, making highly complex the increase of fringe visibility to facilitate the image interpretation in real time.

During recent years, Machine Learning (ML) techniques have been significantly improved in the hardware and software fields, current GPU hardware is able to accelerate usual tensor operations present in ML techniques. In addition, software frameworks like Keras, TensorFlow and Pytorch are able to accelerate the implementation of new Neural Network (NN) models and remove the implementation complexity to make affordable the use of these technologies, reducing its time to market. In this study, we will use these techniques to increase the visibility of interference fringes in speckle pattern interferometry, designing and training a deep convolutional network based in U-NET [3] architecture to denoise speckle interference fringes. We will train the model with a synthetic and heavily noised collection of images, performing all required training activities using public services in the cloud (Google Research Colaboratory) given the current confinement status across countries.

2 The U-Net Architecture

The proposed layout for the denoiser NN is based on the U-NET architecture, well know model broadly used to denoise images [4], and recently used to denoise wrapped phase maps for phase quantitative evaluation and SAR images [5].

The U-Net architecture is a convolutional encoder – decoder with internal connections between encoder and decoder paths (skip connections) with a well-proven performance when it is applied to denoise problems. Is based in one encoder path where 2D-Convolutional (Conv2d) and maxpooling layers are combined to reduce the spatial resolution of the input image capturing the image detail (features) with spatial filters in the Conv2D layers generating feature channels. Once the image has been fully encoded, there is a decoder path reversing the encoding operations.

The significant increase of performance in U-Net architecture over the traditional Convolutional Neural Networks (CNN) is the use (in the decoder path) of skip connections to improve the performance and minimise the training time. The skip connections are connections joining the outputs of the encoding steps with the equivalent decoding layer, introducing part of the encoding information to minimise the information loss in the decoding layers.

2.1 Model Design

The U-Net layout implemented has the following blocks, CD(16)-CD(32)-CD(64)-CD(128)-C(256)- CT(128)- CT(64)-CT(32)-CT(16) where the abovementioned building blocks are the following

* Corresponding author: jose.crespo@me.com

CDXX	Conv2d [XX Filters] + BatchNormalisation + RELU + Conv2d [XX Filters] +BatchNormalisation + RELU + MaxPooling(2,2) + Dropout(0.25)
CXX	Conv2d [XX Filters] + BatchNormalisation + RELU + Conv2d [XX Filters] + BatchNormalisation + RELU
CTXX	Conv2dTraspose [XX Filters, strides=(2, 2)] + Concatenate [CDXX Output] + Conv2d [XX Filters] + BatchNormalisation + RELU + Conv2d [XX Filters] + BatchNormalisation + RELU

2.2 Training

The most relevant aspect designing the applications of NN models to physical problems is the design of the training set. Using Zernike polynomials we followed the method designed in [6] to generate a dataset of 10.000 train images with size 128x128 pixels, based in randomly generate the coefficients for the first 14 Zernike polynomials. This set was the ground true for the inputs and for the noisy images in the training dataset, we included an additive and random noise component based on a normal distribution to align the noisy images to the typical images obtained in speckle pattern interference.

The training procedure for the model was implemented using `ADAM` optimisation [7] and `MEAN_SQUARED_ERROR` as the loss function; the training set was divided into training and validation (80% and 20%). The training process automatically stopped due to the early stop condition after 28 epochs, reaching a loss value of 0.0099 on the validation set.

2.3 Results

The trained model was applied to a new generated set of noisy images, not included in the training nor validation dataset. A sample of the obtained clean images were compared with the results of a denoise procedure using a median filter with a 3x3 kernel (**Figure 1**). The model improved the visibility of the fringes, removing the generated noise and increasing the visibility of the interference fringes. Structural Similarity Index (SSMI) [8] is the selected metric for the quantitative assessment of the perceived increase of quality across the reconstructed image.

(a)	(b)	(c)	(d)
	SSMI 0.15831	SSMI 0.86957	SSMI 0.14955

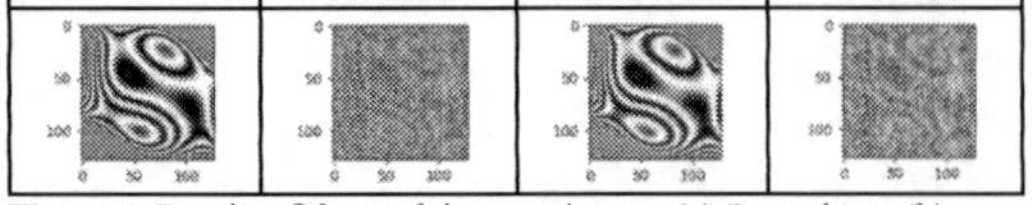

Figure 1. Results of the model on test images.(a) Ground true;(b) Noisy image; (c) Denoised image with trained model; (d) Denoised image using a Median Filter with 3x3 kernel

3 Conclusions and Next Steps

The designed NN, based in U-Net architecture significantly increases the perceived quality of interference fringes, improving fringe visibility under varying noise conditions. The overall process can be implemented in cheap hardware GPU architectures, contributing to decrease the costs of interferometric equipment related to image processing.

Application to real ESPI experimental images will be done after the COVID-19 confinement.

Acknowledgments

This research was supported by the Consolidation and Structuring of Competitive Research Units-Competitive Reference Groups (ED431C 2018/49) funded by Xunta de Galicia, and the Spanish Ministry of Economy, Industry and Competitiveness, R & D National Plan BIA2017-86738-R, with EU FEDER funds.

Authors wish also to acknowledge the financial support of this work by Xunta de Galicia, Conselleria de Educación, Universidades e FP (grant GRC No. ED431C2018/11) and the contract project No.220/01516 USC-FEUGA-ITERA (Merasys).

Data Availability

A test dataset, script and the model definition to check the results with a sample subset have been deposited in the GitHub repository:
`https://github.com/jmcrespoc/uNetFringeDenoise`

References

1. Q. Kemao, *Two-dimensional windowed fourier transform for fringe pattern analysis: Principles, applications and implementations*, Opt. & Las. Eng., vol. 45, no. 12, pp. 1146–1192, 2007
2. H. Wang and Q. Kemao, *Comparative analysis on some spatial domain filters for fringe pattern denoising*, Appl. Opt. 50(12),
3. Ronneberger, O., Fischer, P., & Brox, T. *U-net: Convolutional networks for biomedical image segmentation*. International Conference on Medical image computing and computer-assisted intervention. 2015.
4. Heinrich, M.P., Stille, M. and Buzug, T.M. *Residual U-net convolutional neural network architecture for low-dose CT denoising*. Current Directions in Biomedical Engineering, 4(1), pp.297-300.
5. Lattari, Francesco, et al. *Deep learning for SAR image despeckling*. Remote Sensing 11.13, 2019
6. Ketao Yan, Yingjie Yu, Tao Sun, et al. *Wrapped phase denoising using convolutional neural networks*. Optics and Lasers in Engineering, 2020
7. Diederik P. Kigma, Jimmy BA. *Adam: A method for stochastic optimization*. 2014.
8. Zhou Wang, A. C. Bovik, H. R. Sheikh and E. P. Simoncelli. *Image quality assessment: from error visibility to structural similarity*. IEEE Transactions on Image Processing. 2004

https://doi.org/10.1051/epjconf/202023806010

Characterization progress of a UV-microscope recently implemented at the PTB Nanometer Comparator for uni- and bidirectional measurements

Jan Krüger*, *Rainer* Köning, and *Bernd* Bodermann

Physikalisch-Technische Bundesanstalt (PTB), Bundesallee 100, 38116 Braunschweig

Abstract. In this contribution, we emphasize the need for a sophisticated characterization of measurement devices in particular for optical bidirectional measurements. As an example, the ongoing characterization of the UV-microscope at PTB's linescale comparator is presented. First results of spectroscopic measurements of the employed UV-LED are presented and the impact of deviations in the center wavelength of the LED on the measurements are illustrated by rigorous simulations.

1 Introduction

The steadily increasing demand for precisely manufactured micro- and nanostructured objects stresses the importance of optical dimensional inspections systems. They are significantly faster than competing methods such as tactile probing, are non-destructive, and do not contaminate the sample. For dimensional inspections, unidirectional and bidirectional distance measurements must be distinguished. Distance measurements in between two arbitrary features on an object such as the pitch between two lines are described as unidirectional measurements while distances in between two opposing edges, like line widths or diameters, are referred to as bidirectional measurands. Bidirectional measurements in optical microscopy always require numerical simulations of the microscopes imaging process for further evaluations, because the measured image profiles are influenced by interference and diffraction. The simulations are necessary to locate correctly and accurately, in the measured data, the positions of the physical edges of a structure under test. Therefore, in the most common and easiest way, the intensity level at the position of the edge, subsequently referred to as threshold, is identified in the simulated profile. The transfer of the threshold to the measured line profile allows to determine the correct edge position. However, the simulations are highly dependent on the illumination and imaging parameter of the used microscope and the properties of the measurement object [1]. Hence, investigations of the microscope parameters are inevitable when accurate bidirectional measurements are desired. In section 2, the UV- microscope at PTB's linescale comparator, the so-called *Nanometer Comparator* is introduced. Then, first results of the characterization are presented and discussed in section 3.

2 Characterization setup

The Nanometer Comparator is an interferometry-based length comparator at PTB. Until now, it is used to measure one dimensional, unidirectional distance on objects with length graduations. The Nanometer Comparator performs position measurements on objects up to a size of 550 mm with measurement uncertainties of some nanometers. Typical tasks of the Nanometer Comparator include calibrations of interferometers, line scales, photomasks, and encoders [2]. This field of activity is expanded by bidirectional width measurement through the implementation of a more suited and well characterized UV-microscope. The reflected light microscope originates from a commercial mask metrology tool for the inspection of photomasks. It is equipped with two objectives with a respective NA of 0.55 and 0.9. The total magnification of the microscope is 200x for both objectives. Here an UV-LED is used as light source. The emission is transferred with an optic fiber to the setup. To facilitate the characterization process, the microscope was

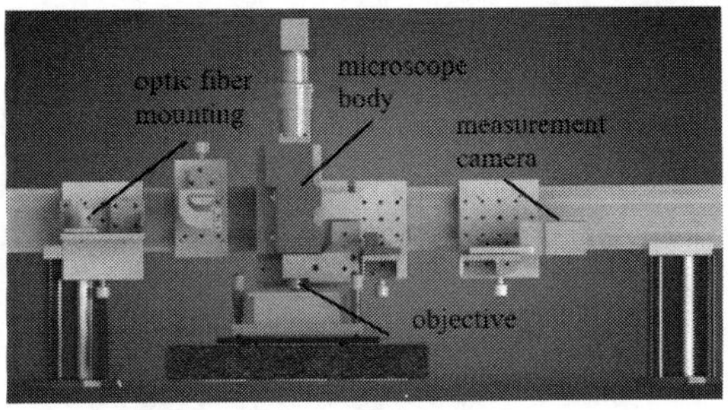

Fig. 1. Illustration of the UV-microscope in its X95 configuration

* Corresponding author: jan.krueger@ptb.de

implemented in a test stand as shown in figure 1. Here, the main advantage is that the open configuration allows for the investigation of single optical planes which would be mechanically inaccessible otherwise.

3 Characterization process

3.1 Overview

For the characterization, we conduct investigations of all microscope parameters which affect the simulations. Therefore, the optical components of the microscope will be tested for their specific impact on the polarization state. Wave front deformations which may be caused by residual optical aberrations and misalignments will be determined by Shack-Hartmann sensors. Especially here, the characterization will benefit from the accessible setup since the sensor has a certain mechanical size which sometimes limits its usage. The optical aberrations obtained through the Shack-Hartmann sensor are represented in form of Zernike polynomials, which can be directly included in the simulations. Interferometric experiments concerning the coherence of the illumination source are performed, too. Here, experiments with a Michelson interferometer indicate that the coherence length of the LED is slightly enlarged by the fiber. Finally, the characterization results will be included in the simulations. The simulation results will be compared to actual microscope measurements.

3.2 Example: UV-LED wavelength

One important parameter is the emission wavelength of the employed LED, because temperature shifts and variable supply currents can influence the wavelength. Therefore, the spectrum of the LED, whose temperature is regulated at 29,5 °C, is measured at three different supply currents with a spectrometer. The results are shown in figure 2.

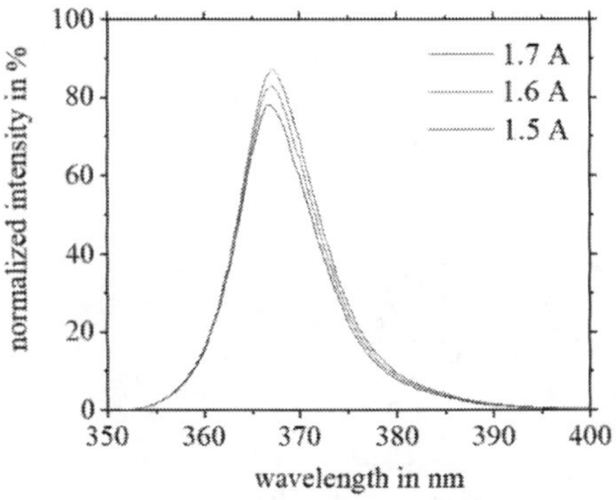

Fig. 2. LED spectrum at 29,5 °C

The center wavelength of the nominal 365 nm UV-LED is measured to be 367 nm. This does not change for the different supply currents. There is only a change in normalized intensity with a linear increase with increasing supply current as expected for a LED. The importance of the knowledge of the correct operation wavelength demonstrated in the simulations shown in figure 3. There, the image profile for one unit-cell of a binary line grating is simulated by the means of a numerical simulation tool which is based on the rigorous coupled wave analysis [3].

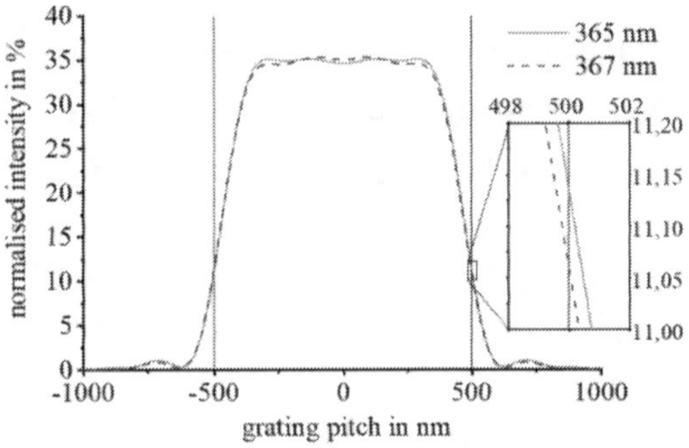

Fig. 3. Simulated profiles for two wavelengths; $NA_{ill.} = NA_{img.} = 0{,}9$; magnification = 200x

The width of the unit-cell is 2 µm. The line consists of chrome, has a width of 1 µm and a height of 100 nm, and is applied on a glass substrate. The illumination and imaging parameters for the simulations are adapted from the microscope.

Even though, the difference between the two wavelengths is small, there are deviations between the two normalized profiles. However, they nevertheless lead to a small but measurable variation in the threshold values derived at the positions of the simulated edges indicated by two vertical lines in the figure. The thresholds at both edges are equivalent for the respective profiles since the line grating is symmetrical. A wavelength of 365 nm leads to a threshold value of 11.14 %, while a wavelength of 367 nm yields a threshold of 11.06 %. This difference in the thresholds is quite small but would already lead to a measurement error of 0.7 nm for both objectives, if not taken into account. In conclusion, it must be acknowledged that the wavelength is just one of many parameters which are considered in the simulation. The difference between the simulations with the assumed parameters based on manufacturer information for example and the simulations with the parameters deduced during the characterization is expected to be more significant. In the final contribution, we will further elaborate on the characterization of UV-microscope at the Nanometer Comparator and on the results shown here.

References

[1] B. Bodermann et al., *Fringe 2013* (Springer-Verlag, Berlin Heidelberg, 2014)

[2] H. Bosse, G. Wilkening, Meas. Sci. Technol.**16** 2155 (2005)

[3] MicroSim, rigorous simulation tool for diffraction problems, developed by ITO, University of Stuttgart: https://www.ito.uni-stuttgart.de/en/research/group-hms/microsim/ (accessed on: 18.03.2020)

https://doi.org/10.1051/epjconf/202023806011

Absolute distance interferometry for subaperture stitching of large freeform optics

Jan Spichtinger[1,*], *Michael* Schulz[1], and *Gerd* Ehret[1]

[1]Physikalisch-Technische Bundesanstalt, Braunschweig, Germany

Abstract. We are developing a form measurement system for the surfaces of freeform optics and large conventional optics. The specifications of the optics are diameters up to 1.5 m and radii of curvature down to 10 m. This includes optics like telescope mirror segments and synchrotron optics. Using a Fizeau interferometer, we propose a subaperture stitching method that involves vertically aligning the interferometer's optical axis to the local surface gradient and measuring the absolute distance from the interferometer's reference flat to the specimen. Experimental results for the absolute distance measurement are shown.

1 Introduction

Measuring the form of an optical surface with a Fizeau interferometer with an aperture smaller than the specimen's diameter makes it necessary to use subaperture stitching [1]. Subaperture stitching is often performed by fitting the offset and the tilt of the subaperture topographies. There are many methods suitable for non-flat optics of different types (e.g. for aspheres, cylinders [2] or freeforms [3]). For arbitrary freeform surfaces with a non-negligible curvature over a large diameter as specified in the abstract, stitching is in general a three-dimensional process. First of all, the interferometer has to be perpendicularly oriented to the local specimen's gradient at each measurement position. By measuring the topography of each subaperture and also the distance from the interferometer's reference flat to the specimen, we will position the topographies from the subaperture measurements in a global coordinate system before stitching. In this way, the stitching process becomes a linearized problem without the need for lateral image registration or computationally intensive multi-dimensional stitching.

2 Methods

2.1 Setup of the form measurement system

Our measurement setup [4] will consist of a Fizeau interferometer with a flat reference surface and a specimen, which are mounted on a motion system (see figure 1). The interferometer is attached to a gimbal, which can be moved up and down with a mechanical z-axis. The z-axis is carried by a linear air-bearing x-axis (1500 mm travel range) on a granite portal. The specimen is placed on a tilt table on top of a rotation table, which is also mounted on an air-bearing linear y-axis of 1500 mm in length. Thereby, the Fizeau interferometer can be perpendicularly oriented to the local specimen surface at any measurement position. The interferometer uses a wavelength shifting laser diode at 633 nm (referenced on an iodine cell) with a scan range of 250 GHz (referenced on an etalon).

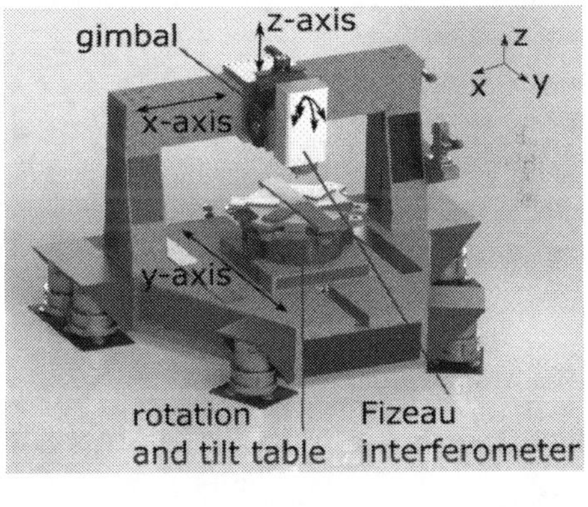

Figure 1. Mechanical setup of the measurement system under development: The arrows represent the possible movements of the Fizeau interferometer and the specimen. The specimen (not shown) is placed on the rotation and tilt table.

2.2 Topography and absolute distance measurement

At each measurement position, a wavelength scan is performed with the laser diode of the Fizeau interferometer. During the scan, the intensity changes are measured for each pixel of the interferometer camera (see figure 2). By means of Fourier analysis, the relative topography is obtained by calculating the phase shift of each pixel and unwrapping the resulting spatially resolved phase map. The absolute distance of the interferometer reference flat to the

*e-mail: jan.spichtinger@ptb.de

specimen, i.e. the cavity length, is calculated from the signal's (temporal) carrier frequency [5]. The resolution of the frequency spectrum obtained by FFT is increased by zero padding. The accuracy of the cavity length measurement is in the order of a few 10 μm (see figure 3), while the accuracy of the relative topography measurement is around 10 nm. Only the central pixels of the interferometer camera are used to determine the cavity length.

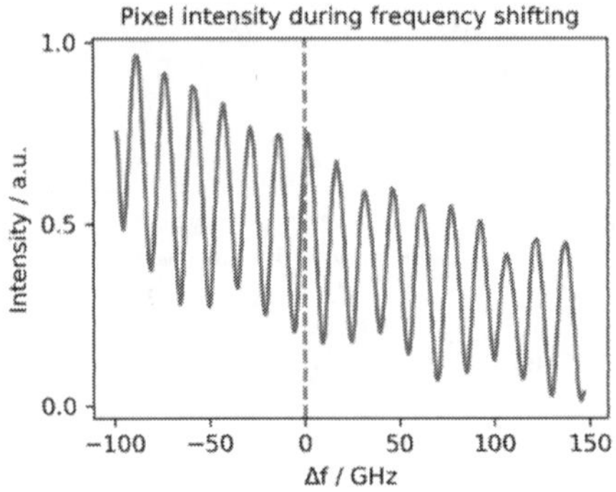

Figure 2. Wavelength shifting interferometry: Measured intensity of a specific pixel in dependence on the frequency shift Δf of the laser diode, relative to the centre wavelength of 633 nm. The carrier frequency corresponds to the cavity length.

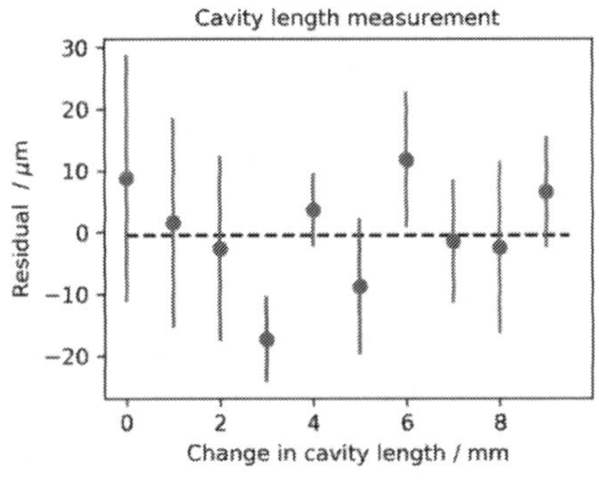

Figure 3. Verifying the absolute distance measurement: A cavity consisting of the Fizeau reference flat and a flat specimen was set up. The cavity length was changed, and the change was measured with a commercial laser interferometer with nanometre precision. The absolute cavity lengths were also measured with the Fizeau interferometer. The figure shows the residuals of the change in cavity length calculated from the absolute distance measurement in comparison to the laser interferometer measurement. Each data point corresponds to the average of 20 measurements with its standard deviation shown as an error bar.

2.3 Using the absolute distance for stitching

With the relative topography measurement only, the subtopography could be located anywhere along the optical axis of the interferometer. Having many subaperture measurements with non-parallel optical axes leads to a non-linear problem in the case of global stitching. In the case of sequential stitching, the challenge is to position the first subaperture.

Positioning the topographies in a global coordinate system and fitting only in the z-direction instead of along the optical axis (see figure 4) leads to a lateral positioning error Δr of the subaperture topography:

$$\Delta r = \Delta z \cdot sin(\psi)$$

with Δz: offset of the subaperture determined by the fitting process and ψ: the total angular displacement of the Fizeau interferometer relative to the z-direction of the fit. For the specimens specified at the beginning, the maximal absolute angle is $\psi = 4.3°$. Therefore, the maximal lateral error will be $\Delta r_{max} = 0.075\Delta z$. Since Δz is in the range of a few 10 μm, Δr is in the range of a few μm.

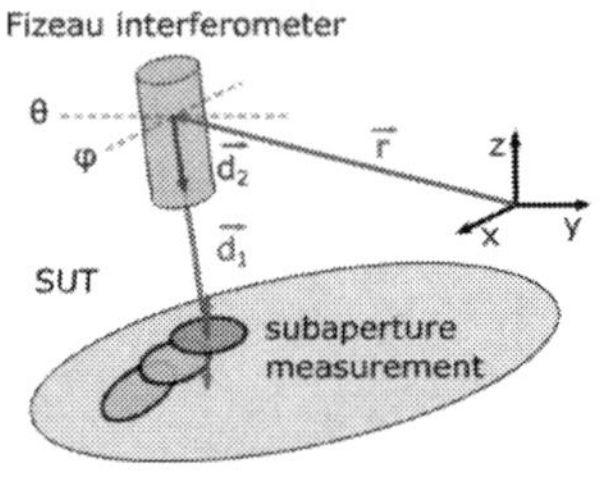

Figure 4. Positioning the topographies from the subaperture measurements: The Fizeau interferometer position $(\vec{r}, \phi, \theta)$ relative to the local surface under test (SUT) is known from the motion system. The subtopographies are positioned with the measured absolute distance $\vec{d_1}$ from the reference flat to the SUT and the distance $\vec{d_2}$ from the pivot point to the reference flat. The green arrow depicts the z-direction in which the topographies are fitted.

3 Outlook

Our next step will be to apply the stitching method with the absolute distance measurement to real measurement data obtained from a measurement system with a smaller movement system than the one presented. The lateral error caused by our approximation of only fitting in the z-direction remains small compared to the typical pixel size of a Fizeau interferometer. This approach is useful as long as the tilt angle range of the interferometer is small enough for the lateral errors to stay within an acceptable limit.

References

[1] P. Murphy et al., Opt. Photon. News 14, 38–43 (2003)
[2] J. Peng et al., Meas. Sci. Technol. 26, 025204 (2015)
[3] C. Supranowitz et al., Proc. SPIE, 1044818 (2017)
[4] J. Spichtinger et al., Proc. DGaO 2019, ISSN:1614-8436
[5] A. Cabral, J. Rebordão, Optical Engineering 46(7), 073602 (2007)

Thinking Camera – Powered by the CAOS Camera Platform

Nabeel A. Riza [1,*]

[1]School of Engineering, University College Cork, Cork, Ireland

Abstract. Introduced is the system level design of a *Thinking Camera* powered by the CAOS camera platform. The proposed camera provides features such as extreme linear dynamic range and full spectrum operations with selectable space-time-frequency CAOS pixel modes using active and passive light. The imager has triple output ports design and is controlled via application-specific classical image processing, machine learning techniques, and human/user feedback. Imaging capabilities cover multi-dimensional mappings allowing diverse full spectrum (350 nm to 2700 nm) applications from industrial metrology to quantitative medical imaging to artefact preserving colour photography.

1 Introduction

Advances in electronic, optical, opto-electronic, opto-fluidic, & MEMS devices coupled with developments in high speed and energy efficient electronic processors have brought about a revolution in compact and handheld sensors impacting both the industrial and commercial worlds. Mass production of these devices and processors along with advances in packaging technologies has brought price points down to consider the realization of a modular design *thinking camera* for multi-function operations empowered by both classical image processing methods as well as modern machine learning algorithms to optimize high efficiency camera controls for a specific application. Hence the unique power of the thinking camera lies both in the raw physics-based image data extraction powers of the modular optical hardware as well as the mathematical basis of the smart camera controls provided by the high speed implementable algorithms optimized per application using training samples. It is important to stress the limitations caused by the intrinsic properties of light and the device-physics of both the light-to-electronic signal conversion as well as the electrical signal flow and processing electronics, all operating with noise levels. Hence an adequate Signal-to-Noise Ratio (SNR) of optimally designed and extracted raw image pixel signal data is required before Digital Signal Processing (DSP) image processing and machine learning can be useful for extraction of robust image data. Recently, such a novel image pixel signal conditioning and extraction technique-based camera has been proposed and demonstrated that works on the principles of an RF wireless mobile phone network where image pixels are treated as time-frequency coded mobile phone-like handsets. This camera is called the Coded Access Optical Sensor (CAOS) camera [1] and forms the foundational modular platform to design the proposed thinking camera system. The CAOS smart camera is currently designed using cost-effective technologies that includes silicon CMOS/CCD image sensors, point detectors, TI Digital Light Processing (DLP) technology Digital Micro-Mirror Device (DMD) chips, lenses, mirrors, etc.

2 Thinking Camera Design

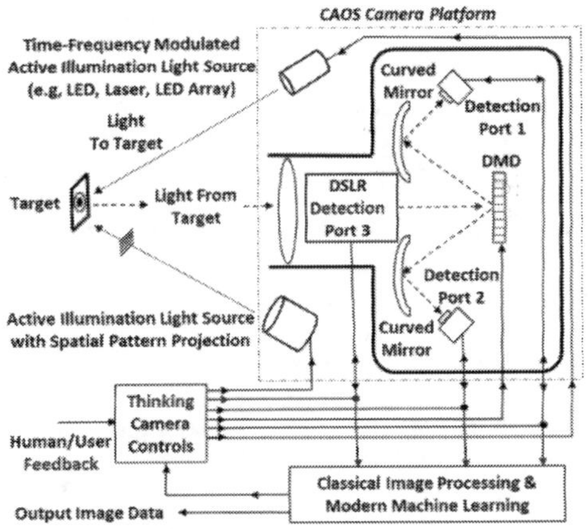

Figure 1 *Thinking Camera* Design empowered by the CAOS Camera Platform.

The CAOS camera 3-port platform used to design the Figure 1 *Thinking Camera* is highly programmable. This extreme programmability also presents a camera system controls challenge, thus requiring a "thinking" camera design for an efficient camera controls solution. For a selected application, specific imaging features of varying weights are desired that can be computed using an application specific image data set produced from both real and synthesized images. Given these weighted features (e.g., with a grey scale zero to 1 weighting where a 1 weight implies feature is extremely important and a zero weight indicating that a feature is not at all

* Corresponding author: n.riza@ucc.ie

important), an optimization process based on classic image processing as well as modern machine learning techniques is deployed to select the best deployed camera mode/modes. The optimization system is trained using a known data set of camera features and modes. In addition, given the mode/modes produced by the optimization process, next a best set of control parameters is required for efficient camera operation. These control mechanisms can be computed using a second optimization process using classical and machine learning techniques and human feedback. Hence pre-processing is implemented for the camera using different application scenario test data given the next described Thinking camera imaging features, modes, and control mechanisms. Specifically, the CAOS camera platform [2-5] can provide powerful *Thinking Camera* **imaging features** such as:

- 177 dB Extreme Dynamic Range (XDR)
- Wide Spectral (350 nm to 2700 nm) Access
- Linear Camera Response Function (CRF) over Full XDR
- Low Contrast Inter-Pixel Recovery within XDR
- Narrow Hyperspectral Access
- 1-D (x), 2-D (x,y), 3-D(x,y,z), 4-D(x,y,z,t), 5-D space-time-wavelength image data mapping modes via coded space-time signalling
- Fast Pixel/s Imaging Speed
- High Adjacent Pixel Isolation
- Bright Light Imaging

CAOS pixel irradiance data harvesting is done using DMD chip pixel-level programming control by implementing various CAOS modes [6]. These CAOS modes using active and/or passive light control are similar in design to the time-frequency modes deployed in the RF wireless multiple access mobile phone network. These CAOS modes are called TDMA, CDMA, FDMA, FM-CDMA, FM-TDMA, CDMA-TDMA, FM-CDMA-TDMA, FDMA-TDMA, FDMA-CDMA, and FDMA-CDMA-TDMA [11]. In addition, there are hybrid camera modes such as CAOS-CMOS mode combining point detector and multi-pixel detector space-time modulation operations. In short, the CAOS camera platform provides powerful *Thinking Camera* **imaging modes** such as:

- Passive Illumination Mode
- Active Illumination modes with Power, Spatial (e.g., Fringe Pattern) and Time/Frequency Controls & Optical Interferometric Imaging
- 10 Time-Frequency Pixel/s CAOS-modes
- Hybrid CAOS-CMOS/CCD/Lidar Pixel Modes
- Super Resolution (Sub-Micromirror)

The CAOS platform control signals connect to the specific hardware components that include DMDs, point detectors, multi-pixel image sensors, DSLR mirror/tilt prism/grating/slit opto-mechanics, and active light sources, etc. To summarize, the CAOS camera platform provides powerful *Thinking Camera* **control mechanisms** such as:

- Pixels of Interest (POI) Location Control
- Region of Interest (ROI) Location Control
- POI & ROI Control in Large Space-Bandwidth-Product, e.g., 1cm^2 area with 10^6 points
- Front Lens-free Imaging Control
- Harvested Pixels Shape Control (circular pinhole, rectangular slit, ellipse, etc)
- Pixel, POI, ROI Acquisition Time Controls
- Moderate Bandwidth Point Photo-Detection Electronic Gain/Bandwidth Control
- Digital Signal Processing (DSP) gain controls
- Multi-pixel full frame image sensor Frame Rate and Gain Controls
- Simultaneous Multi-Port Image Fusion and Noise Cancellation Control
- Pixel Spatial Light Rejection Controls for a Chosen Light Detection Port
- Coherent (with Reference) or Incoherent (Direct) Optical Photo-Detection and Electronic Incoherent/Coherent Signal Detection Controls
- DSLR Mirror for Scanning Grating/Slit Image Sensor & CAOS Hyperspectral Imaging Controls
- DSLR Tilt Prism Option for spatial image dither super-resolution imaging Controls

3. Conclusion

Proposed is the fundamental design and operation of a *Thinking Camera*-based on the CAOS camera platform. This highly programmable camera provides powerful imaging capabilities and requires a 3-faceted "*thinking*" approach to optimize efficient and smart camera operations for specific applications. It relies on classical as well as modern machine learning techniques using processing parameters of desired imaging features, camera modes, and control mechanisms. It also incorporates as needed, human/user feedback for camera controls. Future work includes full design and demonstration of the proposed *Thinking Camera*.

References

1. N. A. Riza, Coded Access Optical Sensor (CAOS), USA Patent 10,356,392 B2, July 16, 2019.
2. N. A. Riza, "The CAOS Camera Platform – Ushering in a Paradigm Change in Extreme Dynamic Range Imager Design," Invited Paper SPIE OPTO, Vol. 10117, Photonics West, Jan. 30, 2017.
3. N. A. Riza, M. J. Amin, and J. P. La Torre, " Coded Access Optical Sensor (CAOS) Imager," *Journal of the European Optical Society (JEOS) Rapid Publications*, 10, pp. 150211-8, 2015.
4. N. A. Riza, "Coded Access Optical Sensor (CAOS) imager and applications," SPIE Photonics Europe Conf. Proc., Vol, 9896, paper No.9, April 2016.
5. N. A. Riza and M. A. Mazhar, "177 dB Linear Dynamic Range Pixels of Interest DSLR CAOS Camera," IEEE Photonics Journal (open access), Volume 11, Number 2, April 2, 2019.
6. N. A. Riza and M. A. Mazhar, "Laser beam imaging via multiple mode operations of the extreme dynamic range CAOS camera," Appl. Opt., Vol. 57, No.E20, 2018.

An optical metrology system for the measurement of the refractive index of glass

Inês Leite[1,2*], and *Alexandre* Cabral[1,2]

[1] Instituto de Astrofísica e Ciências do Espaço, Universidade de Lisboa, Campo Grande 1749-016 Lisboa Portugal

[2] Faculdade de Ciências da Universidade de Lisboa, Departamento de Física, Campo Grande 1749-016 Lisboa Portugal

Abstract. The measurement of the refractive index of parallel plated, optically simple, glass samples is a common and fundamental activity in numerous fields of expertise. This work aimed to optimize a known technique to a simple, cost-effective and reliable system to be implemented in a lab environment, with an accuracy in the results better than 10^{-2}. A setup with a 632.8nm HeNe laser, automatic stage and CMOS camera was used and data was acquired with the help of LabVIEW controlling software. All system components were carefully controlled and optimized with the help of an uncertainty budget. Measurements had an associated uncertainty in the range from 10^{-3} to 10^{-4}.

1 Introduction

The determination of the refractive index depends on simple concepts of optics [1], and there are several well-established methods [2] of measurement of the refractive index (RI) of solid samples. Depending on several factors (type of sample, available equipment, desirable results, and so on) the implemented technique will drastically differ. A state-of-the-art study was conducted on the available methods to implement in order to build a simple, cost-effective and automated system capable of obtaining the refractive index of optically simple samples.

1.1 Analysis of candidate techniques

The analysis of several existing techniques was conducted and some key comparisons were distinguished: the method and technique used, the uncertainty obtained in RI measurement, the complexity of the method and the possibility of optimization – that is, the possibility of altering and optimizing an already existing technique in order to fulfil the set objectives. Based on desired system characteristics, two main methods were considered: Lateral displacement and interferometry.

1.1.1 Lateral displacement vs interferometry

The lateral displacement method relates an incident light beam that crosses a transparent solid sample to the exiting beam [3]. Typically, the incidence angle is known, as well as the wavelength of the beam – to determine the refractive index of the material, usually the exit angle and/or beam displacement are recorded and through notions of geometry and algebra the refractive index of the material can be known using Eq. 1.

$$n_{glass} = \sqrt{\left(\frac{\cos\theta_i}{1-\frac{d}{t\sin\theta_i}}\right)^2 + \sin^2\theta_i \, . \, n_{air}} \qquad (1)$$

Another method for determining the RI of a sample is to employ a simple Michelson interferometer to measure the difference in optical path length created by rotating the glass plate located in one of the arms. Considering a similar setup as the one illustrated in Fig. 1, and replacing the camera with the mirror measurement arm, the refractive index can also be determined through a specific equation [4].

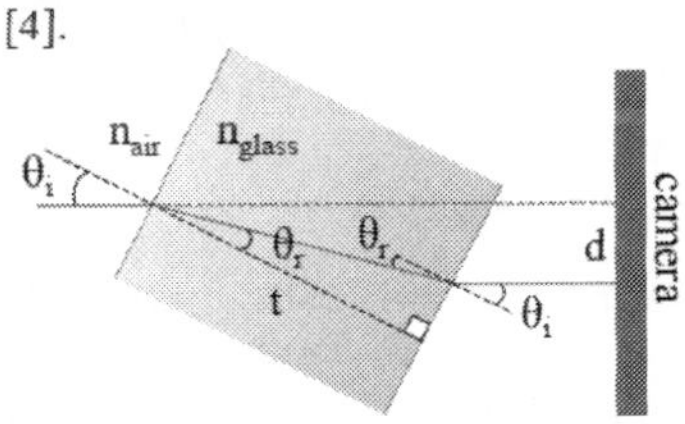

Fig. 1. Lateral displacement *d*, in a camera, of the beam with incident angle θ_i, by a glass sample of thickness *t*.

After building a theoretical model of the performance of both techniques and conducting simple experimentation in a setup with similar conditions it was determined that, while in theory the interferometer, with its higher sensitivity and accuracy is capable of providing results with much better uncertainty, due to its relative measurement principle and instability, experimentally, it proved more complex to implement, which critically deviated from the objectives. The lateral displacement

* Corresponding author: ines.leite995@gmail.com

method, a technique capable of providing absolute measurements, observed to be insensitive to vibrations and of easier implementation, was selected instead.

2 Refractive index of glass by optimized lateral displacement

The final setup of the system is displayed in Fig. 2. The working principle of the system is as follows: The beam that exists the laser will perpendicularly cross the parallel plated sample mounted in the sample holder and a spot image, the zero position, will be captured by the 2D sensor, the camera. The sample holder, mounted on the automatic stage will then rotate symmetrically in the y axis to both sides of the perpendicular position and the controlling software will record the spot position in every rotation step, performing one measurement cycle.

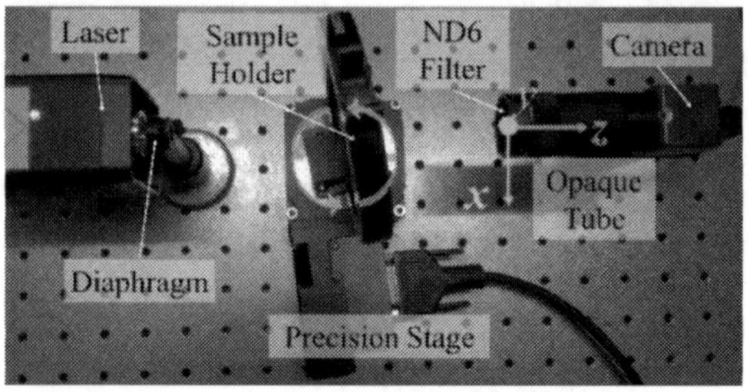

Fig. 2. Final system setup with an HeNe laser (632.8 nm), automatic 1-axis rotation stage and CMOS camera.

It was theorized, and later experimentally verified, that the value of the RI becomes unstable as the incident beam nears the perpendicular position to the sample surface, therefore more measurements statistically improve the average of the obtained results. For every position, the spot centre xy coordinates are determined, which in turn are associated to a specific incidence angle, θ_i. The deviation from the zero position, lateral displacement d, is calculated and by using Eq. 1 and knowing the sample thickness, t, multiple values of the refractive index, for the same sample, are obtained.

2.1. Uncertainty budget

A useful approach, that greatly benefits the effort to optimize the system, is the construction of an uncertainty budget that encompasses every parameter capable of affecting the uncertainty of the refractive index of glass. Each parameter can have more than one source of uncertainty associated to it, and each source often has a different weight in the final uncertainty of each parameter. Fig. 3 gives a global overview of the main parameters, d, θ_i and t, as well as each source origin and weight contribution to the uncertainty in RI measurements.

2.2 Results

By carefully monitoring the sources of uncertainty, the system was now ready to be validated. Two known

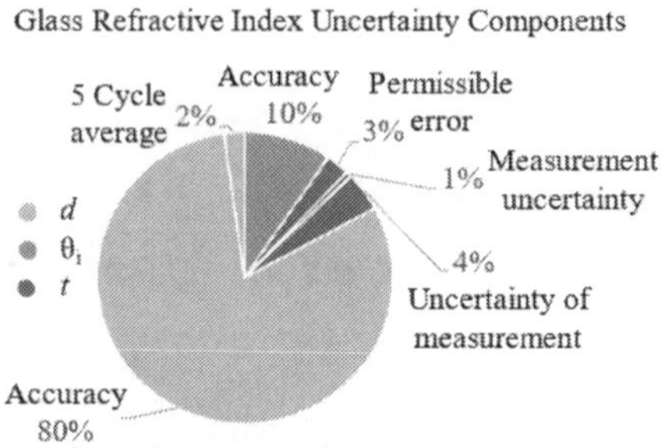

Fig. 3. Uncertainty budget of the built system and every contribution the measurements uncertainty by the three main parameters d, θ_i and t.

calibrated samples, N-BK7 and UVFS, were first used to assess the performance of the system and its reliability. Obtained results achieved an uncertainty in the order of 10^{-4} for calibrated samples and 10^{-3} by two other known samples, of inferior optical quality, as displayed in Table 1.

Table 1. System results of measurements of the RI to four samples: N-BK7 and UVFS (calibration) and B270 glass and a piece of a block of acrylic.

Sample	Exp. RI	Meas. RI	± U	U (%)	|error| (%)
N-BK7	1.5151	1.5151	0.0012	0.08	< 0.01
UVFS	1.4570	1.4571	0.0010	0.07	0.01
B270	1.5211	1.5219	0.0052	0.34	0.05
Acrylic	1.4890	1.4887	0.0016	0.11	0.02

The system works and is ready to be used by the lab, but improvements can always be made. Throughout the work, an interference pattern could not be mitigated due to the nature of the coherent light source – its substitution by an incoherent light source should improve measurement accuracy of the images captured by the CMOS camera, linked to the lateral displacement, the parameter with the greatest contribution to the RI uncertainty.

This work was supported by Fundação para a Ciência e a Tecnologia (FCT) through the research grants UIDB/04434/2020 and UIDP/04434/2020 and the participation was supported by Sociedade Portuguesa de Óptica e Fotónica (SPOF).

References

1. H. Bach, N. Neuroth, *The Properties of Optical Glass* (Springer-Verlag Berlin Heidelberg 1998)
2. S. Singh, *Phys. Scr.* **65**, 167–180 (2001)
3. V. Kolchinskiy, C. H. Shih, I. Lo, R. Romashko, *Phys. Proc.* **86**, 176–180 (2017)
4. I. Leite, A. Cabral, *Proc. SPIE* **11207** (2019)

 https://doi.org/10.1051/epjconf/202023806014

Super-resolution for 2.5D height data of microstructured surfaces using the vdsr network

Stefan Siemens[1,*], *Markus Kästner*[1], and *Eduard Reithmeier*[1]

[1]Leibniz University Hannover, Institute of Measurement and Automatic Control, Nienburger Straße 17, 30167 Hannover, Germany

Abstract. In this work super-resolution imaging is used to enhance 2.5D height data of thermal sprayed Al_2O_3 ceramics with stochastically microstructured surfaces. The data is obtained by means of a confocal laser scanning microscope. By implementing and training a Very Deep Super-Resolution neural network to generate residual images an improvement of the peak signal-to-noise ratio and structural similarity index can be observed when compared to classic interpolation methods.

INTRODUCTION

Advances in manufacturing lead to increasing demands on the quality and durability of technical components. The topography of a components surface such as the microstructure a have major impact on its performance. Surface metrology therefore plays an important role in the research and development of technical components. In recent years, optical measurement methods have increasingly been used for surface metrology as they allow for a contactless and fast measurement of the topography. With regards to measuring microstructures confocal microscopes are among the most widely used measuring instruments [1]. Generally, it is desirable to measure with the highest possible resolution to capture small structures. At the same time a large area should be measured in order to obtain reliable results. These two objectives are in conflict with each other, since lenses with a higher magnification and resolution usually have a smaller field of view. One approach to circumvent this is super-resolution imaging, where low-resolution (LR) images are upscaled to a higher resolution (HR) to obtain fine details whilst keeping a large field of view.

BACKGROUND

Mathematically the degradation of the HR image (denoted by x) to the LR image (denoted by y) can be modelled by the degradation function Φ and the degradation parameters θ_η [3]:

$$y = \Phi(x, \theta_\eta). \tag{1}$$

The goal of the super-resolution algorithm is to find a function $\Phi^{-1}$and parameters θ_ς that invert the degradation function to get an estimate of the HR image $\hat{x}$:

$$\hat{x} = \Phi^{-1}(y, \theta_\varsigma). \tag{2}$$

As Φ and θ_η are unknown and different HR images can result in the same LR image this task is an ill posed problem. There are many approaches to generate a good HR approximation [2, 3]. The most widely used are classic upsampling methods like the nearest-neighbor, bilinear or bicubic interpolation. In recent years different machine learning approaches have been developed, for example linear networks or GAN models [3].
In this work we chose to train a Very Deep Super-Resolution (VDSR) neural network, as it has the advantage of being relatively easy to train whilst yielding good performance metrics. It was published by Kim et al. in 2016 and is inspired by the VGG-Net [4]. The VDSR network uses a pre-upsampling approach to super-resolution. This means that the input is not the LR image but an interpolated low-resolution image (ILR) which is the LR image interpolated to the dimensions of the HR image using classic upsampling methods [4]. The network then learns to generate the residual mapping between the HR and ILR image. This approach simplifies the training objective as the network only has to fill in high frequency information. This in turn allows for the training of deeper networks compared to earlier works like by Dong et al. [5]. With exception to the first layer (which is the input layer) and the last two layers (the reconstruction and output layer) the network consists of 20 pairs of convolution layers each followed by a ReLU activation layer. The convolution layers each consist of 64 filters of the size *3x3*.

DATA & TRAINING SETUP

The examined surfaces are thermal sprayed Al_2O_3 ceramics with a stochastic pore distribution, an average pore area of 15 μm^2 and a porosity of 8 % to 9 %. The confocal laser scanning microscope (CLSM) Keyence VK-210X was used to acquire so called 2.5 D height images where each pixel of the image has a corresponding height value of the measured surface.

* Corresponding author: stefan.siemens@imr.uni-hannover.de

In total 680 height images were taken each with a size of 1024x768 pixels. 600 images were used for training and 80 for testing. The lens used has a magnification of 50x and a numerical aperture of 0.95. An example surface is shown in Figure 1.

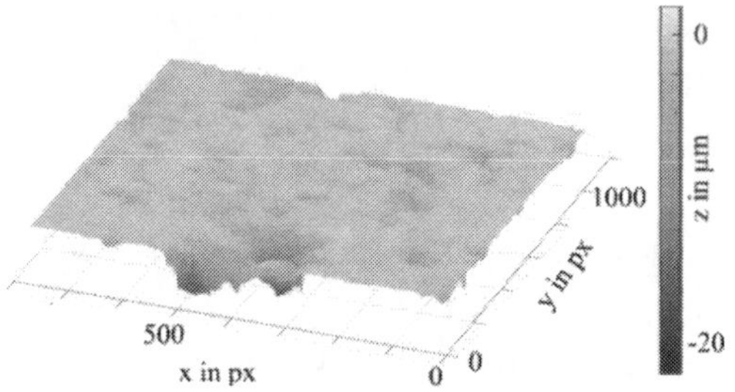

Fig. 1. 2.5D height image of a thermal sprayed Al_2O_3 ceramic.

To train the network 20 patches with a size of 128x128 pixels were extracted from each height image for a total of 12.000 training patches. For this work synthetically degraded data was used. Before being fed into the VDSR the patches were first scaled down by a factor of 2, 3, 4 or 8 and then scaled up by the same factor using bicubic interpolation. Additionally, the training data was augmented with a random rotation by 0°, 90°, 180° or 270° and a random reflection along the x- and y-axis. The network was trained for 150 epochs using a stochastic gradient descent. The hyperparameters were adopted from Kim et al. [4]. The output of the network is the predicted residual image which gets added to the ILR image. The loss function for training is given by half the MSE between ground truth and the predicted residual.

RESULTS

For testing purposes, the image data was also synthetically degraded using bicubic interpolation and scaling factors of 2, 3 and 4. The peak signal-to-noise ratio (PSNR) and the structural similarity index (SSIM) between the prediction and ground truth were calculated and averaged for all testing images. The results where compared to an upscaling of the LR image using only the nearest-neighbor, bilinear or bicubic algorithm. The results are shown in Table 1. The performance of the VDSR network exceeds the other methods in most scenarios, especially for larger scaling factors. An example is shown in Figure 2. The visual impression is consistent with the previous results.

Table 1. Testing results for different methods and scale factors.

Magnification / Method	2x: PSNR / SSIM	3x: PSNR / SSIM	4x: PSNR / SSIM
Nearest-neighbor	27.13 dB / 0.95	22.07 dB / 0.83	22.21 dB / 0.82
Bilinear	28.50 dB / 0.95	23.44 dB / 0.86	23.47 dB / 0.84
Bicubic	30.18 dB / **0.97**	23.71 dB / **0.87**	24,28 dB / 0.86
VDSR	**30.72 dB** / 0.96	**23.92 dB** / **0.87**	**25.14 dB** / **0.88**

While the VDSR results clearly outperform the nearest-neighbor and bilinear approach they are hard to distinguish from the bicubic interpolation.

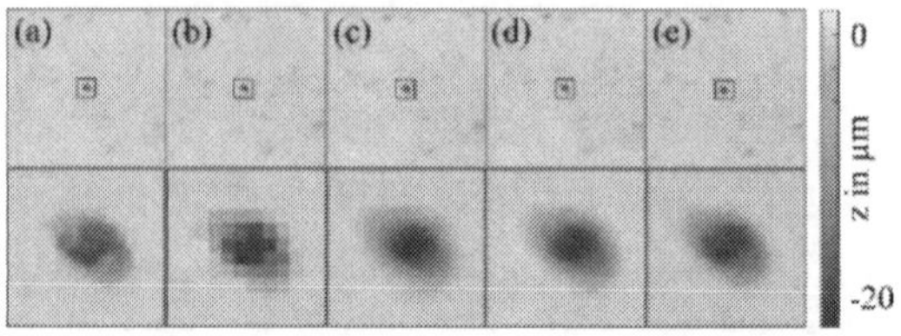

Fig. 2. Overview (top) and detailed view (bottom) of results for a 4x upscaling using different methods. Ground truth (a), nearest-neighbor (b), bilinear (c), bicubic (d) and VDSR (e).

Looking at the residual images (see Figure 3) it can be seen that the VDSR network successfully generates high frequency information which leads to the increase in PSNR and SSIM.

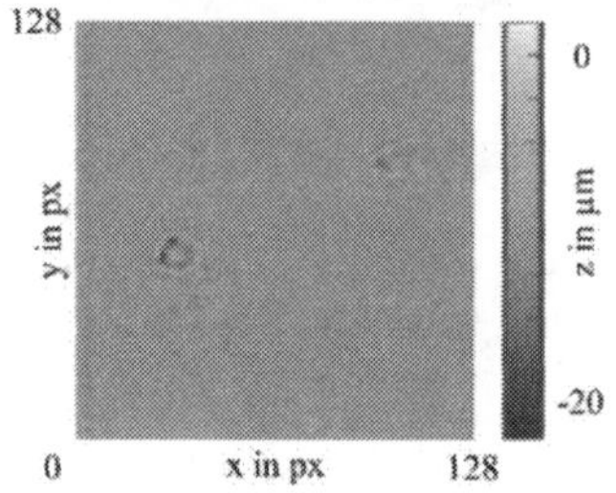

Fig. 3. Example for a residual image generated by the VDSR Net for a 4x upscaling.

CONCLUSION

This work shows that the VDSR network can be used to super-resolve 2.5D height image data acquired by means of a confocal laser-scanning microscope. It is relatively easy to train and in many cases outperforms classic interpolation methods. Limitations of this work are that real degradation processes are unknown and likely not the same as a synthetic down-sampling.

References

1. A.M. Martinez, J. de Vincente Y Oliva, *Materials*, **12(24)**, 4137 (2019)
2. W. Yang, X. Zhang, Y. Tian, W. Wang, J. Xue, Q. Liao, *IEEE Transactions on Multimedia*, **21(12)**, 3106-3121 (2019)
3. S. Anwar, S. Khan, N. Barnes, *ACM Computing Surveys*, **53(3)**, 1-34 (2020)
4. J. Kim, J.K. Lee, K.M. Lee, *2016 CVPR*, 1646-1654 (2016)
5. C. Dong, C.C. Loy, K. He, X. Tang, *Computer Vision – ECCV 2014*, **8692**, 184-199 (2014)

White light intensity stabilization for long term photometric calibrations

João Cachatra[1*], *David* Alves[1, 2], *Manuel* Abreu[1, 2]

1. Departamento de Física, Faculdade de Ciências, Universidade de Lisboa, Edifício C8, Campo Grande, PT1749-016 Lisbon, Portugal
2. Instituto de Astrofísica e Ciências do Espaço, Universidade de Lisboa, Campo Grande, PT1749-016 Lisboa, Portugal

Abstract. This paper presents an overview of the design of a voice coil based white light intensity stabilization tool. A system is created capable of mechanically stabilizing the fluctuations in a beam of light. This is achieved with a use of a voice coil and simple PID based feedback loop, controlled using a LabVIEW interface. Stability results are shown using Allan deviations curves.

1 Introduction

The detection of planetary transits requires sensors with extremely high stability. The calibration of these sensors requires a light source with stability on par or better than the goal stability of the photometer.

As an example, the measurement of the flux impinging on the detector over extended periods of exposure, or the evaluation of the detector gain over time, require the probe beam to be extremely stable so that the relevant information in the signal can be retrieved within the intensity noise of the light source.

However, light sources found in most laboratories have nowhere near the stability that is required for the current generation of photometers used in astrophysics, particularly in the detection and characterization of planetary transits. Typical broadband light sources are listed as having 1-3% fluctuations, or 10000 to 30000 ppm, for extended time periods. For example, the measurements of a Venus transit in a fixed wavelength mode and the relative variation in total irradiance require a precision of approximately 4 parts per million [1].

The main goal of this project is to develop a light unit and an intensity stabilization strategy in order to reduce its long term fluctuations, by at least one order of magnitude, so that it can be used in photometric calibrations.

2 Luminous Power Control System

Most high-precision light source procedures stabilize light at its source, meaning they try to directly control driving current and/or the temperature of the emitter source and circuitry. However, if access to the light source circuitry isn´t possible a different method has to be used. Rather than trying to control primary source parameters like current or temperature, the intended system is based on the creation of an external occlusion system, meaning that intensity and stability of the light is modulated by an attenuator that obscures dynamically the beam of light as function of its variations at source level.

By creating a feedback system we can monitor the light intensity and with that information mechanically adjust the attenuator creating a shadow in the beam path, altering the net flux between the source and the exit port and thus reducing fluctuations in the beam.

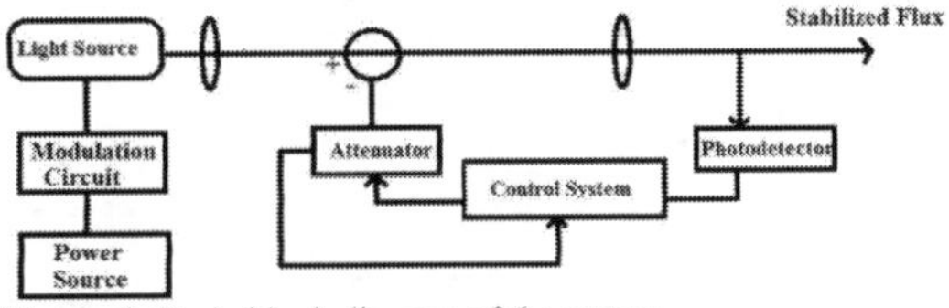

Figure 1: Basic block diagram of the system

The basic layout of this system consists of a photodetector that monitors the light intensity of the source, connected to a control system that tells the attenuator how much it needs to move to perform the necessary corrections.

2.1. Light Source

A current driven white light source is used to emulate a source with noise characteristics. This is achieved through a modulating circuit that connects the lamp to wave generator, enabling the modulation of the beam with a sinus waveform, and on a second stage, with a white noise waveform, mimicking a real light source. This approach allows me to have a good knowledge on the perturbations produced at light source level and thus a more controlled and accurate measure of the amount of ripple really corrected by the obscuration unit.

2.2 Attenuator

The attenuator used is a V-273 PIMag voice coil [2]. This voice coil drives consist of 2 essential components: a permanent magnet and a coil, which is located in the air gap of the magnetic field. When current flows through the coil, it moves in the magnetic field of the permanent magnet. The direction of motion depends on

* Corresponding author: jlcachatra@gmail.com

the polarity. Thanks to their low weight and friction-free drive principle, voice coil drives are particularly suitable for applications, which require high dynamics and high velocities at limited travel ranges.

2.3 Detection and Control System

The measurements from the photodiode are read using a Tektronix_DMM7510 digital multimeter. These readings are used to feed a PID control whose output determines the position of the voice coil. A LabVIEW interface is used to control and monitor the voice coil position and calculate the PID output. The PID gains were optimized using the Ziegler–Nichols tuning method [3].

2.4 Experimental Results

Several measurements were made to provide an overview of the systems capabilities. A long duration test, with a sampling rate of 50mHz, was performed to assess the system's stability. The Allan deviation was chosen has the measurement for stability [4].

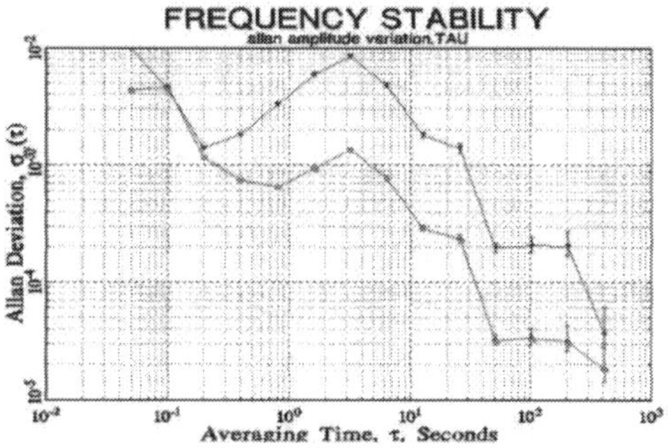

Figure 2: Allan deviation curves for a non-compensated (blue dots) and compensated (red dots) signal, with a frequency of 100mHz and an amplitude of 500mV

Figure clearly shows the added stability provided by the system, increasing with the sampling time, reaching a minimum of 10^{-15}. Further measurements were made to test the systems response to different frequencies and amplitudes of the modulating wave.

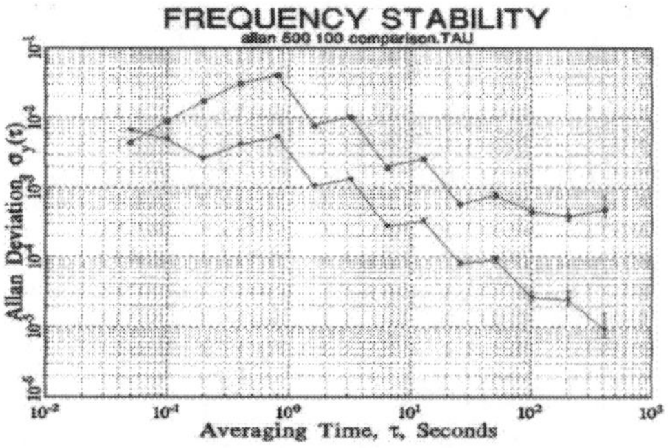

Figure 3: Allan deviation curves for compensated signals with a modulating frequency and amplitude of 300mHz (blue dots) and of 500mHz (red dots)

We can observe from figure 3 that the higher signal frequencies and amplitudes lead to a decrease in the achievable stability. A measurement was also made to test the system when subjected to a 100mV noise signal generated by the wave generator.

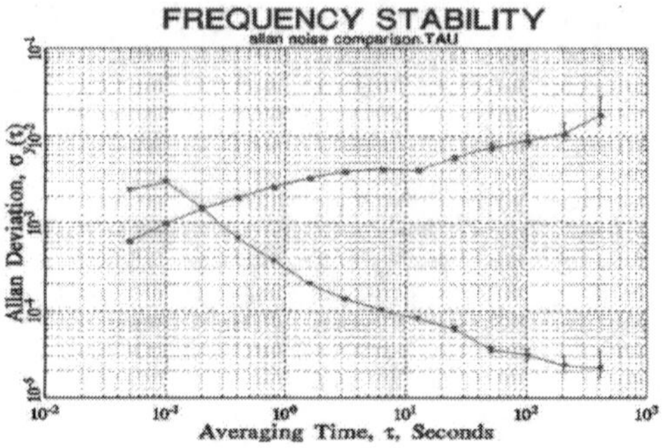

Figure 4: Allan deviation curves for a non-compensated (blue dots) and compensated (red dots) white noise signal, with an amplitude of 100mV

Figure 4 shows how without compensation the system tends to become less stable the longer it runs, but when compensated the opposite occurs, becoming significantly more stable with time.

3 Conclusions

The development of this system requires a study of the noise sources present in the light source and the detection system used, the development of an optical collimating system and the development of a PID control system for the attenuator. Significant stability results are achieved when compared to the non-stabilized results.

This work was supported by Fundação para a Ciência e a Tecnologia (FCT) through the research grants UIDB/04434/2020 and UIDP/04434/2020. We also recognize the support of SPOF on the participation of this conference.

References

1. *https://lasp.colorado.edu/home/sorce/2012/06/10/observations-of-the-venus-transit-total-irradiance/*
2. *https://static.physikinstrumente.com/fileadmin/user_upload/physik_instrumente/files/datasheets/V-273-Datasheet.pdf*
3. W. J. Riley, *Handbook of Frequency Stability Analysis*, NIST, (2007)
4. A. O'Dwyer, *Handbook of PI and PID Controller Tuning Rules*, Imperial College Press, **3**, (2009)

Application of the Vernier method for absolute distance metrology with CW TOF phase shift technique

Nuno M. Gonçalves[1,2,*], *Manuel* Abreu[1,2,**], and *D.* Castro Alves[1,2,***]

[1]Instituto de Astrofísica e Ciências do Espaço, Universidade de Lisboa, Campo Grande, PT1749-016 Lisboa, Portugal
[2]Departamento de Física, Faculdade de Ciências, Universidade de Lisboa, Edifício C8, Campo Grande, PT1749-016 Lisbon, Portugal

Abstract. A phase shift time of flight technique determines a position by comparing the phase angle of a continuously modulated signal in the source and its reflection on a target. However, due to its cyclical properties, the position information is contained within an ambiguity interval. For an absolute measurement, this interval is repeated N times plus a residual part given by the phase shift. In this work we propose an application of the Vernier method to determine N and a setup for mid-range applications (10-20) m with a 3 GHz amplitude modulated source to allow accuracies $\leq 100\ \mu$m.

1 Introduction

Time of flight (TOF) is a very well established optical measuring technique that offer ways to determine absolute or relative distance measurements (ADM and RDM, respectively). In a phase shift (PS) continuous wave TOF (PS CW TOF) method, a coherent light source is continuously modulated in amplitude and directed to a target. The reflected signal will then have a phase shift, $\Delta\phi$, relative to the original, which will be related to the source-target distance by eq. 1.

$$D = \frac{cn}{2f}\left(\frac{\Delta\phi}{2\pi} + N\right) \tag{1}$$

$$\Lambda = \frac{cn}{2f} \tag{2}$$

Where f is the modulation frequency, D the source-target distance, c the speed of light and n the refractive index of the medium. Due to the cyclical properties of the phase, the position information is contained within an ambiguity distance, eq. 2. Therefore, for an absolute measurement, this interval is repeated N times plus a residual part given by the PS. Hence, it's crucial to determine N for absolute measurements. Different methods can be used for this purpose [1]. *Norgia, et al* [2] achieved 100 μm accuracy for a 2 GHz modulation using heterodyne down conversion (HdC), and solved the ambiguity by using a double Vernier method [3].

For a mid-range (10-20) m, by performing a 3 GHz modulation with HdC, one can achieve an accuracy of <100 μm in an RDM, as long as the error of the frequency and PS are in the order of 1 kHz and 0.3 °, respectively. In this work, we propose a different approach to determine N, allowing this level of accuracy for ADM. Although, since N is an integer, it must be determined with an error < 0.5. In order to achieve the distance accuracy goal and N being properly determined, a trade off between both of this components must be done. However, the error analysis isn't in the scope of this document.

*e-mail: fc48080@alunos.fc.ul.pt
**e-mail: maabreu@fc.ul.pt
***e-mail: dmvalves@fc.ul.pt

2 Working Principle

In order to measure the source-target distance, the PS CW TOF technique employs the concept of a "ruler". This ruler has a primary and secondary mark, given by the ambiguity distance and by the PS (as it is referred in eq. 1), as shown by ruler A in figure 1.

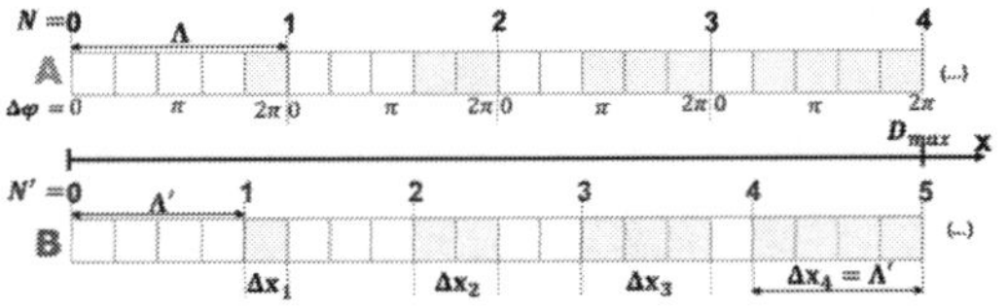

Figure 1. Comparison of two different rulers A and B. In A the distance between marks is determined by f and in B by $f' = f + \Delta f$.

It's possible to determine N by measuring the same distance with two different "rulers". The difference between primary markings of both rulers is defined as Δx_N:

$$\Delta x_N = N\Delta x_1 \quad ; \quad \Delta x_1 = \Lambda - \Lambda' = \frac{cn\Delta f}{2ff'} \tag{3}$$

Where N corresponds to the index of the primary mark of the ruler with biggest spacing between them and $f > \Delta f$ and $\Delta f = f' - f > 0$ since $f' > f$. This represents the difference in frequency for both scales and will rule the variation in position of the different primary marks.

Additionally, by considering the difference in phase shifts between rulers for the same distance, $\Delta\Phi = \Delta\phi' - \Delta\phi$, one gets eq. 4. By rearranging it, the absolute distance can be written as eq. 5.

$$\frac{\Delta\Phi}{2\pi} = D\frac{2\Delta f}{cn} + \Delta N \tag{4}$$

$$D = \frac{cn}{2\Delta f}\left(\frac{\Delta\Phi}{2\pi} - \Delta N\right) \tag{5}$$

Where $\Delta N = N - N' \leq 0$ and an integer, since $f' > f$ then $\Lambda > \Lambda'$ and consequently $N' \geq N$. Additionally, by rearranging eq. 1 we obtain eq. 6. By inserting eq. 5 in eq. 6 and solving for N, eq. 7 is obtained.

$$\frac{\Delta\phi}{2\pi} = \frac{2D}{cn}f - N \tag{6}$$

$$N = \frac{\left(\frac{\Delta\Phi}{2\pi} - \Delta N\right)}{\Delta f}f - \frac{\Delta\phi}{2\pi} \tag{7}$$

Considering a target at a fixed distance D, one can measure a pair of different phase shifts corresponding to known modulation frequencies spaced by Δf, by using eq. 7 it's possible to determine the value of N.

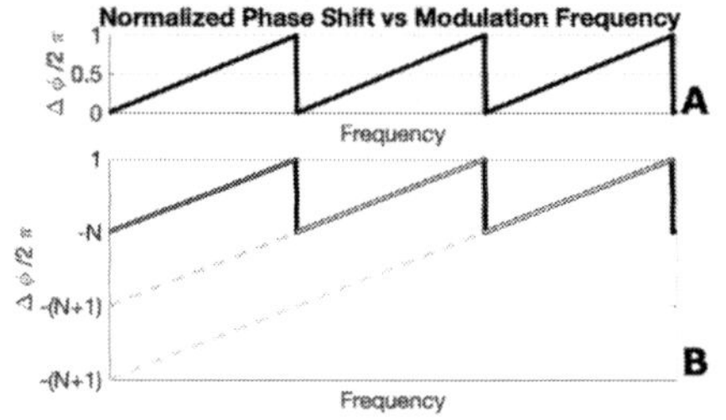

Figure 2. Normalized phase shift for a fixed distance. The colored lines in B are a fit of eq. 6 in the positive slope region of the normalized phase shift.

Very briefly, if one plots eq. 6 for a fixed distance the result is a saw-tooth profile as seen on figure 2A. For a single and arbitrary frequency-PS point in this profile, the corresponding N can be found as the intercept in the normalized PS axis, as shown in figure 2B. However, for a pair of arbitrary and different frequency-PS points, the ΔN between them can be 0,-1,-2,-3... depending on the Δf spacing. More generally, for any given $D \leq D_{max}$, by constraining Δf one can ensure that for a pair of different frequency-PS points, $\Delta N = 0 \vee \Delta N = -1$. The maximum distance at which this condition is fulfilled is when D_{max} is an integer multiple of the ambiguity distances of both frequencies, i.e. the marking of both rulers match and consequently $\Delta\Phi = 0$. However, as seen in figure 1, D_{max} will be reached when $\Delta x_N = \Lambda'$, which mathematically implies that $\Delta N = -1$. By using eq. 5 one can find D_{max}:

$$D_{max} = \frac{cn}{2\Delta f}(0 - (-1)) = \frac{cn}{2\Delta f_{max}} \tag{8}$$

Meaning that for $D \leq D_{max}$ by doing spacings of Δf_{max}, as long as $f > \Delta f_{max}$ for any modulation frequency one can find $\Delta N = 0 \vee \Delta N = -1$. This is consistent with the results of *S. Donati* [1], that with a different approach showed it was possible to measure unambiguously for $D \leq D_{max}$.

Additionally, through mathematical analysis of eq. 5, one can conclude that if $\Delta\Phi < 0 \rightarrow \Delta N = -1$ and $\Delta\Phi > 0 \rightarrow \Delta N = 0$. Hence, by measuring a pair of different PS and modulation frequencies spaced at most by Δf_{max} one can know if $\Delta N = 0$ or $\Delta N = -1$ if the difference in PS between them is positive or negative.

$$slope = \frac{\frac{\Delta\phi' - \Delta\phi}{2\pi}}{\Delta f} = \frac{\Delta\Phi/2\pi}{\Delta f} \tag{9}$$

In conclusion, with eq. 9 and a pair of frequency-PS measurements, one can determine N through eq. 7. Therefore, by knowing N and applying eq. 1 with a GHz modulation frequency it's possible to achieve the desired accuracy.

3 Future work

In order to show that this principle is valid to obtain absolute distances with high accuracy, we intend to build the setup showed in figure 3. One must ensure that given the instrumental limitation, we can measure for a mid-range distance, (10-20) m, with accuracies $\leq$ 100 μm. For that purpose the knowledge of the frequency and PS must be in the order of 1 kHz and 0.3 °, respectively. Additionally, since N is an integer, its measurement error must be smaller than 0.5, to determine the exact integer number of phase cycles to reach the distance D.

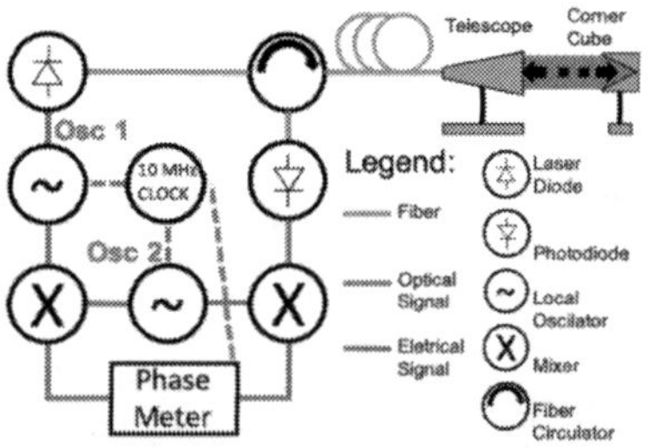

Figure 3. Sensor scheme with *Osc 1* at 3 GHz + Δf and *Osc 2* at 3 GHz.

This work was supported by Fundação para a Ciência e a Tecnologia (FCT) through the research grants UIDB/04434/2020 and UIDP/04434/2020. We also recognize the support of SPOF on the participation of this conference.

References

[1] S. Donati, *Electro-Optical Instrumentation* (Pearson Education, New Jersey, 2004) 61-63

[2] M. Norgia, F. Cavedo, A. Pesatori, *Proc. IEEE MetroAeroSpace*, 408-412 (2017)

[3] M. Rakhmanov, M. Evans, H. Yamamoto, Meas. Sci. Technol. **10**, 190-194 (1999)

Convolutional Neural Network Applied for Nanoparticle Classification using Coherent Scaterometry Data

D. Kolenov[1]*, *D.* Davidse[1], *J.* Le Cam[2], and *S.F.* Pereira[1]

[1]Optics Research Group, Imaging Physics Department, Faculty of Applied Sciences, Delft University of Technology, Lorentzweg 1, 2628 CJ Delft, The Netherlands
[2]Institut d'Optique Graduate School, 2 Avenue Augustin Fresnel, 91120 Palaiseau, France

Abstract. The analysis of 2D scattering maps generated in scatterometry experiments for detection and classification of nanoparticle on surfaces is a cumbersome and slow process. Recently, deep learning techniques have been adopted to avoid manual feature extraction and classification in many research and application areas, including optics. In the present work, we collected experimental dataset of nanoparticles deposited on wafers for four different classes of polystyrene particles (with diameters of 40, 50, 60, 80 nm) plus background (no particles) class. We trained a convolutional neural network, including its architecture optimization, and achieved 95% accurate results. We compared the performance of this network to a existing method based on line-by-line search and thresholding, demonstrating up to a twofold enhanced performance in particle classification. The network is extended by a supervisor layer that can reject up to 80% of the fooling images at the cost of only rejecting 10% of original data.

1 Introduction

In this paper, we propose a method to classify 2D scattered maps of containing isolated nanoparticles on surfaces using CNN. These maps have been generated in a Coherent Fourier Scatterometry (CFS) setup [1–4]. In order to generate the scattered maps to train the data, polystyrene latex (PSL) nanospheres of calibrated sizes were spin-coated on a silicon wafer. The diameters of the particles ranges from 40 to 80 nm. Furthermore, for the classification, we study the areas of the wafer where the nanoparticles were absent, contributing to the "background" class. We also target at a novelty detection, by looking at ways for the network to separate the "unknown" class from the input data, i.e., classes that have been unseen in the training. In order to do that, we rely on a simple approach of baseline and also a more sophisticated approach of introducing the OpenMax layer [5].

We realized an experiment by adding noise to the scattered maps (degrading thus the SNR), fooling images as well as some unfamiliar (reversal) images to the network. As one of the main goals for the CNN is to accurately discriminating between classes, we studied the samples that contain multiple classes of PSL particles with diameters of 40 and 50 nm, 50 and 60 nm, 60 and 80 nm. The results show that our model can successfully discriminate between the 5 classes with accuracies up to 95%. By providing the samples that were unseen

*e-mail: d.kolenov@tudelft.nl

during the training, we highlight the importance of the novelty detection in order to capture the confusing inputs. The results show that the proposed method has superior capabilities compared to the classification with the traditional search algorithm.

2 Closed set classification

In order to see which classes the network struggles to distinguish and to what degree, we built the confusion matrix (Figure 1 A). The horizontal axis represents the particle classes predicted by our model, and the vertical axis represents the true input image labels. It is clearly visible that most misclassification involves the background class. The best model

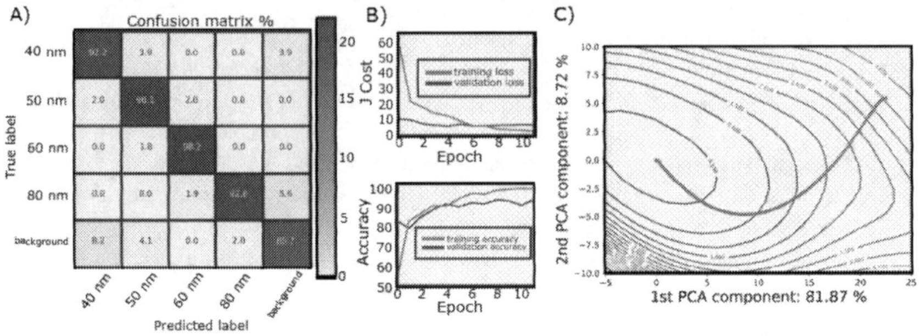

Figure 1. A) The accuracy for the test set in the confusion matrix. B) The training and validation loss - top and accuracy - bottom, is evaluated across different number of epochs based on the optimal architecture of the CNN model. C) 2D visualization of the loss surface of the CNN model with the projected learning trajectories using normalized PCA direction (batch size of 20, Adam optimizer and 15 epochs of training).

has to be selected based on the Accuracy metric calculated on the test data. For the closed set of five classes, we found that after approximately 12 epochs (12 times through all the training examples), the loss no longer decreased significantly. The top performing network ($Accuracy = 95\%$) was stored to be used in the further tasks (see Figure 1 B). Finally, we built the landscape, where we demonstrate the convergence to minima as our learning procedure follows the loss in a gradual manner. The projected learning trajectory is estimated using normalized principal component analysis (PCA) directions. The squared nature of the loss function leads to mostly convex loss landscape (Figure 1 C).

References

[1] S. Roy and A. C. Assafrao and S. F. Pereira and H. P. Urbach," *Opt. Express*, pp. 13250-13262, Jun 2014.

[2] S. Roy and S. F. Pereira and H. P. Urbach and X. Wei and Omar El Gawhary," *Phys. Rev. A*, 96, pp. 013814, Jul 2017.

[3] Omar El Gawhary and Stefan J.H. Petra, "Method and apparatus for determining structure parameters of microstructures," *US Patent*, 2015.

[4] D. Kolenov and R. C. Horsten and S. F. Pereira, "Heterodyne detection system for nanoparticle detection using coherent Fourier scatterometry," *Optical Measurement Systems for Industrial Inspection XI*, 11056, pp. 336-342, 2019.

[5] Abhijit Bendale and Terrance E. Boult, "Towards Open Set Deep Networks," *2016 IEEE Conference on Computer Vision and Pattern Recognition (CVPR)*, pp. 1563-1572, 2015.

https://doi.org/10.1051/epjconf/202023806018

Fully integrated stokes snapshot imaging polarimeter

Carsten Stock[1,2,*], *Thomas* Siefke[1], *Martin* Hubold[3], *Christin* Gassner[3], *Robert* Brüning[3], and *Uwe D.* Zeitner[1,4]

[1]Friedrich Schiller University Jena, Institute of Applied Physics, 07745 Jena, Germany
[2]Fraunhofer Institute of Applied Optics and Precision Engineering, Center of Excellence in Photonics, 07745 Jena, Germany
[3]Fraunhofer Institute of Applied Optics and Precision Engineering, Department Optical and Mechanical System Design, 07745 Jena, Germany
[4]Fraunhofer Institute of Applied Optics and Precision Engineering, Department for Microoptical Systems, 07745 Jena, Germany

Abstract. We present a new concept for a fully integrated nano-optical snapshot imaging polarimeter. It consists of a nano-optical retarder and polarizer array with various orientations combined with a microlens array and crosstalk module. That facilitates a fast, robust, and compact stokes imaging polarimeter without moving parts for applications at 450 nm wavelength.

1 Introduction

Polarized light metrology can be used for many applications to perform non-contact, non-destructive, and fast measurements, e. g. in the semiconductor industry for wafer inspection [1], in tissue characterization [2, 3], chemistry [4], biomedical diagnosis [5] or more.
A full Stokes-polarimeter measures all four terms of the Stokes vector:

$$\vec{S} = \begin{bmatrix} I \\ Q \\ U \\ V \end{bmatrix} = \begin{bmatrix} I_0 + I_{90} \\ I_0 - I_{90} \\ I_{+45} - I_{-45} \\ I_R - I_L \end{bmatrix}, \quad (1)$$

Where I is the total intensity of the input light, Q the intensity ratio of 0° and 90° linear polarized light, U for +45° and -45° polarization, and V for right circular and left circular light. Various methods are established [6] to measure the full stokes vector. In our approach we modified the proposed 2 x 2 x 3 design by Alenin et. al. [7], which consists of a polarizer array with 0° and 90° orientation and retarder array with 126° retardation and 15°, 45° and 75° orientation. To achieve a snapshot polarimeter, we implement all three orientations of the retarder directly on top of the imaging sensor, resulting in a 3 x 4 design. The whole setup consists of a bandpass filter for 450 nm wavelength (design wavelength of the retarder array), nano-optical retarder and polarizer array as well as microlens array and crosstalk module (see figure 1).

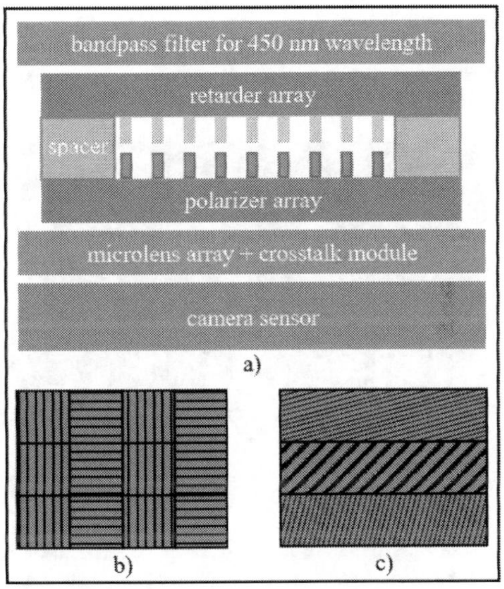

Fig. 1. a) Schematic drawing of the imaging polarimeter setup, b) polarizer array and c) retarder array.

In general, these compact imaging polarimeters consist of only a polarizer array (e.g. the Sony IMX250MRZ sensor) and can only measure linear polarized light (I, Q, V) or use a rotating or liquid crystal retarder, which is bulky and need additional hardware (motors, electronic controller, etc.). Our approach of a full stokes division of focal plane (DoFP) polarimeter offers many advantages: we achieve a robust construction without moving elements and due to the parallelization, a fast measurement becomes possible. In addition, thin element sizes are feasible. Each array (polarizer, retarder, microlens + crosstalk) is fabricated on a 0.5 mm thin substrate and are aligned to and placed directly on top of a CMOS sensor with 2560*1920 pixels, thereby high flexibility and integration in various systems for different metrology tasks is possible.

* Corresponding author: carsten.stock@uni-jena.de

2 Setup and fabrication of the nano-optical elements

The nano-optical retarder array and wire grid polarizer are grating type metasurfaces with periods smaller than the applied wavelength. Their fabrication is well established in our work group [8, 9]. For applications in the visible wavelength regime we use a broadband iridium wire grid polarizer. The nano-optical retarder array consists of a titanium oxide grating, which can be understood as an effective, birefringent medium [10]. The iridium wire grid polarizer array consists of four columns with alternating 0° and 90° orientation. The titanium oxide retarder array consists of three retarder elements with orientations of 15°, 45° and 75°, arranged in three rows. For increased mechanical stability and to protect against environmental influences, the polarizer and retarder arrays are integrated in such a way that the structured areas face each other. A physical contact of the nanostructured areas is prevented by means of spacers (see figure 1). Also, this allows a simplified manufacturing process, as each array can be fabricated individually. During the assembly of all components, they are aligned with the camera sensor and firmly fixed in a holder, so that no adjustment is required after assembly.
A fist image taken by the imaging polarimeter during the calibration is shown in figure 2 and a real-world image is displayed in figure 3.

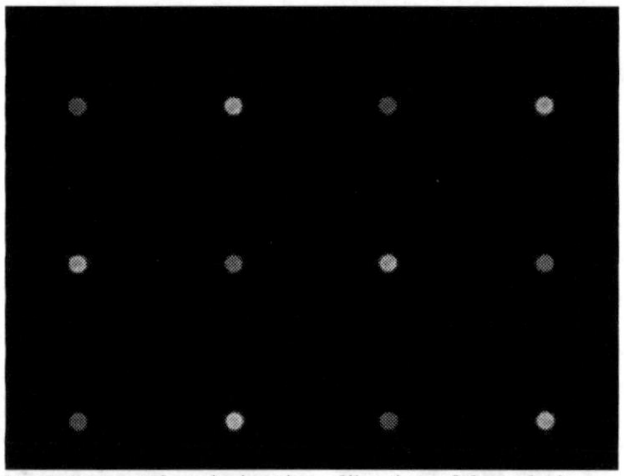

Fig. 2. Image taken during the calibration of the polarimeter. The incoming light is 0° linear polarized. Different intensities are transmitted through the polarizer and retarder array.

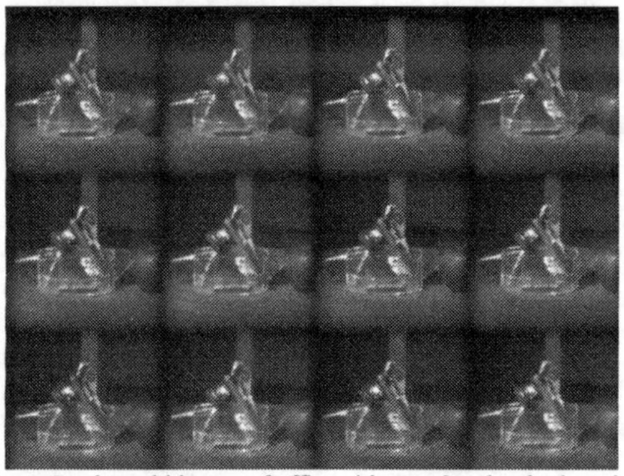

Fig. 3. Real-world image of office objects taken by the imaging polarimeter.

Detailed information about the system design, estimations of the uncertainty budget of the entire polarimeter, and first measurements will be presented at the conference.

3 Conclusion

We presented a concept of a highly integrated full stokes imaging polarimeter using nano-optical elements as analyser and retarder. We fabricate these metasurface gratings via nanostructuring technology and various orientations can be realized within one element. This facilitates parallel measurement by replacing any rotating or electrooptical elements resulting in a compact and robust setup.

References

1. G.C. Giakos, A. Medithe, S. Sumrain, S. Sukumar, L. Fraiwan and A. Orozco, IEEE Transactions on Instrumentation and Measurement, **55**, 2126-2131 (2006)
2. J.C. Ramella-Roman, A. Nayak, S.A. Prahl, Journal of Biomedical Optics, **16**, 047001 (2011)
3. T. Novikova, A. Pierangelo, A. De Martino, A. Benali, P. Validire, Optics and Photonic News, **23**, 26 (2012)
4. D. Hore, A. Natansohn, P Rochon, Journal of Physical Chemistry B, **106**, 9004-9012 (2002)
5. W.L. Hsu, J. Davis, K. Balakrishnan, M. Ibn-Elhaj, S. Kroto, N. Brock, S. Pau, Optics Express, **23**, 4357-4368 (2015)
6. D.H. Goldstein, *Polarized Light*, CRC Press (2010)
7. A.S. Alenin, I.J. Vaughn, J.S. Tyo, Applied Optics **57**, 2327-2336 (2018)
8. T. Siefke, S. Kroker, K. Pfeiffer, O. Puffky, K. Dietrich, D. Franta, I. Ohlídal, A. Szeghalmi, E.B. Kley, A. Tünnermann, Advanced Optical Materials, **4**, 1780-1786 (2016)
9. C. Stock, T. Siefke, U.D. Zeitner, Journal of the Optical Society of America B, **36**, D97-D102 (2019)
10. S.M. Rytov, Soviet Physics JETP, **2**, 466-475 (1956)

EPJ Web of Conferences **238**, 06019 (2020) https://doi.org/10.1051/epjconf/202023806019
EOSAM 2020

Electromagnetic scattering beyond the weak regime: Solving the problem of divergent Born perturbation series by Padé approximants

Thomas van der Sijs[1], *Omar* El Gawhary[1,2,*], and *Paul* Urbach[1]

[1]Optics Research Group, Delft University of Technology, P.O. Box 5046, 2600 GA Delft, The Netherlands.
[2]Dutch Metrology Institute (VSL), Thijsseweg 11, 2600 AR Delft, The Netherlands.

Abstract. Electromagnetic scattering is the main phenomenon behind all optical measurement methods where one aims to retrieve the shape or physical properties of an unknown object by measuring how it scatters an incident optical field. Such an inverse problem is often approached by solving, several times, the corresponding direct scattering problem and trying to find the best estimate of the object which is compatible with a set of measurements. In the direct scattering problem, two regimes can be distinguished depending on the size of the object and the permittivity contrast: the weak-scattering regime and the strong-scattering regime. Generally, the presence of the scatterer alters the form of the incident field inside the scatterer. If that effect is neglected in the physical model, then one speaks of the so-called single-scattering regime or, more often, the Born approximation. The regime in which this approximation is valid is the weak-scattering regime. The corresponding inverse problem, that aims to retrieve the object from scattering data, becomes linear in this case. Linearizing the problem simplifies the method to solve it, but also introduces limitations to the maximum spatial resolution achievable in the reconstruction of the object. In the strong-scattering regime, multiple-scattering effects are not neglected and the inverse problem is treated in its full non-linear nature, which makes finding its solution a far more challenging task.

Despite the existence of numerical methods, a powerful way to solve those direct problems would be to use a perturbation approach where the field is expressed as a series, known as the Born series. The advantage of a perturbation approach stems from the fact that each term of the series has a clear physical meaning and can unveil much more about the scattering process than a purely numerical approach can offer. Unfortunately, the series solution turns out to be strongly divergent in the strong-scattering regime, making it an unpractical approach for problems under these strong-scattering conditions. Thus, despite the fact that multiple scattering could, in principle, allow resolving sub-wavelength details of the unknown object, this possibility is in practice hampered by the divergent nature of the higher-order terms of the Born series.

In this work, we show how to solve this problem by employing Padé approximants and how to treat electromagnetic problems well beyond the weak-scattering regime and provide an accurate evaluation of the scattered field even under strong-scattering conditions. Padé approximants are rational functions that can offer improvements in two ways, namely series acceleration of converging series and analytic continuation of a series outside its region of convergence. In the case of a symmetric approximant of order N, the approximant is calculated from $2N + 1$ terms in the Born series, therefore incorporating multiple-scattering effects to which these higher-order corrections in the Born series correspond. We apply the method to two scalar scattering problems: that of a one-dimensional slab and that of an infinitely long cylinder, which reduces to a two-dimensional problem under normal incidence. In particular, we treat cases in the strong-scattering regime where the Born series diverges, but where Padé approximation retrieves a valuable result. In Fig. 1 the case of a cylinder is shown which is well beyond the weak-scattering regime, but where the most accurate Padé approximant gives a good result for the field.

The presented approach incorporates multiple-scattering effects and can therefore represent an important building block to the application of the Born series to direct and inverse problems, with potential applications in superresolution, optical metrology, and phase retrieval.

*e-mail: o.elgawhary@tudelft.nl

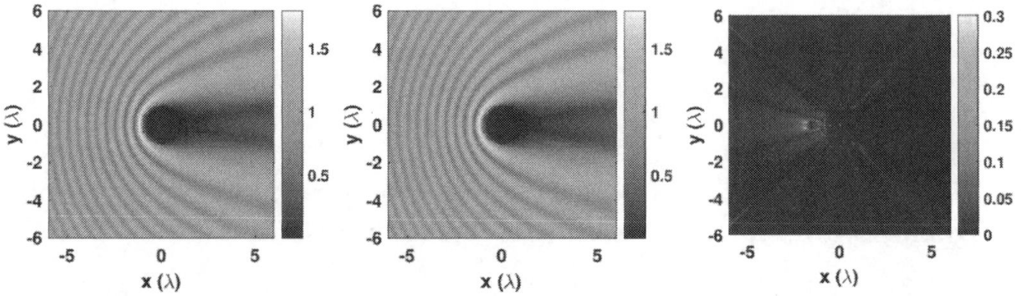

(a) Analytical solution. (b) Padé approximant P_4^4. (c) Error of P_4^4.

Figure 1: Padé approximation in the two-dimensional problem of a cylinder on which a unit plane wave impinges under normal incidence. The cylinder has a radius of 1 λ and has the permittivity of silver $\varepsilon_r = -4.42 + 0.201i$, which is a case that is well outside the weak-scattering regime. The approximant that is shown is P_4^4, the symmetric approximant of order $N = 4$, which was found to give the best approximation to the actual solution.

EPJ Web of Conferences **238**, 06020 (2020) https://doi.org/10.1051/epjconf/202023806020
EOSAM 2020

Two wavelength frequency transfer over an optical fiber link

Carlos Pires[1*], *Manuel* Abreu[2], *Isabel* Godinho[1] and *Rui* Agostinho[2]

[1]Instituto Português da Qualidade, Rua António Gião 2, 2829-513 Caparica, Portugal
[2]Faculdade de Ciências da Universidade de Lisboa, Campo Grande, 1749-016 Lisboa, Portugal

Abstract. In this paper we present the theory and the experimental setup used to transfer a standard frequency, to synchronize two clocks linked by an optical fiber. In order to verify the accuracy on frequency transfer over fiber link, we prepared an experiment that allows testing the performance of the setup for a variable set of environmental conditions, namely associated to temperature and vibration variations. The experimental setup shows the fiber optic link between one laboratory, where the standard frequency is generated, and another laboratory, where the equipment for simulating temperatures and vibrations are installed. The standard frequency, traceable to UTC(IPQ), is used to modulate two lasers with different wavelength, , injected in the optical fiber. At the end of the optical fiber the signals will be out of phase due to the inherent chromatic dispersion, which is also dependent on the temperature of the propagation media. Measuring the phase variations, caused by temperature gradients in the fiber, we can compensate the frequency transfer and synchronize the clocks. Evaluating all uncertainty components in this model, allows the metrological characterization of the synchronization and obtain the associated uncertainty of this quantity.

1 Introduction

Ultra-stable time and frequency sources, primary time standards or even commercial time standards, that contribute to the realization of the Coordinated Universal Time (UTC), have an important role in fundamental physics and in time and frequency metrology, such as clock evaluation, tests in relativity, work in fundamental constants, among others. Currently, the dissemination in long distances of a standard frequency is achieved using the satellites of the Global Navigation Satellite Systems (GNSS), with a stability in the order of 10^{-15} in one day. This value could be overcome replacing the satellites by optical fiber, being a matter of study in the last decade.
The dissemination of frequency, using optical fiber, is being implemented by most of the National Metrology Institutes (NMI) in Europe.
One method to disseminate frequency consists in modulating the frequency into a light signal, and then injecting the light in an optical fiber. At the end of the fiber, the signal is converted to the microwave domain and can be used as a standard frequency.
Replacing the satellites link by an optical fiber link, made possible the comparison of the best time and frequency standards in short time interval with accuracies 2 order of magnitude better than the observed with the GNSS.
The main goal of this work is to study the transfer a frequency signal, from one laboratory, where the standard frequency is generated, to another laboratory where a setup is mounted to measure the delay of the frequency transfer caused by temperature and vibrations variations. Evaluating all the delays and all the uncertainty components in this transfer will allow the synchronization of two clocks at both ends of the fiber, with the best possible uncertainty.
This paper presents the initial work in this experiment, where temperature variation effect on the frequency transfer accuracy is evaluated. For this purpose, a fiber link of 500 m in length was mounted on the roof top of the laboratory and the environment conditions were recorded.

Although considered for this experiment, vibration tests and their impact on the frequency transfer process are not included in this paper
The work produced by the author follows closely the method and experiment performed in 2010, in Sweden, where two light beams of different wavelength injected in a fiber link were used to evaluate the influence of temperature variations in the frequency transfer process [1].

2 Theory

In general, the frequency stability of an oscillator in free mode, with short integration times, is better than the frequency stability of an oscillator located at the end of a means of transport, due to the noise introduced by it. The inherent stability or the accuracy of the oscillator at the remote location turns out to be irrelevant as the greatest source of uncertainty is caused by the means of transport, in this case by the optical fiber link.
From a physical point of view, the group time delay, τ, is given by the slope of the curve that relates the refractive index to the wavelength. This slope is given by the derivative of the index of refraction as function of the wavelength, which multiplied by the length of the fiber allows to obtain the propagation time of a signal in the fiber.

* Corresponding author: carlosp@ipq.pt

$$\tau = \frac{L}{c}\left(n - \lambda\frac{dn}{d\lambda}\right) \quad (1)$$

The delay in the propagation of a beam of light in a fiber can be expressed according to equation 1. The variation in propagation delay can be expressed as a function of temperature (T) using the following expression:

$$\Delta\tau = \frac{L}{c}\frac{\partial n}{\partial T}\Delta T + L\alpha\frac{n}{c}\Delta T + LD\Delta\lambda \quad (2)$$

Where $\alpha = \frac{\frac{\partial L}{\partial T}}{L}$ is the coefficient of thermal expansion of the fiber and $D = \frac{\frac{\partial n}{\partial \lambda}}{c}$ is the chromatic dispersion coefficient of the fiber.

Deriving equation 1 in order of temperature and calculating the difference in propagation time between two different wavelengths, we obtain:

$$\left.\frac{d\tau}{dT}\right|_{\lambda_1-\lambda_2} = \frac{1}{c}\left(\frac{dL}{dT}\left(\left(n_{\lambda_1} - n_{\lambda_2}\right) + \lambda_2\frac{dn_{\lambda_2}}{d\lambda_2} - \lambda_1\frac{dn_{\lambda_1}}{d\lambda_1}\right) + L\frac{d}{dT}\left(\left(n_{\lambda_1} - n_{\lambda_2}\right) + \lambda_2\frac{dn_{\lambda_2}}{d\lambda_2} - \lambda_1\frac{dn_{\lambda_1}}{d\lambda_1}\right)\right) \quad (3)$$

Equation 3 reflects the influence of temperature in the refractive index of two wavelengths, being able to determine the variation in the propagation time.

3 Experimental and reference set up

The experimental setup is presented in the next figure.

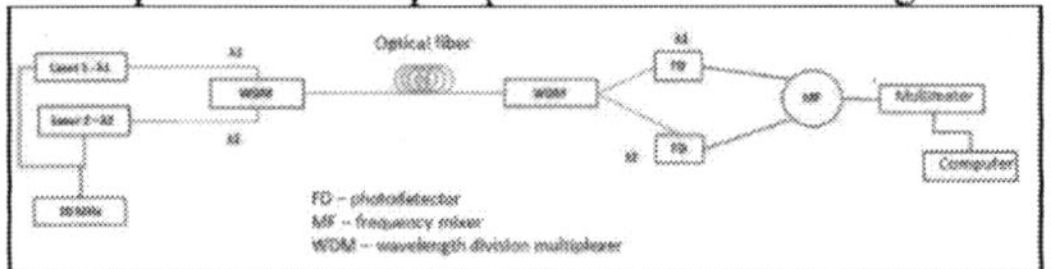

The lasers used are diode lasers with 1310 nm and 1550 nm wavelength, modulated with the 10 MHz reference frequency, from one Caesium clock (Cs8), used in the realization of the national reference time scale, UTC(IPQ). The stability of Cs8 is 6.00×10^{-13} in 100 s. The optical signals from the lasers, after modulation, go to a Wavelength Division Multiplexing (WDM), where they are injected into a 500 m optical fiber. At the end of the fiber both signals go back to a WDM where they are divided by wavelength and sent for detectors, to be converted in microwave. The microwave from the detectors goes to a frequency mixer, to be used as a phase detector [2], and the voltage is measured in a multimeter.

Part of the optical fiber link were installed in the rooftop of the building and is exposed to daily temperature gradients. Several temperature sensors were installed along the optical fiber path, in order to observe the temperature to which the fiber is subjected along the day.

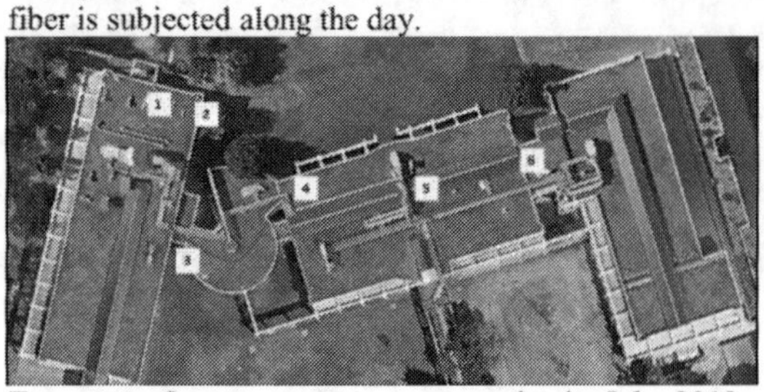

Two stet of measurements were made, in July 2019 and September 2019.

As a result, we have the variation of voltage and temperature across several days.

4 Results

Initially was analysed the stability of the 1550 nm and 1310 nm signals, after passing through the 500 m of optical fiber link. Subsequently the Modified Allan Variation was calculated for both signals, and a decrease about two order of magnitude in stability was observed.

The results from the frequency mixer working as phase detector, are extracted from a dataset of 5 days, collected in September 2019.

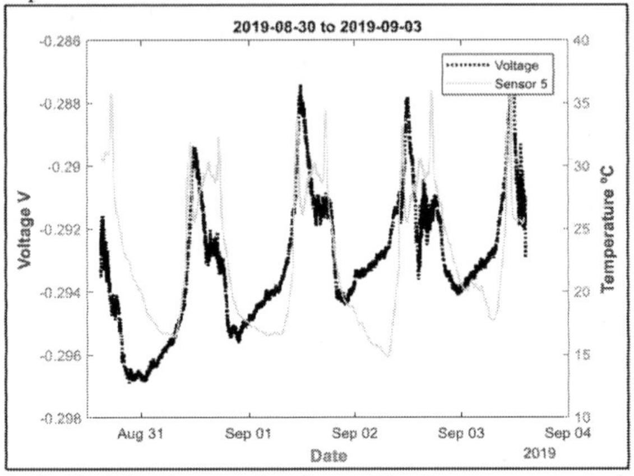

A similar behaviour can be observed by the temperature and voltage (with the information of phase difference between the two channels) which can indicate that changes in the temperature of the fiber in the rooftop can cause a phase variation, in accordance with the theory.

5 Conclusion

There is a decrease in the stability of the reference signal after its transmission in the fiber.

The collected data allows the conclusion that there is an influence of the temperature in the frequency transfer, although it wasn't observed in a categorical way a relation between the variations of the voltage in function of the temperature for this optical fiber path because all sensors presented different values of temperature along the day. It can be considered that, despite being a short path of optical fiber, it is a good example of a real connection, in which different parts of the fiber are at different temperatures.affecting cumulatively the stability of the frequency transfer.

A real implementation, with an optical fiber link of several kilometers is being studied and will allow the synchronization of two clocks, and in this way disseminate the second with the lowest possible uncertainty value.

References

1. S. Ebenhag P. O. Hedekvist, J. Johansson, Proc. 24th Eur Freq and Time Forum, Netherlands (2010)
2. Kurtz Stephan R., WJ Communications, Inc., Vol 5 No 1 (1978)

Membrane Environment Enables Ultrafast Isomerization of Amphiphilic Azobenzene -INVITED

*G. M.释Paternò*1, E. Colombo2,3, V. Vurro1,4, F. Lodola1, V. Sesti5, F. Benfenati2,3, C. Bertarelli5, G. Lanzani1,4*

1Center for Nano Science and Technology, Istituto Italiano di Tecnologia, Via Pascoli 70/3, 20133, Milano, Italy
2Center for Synaptic Neuroscience and Technology, Istituto Italiano di Tecnologia, Largo Rosanna Benzi 10, 16132 Genova, Italy
3IRCCS Ospedale Policlinico San Martino, Largo Rosanna Benzi 10, 16132 Genova, Italy
4Dipartimento di Fisica, Politecnico di Milano, Piazza L. da Vinci 32, 20133 Milano, Italy
5Dipartimento di Chimica, Materiali e Ingegneria Chimica "Giulio Natta", Politecnico di Milano, Piazza L. da Vinci 32, 20133 Milano, Italy

Abstract. We report on the isomerization dynamics of an amphiphilic azobenzene that dwells within the plasma membrane. We found that in aggregates formed in water, the isomerization reaction is hindered, while radiative deactivation is favoured. However, once in the membrane, the molecule reacquires its isomerisation capacity. These data fully account for the recently reported experiments in neurons, showing that the amphiphilic azobenzenes are effective light actuators for the modification of the electrical state of the membrane.

1 Introduction

The precise, fast and efficient spatiotemporal control of cellular signalling via the use of external triggers has become an important topic in chemical biology[1] and photopharmacology.[2] In this regard, light-responsive systems offer the possibility to employ light as a clean, non-invasive and spatio-temporally precise tool for controlling a variety of biological signals both *in vitro* and *in vivo*.[3–5] Very recently a new amphiphilic azobenzene named ZIAPIN2, has been introduced. ZIAPIN2 predominantly localizes in the plasma membrane and potently modulates neuronal firing *in vitro* as well *in vivo* via an opto-mechanical effect.[6] In order to fully characterize the photostimulation mechanism of ZIAPIN2 here we report a detailed investigation of its photophysics and dynamics in different media.[7] By combining steady-state and time-resolved optical spectroscopies, we observed that ZIAPIN2 in water lends itself to a radiative transition likely mediated by the formation of excimers, whereas it reacquires its conformational dynamics in membrane mimicking environments. Characteristics time scales and reaction yields are reported, reconciling the fast response (ms) observed in functional studies and the apparent slow ensemble evolution (s) by determining the molecular conversion time in the tens of picosecond range. Finally, we validate this opto-mechanical phenomenon in a non-excitable cell line that has an intrinsically low concentration of ion channels, to single out the effect on the plasma membrane itself.

2 Results

In order to fully characterize the photostimulation mechanism of ZIAPIN2 here we report a detailed investigation of its photophysics and dynamics in different media. We selected dimethyl sulfoxide (DMSO) that is a good solvent, as a control medium and vector for ZIAPIN2, water that is the main component of extracellular and intracellular media and micelles of sodium dodecyl sulphate (SDS) to mimic the lipid bilayer environment. By combining steady-state and time-resolved optical spectroscopies, we observed that ZIAPIN2in water lends itself to a radiative transition likely mediated by the formation of excimers, whereas it reacquires its conformational dynamics in membrane mimicking environments (figure 1). Characteristics time scales and reaction yields are reported, reconciling the fast response (ms) observed in functional studies and the apparent slow ensemble evolution by determining the molecular conversion time in the tens of picosecond range. Finally, we validate this opto-mechanical phenomenon in a non-excitable cell line that has an intrinsically low concentration of ion channels, to single out the effect on the plasma membrane itself.

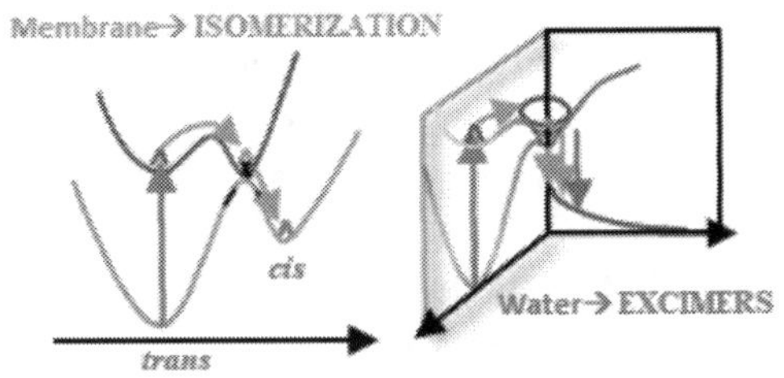

Fig. 1. Sketch of the photodynamics in DMSO, SDS and water as suggested by the TA data.

3 Conclusions

To summarise, the observed aggregation of photochromic molecules in water, where these azobenzenes lose their ability to photoisomerise, might represent an impediment to their

* Corresponding author: giuseppe.paterno@iit.it

 https://doi.org/10.1051/epjconf/202023807001

application in biology as water is the main component of biological environments. However, here we find that it is not true if the molecules have a preferential affinity for the cell membrane. In particular, we show that an amphiphilic azobenzene (ZIAPIN2) with non-covalent affinity for cell membranes can act as an efficient, fast and reversible modulator of the membrane potential via visible light. By applying steady-state and time-resolved ultrafast spectroscopies, we note that in water a strong aggregation occurs, effectively hampering photoisomerisation and opening up a deactivation path likely mediated by excimer formation. Alternatively, internalisation in the cell membrane prevents self-aggregation and restores the intrinsic ultrafast response of the photoswitch. This, ultimately allows light-triggering a large hyperpolarization in HEK293 cells whose magnitude follows the absorption profile of the trans state in membrane and, remarkably, permits to evoke biological photoinduced effects in vivo.[34] Such actuators are new tools that could contribute to the developing of opto-neuroscience, biophotonics and photopharmacology.

References

1. Beharry, A. A. & Woolley, G. A. *Chem. Soc. Rev.* **40**, 4422–4437 (2011).
2. Cabré, G. *et al. Nat. Commun.* **10**, 907 (2019).
3. Parameswaran, R. *et al. Nat. Nanotechnol.* **13**, 260–266 (2018).
4. Acarón Ledesma, H. *et al. Nat. Nanotechnol.* **14**, 645–657 (2019).
5. Chen, S. *Science (80-.).* **365**, 456–457 (2019).
6. DiFrancesco, M. L. *et al. Nat. Nanotechnol.* (2020). doi:10.1038/s41565-019-0632-6
7. Paternò, G. M. *et al. Adv. Sci.* **7**, 1903241 (2020).

Hybrid plasmonic/photonic crystals for optical detection of bacterial contaminants

Giuseppe M. Paternò1, Liliana Moscardi1,2, Stefano Donini1, Aaron M. Ross2, Silvia M. Pietralunga1,3, Nicholas Dalla Vedova1, Simone Normani1, Ilka Kriegel4, Guglielmo Lanzani1,2 and Francesco Scotognella*

1Center for Nano Science and Technology, Istituto Italiano di Tecnologia (IIT), Via Pascoli 10, 20133, Milano, Italy
2Physics Department, Politecnico di Milano, Piazza L. da Vinci 32, 20133 Milano, Italy
3Institute for Photonics and Nanotechnologies (IFN), Consiglio Nazionale delle Ricerche (CNR), Piazza L. da Vinci 32, 20133 Milano, Italy
4Department of Nanochemistry, Istituto Italiano di Tecnologia (IIT), via Morego, 30, 16163 Genova, Italy

Abstract. Here, we show that a hybrid plasmonic/photonic crystal consisting of a thin layer of bioactive plasmonic material (i.e. silver) deposited on top a 1D PhC can detect one of the most common bacterial contaminant, namely Escherichia coli. We speculate that the change in the plasmon charge density brought about by metal/bacterium interaction results in a variation of the plasmon resonance that, in turns, translates in a shift of the photonic structural color.

1 Introduction

Photonic crystals (PhCs) have been largely employed as detection/sensing devices in recent years, since the photonic stop-band can be tuned by applying a number of external stimuli, such as chemical[1], thermal[2] and mechanical triggers[3]. In this context, we have recently proposed porous 1D photonic structures exhibiting electro-optical tunability, due to the incorporation of optoelectronically-active plasmonic nanoparticles in the photonic structures.[4–6]
Here, we show that a hybrid plasmonic/photonic crystal consisting of a thin layer of bioactive plasmonic material (i.e. silver) deposited on top a 1D PhC can detect one of the most common bacterial contaminant, namely *Escherichia coli.*[7] We speculate that the change in the plasmon charge density brought about by metal/bacterium interaction results in a variation of the plasmon resonance which, in turns, translates in a shift of the photonic structural color.

2 Results

We integrated the bio-responsive silver layer on top of a 1D PhC consisting of alternating layers of SiO_2 and TiO_2 nanoparticles (5 bilayers). In this context, the plasmonic metal can be seen as defective layer of the photonic crystal that affects the optical response of the PhC via its free carrier density.> Thus, the main goal is to modify the dielectric properties at the PhC/metal interface and, thus, the overall optical read-out by exploiting the possible change in the silver complex dielectric function caused by Ag/bacteria interaction. To localise strongly the plasmonic response at the interface of the photonic structure, we selected the minimum Ag thickness attainable with our deposition apparatus. Interestingly, while interaction with the culture medium (pristine) does not alter appreciably the photonic read-out (Fig. 1), when the hybrid structure is interfaced with the bacterial colonies we noticed a 40 nm blue-shift of the photonic band-gap.

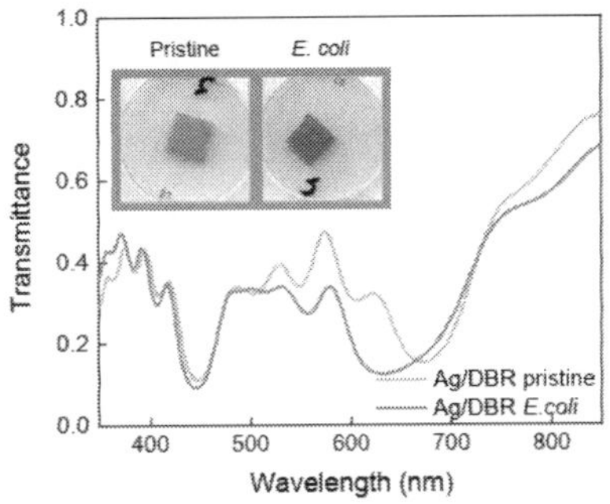

Fig. 1. Transmission spectrum of the pristine and E.coli-contaminated Ag/DBR. The inset shows the visual change in the reflection properties of the sample.

3 Conclusions

Here, we show that a novel hybrid plasmonic–photonic device consisting of a thin layer of silver deposited on top of a solution-processed BS is responsive to bacterial contamination. In particular, we assessed the validity of our approach by detecting one of the most hazardous Gram-negative bacterial contaminants in food and water, Escherichia coli.

References

1. Hong, W., Hu, X., Zhao, B., Zhang, F. & Zhang, D. **22**, 5043–5047 (2010).
2. Kanai, T., Lee, D., Shum, H. C., Shah, R. K. & Weitz, D. A. *Adv. Mater.* **22**, 4998–5002 (2010).
3. Fudouzi, H. & Sawada, T. *Langmuir* **22**, 1365–1368 (2006).

* Corresponding author: giuseppe.paterno@iit.it

4. Aluicio-Sarduy, E. *et al. Beilstein J. Nanotechnol.* **7**, 1404–1410 (2016).

5. Paternò, G. M. *et al. Sci. Rep.* **8**, 3517 (2018).

6. Paternò, G. M., Moscardi, L., Kriegel, I., Scotognella, F. & Lanzani, G. *J. Photonics Energy* **8**, 1 (2018).

7. Paternò, G. M. *et al. J. Phys. Chem. Lett.* **10**, 4980–4986 (2019).

8. Drude, P. *Ann. Phys.* **312**, 687–692 (1902).

https://doi.org/10.1051/epjconf/202023807003

Control of the emission properties of organic light emitting diodes using plasmonic gratings of Ag nanoparticles

***Sarah.* Hamdad, *Amadou. Thierno.* Diallo, *Mahmoud.* Chakaroun, and *Azzedine.* Boudrioua**
Laboratoire de Physique des Lasers CNRS UMR 7538, Université Paris 13, Sorbonne Paris Cité, Villetaneuse, France

Abstract. We report the influence of Ag nanoparticle arrays on the emission properties of organic light emitting diodes (OLED). First, we studied the features of Ag arrays related to the periodicity. The obtained results emphasize the existence of two coupling regimes dominated by near or far field effects, respectively. Thereafter, we studied the influence of these gratings on the performances of OLED devices. Our results reveal the benefit of the plasmonic structures which allow a precise tuning of the OLED properties.

1. Introduction

In recent years, several works have been reported on the plasmonic effect of nanoparticle arrays [1]. Besides, numerous studies concerning the interaction of metallic nanoparticles nearby active molecules have been, also, reported [2]. Nevertheless, a deep understanding of the influence of metallic nanoparticle gratings on the optical responses of organic emitters is still needed, in order to efficiently control their emission features. Many questions are still under investigation and need to be thoroughly studied. For instance, the plasmonic mode contribution to the diffractive orders has not been clarified. Moreover, almost the studies, particularly, focus on the optical excitation of emitter-nanoparticle system. Thereby, understanding this phenomenon under electrical excitation might reveal new coupling features and bring additional insight to the effects induced by plasmonic structures. The careful study of these processes are particularly crucial to develop efficient and new organic light emitting diodes (OLED), which may pave the way to new organic laser diode.

In this context, we report the investigation of the influence of Ag nanoparticle arrays on the emission of OLED devices. Specially, we aim to study the interaction range of plasmonic nanoparticles on the injection, recombination and extraction mechanisms of organic hetero-structures. For that purpose, we performed a study of the optical responses of Ag arrays by varying their periodicities in order to clarify the existence of two coupling regimes, from short periodicities dominated by near field effect to the longue ones exhibiting merely far field features. Thereafter, we studied the influence of these regimes on the radiative lifetimes of nearby emitters. Finally, the electrical and optical characteristics of OLEDs containing Ag nanoparticle arrays have been, carefully, studied.

2. Results and discussion

The studied structures are regular square arrays of silver nano-cylinders of 100nm diameter on top of glass/ITO substrate, covered by 100nm of organic material. Various Ag arrays with periods (p) ranging from 180nm to 480nm by a step of 20nm, fabricated by using electron beam lithography technique, have been simulated and experimentally studied. For instance, fig.1.a shows the experimental map of the extinction intensity as a function of the Ag grating period at a normal excitation.

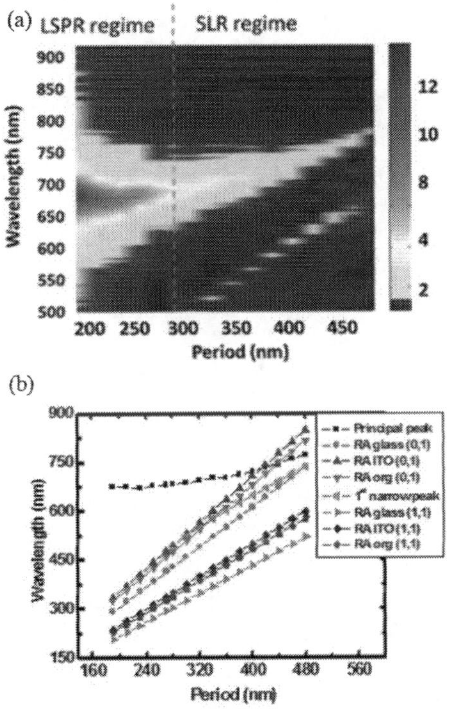

Fig. 1. (a) Extinction map of the Ag arrays as a function of the period and wavelength. (b) Variation of the LSP resonance, the first excited photonic mode and the appeared Rayleigh Anomalies as a function of the period.

Depending on the grating period, the obtained results emphasize the existence of two coupling regimes, localized surface plasmon resonance (LSPR) regime [3] and the surface lattice resonance (SLR) regime [4]. The former is associated to short period gratings of $p < 280nm$, for which a single resonance characterized by a high damping rate is obtained. In this case, each metallic nanoparticle presents intense confined (LSPR) modes. The latter regime is related to inter-particle distance $p \geq 280nm$, for which we observe a set of narrow resonances appearing on the extinction spectra that reveals the excitation of collective modes (SLR). Figure 1.b reports the spectral positions of the experimental LSP resonance, the first narrow peak and the associated Rayleigh anomalies [5]. As shown in this figure, the excited sharp resonances closely follow the dispersion of Rayleigh anomalies. Thereby, in this particular case, the coupling between the nanoparticles is mediated *via* delocalized modes, highly dominated by dipolar far-field effects.

In addition to that, we also performed fluorescence lifetime measurements by using a time correlated single photon counting technique (TCSPC). The obtained results show significant lifetime reductions of the radiative excited states for the short grating periods from 1.57ns to 1.09ns. In contrast to that, no significant reductions have been obtained in the case of the SLR regime according to their delocalized photonic features.

This part of our work shows that, by tuning the nanoparticle grating parameters, it is possible to select specific optical responses for emitters at near or far field zones.

(a)

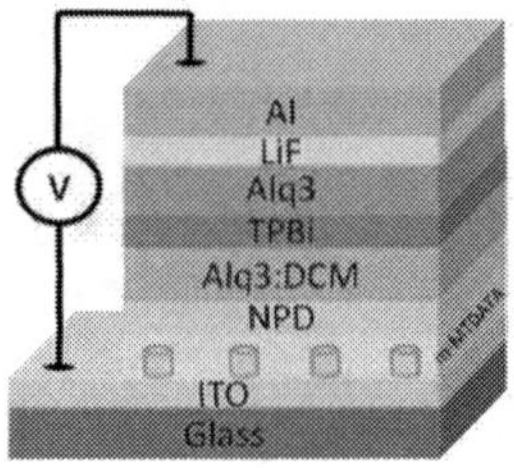

(b)

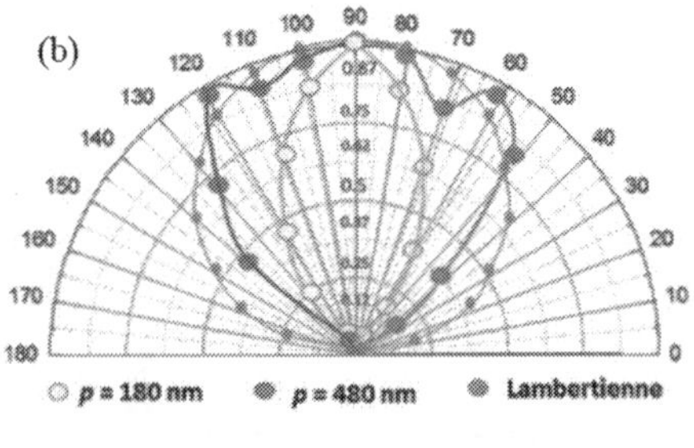

Fig. 2. (a) Scheme of the plasmonic organic light emitting diode. (b) Angular emission diagram for a conventional, plasmonic short and long period Ag-grating based OLEDs.

Based on the previous study, we investigate the plamonic effect of the studied Ag gratings on the optical and electrical properties of OLED devices. The studied OLED is depicted in fig.2.a. As an example, two devices containing gratings of $p = 180nm$ and $p = 480nm$ have been fabricated and studied (fig.2.a). *IVL* measurements and angular electro-luminance pattern have been measured and compared to a reference device without nanoparticles.

In addition to the *IVL* enhancement properties of the plasmonic OLEDs, the obtained emission patterns (fig.2.b) show various features depending on the array period. In the case of the OLED containing the short-period grating (p = 180nm), the emission is modified to a more directional one, fully contained in a cone aperture of 30 degrees, as compared to the well-known Lambertien pattern of the reference device. However, in the case of the OLED containing a long-period array (p = 480nm), the pattern shows a maximum of emission in preferred directions, at ± 30° from the normal to the substrate. Therefore, the redirection of the emission can be attributed to the excitation of hybrid modes (SLR); the metallic nanoparticles interacting *via* a dipole-dipole coupling are involved in constructive interference processes, leading to a significant re-emission of the light into diffractive modes at well-defined directions.

3. Conclusion

In this work, we report the investigation of the influence of Ag nanoparticle gratings on the OLED features. Our results demonstrate the existence of two coupling regimes dominated by near or far field effects, respectively. These plasmonic structures can strongly modify the optical and electrical performances of OLED devices. Specifically, the emission pattern can be drastically changed to a more directed one. Therefore, plasmonic structures can be used as an integrated nano-optical compound to spectrally and spatially shape and control the emission in terms of intensity, efficiency and directionality.

4. References

1. B. Lamprecht, G. Schider, R. T. Lechner, H. Ditlbacher, J.R. Krenn, A. Leitner, and F. R. Aussenegg, "*Metal Nanoparticle Gratings: Influence of Dipolar Particle Interaction on the Plasmon Resonance,*" PRL, **84** (20), 4721-4724 (2000).
2. T. D. Neal, K. Okamoto, and A. Scherer, "*Surface plasmon enhanced emission from dye doped polymer layers,*" Optics Express , **13**(14), 5522-5527 (2005).
3. M. A. Garcia, "*Surface plasmon in metallic nanoparticles: fundamentals and applications,*" Journal of Physics D: Applied Physics, **44**(28), 283001-283020 (2011).
4. D. Khlopin, F. Laux, W. P.Wardley, J. Martin, G. A.Wurtz, J. Plain, N. Bonod, A. V. Zayats, W. Dickson and D. Gerard, "*Lattice modes and plasmonic linewidth engineering in gold and aluminum nanoparticle arrays,*" JOSA B, **34**(3), 691-700 (2017).
5. A. Maradudin, I. Simonsen, J. Polancoand R. M. Fitzgerald, "*Rayleigh and Wood anomalies in the diffraction of light from a perfectly conducting reflection grating,*" Journal of Optics, **18**(2), 024004-024014 (2016).

A new route for high quality nanometric films of inorganic halide perovskites

Naomi Falsini[1], *Nicola* Calisi[2,3], *Giammarco* Roini[1,4], *Andrea* Ristori[5], *Francesco* Biccari[1], *Paolo* Scardi[6], *Claudia* Borri[2], *Stefano* Caporali[2,3], and *Anna* Vinattieri[1,5,7]

[1] Department of Physics and Astronomy, University of Florence, via G. Sansone 1, I-50019 Sesto Fiorentino (FI), Italy
[2] DIEF-Industrial Engineering Department, University of Florence, via S. Marta 3, I-50139 Florence, Italy
[3] INSTM-Interuniversity National Consortium for Material Science and Technology, via Giusti 9, I-50121 Florence, Italy
[4] Department of Information Engineering, University of Brescia, via Branze, 38, I-25123 Brescia, Italy
[5] LENS-European Laboratory for Non-Linear Spectroscopy, via Nello Carrara 1, I-50019 Sesto Fiorentino (FI), Italy
[6] Department of Civil, Environmental and Mechanical Engineering, University of Trento, via Mesiano 77, I-38123 Trento, Italy
[7] INFN-Firenze, via G. Sansone 1, I-50019 Sesto Fiorentino (FI), Italy

Abstract. Successful deposition of $CsPbX_3$ (X = Br, Cl) thin films (50-500 nm) on several kind of substrates has been realized by Radio-Frequency Magnetron Sputtering. The technique allows for high homogeneity of the samples on large areas (several cm^2) not only in terms of morphology but also referring to the optical properties. In particular, high resolution (in space, spectrum and time) photoluminescence studies in a wide temperature range (10-300 K) reveal that the low inhomogeneous broadening comes from a submicron size disorder while no significant contribution arises from a micrometric or even larger disorder. Given the relevance of inorganic halide perovskites for innovative optoelectronic devices, our results prove the scalability of this technique. Moreover, the successful deposition on several different substrates open the route for an easy integration in multi-layered structures.

1 Introduction

In the last decades fully inorganic halide perovskites have emerged as one of the most interesting class of semiconductor materials for applications in the field of energy harvesting, photonics and optoelectronics [1]. In particular, caesium-based halide perovskites are of extreme relevance for efficient incoherent (LED) and coherent (Laser) light emitters in the visible spectral region, being direct-band gap semiconductors, easily tunable by controlling the chemical composition. $CsPbCl_3$ shows a blue emission of interest for white light generation, when coupled with a converting phosphor; $CsPbBr_3$, having a bandgap at 2.3 eV, helps in overcoming the green gap problem of nitrides [2]. Moreover, white light generation can be obtained by multilayer structures of $CsPbI_3$, $CsPbBr_3$ and $CsPbCl_3$.

However, one of the main problems that limits the scalability of the material for innovative devices comes from the lack of homogeneity over large sample area (several cm^2). In this work we demonstrate the first successful deposition of thin films of fully inorganic perovskite of $CsPbBr_3$ and $CsPbCl_3$ by the Radio-Frequency (RF) Magnetron Sputtering technique. In this way we succeeded in obtaining high quality samples with a very good uniformity and homogeneity concerning the morphological and optical properties.

2 Results

$CsPbBr_3$ and $CsPbCl_3$ thin films were grown by RF Magnetron Sputtering on several substrates (soda lime glass - SLG, quartz, TiO_2, Silicon, etc.) with thickness ranging between 50 and 500 nm [3, 4].

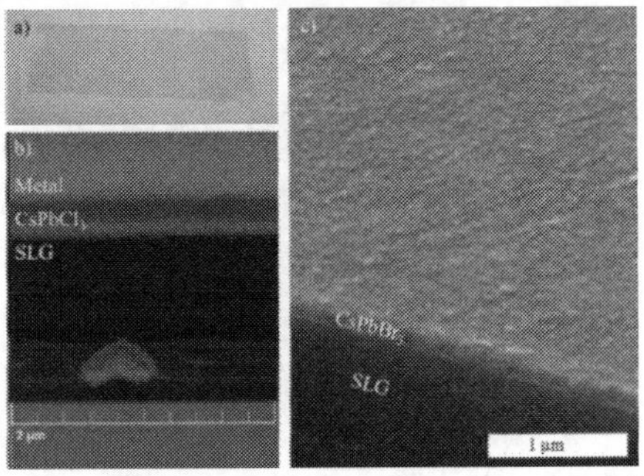

Fig. 1. a) Image of a 70 nm $CsPbCl_3$ film on quartz. b-c) SEM micrographs of a 150 nm $CsPbCl_3$ film and of a 140 nm $CsPbBr_3$ film on glass.

* Corresponding author: vinattieri@fi.infn.it

In Figure 1 a typical image of a 70 nm $CsPbCl_3$ film on quartz is shown along with two SEM micrographs of a $CsPbCl_3$ and a $CsPbBr_3$ thin film. The samples appear to be highly uniform and homogeneous at the macro-scale (several cm^2) and compact as from the SEM analysis. XPS measurements were performed, providing a substantial agreement with the expected stoichiometry, whereas XRD spectra clearly indicate the presence of the correct crystalline phase.

A detailed photoluminescence (PL) spectroscopy study was conducted at the macro and micro scale in a wide temperature range (10-300 K) to assess the origin of the inhomogeneous broadening and to quantify the PL quantum yield quenching. In figure 2a the PL spectrum at 10 K shows an inhomogeneous broadening less than 20 meV (FWHM); the comparison between macro-PL (excited area ~ 10^4 μm^2) and micro-PL (excited area ~ 1 μm^2) spectra (Fig. 2b) indicate that the origin of the inhomogeneous broadening comes from a sub-micrometric scale disorder. A significant decrease of the PL intensity is observed starting from ~150 K in both kind of samples and it originates from a reduced number of carriers and excitons that initially populate the radiative states [4].

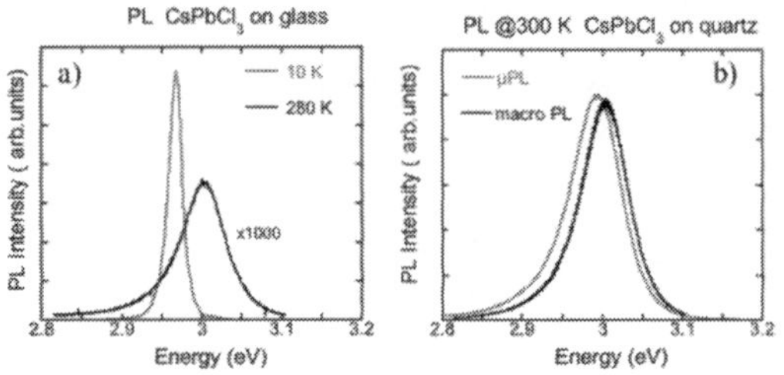

Fig. 2. a) PL spectra at 10 K and 280 K of a $CsPbCl_3$ thin film on glass. b) Comparison between micro and macro-PL spectra at room temperature of a $CsPbCl_3$ thin film on quartz.

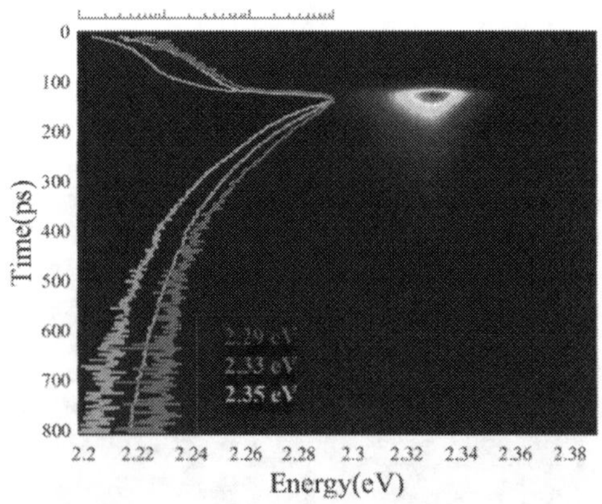

Fig. 3. Streak camera image of a $CsPbBr_3$ film along with the normalized PL decays (Log scale) extracted from the image at three different energies: 2.35 eV free carrier recombination, 2.33 eV localized exciton emission, and 2.29 eV low energy side due to deeper exciton localization.

Time-resolved (TR) spectra were acquired by a streak camera after excitation of the sample at 365 nm with 1.2 ps mode-locked laser (Figure 3).

Similar results are found for $CsPbBr_3$ and $CsPbCl_3$: non exponential decays are commonly found in this class of materials with evidence of an exciton localization at low temperature and the presence of a thermalized high energy tail due to free carrier recombination [5].

Moreover, high satisfactory is the stability of PL signal in the whole sample area despite sample ageing. In fact, the loss of PL signal is of the order of few tens of percent after 2-3 months under ambient conditions, which, to our knowledge, is a considerable stability when compared with literature data.

3 Conclusions

We have demonstrated that high quality thin films of $CsPbX_3$ (X = Br, Cl) can be realized by RF-Magnetron sputtering with state-of-the art homogeneity in terms of morphology and optical properties on sample areas as large as several square centimeters. In particular, this technique turns out to be highly promising for $CsPbCl_3$ deposition because of limitations in realizing films by standard solution-based techniques.

References

1. Y. Kanemitsu, T. Handa, Jpn. J. Appl. Phys. **57**, 1 (2018) https://doi.org/10.7567/JJAP.57.090101
2. K. Sim, T. Jun, J. Bang, H. Kamioka, J. Kim, H. Hiramatsu, H. Hosono, Appl. Phys. Rev., **6**, 031402 (2019) https://doi.org/10.1063/1.5098871
3. C. Borri, N. Calisi, E. Galvanetto, N. Falsini, F. Biccari, A. Vinattieri, G. Cucinotta, S. Caporali, Nanomaterials, **10**, 60 (2020) https://doi.org/10.3390/nano10010060
4. N. Falsini, N. Calisi, G. Roini, A. Ristori, F. Biccari, P. Scardi, S. Caporali, A. Vinattieri, Adv. Mater. (under submission)
5. F. Gabelloni, F. Biccari, N. Falsini, N. Calisi, S. Caporali, A. Vinattieri, Nanophotonics, **8**, 1447 (2019) https://doi.org/10.1515/nanoph-2019-0013

Room-temperature continuous-wave upconverting micro- and nanolasing for bio-optofluidics

Angel Fernandez-Bravo[*1], *Liliana Moscardi*[1], *Aaron M. Ross*[1], *Guglielmo Lanzani*[1], *Emory M. Chan*[2], *Teri W. Odom*[3], *P. James Schuck*[4], *and Francesco Scotognella*[1]

[1]Dipartimento di Fisica - Politecnico di Milano P.zza Leonardo da Vinci 32, 20133 Milano
[2]The molecular Foundry, Lawrence Berkeley National Laboratory, 94720, Berkeley, California
[3]Department of Chemistry, Northwestern University, Evanston, IL, USA. 5
[4]Department of Mechanical Engineering, Columbia University, New York, NY, USA

Abstract. Nanolasers that operate under the continuous-wave pump and are robust in diverse environments will make possible compact optoelectronic devices, biomedical imaging, and large-scale quantum photonics. However, current nanolasers require low temperatures or pulsed excitation because their small mode volumes severely limit gain relative to cavity loss. Here, I will present continuous-wave upconverting micro- and nanolasing at room temperature with record-low thresholds and high photostability. I will explore the future implications of such a low-threshold laser for optofluidics.

Nanolasers that operate under continuous-wave pump and are robust in diverse environments will make possible compact optoelectronic devices, biomedical imaging, and large-scale quantum photonics. However, current nanolasers require low temperatures or pulsed excitation because their small mode volumes severely limit gain relative to cavity loss. Here, I will present continuous-wave upconverting micro- and nanolasing at room temperature with record-low thresholds and high photostability. I will talk about the coupling between Tm^{3+}-based energy-looping nanoparticles (ELNPs) and stand-alone optical microcavities with implications in deep-tissue imaging[1,2].

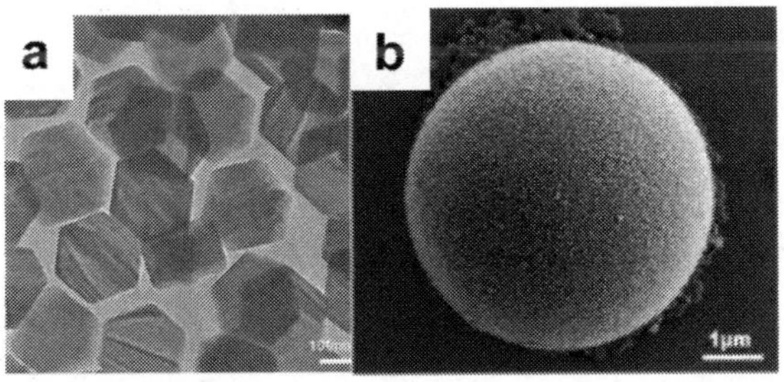

Figure 1 – **a.** TEM image of upconversion nanoparticles used to coat microresonators. **b.** SEM image of a microsphere.

I will also cover how to achieve directional lasing to enhance interactions at the nanoscale and selective single-mode lasing from Yb^{3+}/Er^{3+}-co-doped upconverting nanoparticles conformally coated on silver nanopillar arrays supporting sharp lattice plasmon modes, Figure 2.

The intense electromagnetic nearfields resulted in a threshold of 29 W/cm2, orders of magnitude lower than other small lasers[3]. Micron-sized lasers fabricated from upconverting nanoparticles (UCNP) coupled to whispering gallery mode (WGM) microresonators can exhibit continuous-wave anti-Stokes lasing useful for tracking cells, environmental sensing, and coherent stimulation of biological activity. The integration of these microlasers into organisms and microelectronics requires even smaller diameters, however, which raises threshold pump powers beyond practical limits for biological applications. To meet the need for low lasing thresholds and high-fidelity fabrication methods, we use correlative optical and electron microscopy, Figure 2, to uncover the nanoparticle assembly process and structural factors that determine efficient upconverted lasing.

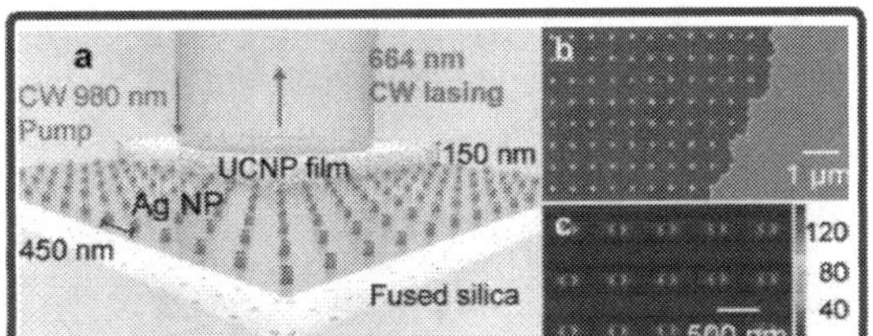

Figure 2 – Sub-wavelength plasmon nanolaser. a. Schematic. **b.** SEM of nanocavity arrays. **c.** FDTD simulation.

* Corresponding author: angel.fernandez@polimi.it

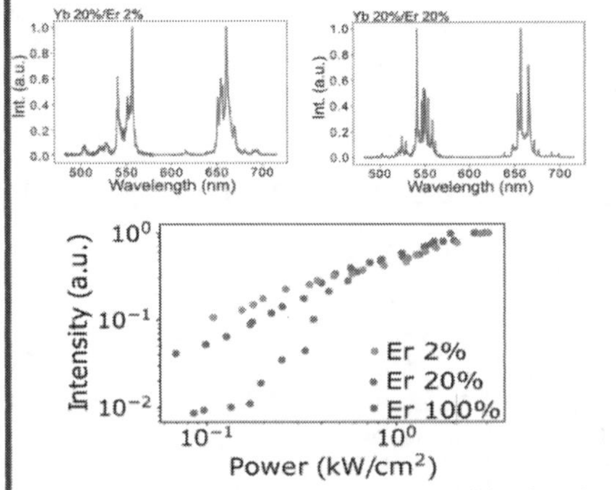

Figure 3 – Upper panels: Spectra with changing concentrations of Er^{3+} ions in the crystal lattices use for lasing at low thresholds. Lower panel: Light-Light curves for 3 lanthanide compositions with changing spontaneous emission coupling (beta) factors, showing thresholdless lasing behavior.

I will explore the future implications of such a low-threshold laser and the potential paradigm change they can bring to, deep-tissue interrogation, microfluidic design and bio-optofluidics. Here we show upconverting microlasing with high photostability under continuous wave (CW) pump enabled by photonic microcavities. Solution-processed UCNPs at a submonolayer thickness can be coated on biocompatible polymeric (polystyrene) cavities. After pumping with a CW laser at NIR wavelength, we observed lasing in the visible regime ca. 550 and 650 nm at room temperature. We observe lasing at thresholds in the order of 100's of W/cm^2 which helps preserve biological tissues and species, Figure 3. This, along with the exceptionally large spontaneous emission coupling and low lasing threshold will be critical in the development of future biologically compatible optomechanical traps, and bio-integrated sensors.

We demonstrate that the WGM signatures of these upconverting microlasers can be imaged and distinguished through tissue-mimicking phantoms[2,4]. These advances will enable the fabrication of more efficient upconverting lasers for imaging, sensing, and actuation in optically complex environments.

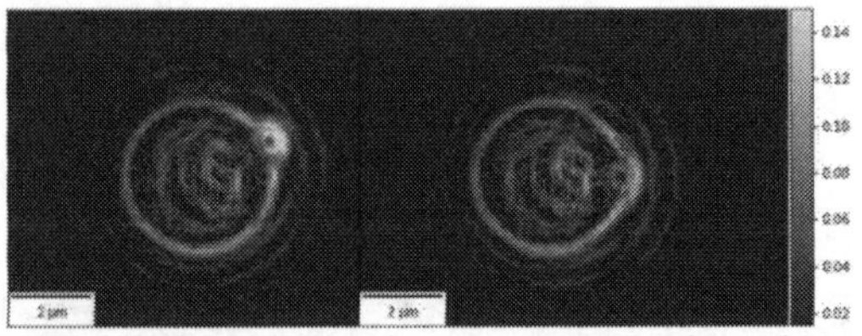

Figure 4 – Segmented images show proteins orbiting around a stand-alone micrometer size optofluidic trap.

Additionally, the solid-state laser showed high photostability that allows for long experiments with biological samples with no observed degradation. I will extend the Yb^{3+}/Er^{3+}-co-doped lasers to stand alone microcavities to fabricate optomechanical traps for biologically relevant species, shown in Figure 4. The result of these microscopic optical traps is a fine control of nanometer scale species in a reduced volume, fully controlled by light.

References

1. Levy, E. S. *et al.* Energy-Looping Nanoparticles: Harnessing Excited-State Absorption for Deep-Tissue Imaging. *ACS Nano* **10**, (2016).
2. Fernandez-Bravo, A. *et al.* Continuous-wave upconverting nanoparticle microlasers. *Nat. Nanotechnol.* **13**, (2018).
3. Fernandez-Bravo, A. *et al.* Ultralow-threshold, continuous-wave upconverting lasing from subwavelength plasmons. *Nat. Mater.* **18**, 1172–1176 (2019).
4. Liu, Y. *et al.* Controlled Assembly of Upconverting Nanoparticles for Low-Threshold Microlasers and Their Imaging in Scattering Media. *ACS Nano* **14**, 1508–1519 (2020).

Tuning the transient opto-electronic properties of few-layer MoS_2 nanosheets via substrate nano-patterning

Andrea Camellini[1, 6*], *Carlo* Mennucci[2], *Andrea* Mazzanti[3], *Christian* Martella[4], *Alessio* Lamperti[4], *Alessandro* Molle[4], *Francesco* Buatier de Mongeot[2], *Giuseppe* Della Valle[3, 5], and *Margherita* Zavelani-Rossi[1, 5]

[1]Dipartimento di Energia, Politecnico di Milano, via G. Ponzio 34/3, I-20133 Milano, Italy
[2]Dipartimento di Fisica, Università di Genova, Via Dodecaneso 33, I-16146 Genova, Italy
[3]Dipartimento di Fisica, Politecnico di Milano, Piazza L. da Vinci 32, I-20133 Milano, Italy
[4]IMM – CNR, Unit of Agrate Brianza, via C. Olivetti 2, I-20864 Agrate Brianza, Italy
[5]IFN – CNR, Piazza L. da Vinci 32, I-20133 Milano, Italy
[6]present address: Functional Nanosystems, Istituto Italiano di Tecnologia,via Morego 30, I-16163 Genova, Italy

Abstract. We study the optical properties of Molybdenum Disulphide nanosheets deposited by chemical vapor deposition onto a nanopatterned substrate endowed with a uniaxial corrugation. The uniaxial nanocorrugation leads to a polarization dependence in steady state and an anisotropic relaxation around the C resonance. Finite element numerical simulations allow to better disclose the scenario. Our findings point towards new possibilities for the manipulation of the optical properties of two-dimensional transition metal dichalcogenides via substrate nanopatterning.

Nowadays two-dimensional transition metal dichalcogenides (TMDs) constitute one of the most promising class of materials for novel opto-electronic devices due to their peculiar physical properties including strong light-matter interaction and excitonic effects [1]. As recently demonstrated, their properties can be modified via substrate pattern design by introducing an anisotropic degree of freedom in the spatial arrangement of TMD nanosheets [2, 3].

In this context, a new scalable deposition methodology which allows chemical vapour deposition growth over macroscopic area (cm^2 range) of Molybdenum Disulfide (MoS_2) has been applied to a self-organized nano-patterned substrate with sub-wavelength rippled shaped corrugation [4].

So far, the effect of nano-patterning on the opto-electronic properties of TMDs has been considered only in terms of Raman modes, photoluminescence, absorption whereas the effects of morphology induced anisotropy in the time domain has been largely overlooked.

In this work, we present a detailed analysis of the linear and the transient optical properties of a few layer MoS_2 nanosheets, deposited on top of a SiO_2 substrate endowed with a uniaxial nano-corrugation [5]. Our study of the morphology dependent behavior is based on polarization resolved optical extinction and broadband ultrafast pump-probe experiments with 100fs temporal resolution. The experimental evidences, supported by full-wave finite element numerical simulations, show a rich and complex scenario. This is characterized by: (i) a giant optical anisotropy in the spectral region of the four main excitonic resonances (A, B, C, D) with peculiar features dictated by of the out-of-plane component of MoS_2 dielectric permittivity, (ii) a polarization dependence of the relaxation dynamics around the C resonance (Fig. 1) where the optical response is dominated by a van Hove singularity in the joint density of states. We argue that the polarization dependent dynamics could involve different electron-phonon scattering mechanisms enhanced by the substrate nano-patterning thus allowing the possibility of anisotropy-related exciton engineering.

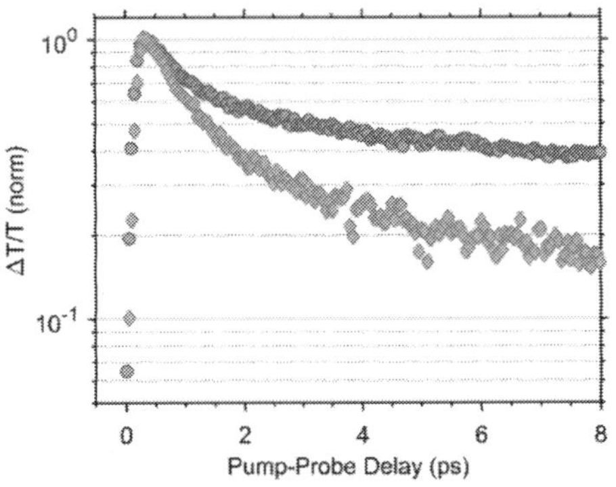

Fig. 1. Normalized transient transmission ($\Delta T/T$) dynamics around the spectral position of the C exciton for different probe polarization direction: parallel (circles) and orthogonal (diamonds) to the corrugation axis.

* Corresponding author: andrea.camellini@iit.it

The same nano-corrugation of the SiO_2 substrate can also be exploited for the deposition of an array of metal nanowires in close contact with MoS_2 nanosheets [2]. This hybrid metal/semiconductor interface can be exploited for increasing the photon harvesting properties of TMDs and could potentially enable hot carrier based photodetection over cm^2 areas.

Our approach show that substrate nano-patterning is a viable tool for tailoring the opto-electronic properties of two-dimensional materials and provide a suitable platform for new applications in the field of hot carrier technology.

References

1. Q.H. Wang, K. Kalantar-Zadeh, A. Kis, J. N. Coleman, M. S. Strano, Nature Nanotech. **7**, 699-712, (2012)
2. C. Martella, C. Mennucci, A. Lamperti, E. Cappelluti, F. Buatier de Mongeot, A. Molle, Adv. Mater. **30**, 1705615, (2018)
3. C. Martella, L. Ortolani, E. Cianci, A. Lamperti, V. Morandi, A. Molle, Nano Research **12**, 1851–1854 (2019)
4. C. Martella, C. Mennucci, E. Cinquanta, A. Lamperti, E. Cappelluti, F. Buatier de Mongeot, A. Molle, Adv. Mater. **29**, 1605785, (2017)
5. A. Camellini, C. Mennucci, E. Cinquanta, C. Martella, A. Mazzanti, A. Lamperti, A. Molle, F. Buatier de Mongeot, G. Della Valle, M. Zavelani-Rossi, ACS Photonics **5**, 3363 (2018)

EPJ Web of Conferences **238**, 07007 (2020) https://doi.org/10.1051/epjconf/202023807007
EOSAM 2020

Revealing excited states dynamics in cross-linked covalent hybrids of graphene and diketopyrrolopyrrole oligomers via ultrafast transient absorption spectroscopy

Aaron M. Ross,[1*] *Meng Zheng,*[2] *Michele Maggini,*[2] *Fabio Marangi,*[1,3] *Giulio Cerullo,*[1,3] *Teresa Gatti,*[4] *Francesco Scotognella*[1,3]

[1] Department of Physics, Politecnico di Milano, Piazza L. da Vinci 32, 20133 Milano, Italy
[2] Department of Chemical Sciences, University of Padova, via Marzolo 1, 35131, Padova, Italy
[3] Center for Nano Science and Technology@PoliMi, Istituto Italiano di Tecnologia (IIT), Via Giovanni Pascoli, 70/3, 20133, Milan, Italy
[4] Center for Materials Research and Institute of Physical Chemistry, Justus Liebig University Giessen, Heinrich Buff Ring 17, 35392, Giessen, Germany.

Abstract. We report on ultrafast dynamics in cross-linked covalent hybrids of graphene and diketopyrrolopyrrole (TDPP) oligomers. Comparisons between non-polymerizing (EXG-TDPP) and polymerizing (c-EXG-TDPP) hybrids show that charge/energy transfer occurs faster than 50 ps in EXG-TDPP, while the c-EXG-TDPP system shows quenching within 18 ps. The transient response of c-EXG-TDPP is completely quenched (>99%), while in EXG-TDPP full quenching occurs after 1.39 ns. Photocurrent studies of c-EXG-TDPP films show IPCE > 2%, indicating ultrafast charge transfer between TDPP and graphene.

1 Introduction

One promising class of synthetic systems in which efficient energy/charge transfer may be achieved is graphene-based materials (GBMs). Graphene monolayers exhibit exquisitely high carrier mobilities, broadband absorption (2.3% per monolayer), and ultrafast optical non-linearities such as saturable absorption [1].

The utility of graphene can be enhanced further via functionalization, either with nanoparticles or molecules, and via covalent or non-covalent linkages [2]. Sensitization of graphene via attachment of strongly-absorbing chromophores may provide a new avenue for the development of cheap and non-toxic solar cells. Additionally, the presence of graphene, whether via covalent or non-covalent linkage, has been shown to significantly modify the electronic properties of chromophores [3].

Here we investigate the ultrafast dynamics of mechanically-exfoliated few-layer graphene and diketopyrrolopyrrole oligomers via ultrafast transient absorption spectroscopy with 70 fs time resolution. Two hybrids are studied in comparison to the isolated Ph_2TDPP chromophore, one of which (c-EXG-TDPP) cross-links graphene flakes via long polymeric backbone chains of TDPP, and the other of which (EXG-TDPP) does not polymerize (Figure 1). While quenching of the transient optical absorption is observed in both samples, the efficiency of this quenching process is shown to be much higher in c-EXG-TDPP compared to EXG-TDPP.

2 Results

We examined a recently developed cross-linked covalent hybrid between few-layer graphene flakes and bifunctional diphenyl-dithiophenedikethopyrrolopyrrole (Ph_2TDPP) units [3,4]; the resulting cross-linked structure is referred to as c-EXG-TDPP. A similar model hybrid unit that does not cross-link is referred to EXG-TDPP (Figure 1). These samples are dissolved in toluene and placed in 1 mm path length cuvettes. It was confirmed here via steady-state linear absorption and emission experiments that the electronic properties of the cross-linked system c-EXG-TDPP are modified significantly compared to both Ph_2TDPP and EXG-TDPP: the main absorption band of Ph_2TDPP is red-shifted by 66 nm in c-EXG-TDPP, indicating a large increase in conjugation length along the polymeric backbone chain. These changes in electronic properties may explain the differences in charge/energy transfer processes between the two hybrids.

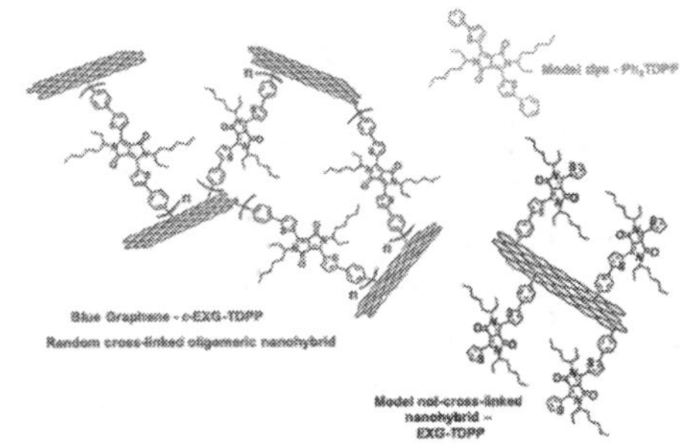

Figure 1: Structure of the cross-linked graphene-dithiophenediketopyrrolopyrrole hybrid c-EXG-TDPP, of the model Ph_2TDPP chromophore unit constituting its cross-linking oligomers, and of the analogous hybrid system lacking cross-linking EXG-TDPP.

The ultrafast dynamics of both the base unit Ph_2TDPP, and the two hybrids, are studied via ultrafast transient absorption spectroscopy. A non-collinear optical parametric amplifier (NOPA) is pumped with 800 nm,

* Corresponding author: aaronmichael.ross@polimi.it

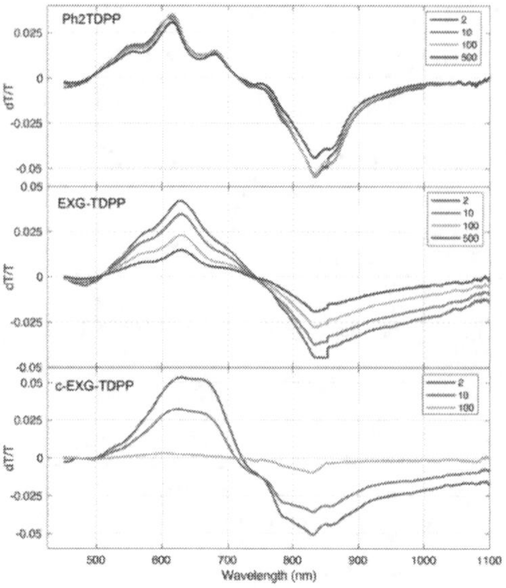

Figure 2: Transient absorption spectra for Ph_2TDPP, EXG-TDPP, and c-EXG-TDPP at selected pump-probe delay times (labeled in picoseconds in legend). Discontinuity at ~850 nm is due to stitching between different experimental runs.

150 fs pulses generated by a chirped pulse amplified (CPA) Ti:sapphire laser at a 1 kHz repetition rate, yielding sub-100 fs pulses at 575, 625, and 645 nm [5], used to pump Ph_2TDPP, EXG-TDPP, and c-EXG-TDPP, respectively. Broadband white light is generated in a sapphire crystal as a probe, and differential transmission ($\Delta T/T$) is measured by comparing the transmission of the probe through the sample when the pump is blocked/unblocked.

Global analysis of $\Delta T/T$ spectra (Fig. 2) using the sum of three exponentials allowed for the elucidation of energy/charge transfer processes. In EXG-TDPP, a process faster than 50 ps is observed, followed by a much longer decay with a lifetime of 1.39 ns, in agreement with previous TRPL measurements [3]. Ground state bleaching (GSB) features are observed at 575 and 645 nm, stimulated emission (SE) signals are observed at 700 nm; these results are in agreement with linear absorption and steady-state emission. A broad excited-state absorption (ESA) band is observed starting at 725 nm, and extending out into the NIR past 950 nm. The decay-associated spectrum (DAS) extracted via global fitting (not shown) associated with the 47 ps lifetime is broadened out into the NIR farther than both Ph_2TDPP ESA features and the longer EXG-TDPP DAS with lifetime of 1.39 ns; this may indicate the presence of a charge-transfer state, which is followed by back electron transfer into the TDPP moiety.

However, much more efficient and rapid quenching is observed in c-EXG-TDPP: a broad DAS with lifetime 3.8 ps is observed at longer wavelengths than the main GSB peaks, indicating an in-filling of the excited singlet states from higher-lying states. The full transient absorption signal is quenched within 18 ps. Since the GSB features observed from 600-660 nm in the 18 ps DAS are also observed in linear absorption, and the ESA features are similar to those associated with the excited singlet states of Ph_2TDPP, we believe preliminarily that charge separation between the TDPP moieties and covalently-bound graphene occurs on the 18 ps timescale, and that the charge separated state may not be optically active. This phenomenon may be explained by the increased conjugation length in c-EXG-TDPP, which could lead to a more stable charge separated state than in EXG-TDPP. It is also expected that there is an interplay between charge transfer from nearest TDPP moieties attached directly to graphene, energy transfer to graphene from more distant TDPP moieties, and increased conjugation length in c-EXG-TDPP compared to oligomeric EXG-TDPP.

Photocurrent measurements of c-EXG-TDPP films indicate that a charge transfer process is taking place, with incident photon to electron conversion efficiencies (IPCE) > 2%. This result may confirm the rapid charge separation process occurring in c-EXG-TDPP; photocurrent collection is further enhanced likely due to transport of the extracted electron along the graphene scaffolding. Photocurrents were not detected for Ph_2TDPP or non-film-forming EXG-TDPP drop-cast samples.

We conclude that energy/charge transfer processes in these hybrid graphene-dye systems that occur on sub 50 ps timescales are observed using ultrafast transient absorption spectroscopy, with charge transfer confirmed by photocurrent measurements.

References

1. D. R. Cooper, B. D'Anjou, N. Ghattamaneni, B. Harack, M. Hilke, A. Horth, N. Majlis, M. Massicotte, L. Vandsburger, E. Whiteway, V. Yu, ISRN Condens. Matter Phys. **2012**, 501686 (2012).
2. G. Bottari, M. Ángeles Herranz, L. Wibmer, M. Volland, L. Rodríguez-Pérez, D. M. Guldi, A. Hirsch, N. Martín, F. D'Souza, T. Torres, Chem. Soc. Rev. **46**, 4464 (2017).
3. M. Zheng, F. Lamberti, L. Franco, E. Collini, I. Fortunati, G. Bottaro, G. Daniel, R. Sorrentino, A. Minotto, A. Kukovecz, E. Menna, S. Silvestrini, C. Durante, F. Cacialli, G. Meneghesso, M. Maggini, T. Gatti, Synth. Met. **258**, 116201 (2019).
4. P. E. Hartnett, E. A. Margulies, C. M. Mauck, S. A. Miller, Y. Wu, Y. L. Wu, T. J. Marks, M. R. Wasielewski, J. Phys. Chem. B **120**, 1357 (2016).
5. C. Manzoni, D. Polli, and G. Cerullo, Rev. Sci. Instrum. **77**, 1 (2006).

Chalcogenide Optical Fiber Components -INVITED

Martin Rochette
Department of Electrical and Computer Engineering, McGill University, Montréal (QC), Canada, H3A 2A7
Author e-mail address: martin.rochette@mcgill.ca

Abstract: This talk features recent achievements of the Nonlinear Photonics Group at McGill University towards the fabrication of chalcogenide-based optical fiber components such as nonlinear gain fibers, optical fiber couplers, filters, and saturable absorbers.

OCIS codes: (060.2320) Fiber optics amplifiers and oscillators; Nonlinear optics, fibers; (140.3070) Infrared and far-infrared lasers.

Introduction

In a quest for the development of optical fiber sources that operate in the mid-infrared, the availability of optical components that are all-fiber and compatible with the mid-infrared is of utmost importance. Fiber components made from chalcogenide glass raise attention in this context, owing to their transparency in the 1-13 μm wavelength range and their high nonlinearity that enables nonlinear processing. In this talk, I will present recent achievements of the Nonlinear Photonics Group at McGill University towards the fabrication of all-fiber chalcogenide components and devices that are compatible with the mid-infrared. These include gain media, optical fiber couplers, optical fiber filters, all-fiber saturable absorbers and emerging devices build from those building blocks.

References

Gain media

1. Alamgir, I., et al., "All-fiber nonlinear optical wavelength conversion system from the C-band to the mid-infrared," Optics Letters 45(4), 857-860 (2020).
2. Alamgir, I., et al., "In situ fabrication of far-detuned optical fiber wavelength converters," Optics Letters 44(18), 4467-4470 (2019).
3. Rochette, M., "Design and fabrication of nonlinear optical waveguides for far-detuned wavelength conversion to the mid-infrared," Advanced Optical Technologies 7(5), 321-326 (2018).
4. Li, L., et al., "Mid-infrared wavelength conversion from AsSe microwires," Optics Letters 42(3), 639-642 (2017).
5. Li, L., et al., "Chalcogenide optical microwires cladded with fluorine-based CYTOP," Optics Express 24(17), 18931-18937 (2016).
6. Li., L., et al., "Design, fabrication and characterization of PC, COP and PMMA-cladded As2Se3 microwires," Optical Material Express 6(3), 912-921 (2016).
7. Godin, T., et al., "Far detuned mid-infrared frequency conversion via normal dispersion modulation instability in chalcogenide microwire," Optics Letters 39(07), 1885-1888 (2014).
8. Al Kadry, A., et al., "Mid-infrared sources based on the soliton self-frequency shift," Journal of the Optical Society of America B 29(6), 1347-1355 (2012).

Optical fiber couplers

9. Rezaei, M., et al., "All-Chalcogenide Single-Mode Optical Fiber Couplers," Optics Letters 44(21), 5266-5269 (2019).
10. Tavakoli, F., et al.,"Broadband and Wavelength-Dependent Chalcogenide Optical Fiber Couplers," IEEE Photonics Technology Letters 29(9), 735-738 (2017).

Filters

11. Zhang, K., et al., "Mid-infrared Compatible Tunable Bandpass Filter Based on Multimode Interference in Chalcogenide Fiber," Journal of Lightwave Technology 38(4), 857-863 (2020).
12. Zhang, K., et al., "Chalcogenide Fabry–Perot Fiber Tunable Filter," IEEE Photon. Technol. Lett. 30(23), 2013-2017 (2018).
13. Ahmad, R. et al., "Photosensitivity at 1550 nm and Bragg grating inscription in As2Se3 microwires," Applied Physics Letters 99 061109 (2011).

Pulse measurement devices

14. N. Abdukerim et al., "All-fiber frequency-resolved optical gating pulse characterization from chalcogenide glass," Optics Letters 43(14), 3228-3231 (2018).
15. N. Abdukerim et al., "Pulse characterization by cross-phase modulation in chalcogenide glass," Optics Letters 43(19), 4771-4774 (2018).

EPJ Web of Conferences **238**, 08002 (2020) https://doi.org/10.1051/epjconf/202023808002
EOSAM 2020

Crossing the exceptional point in a fiber-plasmonic waveguide -INVITED

Alessandro Tuniz[1,*], *Torsten* Wieduwilt[2], and *Markus* A. Schmidt[1]

[1] The University of Sydney Nano Institute (Sydney Nano), Institute of Photonics and Optical Science (IPOS), School of Physics, The University of Sydney, NSW 2006, Australia

[2]Leibniz Institute of Photonic Technology (IPHT Jena), Albert-Einstein-Str. 9, 07745 Jena, Germany

Abstract. We experimentally demonstrate a hybrid plasmonic fiber with tuneable Eigenmode interactions near the exceptional point. We experimentally observe a transition through the exceptional point in a fiber-plasmonic system: transmission experiments reveal fundamental changes in the underlying Eigenmode interactions as the environmental refractive index is tuned due to a crossing through the plasmonic exceptional point. These results extend the design opportunities for tunable non-Hermitian physics to plasmonic waveguide systems.

1 Introduction

Photonics is the ideal landscape for investigating non-Hermitian physics, where tailored amounts of optical gain and loss can be introduced into a photonic system to unlock otherwise inaccessible Eigenmode topologies [1]. Recent technological advances have shown many far-reaching applications [2], which rely on the unique characteristics of the Eigenstates close to exceptional point (EP) – where the system's Eigenvalues coalesce to a degenerate point that is highly sensitive to the environment – including sensitive modal manipulation, topological energy transfer, unidirectional propagation, and polarization conversion. These advances fundamentally rely on the unique characteristics of the eigenstates close to the EP: for example, nanoscale events can be detected by measuring their impact on macroscopic eigenstates [3].

Surface plasmon polaritons (SPP) [4], i.e., propagating surface waves at a metal-dielectric interface, can also be harnessed in the context of non-Hermitian photonics due to their dissipative nature [1]. The high sensitivity of SPPs to refractive index (RI) changes of the environment makes them attractive for sensing applications, leading to intensive research efforts in hybrid dielectric-plasmonic waveguides that can be straightforwardly interfaced with commercial light sources [5]. Such waveguides are inherently non-Hermitian and can therefore provide a platform for accessing EP-based functionality. A recent report presented nanostructures with localized plasmon resonances that show discrete frequency spectra [6], but experimental observations of EPs in plasmonic waveguides have so far been limited. Here, we present measurements of the transition across the plasmonic EP in a waveguide system, namely a tuneable hybrid plasmonic fiber [7].

2 Device Design and Characterization

A schematic of the experiment is shown in Fig. 1(a). The system is formed by a cylindrical SiO_2 waveguide (diameter: 20 μm) coated with a 30 nm gold nanofilm, surrounded by a liquid of a certain RI n_{ID}. Figure 1(b) show the properties of the real- and imaginary- parts of the two supported Eigenmodes (EM_j, j=1,2) that can be tuned by changing n_{ID}: here the real part of the effective indices either crosses or anti-crosses at a resonant wavelength λ_R, depending on n_{ID}. A transition between the two regimes occurs at the EP.

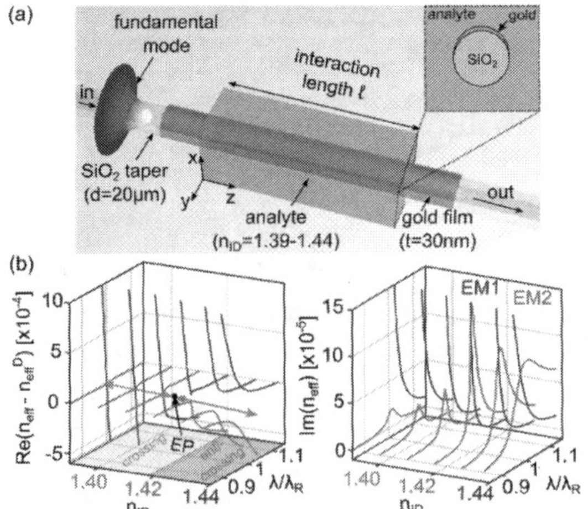

Fig. 1. (a) Hybrid fiber-plasmonic waveguide schematic. A silica fiber coated by a gold nanofilm is immersed in a liquid with RI of n_{ID} and length ℓ. (b) Calculated spectral distribution of $\mathrm{Re}(n_{eff,j} - n_{eff,D})$ (j=1,2, $n_{eff,D}$: equivalent dielectric waveguide without gold film) and $\mathrm{Im}(n_{eff})$ for EM1 and EM2 as a function of λ/λ_R

* Corresponding author: alessandro.tuniz@sydney.edu.au

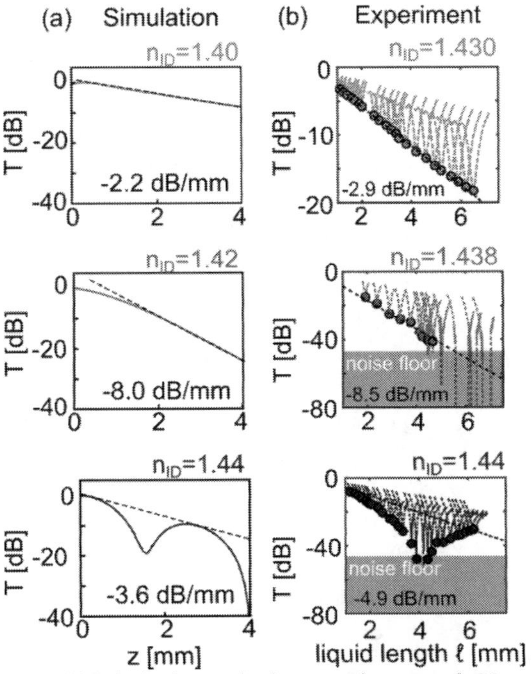

Fig. 2. (a) Calculated transmitted power at three example RIs. (b) Circles: measured on-resonance transmission as a function of ℓ for different analytes as labeled, showing the a crossing of the EP. Dashed lines show the full resonance spectrum centered in each transmission minimum (i.e., the horizontal axis of each spectrum is scaled by a factor ℓ/λ_R [7]). Dashed lines: fitted loss as labeled

Calculations show that the transition across the EP can be inferred from the transmission properties of this waveguide system by changing n_{ID}. Figure 2(a) shows the calculated transmitted power due to the modal evolution of the two hybrid EMs excited by the fundamental mode of the dielectric waveguide [8] at λ_R. The anti-crossing regime n_{ID}=1.44 shows an oscillating damped spatial energy distribution, [Fig. 2(a) bottom], while within the crossing regime (n_{ID}=1.40), no such oscillations are observed [Fig. 2(a), top]. At n_{ID}=1.42 we find that the transmission monotonically decays with the highest overall loss [Fig. 2(a), middle].

We perform a comprehensive series of transmission experiments on the above system [7] – different interaction lengths are realized by liquid columns of different length. Figure 2(b) shows measurements of the on-resonance power T vs. λ_R under different conditions: for n_{ID}=1.440 we measure a periodic energy exchange between the waveguides (i.e., the anti-crossing regime); a lower n_{ID} ≤1.438 yields an exponentially decreasing power with no oscillations (i.e., the crossing regime). Finally, an intermediate refractive index n_{ID}=1.438 shows no oscillations and the highest loss – a confirmation that the EP was locked via the additional degree of freedom of tuning n_{ID}. Since this device operates near the plasmonic cutoff, this sensor has a sensitivity >50µm/RIU for n_{ID}>1.43 [7], which, to the best of our knowledge, is one of the highest values measured for a plasmonic device to date.

3 Conclusions and Future Directions

In conclusion, we have presented a plasmonic fiber operating close the exceptional point. Our platform extends the capabilities of tuneable non-Hermitian photonics near the EP to plasmonic waveguides. The results presented here apply to any non-Hermitian waveguide, and immediately provide design tools for future EP-based photonic devices. Preliminary calculations show that this design can be easily modified to operate in the visible regime with a reduced footprint. The unique dispersion characteristics near the EP widen the scope of plasmonics-related applications. For example, this structure can be easily modified for compact operation in the visible regime and at a biologically relevant refractive index (n_{ID} ~1.33): preliminary calculations reveal that a scaled-down version of the above structure in contact with a gold nanofilm possesses an exceptional point at n_{ID} = 1.330, by phase-matching with the short-range surface plasmon polariton [8] at λ ~ 560nm. This device possesses a significantly smaller footprint, both in terms of lateral dimensions (one order of magnitude smaller) and interaction length (four orders of magnitude shorter), while operating in the visible regime. The same principles could be further applied to chip-scale devices with dimensions further reduced by one order of magnitude, e.g. using hybrid plasmonic waveguides modules [9]. These devices could find important applications in convenient, monolithic plasmonic biosensors of single-molecule events [6].

References

1. M.-A. Miri and A. Alù , Science **363**, eaar7709 (2019).
2. Ş.K. Özdemir, S. Rotter, F. Nori, L. Yang, Nat. Mater. **18**, 783 (2019).
3. W. Chen, Ş. K. Özdemir, G. Zhao, J. Wiersig, L. Yang , Nature 548, 192 (2017).
4. Lukas Novotny and Bert Hecht, *Principles of Nano-Optics* (Cambridge University Press, Cambridge, England, 2012).
5. T. Wieduwilt, A. Tuniz, S. Linzen, S. Goerke, J. Dellith, U. Hübner, M. A. Schmidt, Sci. Rep. **5**, 17060 (2015).
6. J.-H. Park, A. Ndao, W. Cai, L. Hsu, A. Kodigala, T. Lepetit, Y.-H. Lo, B. Kanté, Symmetry-breaking-induced plasmonic exceptional points and nanoscale sensing. Nat. Phys. **16**, 462(2020)
7. A. Tuniz, T. Wieduwilt, M. A. Schmidt, Phys. Rev. Lett. **123**, 213903 (2019).
8. A. Tuniz and M. A. Schmidt, Optics Express **24**, 7507 (2016).
9. A. Tuniz, O. Bickerton, F. J. Diaz, T. Käsebier, E.-B. Kley, S. Kroker, S. Palomba, C. M. de Sterke, Nat. Commun. **11**, 2413 (2020)

Micro/Nanofiber-coupled low-dimensional structures for nanophotonic applications -INVITED

Xin Guo[1,2*]

[1]State Key Laboratory of Modern Optical Instrumentation, Zhejiang University, Hangzhou 310027, China
[2]College of Optical Science and Engineering, Zhejiang University, Hangzhou 310027, China

Abstract. Optical micro/nanofiber, with diameter close to the wavelength of the waveguided light, leaves a considerably large fractional evanescent fields for coupling with low-dimensional functional nanostructures. Here I introduce our works on micro/nanofiber-coupled low-dimensional structures for nanophotonic applications including gas sensing and ultrashort pulse measurement.

1 Introduction

An optical micro/nanofiber (MNF) can guide light with tight optical confinement and strong evanescent waves, making it possible to efficiently couple with functional nanostructures [1-3]. Here I introduce our progress on micro/nanofiber-coupled low-dimensional structures, including metal nanoparticles, metal nanowires and semiconductor microwires, for nanophotonic devices and applications [4-8].

2 MNF-coupled low-dimensional structures and applications

2.1 Coupling of MNF with metal nanoparticles

Owing to their fascinating optical properties originated from localized surface plasmon resonance, metal nanoparticles are attracting increasing attentions as it is possible to confine light into the deep-subwavelength scale in all three dimensions. By attaching metal nanoparticles to the inner edge of a bent nanofiber, we theoretically show a hybrid structure to reduce the bending loss of a sharply-bent optical nanofiber [4]. Due to the near field coupling between the metal nanoparticles and the nanofiber, a considerable fraction of energy can be confined to the interface of the metal nanoparticles and the optical nanofiber, resulting in a significant reduction of the bending loss in the sharply-bent optical nanofiber. It shows the merits of easy operation, high integration, and a very small bending radius down to 1-μm level.

2.2 Coupling of MNF with micro/nanowires

Chemically synthesized metal nanowires or semiconductor microwires, with atomically smooth surface and well-developed crystal structure, have favorable properties including ultra-tight optical confinement, large surface-to-volume ratio, and high optical nonlinearity for nanophotonic applications such as sensors and optical correlators.

Via ultracompact near-field interaction, we realize effective coupling by directly contacting optical micro/nanofibers with metal nanowires [5,6]. A photon-to-plasmon conversion efficiency higher than 80% is experimentally realized in a microfiber-coupled Ag nanowire. Based on highly efficient coupling between optical fibers and Ag nanowires, we have assembled all-fiber hybrid photon-plasmon components such as hybrid Mach-Zehnder interferometer and ring resonators [6]. For practical applications, we further demonstrate a nanowire plasmonic sensor for detecting ammonia gas (NH_3) with detection limit lower than 100 ppm and response time of about 300 ms, which offers a promising route to bridge light from macroscopic fiber systems to microscopic nanowire plasmonics for highly localized and flexible sensing applications [7].

Moreover, by counterpropagating ultrafast pulses in a CdTe microwire evanescently coupled to microfibers at both ends, and converting the temporal profile of the pulses into the transverse second harmonic spatial image, we demonstrate a single-microwire optical correlator for femtosecond pulses at 2-μm wavelength [8]. The measured pulse duration from the microwire-based auto-correlator agrees well with that extracted by the standard auto-correlator. Due to the low-loss transmission window for the CdTe microwire in the mid-infrared region, the microwire-based auto-correlator shows great potential for mid-infrared ultrashort pulse measurement.

3 Conclusions

Our results above suggest that the micro/nanofiber-coupled low-dimensional structures may offer a versatile platform for nanophotonic devices and applications.

* Corresponding author: guoxin@zju.edu.cn

 https://doi.org/10.1051/epjconf/202023808003

This research was supported by the National Key Research and Development Project of China (2018YFB2200404), and the Fundamental Research Funds for the Central Universities.

References

1. L. M. Tong, R. R. Gattass, J. B. Ashcomv, S. L. He, J. Y. Lou, M. Y. Shen, I. Maxwell, E. Mazur, Nature **426**, 816 (2003)
2. L. M. Tong, F. Zi, X. Guo, J. Y. Lou, Opt. Commun. **285**, 4641 (2012)
3. L. M. Tong, Sensors **18**, 903 (2018)
4. Y. X. Yang, J. Gong, Y. X. Gao, H. Wu, X. Guo, P. Wang, L. M. Tong (to be submitted)
5. X. Guo, M. Qiu, J. M. Bao, B. J. Wiley, Q. Yang, X. N. Zhang, Y. G. Ma, H. K. Yu, L. M. Tong, Nano Lett. **9**, 4515 (2009)
6. X. Y. Li, W. Li, X. Guo, J. Y. Lou, L. M. Tong, Opt. Express **21**, 15698 (2013)
7. X. Guo, Y. B. Ying, L. M. Tong, Acc. Chem. Res. **47**, 656 (2014)
8. J. B. Zhang, Z. X. Shi, L. Yang, C. G. Xin, Y. H. Li, Y. Kang, L. Z. Wang, J. Gong, X. Guo, L. M. Tong (to be submitted)

Strain Sensor from Tapered Fibres

Celaschi Sergio[1]*, *Guerra* Christiano P.[1], *Biazoli* Claudecir [1], *Cordeiro* Cristiano B.[2], and *Grégoire* Nicolas [3]

[1]CTI Renato Archer, Campinas 13069-901, Brazil

[2] Dept. Eletrônica Quântica - IFGW, UNICAMP, Campinas 13083-859, Brazil

[3]Centre d'Optique, Photonique et Laser, Université Laval, Québec, Canada, G1V 0A6

Abstract. A new all fibre, and low transmission loss, digital optical strain sensor is proposed. This sensor behaves as a Coaxial Mach-Zehnder interferometer. Special depressed cladding single-mode fibre DCF was tapered down to the micrometer scale presenting FSR in the nm range. The sensor is modelled to probe up to ± 0.2% strain when under expansive or compression stresses, returning 20 optical Power Transfer Turning Points (PTTP) at 1575 nm transmitted wavelength.

1 Introduction

Devices fabricated by tapering optical fibres, enable additional effects and applications other than those reported using bulky optical fibres [1]. Analogue fibre optics sensing devices based on interferometer`s principles are crucial for many applications. Unfortunately, these sensors have as disadvantage a low linear dynamic range. On the other hand, the ease of fabrication of tapered fibre-based devices contributes to their use in applications such as: Brillouin optical sensors [2], tuneable acoustic-optical filters, super continuous radiation generation, among others applications [3], useful in Internet of Things (IoT). For these reasons, the concept of multimodal interference has been studied aiming at the development of different devices. Devices intrinsically manufactured from tapering single-mode fibres (SMF) allow the generation and propagation of selected super-modes, causing interference at the output of the multimode (MM) tapered section. As a result, low loss transmitted optical power may be observed at specific wavelengths. In addition, when this SMF has a Depressed Refractive Index profile (DCF), and the adiabatic condition holds at the tapered section, this structure behaves as a Coaxial Mach-Zehnder interferometer (CMZI) [3]. The interrogation of such sensor occur acting externally on either the MM optical path, or on the super-modes speed (external refractive index).

In this work, an all-fibre digital strain sensor based on a taper from a DCF is proposed. Our fabricated tapered-based CMZI senses external force fields acting on the optical path of the excited super-modes.

1.1 Materials, principles and methods

The DCF has been specified, and manufactured to enable HE_{11} HE_{12} mode coupling at the adiabatic tapering section [4]. The core and cladding radius and refractive index parameters of the DCF is shown in Figure 1. Determining which profile results from stretching a fibre into a given heat source with its own temperature distribution is a complex fluid mechanics problem.

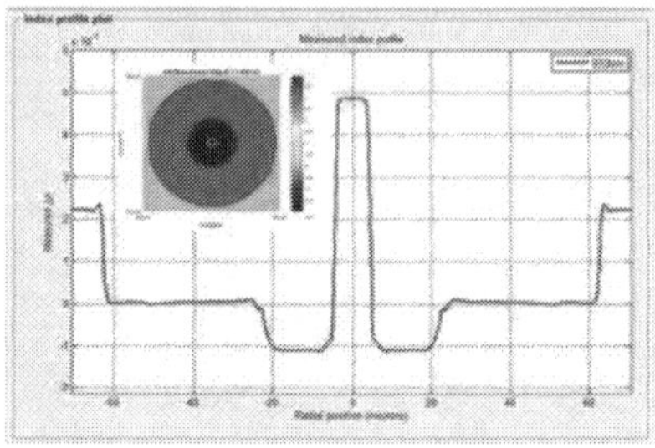

Fig.1. Refractive index profile of the DCF. Inset 2-D measurement data at 633 nm.

This problem was previously modelled by a coupled of partial differential equations and solved numerically [2]. However, no knowledge of fluid mechanics other than mass conservation is required to describe a simple model in which a fixed length of fibre is heated to a constant/uniform temperature and stretched. In this approach [4], simple mathematical equations (Eq. 1 and 2) are obtained for the shape of the profile,

$$\rho(\alpha, Lo, z) = \rho_0 \left[1 + \frac{2\alpha(z_f - z)}{(1-\alpha)L_O}\right]^{-1/2\alpha} \quad (1)$$

$$L_w = L_o + 2.\alpha.Z_f \quad (2)$$

Where ϕ *(α, Lo, z)* = *2ρ(α, L_o, z)* is the diameter of the fibre in the profile region. ρ_o is the radius of the fiber outside the tapered region, α is the linear rate of change in the length of the heated zone with the fiber extension. L_o is the initial length of the heated zone, $Z_f = Z + L_w/2$ the final extension of that region on the axial axis z of the fibre, and L_w the length of the waist region.

2 Experimental data and discussion

Three fibre tapers with different parameters, as shown in Table 1 were modelled and manufactured. A micro-torch, displaced in the longitudinal direction was used. The DCF was pulled using step motors.

* Corresponding author: sergio.celaschi@cti.gov.br

https://doi.org/10.1051/epjconf/202023808004

Table 1. Tapers parameters, model and experiments

Sample	*α*	*Lo (mm)*	*D_w (μm)*	*L_w (mm)*	*Z (mm)*	*L_{total} (mm)*
1	*- 0.1*	*14*	*3.9*	*6.7*	*38.5*	*84*
2	*- 0.1*	*14*	*10.6*	*8.5*	*30*	*68.5*
3	*- 0.1*	*14*	*27.8*	*10.4*	*20*	*50.4*

The resulting taper profiles, and fitting curves, according to Eqn. (1) and (2), are presented in Fig.2.

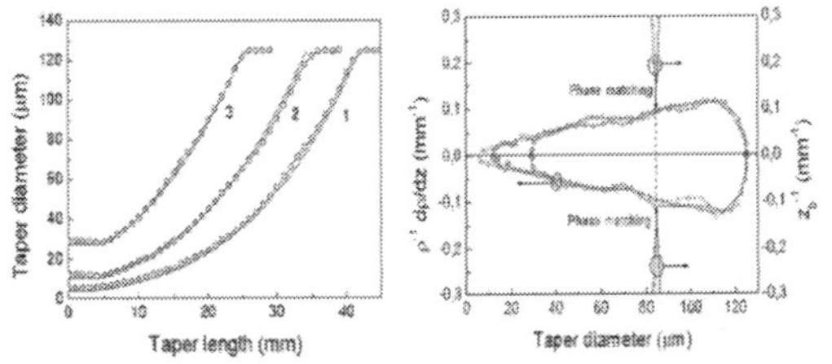

Fig.2. Left: Half taper profiles for three different heat-and-pull processes. Open circles are measured data. The solid lines are fittings to the analytic model, using parameters from Table 1. Right: Adiabatic condition $(1/\rho)d\rho/dz < 1/z_b$ where z_b is the beat length between HE_{11} and HE_{12}.

During the heat-and-pull process a number of oscillations (inset of Figure 3) is observed in the transmitted optical power, and defined as Power Transfer Number (PTN). PTN presents an exponential dependence for different taper elongations. According to mode coupling theory, under adiabatic condition [6], $PTN(\propto, L_o, Z_f) = (2/\pi)\int_0^{Z_f} \Delta\beta(\alpha, L_o, z).dz$ where $\Delta\beta$ is the difference for HE_{11} and HE_{12} propagation constants.

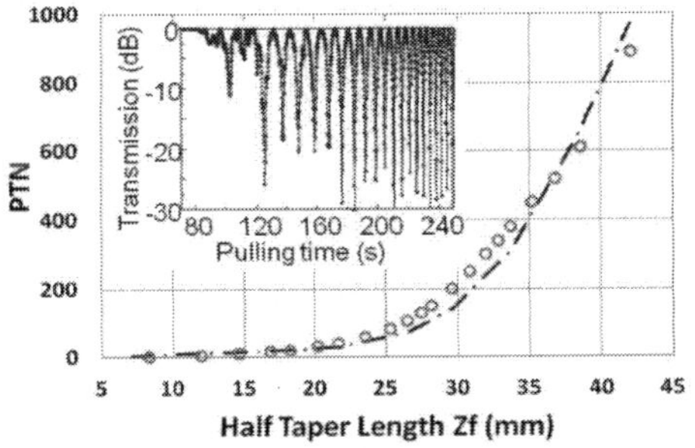

Fig.3. Power transfer number PTN versus ½ taper length for different taper elongations. Broken line, mode coupling theory fitting. Inset - transmitted power at 1575 nm as a function of the first 240 seconds of the heat-and-pull process.

The intrinsic sine square response in transmitted power (@ 1575 nm) to external mechanical loads of sample #1 can be seen in the inset of Figure 4. The output power experiences a number of Power Transfer Turning Points (PTTP) when under stress, and this PTTP is directly proportional to $\Delta L/L_o$ (Figure 4). An intrinsic sine square transfer function has two PTTP per cycle. Submitted to 0.02 Newton it elongates by ΔZ = 180 μm, which is equivalent to an increase ΔPTN = 10, and a number of PTTP = 2.ΔPTN = 20. It can be clearly noticed from Figure 4 that the digital sensor operates counting the PTTP when under traction or compression along its longitudinal direction. Negligible hysteresis was observed between traction and compression. Transmission and detection of two distinct wavelengths spectrally spaced by an odd multiple of a quarter of the FSR are necessary for sensing compression and traction. Finally, the proposed sensor can be a potential candidate for monitoring strains. In special, in environments under high electromagnetic interference or under risk of explosion where conventional electric strain sensors might fail. Moreover, several of these sensors can be distributed along one fiber making it possible to cover large areas.

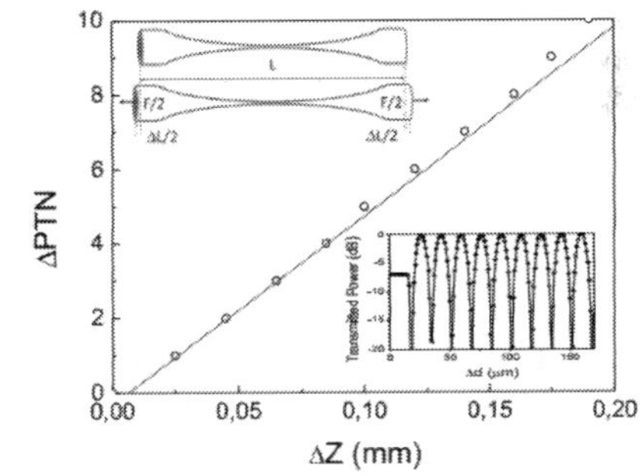

Fig.4. Experimental ΔPTN as a function ΔZ (open circles). Red line represents the analytic model for the digital response of the sensor for the strain range $\Delta L/L_0$. Inset shows 20 PTTP of the transmitted power as a function of taper elongation. A negligible hysteresis value of - 0.4 μm was observed.

S. Celaschi acknowledges FINEP/MCTIC financial support (01.16.0053.01) under project Plat IoT, granted by FUNTTEL. CNPq sponsors authors C.P. Guerra, and C. Biazoli under PCI scholarships (300473/2018-4). This paper reflects only the author's views.

References

1. S. Celaschi, G. N. Malheiros-Silveira, J Lightwave Tech. **35** (24), pp. 5381, (2017)
2. J. Beugnot, S. Lebrun, G. Pauliat, et al. Nature Commun. **5**, 5242 (2014)
3. G. Brambilla et al., Advances in Optics and Photonics **1**, 107–161 (2009)
4. S. Celaschi, G. N. Silveira, et al., in OSA Latin Am. Optics and Photonics Conf, Lima, p. W3B.7, (2018)
5. T.A. Briks, Y.W. Li, " J Lightwave Tech. **10** (4), 432, (1992)
6. A. W. Snyder and J. D. Love, *Optical Waveguide Theory*, Boston, MA: Springer US, (1984).

Transmittance of tapered photonic crystal fibers with absorbing coatings

Mauricio Salazar Sicacha[1], *Vladimir P.* Minkovich[1,*], *Alexander B.* Sotsky[2], *Artur V.* Shilov[2], and *Luidmila I.* Sotskaya[3]
[1]Centro de Investigaciones en Optica, A.C., 37150, Mexico
[2]A.A. Kuleshov Mogilev State University, 212009, Mogilev, Belarus
[3]Belarussian-Russian University, 212000, Mogilev, Belarus

Abstract. The interaction of the adiabatically tapered photonic crystal fiber fundamental mode with a thin-film absorbing coating, deposited on a surface of a taper waist, on transmission of a tapered fiber is studied. Examples of using this interaction in refractive index sensors and for detection of an adsorption layer with ammonia molecules upon contact of the absorbing coating with a liquid medium are presented. It is obtained that a pronounced sensory effect occurs in the case of a resonant coupling between the fundamental fiber mode and cladding modes localized between photonic crystal fiber air channels and the absorbing coating.

It is known that the fundamental mode of an index guiding photonic-crystal fiber (PCF) can be resonance coupled with leaky modes of a thin film PCF coating (cover) [1]. If the coating is absorbing, the coupling leads to a quasi-periodic spectral transmittance of the PCF that can be modified by interaction of coating modes with environment. However, at using the untapered PCF's mentioned mode, the coupling is rather weak. As a result, it can be only observed when sufficiently long stretches (tens of centimeters) of coated PCFs are used. Besides, to realize corresponding sensory effects it is necessary to keep a coating thickness d constant along the entire sensitive stretch of the PCF with a tolerance of $0.075\lambda/\sqrt{n_c^2-n_s^2}$, where λ is the light wavelength, n_c and n_s are refractive indexes of the coating and a fiber cladding, respectively [1]. For long PCF stretches this requirement is difficult to fulfil. It is hoped that the use of adiabatic PCF tapers with a length of a sensitive region of a centimeter order will overcome these difficulties. Due to a small transverse size of a taper waist, the interaction of the fundamental PCF mode with an absorbing coating of the taper waist has some attractive features and therefore are considered in this report. Corresponding estimates are presented for transmission spectra T(λ) of PCF tapers at sensing the refractive index of a liquid, and for detection of an adsorption molecular layer. Numerical simulations are performed with the use of the exact method of integral equations [1].

Fig.1 illustrates a resonance mode coupling in an adiabatic PCF taper used as a liquid refractive index sensor. In modelling we used a home-made silica glass PCF with 3 hexagonal rings of air holes in its cross section. In our untapered PCF, cladding diameter D_0=123.4μm, diameter of air holes a_0=3μm, photonic crystal fiber pitch Λ_0=8.8μm [2]. At the adiabatic taper fabrication, the diameter of holes a and the pitch Λ within the taper waist modify by the rule [3] $a=(D/D_0)a_0$, $\Lambda=(D/D_0)\Lambda_0$, where D=44μm is the taper waist diameter; the length of the waist is 3cm [2].

Fig.1 refers to the waist coated with the absorbing coating of a multifunctional butyl acrylate with a thickness d=5μm and the refractive index n_c=1.54-i0.00002. The left and right transmittance minima in Fig.1 correspond to a coupling of the fundamental PCF mode with TM and TE leaky modes of the coating, respectively, that realizes at the phase matching of the modes [1]. This coupling leads to a resonance energy capture of the fundamental PCF mode by the absorbing coating and a release of heat in it.

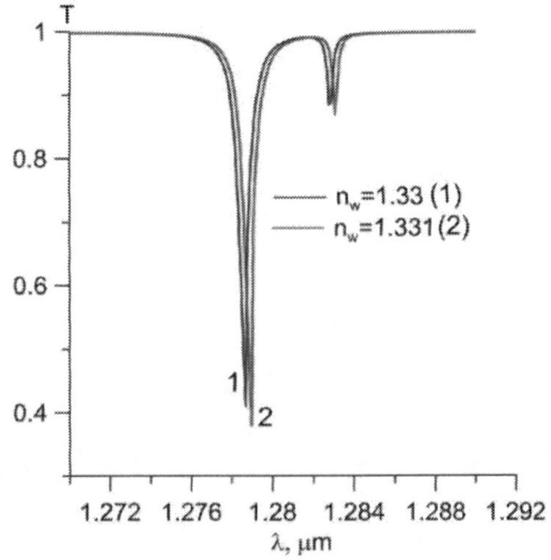

Fig.1. Shift of the taper transmittance spectrum with a change in the refractive index of a surrounding liquid.

*Corresponding author: vladimir@cio.mx

The sensory effect is explained by increase in effective mode indices of cover modes with increase in the refractive index of a liquid contacted with the absorbing coating.

Fig.2 shows a modification of the taper transmission spectrum caused by formation of an adsorption layer of ammonia molecules with a thickness of 0.0085μm on a surface of the absorbing coating which consists of a tin dioxide doped with antimony (7%) contacting with an aqueous ammonia. In this case d=0.1μm, n_c=1.9082-i0.00042. Due to a small d, the absorbing coating does not have leaky modes that can be in phase synchronism with the fundamental PCF mode in the spectral range of Fig. 2. In this case a sensory effect is explained by a resonance coupling of the fundamental PCF mode concentrated in the fiber core with a mode of the fiber cladding that concentrated in the area between air channels and coating [3]. This coupling mechanism is confirmed by Fig. 3 that illustrates a working mode intensity distribution at the wavelength corresponding to the transmission minimum for the curve 1 in Fig. 2. Cladding modes of the tapered PCF interact with the absorbing coating much more effective than the fundamental mode. This explains more pronounced sensory effect in Fig. 2 in comparison with Fig. 1. Calculations show, that necessary conditions for its realizations are high refractive index of the cover (coating) and rather small diameter of the taper waist. Only under these conditions and the mentioned mode coupling, the expressed standing waves are realized in the PCF cladding (see Fig. 3).

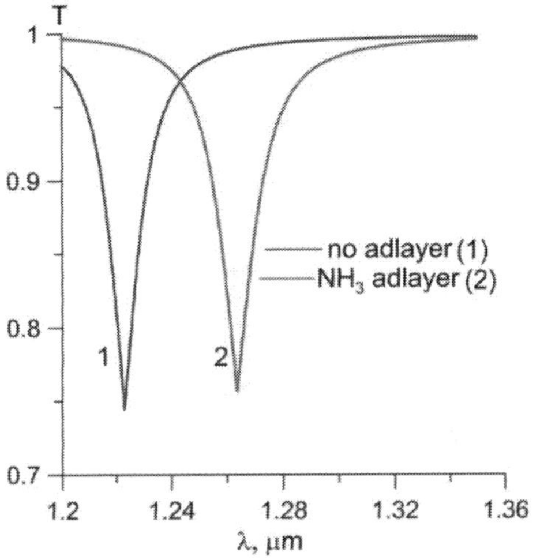

Fig.2. Shift of the PCF taper transmission spectrum at a formation of an adsorption layer of ammonia molecules on a surface of a tin dioxide coating.

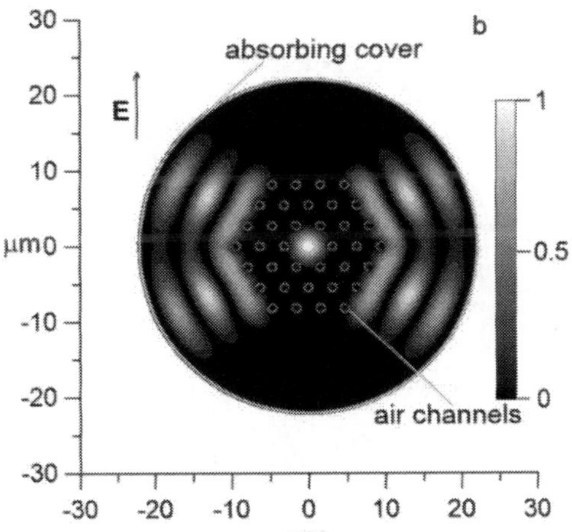

Fig.3. The optical density of the taper waist working mode at a minimal taper transmission in the case of the curve 1 in Fig. 2. The arrow **E** indicates the main orientation of a mode electric field vector.

This work was partially supported by the State Scientific Research Program "Photonics, Opto- and Microelectronics 1.3.03" of the Belarus Republic.

References

1. A.B. Sotsky, L.I. Sotskaya, V.P. Minkovich, D. Monzon-Hernandez, Tech. Phys., **54**, 6 (2009)
2. G.A. Cardenas-Sevilla, D. Monzon-Hernandez, V. P. Minkovich, Appl. Phys. Express, **5**, 022502 (2012)
3. V.P. Minkovich, D. Monzon-Hernandez, J. Villatoro, A.B. Sotsky, L.I. Sotskaya, J. Lightw. Technol., **24**, 11, (2006)

Monolayer 2D material lasing from photoactivation-enhanced photoluminescence at room temperature

Zhaoqi Gu, Feng Liao, Fuxing Gu

Laboratory of Integrated Opto-Mechanics and Electronics, Shanghai Key Laboratory of Modern Optical System, Engineering Research Center of Optical Instrument and System (Ministry of Education), University of Shanghai for Science and Technology, Shanghai 200093, China

gufuxing@usst.edu.cn

In recent years, monolayer transition metal dichalcogenides (TMDs) with unique photoluminescence (PL) properties have received extensive attention [1]. However, the defects introduced during the fabrication and processing steps provide nonradiative recombination pathways. In addition, as the pump intensity increases, the generated high-density carriers in this atomic thickness enhance many-body interactions such as nonradiative exciton-exciton annihilation.

In this work, we first grow single-crystal monolayer MoS_2 on tapered silica MNFs (micro/nano fibers)[2]. We then use a simple photoactivation strategy to obtain strongly enhanced and highly stable PL quantum yields (QYs) in a wide pump dynamic range at room temperature.

Our work demonstrates substantially enhanced and stable room temperature PL QYs of monolayer MoS_2 grown on silica MNFs in a wide pump dynamic range, using a simple photoactivation method. The taper-drawing process greatly lowers the photoactivation energy and allows fast completion of the entire photoactivation process in a few minutes with a CW laser. The high-density oxygen dangling bonds released from the tapered micro/nanofiber surface are the key to this strong enhancement of QYs. As the pump intensity increases from 10^{-1} to 10^{4} W cm^{-2}, our photoactivated monolayer MoS2 exhibits QYs from ~30 to 1% while maintaining high environmental stability, allowing direct lasing with greatly reduced thresholds down to 5 W cm^{-2}. By using a monolayer/microbottle structure with an outer diameter of 5.5 μm, the lasing threshold around 5 W cm^{-2} (Fig 1.c) for a best fitted β of 0.5 is observed at a wavelength of 682.5 nm (TM^{1}_{31}).[3]

We foresee that our photoactivation method for single-crystal TMD monolayer/MNF structures provides an exciting direction for various integrated applications ranging from low-threshold coherent light sources and nonlinear optics to optoelectronics. Since microfiber and microbottle structures themselves can serve as good optical microcavities [3,4], the high-QY and stable emission make as-activated monolayers good choice for lasing generation at room temperature.

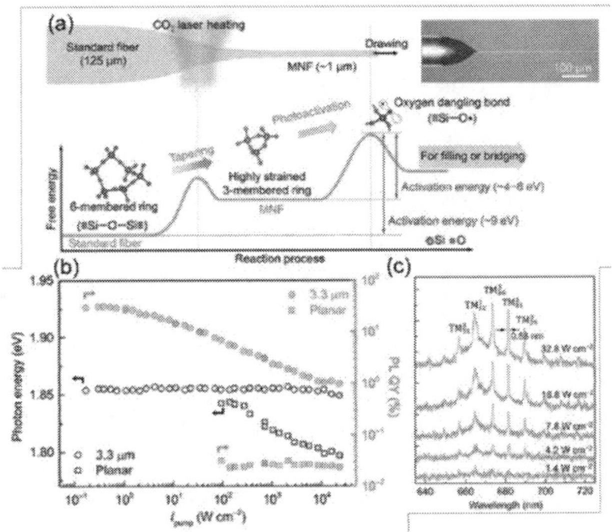

Fig.1 (a) Schematic illustration of generation of oxygen dangling bands from a tapered fiber. The tapering process greatly lowers the activation energy from ≡Si—O—Si≡ to ≡Si—O·. Inset: Optical image of a MNF tapered from a standard fiber. **(b)** Pump-dependent peak positions (black data) and PL QYs (red data) in the as-photoactivated monolayer MoS_2 (solid circles). A planar sample is also provided for reference (open squares). **(c)** Steady-state PL spectra in an as-photoactivated MoS_2 monolayer/microbottle (outer diameter,5.5 mm) with increasing $\mathbf{I_{pump}}$, in which WGM oscillations are observed.

References:

[1] F. Xia, H. Wang, D. Xiao, et al. Two-dimensional material nanophotonics. Nat. Photonics 8, 899–907 (2014).

[2] F. Liao, Z. Gu, J. Yu, et al. Enhancing monolayer photoluminescence on optical micro/nanofibers for low-threshold lasing. Sci. Adv. 5, eaax7398 (2019).

[3] Gu, F., Xie, F., Lin, X. et al. Single whispering-gallery mode lasing in polymer bottle microresonators via spatial pump engineering. Light Sci. Appl. 6, e17061 (2017)

Optical scattering field imaging along a nanofiber in operando

Yosri Haddad *, *Jacques* Chrétien, *Samuel* Margueron, *Jean-Charles* Beugnot, & *Gil* Fanjoux

FEMTO-ST Institute, optics department, UMR 6174 CNRS / Université Bourgogne Franche-Comté, 25030 Besançon - France

Abstract. We present a non-destructive and non-invasive imaging spectroscopic technique with a high spatial and spectral resolution based on the detection of the Rayleigh scattering field radiated out of a centimetric length and micrometric size optical waveguide in operation.

1 Introduction

The linear or nonlinear optical propagation behaviour lengthwise of micro- and nano- waveguides in operation is difficult to experimentally evaluate, although this information is crucial for their optimal development. Previously, apart from the destructive cut-back method, far-field qualitative imaging of Rayleigh scattering (RS) have been used for rapid analysis of the propagation of light in few-mode optical nanofibers (ONF) [1]. In the present work, we demonstrate a novel method, non-destructive and non-invasive, with a high spatial and spectral resolution, to analyse the light propagation in an optical waveguide in operation. By analysing with a confocal spectrometer the RS radiated out of an ONF in operation, we show that our method can spatially characterize the entire ONF in the linear regime with a micrometric spatial resolution. This approach provides direct observation of the higher-order modes (HOMs) leaking from the ONF surface along the taper region and the transition from core–cladding guidance to cladding–air guidance. This method can also be used to characterize the linear loss and provides the localization of defects along an optical waveguide.

2 Experimental Setup

Our measurement setup, illustrated in Fig. 1, consists of a high-precision motorization for longitudinal X and transverse Y and Z translation (with a step size < 10nm) of the component under study, allowing the generation of an accurate recordable 3D spatial trajectory point-by-point. RS of the guided light emitted from the outside of an optical waveguide is collected in the direction orthogonal to the propagation direction (Fig. 1) with a ×10 microscope objective (NA=0.3) and recorded through a confocal spectrometer. The spectrometer monochromator is equipped with three distinct diffraction gratings (300, 1500 and 2400 lines/mm), and an ultra-sensitive cooled CCD detector (−80°C) to have a low detection threshold. Alternatively, a CCD image camera is used to image the waveguide for positioning. The waveguide studied in this work corresponds to an ONF fabricated from telecom single mode fiber (SMF@1550nm) as described in [2]. After tapering process, the ONF has a uniform waist of 1µm over around 17mm accurately measured by Brillouin scattering methods [2], and the two transition tapers are around 37mm long (Fig. 1). The laser beam injected in the ONF comes from a wavelength and power stabilized cw laser diode with a wavelength of 653nm. The fiber before stretching is therefore strongly multimode at this wavelength. The ONF is then placed inside a closed box to protect it from external air vibrations and to reduce any kind of contamination. Moreover, we monitored the transmitted power during the RS measurement and we did not observe any significant variation (Fig. 1).

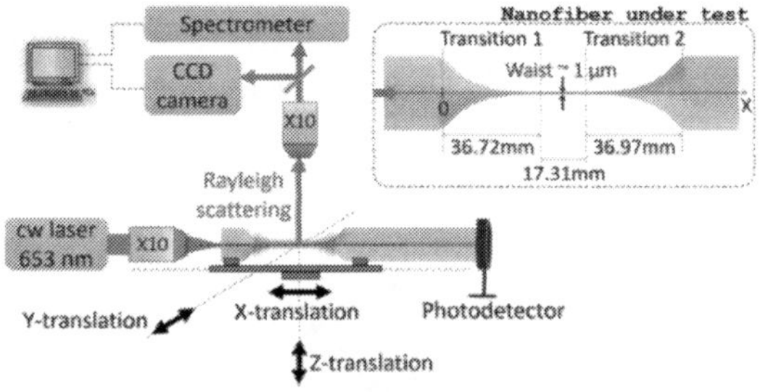

Fig. 1. Schematic experimental setup, and geometrical data of the ONF studied in this work (in dashed box).

3 Experimental results and discussion

First, CCD camera pictures of the fiber allow the measurement of the evolution of the fiber taper radius with the distance X (black crosses in Fig. 2a). This analyzed fiber part includes the end of the non-stretched SMF fiber (62.5 µm), the taper region, and the ONF.
Fig.2b shows an XY-2D mapping in logarithm scale of the RS intensity along the fiber input side before the ONF. Note that no normalization in intensity has been realized. Therefore, the evolution of the RS intensity directly represents the evolution of the scattering process. For the first two millimeters, Fig. 2b shows that the light is only confined in the core of the fiber and thus the fluctuations in refractive index of the core dominate the scattering process. The RS signal shows a uniform trace and the

* Corresponding author: yosri.haddad@femto-st.fr

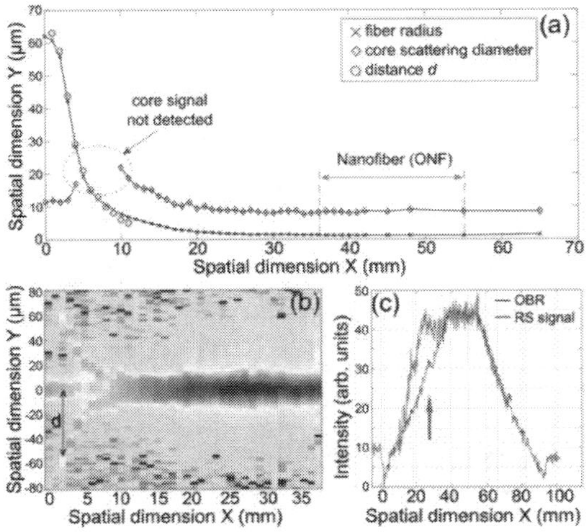

Fig. 2. **(a)** Evolution vs. the longitudinal distance X of the fiber radius (**×**), the linewidth (FWHM) of the scattering signal from the core (◇), and the distance *d* between the scattering traces symmetrically shifted from the core (○). **(b)** 2D XY mapping of RS intensity in the transition 1 (logarithm scale). **(c)** Evolution of RS intensity from OBR trace (blue) and our measurements (red).

linewidth of the core signal is almost stable (blue diamond in Fig. 2a). After 2mm, the linewidth clearly increases meaning that light progressively escapes the core to the cladding along the taper. Therefore, a substantial portion of the optical power is propagating within the cladding. In addition, Fig. 2b shows that the scattering light presents three distinct traces, one central peak corresponding to the core, and two scattering traces symmetrical with respect to the core by a distance *d*. In fig. 2a, the green circles show the evolution of the distance *d* along X. These match clearly the evolution of the radius in the taper region. This mean that these two scattering traces originate from the surface. Thereafter, between 5mm and 10mm, as the fiber core signal is negligible, the central scattering peak is no more measurable because of its superposition with the two lateral peaks. This shows that the amount of light remaining in the core is now extended into the cladding. Here, the refractive-index fluctuations of the cladding dominate the scattering process. Finally, after 10mm, the RS signal show a single trace again which increases in intensity with the propagation distance as it approaches the nanofiber as we will explain thereafter. In this region, the surrounding air becomes the cladding and the whole fiber the core. Moreover, the taper surface roughness dominates the scattering process [3]. On the other hand, the signal linewidth decreases up to a limit of about 8µm. As the nanofiber has a waist diameter equal to 1µm, this threshold corresponds to the spatial resolution of our system. This resolution can reach 5µm by reducing the width of the entrance slit of the spectrometer to 10µm.

Let us analyze the evolution of the longitudinal RS signal intensity with the distance X. We have therefore compared our experimental results with the state-of-art method, i.e., an OBR system from Luna® which uses an internal 1566nm laser source. Fig. 2c shows the OBR trace (blue curve) and our scattering trace (red curve). Fig. 2c shows that the scattering signal shape obtained with our technique with a wavelength of 653nm is in good agreement with the OBR signal shape. Nevertheless, our signal in transition 1 is strongly blurred by a large increase (red arrow), contrary to the RS signal in transition 2. As the fiber is multimode and the tapered transition is not adiabatic, coupling of guided modes with lossy modes induces the spatially filtering by the transition of several high order modes. Therefore, the lossy modes cross the interface glass-air and consequently undergo large scattering by surface roughness.

To confirm this observation, we have analyzed an ONF with 1µm of diameter and 2cm long, fabricated from a single mode fiber in the visible range SM450. Fig.3 shows an XY mapping RS intensity (in logarithm scale) for this ONF under the same conditions as Fig 2b. The RS intensity shows the absence of the surface scattering in the taper region (transition 1). This confirms that the surface scattering obtained for the multimode ONF (Fig. 2) come from the no longer guided HOMs modes, which go out the fiber in the transition 1. Accordingly, when reach cutoff, those modes cross the surface of the fiber presenting a high scattering efficiency. Fig. 3 also clearly shows that the guiding process drastically change near X = 10mm of propagation, from a core-cladding to a fiber-air guidance, in perfect agreement with numerical prediction.

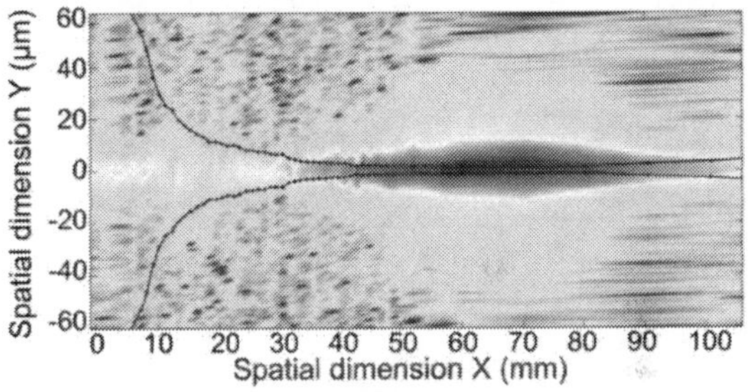

Fig. 3. 2D XY mapping of RS intensity for a single mode fiber @653nm with adiabatic transition (logarithm scale). Black curves correspond to the fiber diameter from CCD camera.

4 Conclusion

Preliminary results have shown that our technique is a powerful tool to characterize the evolution of the field propagating along a photonic waveguide with a micrometric spatial resolution. For a nanofiber, RS signal provides direct observation of the transition from core–cladding to cladding–air guidance. Moreover, we have shown that the presence of the surface scattering indicates non-adiabaticity in the tapering part.

This work has been supported by the EIPHI Graduate School (contract ANR-17-EURE-0002) and by the Conseil Régional de Bourgogne - Franche-Comté (DECAP and CIPPIC project).

References

1. J. E. Hoffman *et al.*, Optica 2, 416-423 (2015).
2. A. Godet *et al.*, Optica 4, 1232–1238 (2017).
3. Y.-H. Lai *et al.*, Opt. Exp. 25, 22312 (2017).

Evanescent Kerr effect using an optical nanofiber in acetone

Gil Fanjoux*, Jacques Chrétien, Adrien Godet, Kien Phan-Huy, Jean-Charles Beugnot, & Thibaut Sylvestre

FEMTO-ST Institute, Optics Department, Université Bourgogne Franche-Comté, CNRS, Besançon – France

Abstract. We report on a theoretical and experimental investigation of the optical Kerr effect in the evanescent field of silica nanofibers immersed in several highly nonlinear liquids such as ethanol, acetone and water and we further compare them with air cladding. We provide formula of the effective nonlinear coefficients including the contribution of the nanofiber silica core and of the evanescent field for varying nanofiber diameter and for different surrounding media.

1 Introduction

Optical nanofibers are ultra-thin tapered optical fibers that feature unique optical properties, including a strong field confinement in the core, enhanced nonlinear effects, and a wide evanescent field that can interact with the surrounding medium and can therefore modify or detect chemical elements [1]. Among third-order nonlinear effects, stimulated Raman scattering (SRS) has recently been evidenced in the evanescent field of tapered silica nanofibers dipped in nonlinear liquids such as ethanol and toluene [2]. It has been shown in particular that the evanescent Raman effect in the highly nonlinear liquid is stronger than the standard Raman effect in the silica nanofiber itself. In this paper, we theoretically demonstrate that the evanescent Kerr effect can also be stronger than that in the silica taper. We further report the experimental observation of the evanescent Kerr effect through measurements of the stimulated Raman-Kerr scattering in a silica nanofiber immersed in acetone. The evanescent Kerr effect is shown to give rise to a strong asymmetric spectral broadening of the first Raman order generated in the nonlinear liquid.

2 Theory and experiments

The optical Kerr effect in a nonlinear optical waveguide is usually defined using the following nonlinear parameter

$$\gamma = \frac{2\pi n_2}{\lambda A_{eff}} \quad (1)$$

where n_2 is the nonlinear Kerr index of the nonlinear medium, A_{eff} the effective area of the guided mode, and λ the wavelength of the field propagating in the waveguide. For a tapered fiber with a waist close to the optical wavelength, the evanescent field of the fundamental mode interacting with the external environment becomes more important than that the electric field guided into the core. To include the evanescent field into the Kerr effect, we can write the effective nonlinear parameter γ_{eff} as

$$\gamma_{eff} = \frac{2\pi}{\lambda}\frac{\iint_{-\infty}^{+\infty} n_2(x,y) S_z^2 \, dx\, dy}{\left(\iint_{-\infty}^{+\infty} S_z^2 dx\, dy\right)^2} \quad (2)$$

where S_z is the Poynting vector component in the waveguide direction. This equation is an approximation as it does not assume the z field component, which is no more negligible for very small diameter [3]. However, although the propagating modes of a waveguide are not fully transverse in strong guidance regime, the full vectorial formalism of γ_{eff} and the simplified expression of Eq. (2) gives the same results of γ_{eff} for silica nanofiber in air even for small core diameter down 0.3 nm [3]. If now the core and cladding have different Kerr coefficients, the previous Eq. (2) is written as

$$\gamma_{eff} = \frac{2\pi n_{2,co}}{\lambda}\frac{\iint_{co} S_z^2 \, dx\, dy}{\left(\iint_{-\infty}^{+\infty} S_z^2 dx\, dy\right)^2} + \frac{2\pi n_{2,cl}}{\lambda}\frac{\iint_{cl} S_z^2 \, dx\, dy}{\left(\iint_{-\infty}^{+\infty} S_z^2 dx\, dy\right)^2} \quad (3)$$

with $n_{2,co}$ and $n_{2,cl}$ the core and cladding Kerr coefficient, respectively, considered as uniform. Then, the first term of Eq. (3) corresponds to the contribution to the Kerr effect of the silica core, while the second one is the contribution of the cladding, i.e., $\gamma_{eff} = \gamma_{co} + \gamma_{cl}$. To get the effective Kerr coefficient, we first computed the S_z Poynting component of the fundamental mode using a finite element method. The studied system corresponds to a silica nanofiber with a various diameter, and surrounding by air or liquid. Parameters used are laser wavelength of 532 nm, refractive index of 1.4606, 1.000, 1.3337, 1.3637 and 1.3614 for silica, air, water, ethanol and acetone, respectively, and the corresponding nonlinear Kerr indices n_2 are 2.6, 3.10^{-3}, 4.1, 7.7 and 24 ($\times 10^{-20}$ m²/W).

Figs. 1(a-d) show the nonlinear parameters of the silica core γ_{co} and of the cladding γ_{cl}, the effective nonlinear coefficient γ_{eff} and the evanescent Kerr rate, all as a function of the nanofiber diameter. Concerning the silica core, γ_{co}, it first increases until reaching a maximum value for diameters close to those of the minimal effective area. This is because if the effective mode area is very weak, the field intensity and therefore the Kerr effect in the silica

* Corresponding author: gil.fanjoux@univ-fcomte.fr

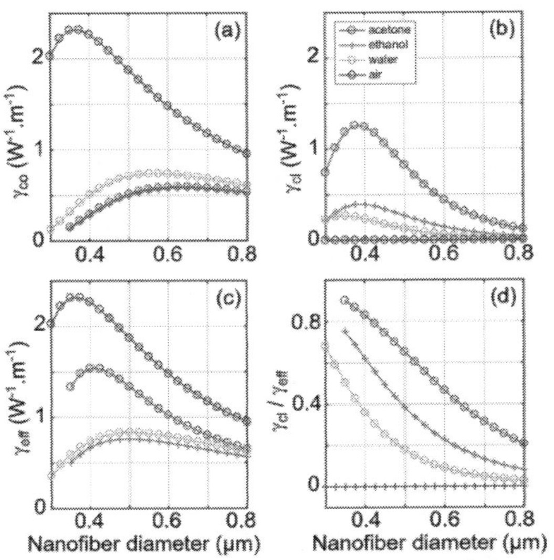

Figure 1. Evolution in function of the nanofiber diameter and for different surrounding media of (a) the nonlinear parameter contribution of the silica core γ_{co} (b) the nonlinear parameter contribution of the cladding γ_{cl}, (c) the whole effective nonlinear parameter γ_{eff}, and (d) the γ_{cl}/γ_{eff} ratio. Wavelength $\lambda = 532$nm.

core are maximum. Consequently, it is clear that γ_{co} is strongly greater for nanofiber in air than in liquids (Fig. 1a). For small diameters, the core contribution drastically decreases and tends to zero since the fundamental mode spreads out spatially with increasing mode effective area. Concerning the cladding contribution (Fig. 1b), the evolution is nearly similar. Since the intensity fraction in the evanescent field increases with decreasing nanofiber diameter, γ_{cl} increases up to reach a maximum and to decrease thereafter. This peak value corresponds to an optimal balance between the increase of the evanescent field that gives rise to the Kerr effect enhancement and to the decrease of the field intensity. For a given nanofiber diameter, the dependence of γ_{cl} on the cladding materials follows the order of their Kerr indices.

Figure 1(c) shows the effective Kerr parameter γ_{eff}. First, it is clear that the Kerr parameter γ_{eff} remains maximum for a nanofiber in air, even if the contribution of the evanescent field to the Kerr effect is null, and even compared to a nanofiber immersed in acetone which presents the highest Kerr coefficient n_2. This is because the effective area of the fundamental mode for a nanofiber in air remains the smallest, leading to high field intensity and Kerr effect. Although γ_{eff} is lower for a nanofiber immersed in a liquid, the contribution of the evanescent field to the Kerr effect may be comparable or even higher than that from the silica core. For acetone, γ_{eff} has a maximal value for a core diameter of d = 410nm, and almost 80% of the Kerr parameter is in the evanescent field. If we now compare the role of ethanol and acetone that have the same effective mode area, the contribution γ_{co} is therefore equivalent for both liquids (Fig. 1b). However, as the Kerr coefficient n_2 of acetone is higher than that of ethanol, the contribution γ_{cl} is clearly greater for acetone. Thus, the effective Kerr coefficient is about 2 times larger for acetone than ethanol. The same analysis can be performed when comparing water and ethanol. A nanofiber in water will have a lower evanescent field than in ethanol due to the different step refractive index. However, water has a lower Kerr coefficient than ethanol, leading to a highest cladding contribution γ_{cl} in ethanol. These two contributions have opposite behavior and give rise to similar nonlinear parameters γ_{eff} for both liquids (Fig. 1c). By comparing the outer fraction of field and the γ_{cl}/γ_{eff} ratio, the contribution of the evanescent field to the Kerr effect is higher than the outer fraction of the field: the Kerr effect is not only related to the spatial extent of the electric field but also to the Kerr coefficient of the nonlinear liquid. Thus, for a nanofiber in acetone, the ratio γ_{cl}/γ_{eff} reaches 50% for a nanofiber diameter of 550 nm, while the outer fraction of the field is only 39 %.

Our experimental approach for the observation of the evanescent Kerr effect is based on using stimulated Raman-Kerr scattering (SRKS) [4]. This inelastic light scattering comes from the inertial nature of reorientation motion of anisotropic molecules of many liquids, and gives rise to a broad asymmetric broadening of the Raman Stokes spectrum towards the red. By immersing a nanofiber in acetone, we indeed observe a large broadening of the Raman spectral components on the Stokes side, and conclude to the SRKS process, and thus the conjugated effect evanescent Raman and Kerr effects, to explain this phenomenon.

3 Conclusion

We theoretically and experimentally investigated the optical Kerr effect in the evanescent field of a silica nanofiber surrounded by nonlinear liquids such as water, ethanol and acetone. We have shown that the effective Kerr coefficient has two contributions: one from the silica core, which depends on the fiber diameter and follows the inverse of the effective area, the other from the cladding via the evanescent field, which depends on the diameter of the nanofiber and the nonlinear properties of the external environment

This project has received funding from the Agence Nationale de la Recherche (ANR) (ANR-16-CE24-0010; ANR-15-IDEX-0003, EIPHI graduate school ANR-17-EURE-0002) and the European union H2020 program MSCA-ITN-SUPUVIR under grant agreement 722380, and the Conseil Regional de Bourgogne Franche-Comte.

References

1. L. Tong and M. Sumetsky, Springer, 978-3-642-03362 (2011).
2. L. Shan, G. Pauliat, G. Vienne, L. Tong, and S. Lebrun, Appl. Phys. Letters 102, 201110 (2013)
3. S. Afshar V., and T. M. Monro, Opt. Express 17(4), 2298-2318 (2009)
4. G. S. He, R. Burzynski, and P. N. Prasad, J. Chem. Phys. 93, 7647–7655 (1990)

Tapered optical fiber for Micro-Newton force sensor

Adrien Godet, Jacques Chrétien, Kien Phan Huy, Jean-Charles Beugnot
Institut FEMTO-ST, UMR 6174 CNRS, Université Bourgogne Franche-Comté, Besançon, France

Abstract. We used Brillouin scattering in silica nanofiber to demonstrate a microNewton force sensor having weak optical losses and using only one access of the nanofiber. The measurements are in good agreement with the theoretical model.

Fiber force sensor, offering both a broad detection range and a high sensitivity, is essential for several applications from earthquake monitoring to automation system [1]. Due to the small transverse dimension, the tapered optical fiber have a number of optical and mechanical properties that make them very attractive for both fundamental physics and technological applications [2-5]. Contrary to standard telecom fiber where the Brillouin scattering effect is characterized by a single Lorentzian resonance centred at 10.86 GHz (@ 1550nm), in tapered silica fiber, we identified several Brillouin resonances at different frequencies from 5 GHz to 10 GHz coming from surface, shear and compression elastic waves [6]. In this work, we used the high sensitivity of Brillouin scattering to longitudinal strain and the nanoscale cross-section of nanofiber (the waist part of the tapered fiber) in order to design a very sensitive force at the 10^{-6} Newton range. We report a point fibre force sensor with sensitivity up to 0.66 µNewton/MHz by using a nanofiber diameter of 630 nm.

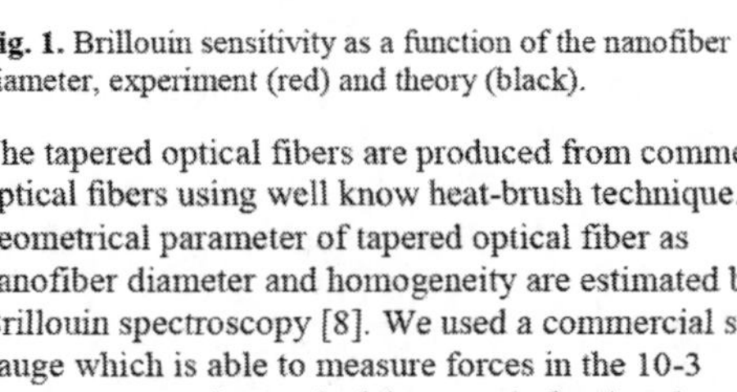

Fig. 1. Brillouin sensitivity as a function of the nanofiber diameter, experiment (red) and theory (black).

The tapered optical fibers are produced from commercial optical fibers using well know heat-brush technique. The geometrical parameter of tapered optical fiber as nanofiber diameter and homogeneity are estimated by Brillouin spectroscopy [8]. We used a commercial strain gauge which is able to measure forces in the 10-3 Newton range. One end of the taper is fixed to the translation stage while the other end is fixed to the force sensor. After benchmarking of force sensor, we moved the translation stage by step of 100µm and measured the spontaneous backscattering Brillouin spectrum using heterodyne detection.

In figure 1, the experimental results of BFS sensitivity as a function of nanofiber diameter are represented. The analytical model [8] is in good agreement with experimental measurement.

To conclude, we have experimentally investigated the force sensing potential of Brillouin scattering in tapered silica optical fiber. The linear dependence of Brillouin resonance from compression elastic wave is in good agreement with analytical model. Sensibility (1 MHz/µN) and dynamic ratio (1:2000) can be tune with diameter of taper and the total losses is not modified and below 0.1dB. These results demonstrate the strong potential of tapered optical fiber for vibrational acoustic sensor.

The authors would like to acknowledge the financial support of Agence Nationale de la Recherche (FUNFILM-ANR-16-CE24-0010-03).

References

1. N. J. Lindsey, T. D. Craig and J. B. Ajo-Franklin "Illuminating seafloor faults and ocean dynamics with dark fiber distributed acoustic sensing," Science 366, 1103-1107 (2019).
2. W. Luo, J.-L. Kou, Y. X. Chen, F. Xu, and Y. Lu. "Ultra-highly sensitive surface-corrugated microfiber Bragg grating force sensor," Applied Physics Letters 101, 133502 (2012).
3. T. Wieduwilt, S. Brückner and H. Bartelt., "High force measurement sensitivity with fiber Bragg gratings fabricated in uniform-waist fiber tapers," Meas. Sci. Technol. 22 075201 (2011).
4. Y. Chen, S-C. Yan, X. Zheng, F. Xu, and Y.-Q. Lu, "A miniature reflective micro-force sensor based on a microfiber coupler," Optics Express 22, 2443-2450 (2014).

* Corresponding author: adrien.godet@femto-st.fr

https://doi.org/10.1051/epjconf/202023808009

5. L. Tong, "Micro/Nanofibre Optical Sensors: Challenges and Prospects," Sensors 18, 903, (2018).
6. J-C. Beugnot, S. Lebrun, G. Pauliat, H. Maillotte, V. Laude, and T. Sylvestre, "Brillouin light scattering from surface acoustic waves in a subwavelength-diameter optical fibre," Nat. Commun. 5, 5242 (2014).
7. A Godet, T Sylvestre, V Pécheur, J Chrétien, J-C. Beugnot, and K. P. Huy., "Nonlinear elasticity of silica nanofiber," APL Photonics 4, 080804 (2019).
8. A. Godet, A. Ndao, T. Sylvestre, V. Pecheur, S. Lebrun, G. Pauliat, J-C. Beugnot, and K. P. Huy," Brillouin spectroscopy of optical microfibers and nanofiber," Optica 4, 1232-1238 (2017).

Laser-induced breakdown measurements of silica nanofibers in air and immersed in water, ethanol and isopropanol

Maha Bouhadida[1*] and *Sylvie Lebrun*[1]

[1]Laboratoire Charles Fabry, Institut d'Optique, CNRS, Université Paris-Saclay, 91127 Palaiseau cedex, France

Abstract. In this paper we present the first study to our knowledge of Laser-Induced Breakdown (LIB) measurements of silica nanofibers in air and in three liquids for different radii in the ns regime. The experimental protocol is described. A significant number of samples are tested and the results are highly repeatable. We showed that immerging a nanofiber in a liquid substantially enhanced the LIB, the most salient increase having been obtained for nanofibers immersed in water for which the LIB has been almost multiplied by a factor of 2 compared with air. This property offers a new degree of freedom to widen the field of applications of nanofibers, where high peaks powers are needed.

1 Introduction

Optical tapered nanofibers are fabricated by pulling standard silica fibers until their diameters become typically smaller than 1 µm. The nanofibers are linked to the untapered end-fibers via two flared sections called the tapers and can be inserted in fibered networks with very low losses, making them highly attractive for many applications. At such small diameters, nanofibers exhibit a strong confinement of light inside the silica which enables the generation of nonlinear effects [1] such as, for example, the generation of supercontinuum [2]. Nanofibers can also exhibit an intense evanescent field which has been exploited for optical sensing [3] or optical traps for quantum applications [4,5]. The excitation of optical non linearities such as stimulated Raman effect [6,7] or Kerr effect in the evanescent field of nanofibers immersed in liquids has also been demonstrated [8]. In all the above-mentioned applications the optical mode is guided by total internal reflection at the interface between the silica nanofiber and the surrounding medium that can be vacuum, a gas or a liquid. The Laser-Induced Breakdown (LIB) for applications in general and consequently for nanofibers is often a limiting issue. For pulsed regimes, to our knowledge, there is a lack of experimental data on the LIB in nanofibers in the literature.

In this work our aim is to give experimental data and reliable trends on the LIB of silica nanofibers in ambiant air and immersed in different low-index common liquids. We present the experimental protocol we have followed to measure the LIB. The pump source is a sub-ns frequency doubled Nd:YAG microlaser emitting at 532 nm with a frequency rate repetition of 4.7 kHz. The results we obtained are highly repeatable and physical interpretations will be proposed during the conference. We show that immerging a nanofiber in water can increase the LIB obtained in air by more than a factor of 2, offering a simple solution to increase the LIB for many applications in which high peak powers are needed (Raman converters, sensors etc …).

2 Experimental setup and protocol

The experimental setup is depicted in Fig. 1. The preparation and the drawing of the nanofibers were performed in a class-5 clean room. In the following experiments, we have used the SMF28 fiber (ref. Corning). As this fiber is highly multimode at 532 nm, high-order modes were filtered beforehand by pulling the input connecting fiber over 1 cm until reaching a core radius of 1.3 µm and a cladding radius of 20 µm allowing the propagation of the fundamental mode only (single mode filter on Fig. 1). We have used short and straight connecting fibers to remain on a single mode propagation from the input (indicated by E_I) to the output (indicated by E_{out}). Then, before being pulled the fiber was removed from its polymer cladding with a mechanical stripper and cleaned in a single pass with an optical paper wet by a drop of ethanol. The nanofibers were drawn with a pulling rig using the "pull and brush" technique and a simple butane flame to soften the glass [6]. Our pulling rig enabled to routinely draw high optical quality nanofibers with an overall optical transmission T defined as the ratio of the output energy E_{out} versus the input energy E_I always higher than 80%. Once drawn, the nanofiber was transported out of the clean room and put in the experimental setup in a tank with a removable protection glass cover. The pump source beam, which was linearly polarized, (ref. Horus HLX-G-F004) was sent on a half-wavelength plate associated with a polarizer cube to control the incident energy on the nanofiber. The rejected beam served as a reference for the input energy (E_{ref}). The direct beam was

* Corresponding author: maha.bouhadida@institutoptique.fr

launched in the SMF28 fiber through a microscope objective, propagated across the single mode filter, the nanofiber and was collected at the output for energy measurements. The input energy E_1 just after the single mode filter was calibrated with respect to E_{ref}. The nanofiber was a little bit stretched to be straight but not constrained and kept in the position thanks to two smooth magnets. The pump beam was injected in the fiber at low energy and the optical transmission was optimized. For experiments with liquids, the tank was filled carefully and immediately afterwards with a micro-syringe until the nanofiber was fully immersed. The optical transmission was optimized again and the laser input energy was slowly increased until the nanofiber breakdown. Input and output energies were recorded during all the process.

Four external media were used : ambiant air, tap water, ethanol and isopropanol. The radii of the three nanofibers we tested (NF1, NF2 and NF3) were respectively 220 nm, 300 nm and 350 nm. For each medium we have drawn 10 nanofibers of each radius. The length of each nanofiber was 2 cm in order to avoid non linear effects such as Raman effect in the liquid and hence to study the LIB alone [6,7].

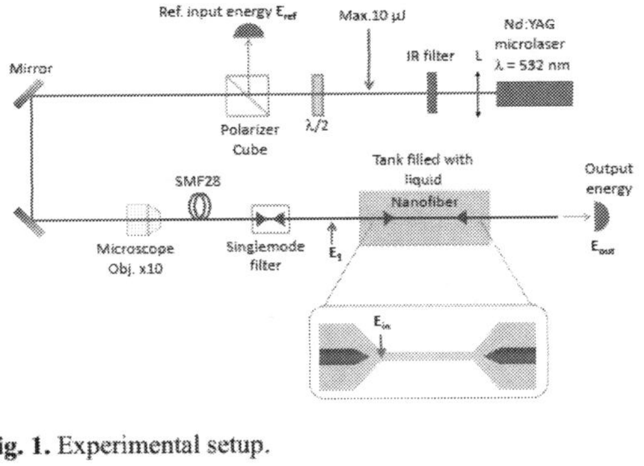

Fig. 1. Experimental setup.

3 Results

The overall transmission T of the immersed nanofibers remained very stable during the whole experiment until the LIB, respectively 90% in water, 92% in ethanol and 95% in isopropanol. This was not the case when they were in air : the average transmission decreased from 80% to 57% during the experiment. This first systematic observation showed that whatever the liquid we used it had a protection effect from additional dust. It also enabled to mechanically stabilize the nanofiber, avoiding the fluctuations due to micro-turbulences.

We have plotted in Fig. 2 the energy inside the nanofiber at LIB, named E_{in}, for each radius r_{NF} and each external medium :

$$E_{in}(r_{NF}, medium) = \eta\sqrt{T}E_1(r_{NF}, medium) \qquad (1)$$

η is the fractional energy inside the nanofiber. The standard deviation on the incident energies (not shown on the graphic) was estimated to be 0.01 µJ in all cases. In this range of radii, the uncertainty on the radii was below 15 nm. The analyze of the results confirmed that surface effects were predominant compared to bulk effects. Fig. 2 shows that immerging a nanofiber in a liquid substantially enhanced the LIB, the most salient increase having been obtained for nanofibers immersed in water for which the LIB has been almost multiplied by a factor of 2 compared with air.

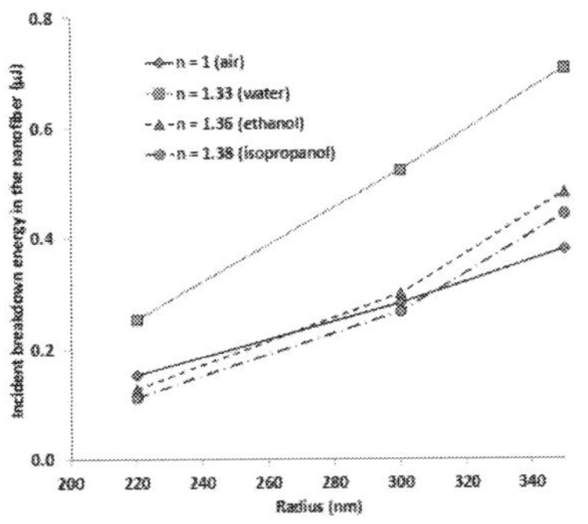

Fig. 2. LIB energy measurements for three nanofiber radii and four external media.

4 Conclusion

In this paper we have presented the first study to our knowledge of LIB of nanofibers in air and in several liquids for different radii in the ns regime. The same protocol was applied to a significant number of samples and showed highly repeatable results. We showed that immerging a nanofiber in a liquid substantially enhanced the LIB, offering a new degree of freedom to widen the field of applications of nanofibers, where high peaks powers are needed.

This work is supported by the French National Research Agency (ANR) (FUNFILM-ANR-16-CE24-0010- 03).

References

[1] M. A. Foster, A.C. Turner, M. Lipson, A.L. Gaeta, Opt. Express 16, pp. 1300-1320 (2008).
[2] S. Leon-Saval, T. Birks, W. Wadsworth et P. Russell, Opt. Express, 12 (13), pp. 2864-2869 (2004).
[3] L. Tong, Sensors, 18, 903 (2018).
[4] B. Gouraud, D. Maxein, A. Nicolas, O. Morin, and J. Laurat, Phys. Rev. Lett. 114, 180503, May 2015.
[5] K. S. Rajasree, T. Ray, K. Karlsson, J. L. Everett, and S. Nic Chormaic, Phys. Rev. Research 2, 012038(R) February 2020.
[6] L. Shan, G. Pauliat, G. Vienne, L. Tong, and S. Lebrun, Applied Physics Letters, 102:201110, May 2013.
[7] M. Bouhadida, J. C. Beugnot, P. Delaye, K. Phan Huy and S. Lebrun, Applied Physics B, Volume 125, Issue 12, pp.1-7 (sept. 2019).
[8] G. Fanjoux, J. Chrétien, A. Godet, K. Phan-Huy, J.-C. Beugnot, T. Sylvestre, Opt. Express 27 (20), pp. 29460-29470 (2019).

https://doi.org/10.1051/epjconf/202023808011

Long-time optical transmittance measurements of silica nanofibers

Maha Bouhadida[1*] and *Sylvie Lebrun*[1]

[1] *Laboratoire Charles Fabry, Institut d'Optique, CNRS, Université Paris-Saclay, 91127 Palaiseau cedex, France*

Abstract. We perform long-time measurements of the optical silica transmittance during several months in different environments and with different nanofiber lengths. These measurements are repeatable and give guidelines to control and to improve the lifetime and the performances of the nanofiber. The dust particles on the nanofiber surface is the fundamental reason behind its degradation. Enhancing the cleanness conditions of the nanofiber environment makes its lifetime increases significantly (from some hours to some months) and enables to avoid the dramatic decrease of its transmittance even after months. The nanofiber length does not contribute to the nanofiber transmittance degradation. Stabilizing the nanofiber transmittance after its decrease is possible by putting in in a dust free box.

1 Context and motivation

The optical nanofiber is created by stretching a standard silica fiber until its diameter becomes typically comparable to the light wavelength. It is composed by a central part called the waist having a sub-micrometric diameter linked to the untapered end-fibers via two transition sections called tapers. The nanofiber can be easily inserted in fibered networks with very low losses which makes it significantly explored for many applications. Thanks to its small diameter, this component has a strong confinement of light inside the silica enabling the generation of nonlinear effects such as the generation of supercontinuum [1]. It has also an intense evanescent field exploited for optical sensing [2], optical traps for quantum applications [3] and the excitation of non linearities when immersed in liquids [4,5,6]. Despite all these attractive applications, the nanofiber performances are limited by the degradation of its transmittance during time and by its short lifetime. It has been previously demonstrated that the nanofiber transmittance degradation is mostly due to the dust particles deposited on its surface [7]. Just after pulling, the lifetime of a nanofiber placed in an environment with no particular attention to cleanliness is only some hours. In [7] the investigation of this degradation was realized using rooms with different cleanliness degrees and nanofibers with different diameters varying from 340 nm to 460 nm. However, the measurements were not repeated and were performed only for one week. Besides, the possibility of stabilizing the transmittance of an already degraded nanofiber was not discussed. Only one nanofiber waist length of 1 cm was tested. In this work, we perform long-time measurements (over several months) of the nanofiber transmittance in environments with different levels of cleanliness. Two nanofiber waist lengths are tested (2 cm and 7 cm). We also give a simple solution to the nanofiber transmittance stabilization after its degradation. Firstly we describe the experimental setup and protocol. Then, we present the results and we discuss them giving some guidelines to preserve the nanofiber transmittance stability and to extend its lifetime.

2 Experimental setup and protocol of measurements

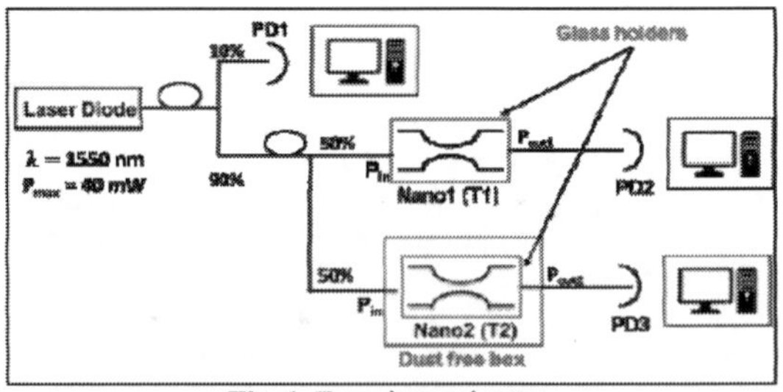

Fig. 1. Experimental setup

Fig.1 illustrates the experimental setup we build to perform the long-time measurements of nanofibers transmittance stability. This setup is totally fibered. A laser diode delivering a maximal power of 40 mW emits a light beam at 1550 nm in the cw regime. The beam is split by a 90/10 coupler. The 10% side is used as a reference and detected by photodiode PD1. The other 90% side is split by a 50/50 coupler, delivering two beams of the same power P_{in}. The first beam is coupled

* Corresponding author: maha.bouhadida@institutoptique.fr

to a nanofiber (Nano1) placed in a partially protective glass holder. The second beam is coupled to a nanofiber (Nano2) placed in an identical glass holder and stored in dust free plexiglas box. Only the end-fibers connectors are out of this box. The nanofiber pulling process is described in [4]. A butane flame softens the fiber central part while two computer-controlled translation stages elongate it following the classical "pull and brush" technique. This fabrication process is performed in a class-5 clean room. The resulting nanofiber has a radius of 500 nm. Two waist lengths are fabricated for these measurements (2 cm and 7 cm). The standard fiber is the SMF28 (ref. Corning). It is connectorized at both sides before its pulling. After fabricating the nanofiber, we partially protect it by putting it in the glass holder and transport it to the experimental setup. Our setup is built in normal lab cleanliness and humidity conditions. We check that the humidity conditions of the dust free box are the same as those of the lab. The light transmittance of the nanofiber T is defined as the ratio between the output transmitted power P_{out} and the input power P_{in}. We denote respectively by T1 and T2 the transmittances of Nano1 and Nano2. The initial nanofibers transmittances are measured after their installation in this setup. P_{out1} and P_{out2} are detected by photodiodes PD2 and PD3. All the photodiodes acquisitions are controlled by computer. We tested two different environments of the nanofibers : the partially protective glass holder directly held in the normal lab cleanliness conditions (Environment1) and the same holder overprotected in a dust free box (Environment2). The measurements duration is about 150 days. The output power values are taken every minute, which is equivalent to 1440 samples per day. The transmittance measurements are divided in two phases : in the first phase, Nano1 and Nano2 are respectively placed in Environment1 and Environment2. In the second phase, Nano1 and Nano2 are swapped. We repeat these long-time measurements several times (3 times until now) and for the two different waist lengths.

3 Results and discussion

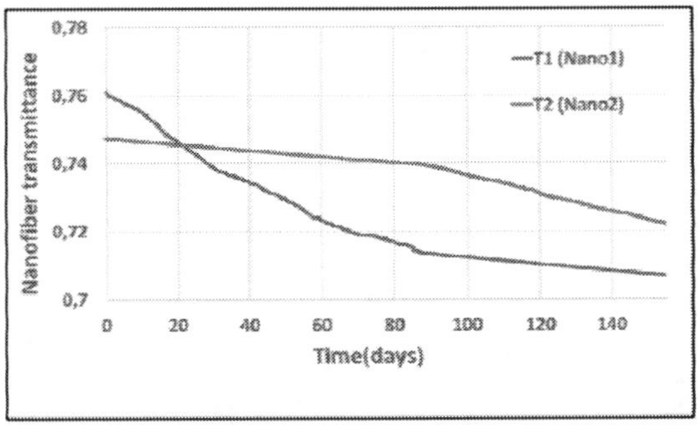

Fig. 2. Typical transmittance measurement

Fig.2 shows a typical continuous long-time transmittance measurement (about 150 days) of nanofibers having a waist length of 2 cm. The initial transmittances are T1 = 0.76 and T2 = 0.745. The first phase lasted about 85 days and the second one about 65 days. The nanofibers having 7 cm of length give the same trends, which proves that the waist length doesn't affect the transmittance degradation. This result is repeatable (3 measurements for the two tested nanofibers of different lengths). We observe that the transmittances of both Nano1 and Nano2 are not dramatically degraded after 150 days as it is the case when the nanofiber is put directly in the normal lab conditions without any protection. This confirms the protection efficiency of our glass holders even if it is partial. However, the decrease of T1 is faster than the decrease of T2 in the first phase of the measurement. In environment1, T1 goes from 0.76 to 0.715 ($-5x10^{-4}$/day) but T2 is only reduced from 0.745 to 0.74 ($-5x10^{-5}$/day). This slighter slope degradation illustrates the impact of the additional protection offered by the dust free box. In the second phase, we swap Nano1 and Nano2. We notice that the slope of the degradation is decreasing : T1 decreases from 0.715 to 0.71 in 65 days, i.e. with a slope of $-2x10^{-4}$/day (at the same time T2 decreases by $3x10^{-4}$/day). Then, by placing an already degraded nanofiber in a dust free box, it is possible to slow down the degradation of its transmittance. Our results also enable to predict the value of the transmittance of a nanofiber over several months, which is an important parameter for experiments.

4 Conclusion and perspectives

In this paper, we present long-time and repeatable measurements of nanofiber transmittances during several months, performed in different environments and with different nanofiber waist lengths. We show that the length of the nanofiber does not contribute to the transmittance degradation. By protecting the nanofiber from dust just after pulling, we are able to increase its lifetime from some hours to several months and to preserve its optical performances. Placing the nanofiber in a cleaner environment enables to slow down the degradation. These guidelines are useful to control and to enhance the lifetime and the performances of an optical nanofiber.

This work is supported by the French National Research Agency (ANR) (FUNFILM-ANR-16-CE24-0010- 03).

References

[1] M. A. Foster, A.C. Turner, M. Lipson, A.L. Gaeta, Opt. Express **16**, pp. 1300-1320 (2008).

[2] L. Tong, Sensors, **18**, 903 (2018).

[3] B. Gouraud, D. Maxein, A. Nicolas, O. Morin, and J. Laurat, Phys. Rev. Lett. **114**, 180503(2015).

[4] L. Shan, G. Pauliat, G. Vienne, L. Tong, and S. Lebrun, Appl. Phys. Lett., 102:201110 (2013).

[5] M. Bouhadida, J. C. Beugnot, P. Delaye, K. Phan Huy and S. Lebrun, Appl. Phys. B, **125**, 228 (2019).

[6] G. Fanjoux, J. Chrétien, A. Godet, K. Phan-Huy, J.-C. Beugnot, T. Sylvestre, Opt. Express, **27**, 20 (2019).

[7] M. Fujiwara, K. Toubaru, S. Takeuchi, Opt. Express, **19**, 9 (2011).

EPJ Web of Conferences **238**, 09001 (2020) https://doi.org/10.1051/epjconf/202023809001
EOSAM 2020

Procedure to optimize the intercarrier spacing in superchannels impaired by the cascading of WSS-based ROADMs

Paulo J. Pereira[1,*], *João L.* Rebola[1,2], and *Luís G.* Cancela[1,2]

[1]Department of Information Science and Technology, ISCTE-Instituto Universitário de Lisboa, Lisboa, Portugal
[2]Optical Communications and Photonics Group, Instituto de Telecomunicações, Lisboa, Portugal

Abstract. A new procedure is proposed to optimize the intercarrier spacing and wavelength selective switch (WSS) bandwidth for superchannels, using a Monte Carlo simulation. We perform an exhaustive assessment of the optical signal-to-noise ratio (OSNR) penalties due to the optical filtering and intercarrier crosstalk, and concluded that the optimum intercarrier spacing is at most 1.1 GHz larger than the Nyquist bandwidth and the results show how the number of cascaded WSSs influences the intercarrier spacing.

1 Introduction

In superchannels, the spectrum of the subchannels is tightly packed and the crosstalk between subchannels can become a performance constraint. Furthermore, along its path in the optical network, a superchannel passes through several reconfigurable optical add-drop multiplexers (ROADMs). The cumulative filtering reduces the available bandwidth and leads to signal distortion. The performance degradation induced by the wavelength selective switch (WSS) filtering and the intercarrier crosstalk have been studied jointly [1] but a separate quantification of their performance degradation is useful [2,3]. In this work, we perform an exhaustive assessment of the optical signal-to-noise ratio (OSNR) penalties due to the optical filtering and intercarrier crosstalk between subchannels using different *M*-ary quadrature amplitude modulation (*M*-QAM) formats, symbol rates, number of traversed WSSs, WSS bandwidths, number of subchannels, intercarrier spacings and roll-off factors. We also provide a procedure to obtain optimized values of the intercarrier spacings and WSS bandwidths that guarantee an OSNR penalty below 1.5 dB [1] for all subchannels after traversing 20 WSSs.

2 Superchannels with 2 up to 10 carriers

We consider the WSS analytical model described in [4] and an optical transfer function with a -3 dB bandwidth BW_{OTF}=8.5 GHz. For the Monte Carlo (MC) simulation, we use 28 and 32.5 GBd signals considering two modulation formats, 4-QAM and 16-QAM. The target BER is 4×10^{-3} for 28 GBd and 10^{-2} for 32.5 GBd [5]. The transmitted symbols in each simulation run are 131072 for the two modulation formats. Two root raise cosine filters are considered, one at the transmitter and the other at the receiver, with the same roll-off factor ρ=0.1. First, we perform the simulation for an edge subchannel, then for a center subchannel and, finally, we combine the results from the two subchannels and find the optimum intercarrier spacing and WSS bandwidth. Hence, the procedure has 3 steps:

1. The WSS bandwidth is set equal to the superchannel bandwidth. For an edge subchannel, by varying the intercarrier spacing, we examine if the OSNR penalty is below 1.5 dB after traversing 20 WSSs. We can increase the WSS bandwidth by 12.5 GHz, until we obtain the lowest WSS bandwidth that ensures less than 1.5 dB OSNR penalty.
2. For the WSS bandwidth found in step 1 and for a center subchannel, we change the intercarrier spacing and obtain the interval of acceptable intercarrier spacings that provide OSNR penalties below 1.5 dB after 20 WSSs.
3. The intersection of the two intervals obtained in steps 1 and 2 provides a new interval. Then, and only for intercarrier spacings in this new interval, we add the OSNR penalties from the edge and center subchannels to obtain the total OSNR penalty and then find the intercarrier spacing that provides the minimum total OSNR penalty.

With the proposed procedure, we find the intercarrier spacing that provides the lowest OSNR penalty, below 1.5 dB, for all subchannels after 20 traversed WSSs and the superchannel uses the lowest WSS bandwidth and, subsequently, has the highest spectral efficiency. In Fig. 1, the total OSNR penalty is presented as a function of the intercarrier spacing, for a superchannel with 10 subchannels, 28 GBd, ρ=0.1, a -6 dB WSS bandwidth of 300 GHz and a) 4-QAM and b) 16-QAM. We can see that, for low intercarrier spacings, the OSNR penalty increases due to higher crosstalk and the OSNR degradation is practically independent of the number of WSSs traversed. On the other hand, the OSNR penalty increases for higher intercarrier spacings due to WSS filtering-induced distortion and becomes dependent on the number of WSSs traversed. The OSNR degradation is higher for the 16-QAM modulation format than for the 4-QAM. An optimum intercarrier spacing for the subchannels is found at the lowest total OSNR penalty due to a balance between these two effects, for a specified number of cascaded WSSs.

* paulo_jose_pereira@iscte-iul.pt; joao.rebola@iscte-iul.pt; luis.cancela@iscte-iul.pt

Table 1. Intercarrier spacing and the lowest OSNR penalty after 20 cascaded WSSs for 28 GBd with ρ=0.1, 4-QAM and 16-QAM modulation formats, for a different number of subchannels in the superchannel.

Number of subchannels	WSS bandwidth [GHz]		Intercarrier spacing [GHz]		Center subchannel OSNR penalty [dB]		Edge subchannel OSNR penalty [dB]	
	4-QAM	16-QAM	4-QAM	16-QAM	4-QAM	16-QAM	4-QAM	16-QAM
2	75	87.5	28.3	28.9	-	-	0	0
4	125	137.5	28.6	29.0	0.1	0	0.1	0
5	150	162.5	28.3	29.1	0.3	0	1.2	0
6	187.5	187.5	28.7	29.0	0	0	0	0
8	237.5	250	28.6	29.0	0.1	0	0.6	0
10	300	300	28.9	28.6	0	0.8	0	0.8

Table 2. Intercarrier spacing and the lowest OSNR penalty after 20 cascaded WSSs for 32.5 GBd with ρ=0.1, 4-QAM and 16-QAM modulation formats, for a different number of subchannels in the superchannel.

Number of subchannels	WSS bandwidth [GHz]		Intercarrier spacing [GHz]		Center subchannel OSNR penalty [dB]		Edge subchannel OSNR penalty [dB]	
	4-QAM	16-QAM	4-QAM	16-QAM	4-QAM	16-QAM	4-QAM	16-QAM
2	75	87.5	33.1	33.2	-	-	0.3	0
4	137.5	150	32.6	33.5	0.6	0	1.4	0
5	175	187.5	32.8	33.5	0.1	0	0.1	0
6	212.5	212.5	33.1	33.5	0	0	0	0.5
8	275	287.5	32.9	33.4	0.1	0	0	0
10	337.5	350	32.7	33.5	0.3	0	0.4	0

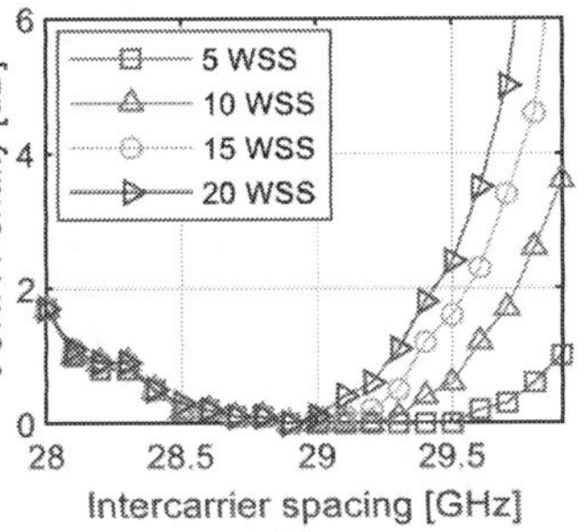

(a)

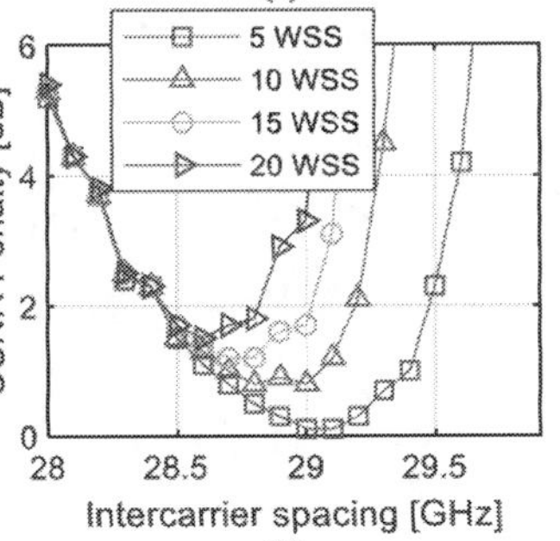

(b)

Fig. 1. Total OSNR penalty dependence on the intercarrier spacing after 5, 10, 15 and 20 WSSs, 10 subchannels, 28 GBd with ρ=0.1 and a -6 dB WSS bandwidth of 300 GHz, for each WSS in the cascade, for (a) 4-QAM and (b) 16-QAM.

The results obtained from thorough superchannels optimization with a different number of subchannels are presented in Tables 1 and 2, showing the optimum intercarrier spacings and the -6 dB WSS bandwidths with the corresponding lowest OSNR penalties for a center and an edge subchannels, for the 4-QAM and 16-QAM modulation formats. For 16-QAM, due to the lower tolerance of higher modulation formats to WSS bandwidth narrowing, we observe almost always a higher intercarrier spacing and almost always a larger WSS bandwidth. We can see that the optimized intercarrier spacing is at most 1.1 GHz larger than the Nyquist bandwidth. If we increase the number of WSSs, it is necessary to reduce the intercarrier spacing and allow a higher OSNR penalty. If the roll-off factor is increased from 0.1 to 0.3, we have to increase the WSS bandwidth by one or two slots of 12.5 GHz and adjust the intercarrier spacing to ensure OSNR penalty lower than 1.5 dB in all subchannels after traversing 20 WSSs.

3 Conclusions

In this work, a new method to optimize the WSS bandwidth and intercarrier spacing in superchannels has been proposed and implemented using a MC simulation. The optimization has been performed for different symbol rates, modulation formats, number of subchannels, number of cascaded WSSs and roll-off factors. This method guarantees an OSNR penalty below 1.5 dB and a superchannel with the highest spectral efficiency.

This work was supported in part by Fundação para a Ciência e a Tecnologia (FCT) from Portugal under the project of Instituto de Telecomunicações UIDB/EEA/50008/2020.

References

1. G. Bosco, *et al.*, JLT, **29**, 56-57, (2011)
2. B. Clouet, , *et al.*, JOCN, **8**, A122-A124, (2016)
3. J. Pedro, JOCN, **9**, C39-C40, (2017)
4. C. Pulikkaseril, *et al.*, Optics Express, **19**, 8460 (2011)
5. A. Morea, *et al.*, JOCN, **7**, A293-A295, (2015)

EPJ Web of Conferences **238**, 09002 (2020) https://doi.org/10.1051/epjconf/202023809002
EOSAM 2020

Autocompensating Measurement-Device-Independent quantum cryptography in few-mode optical fibers

Jesús Liñares-Beiras[1,*], *Xesús* Prieto-Blanco[1], *Daniel* Balado[1] and *Gabriel M.* Carral[1,**]

[1]Quantum Materials and Photonics Research Group, Department of Applied Physics, Faculty of Physics / Faculty of Optics and Optometry, University of Santiago de Compostela, Campus Vida s/n, E-15782, Santiago de Compostela, Galicia, Spain.

Abstract. We present an autocompensating quantum cryptography technique for Measurement-Device-Independent quantum cryptography devices with different kind of optical fiber modes. We center our study on collinear spatial modes in few-mode optical fibers by using both fiber and micro-optical components. We also indicate how the obtained results can be easily extended to polarization modes in monomode optical fibers and spatial codirectional modes in multicore optical fibers.

1 Introduction

Quantum cryptography is based on the properties of quantum mechanics to obtain secure quantum key distribution (QKD) by using different protocols. The seminal protocol has been the so-called BB84 one in which four states define a set of two mutually unbiased basis. The security in line is guaranteed by the laws of physics, but side channel attacks are still posible. However, the Measurement-Device-Independent (MDI) quantum cryptography [1] has solved the detector side channel attacks. As to the different emisor channel attacks they can be solved *ad hoc*. On the other, different optical fiber systems have been proposed to implement QKD cryptography. Such systems can use different kind of modes, for instance, polarization modes in monomode optical fibers, spatial collinear modes in few-mode optical fibers (FMF) and spatial codirectional modes in multicore optical fibers. However, a common problem is that at least four linear combinations of two modes have to be used and therefore these modes need to keep stable over large distances of optical fibers. Mode instability is due to the modal coupling undergone by the modes in their propagation along real optical fibers with small imperfections or (slow) temporal perturbations. To overcome this drawback polarization autocompensating techniques have been proposed in cryptography with single-photon quantum states [2]. In this work we propose an autocompensating quantum cryptography technique for two-photon quantum states, and in particular Autocompensating MDI-QKD (A-MDI-QKD). We study in detail two collinear modes of a few-mode optical fiber by using both fiber and micro-optical components. The results can be easily extended to both polarization and codirectional modes. The photonic devices are analogous to those ones used in 1-qudit cryptography [3].

*e-mail: suso.linares.beiras@usc.es
**e-mail: gabrielmaria.carral@rai.usc.es

2 Basis and Bell states for A-MDI-QKD

A FMF can support several collinear modes. Let us consider the horizontal and vertical Hermite-Gaussian (HG) modes, which are denoted as X and Y modes. Likewise, we assume a biphoton quantum source, or alternatively, two sources of weak coherent pulses together with decoy states for security purposes. Moreover, let us consider that the initial state is excited in HG modes rotated $\pi/4$,

$$|L_i\rangle = |L_{hi}\rangle\,|L_{vi}\rangle =\rangle \frac{1}{2}(|1_{hX}\rangle + |1_{hY}\rangle)\,(|1_{vX}\rangle + |1_{vY}\rangle) \quad (1)$$

where subindices h, v stand for modes in Alice and Bob paths. The basic setup of the quantum cryptographic system is shown in figure 1. The transmission between Alice (A) and Charlie (C) and Bob (B) and C is made by means of FMFs. The state (1) is produced by the initial states generator (ISG) (a SPDC source or two lasers). Moreover, the systems A-(A) and B-(A) contain autocompensating optical devices and a quantum states generator device by

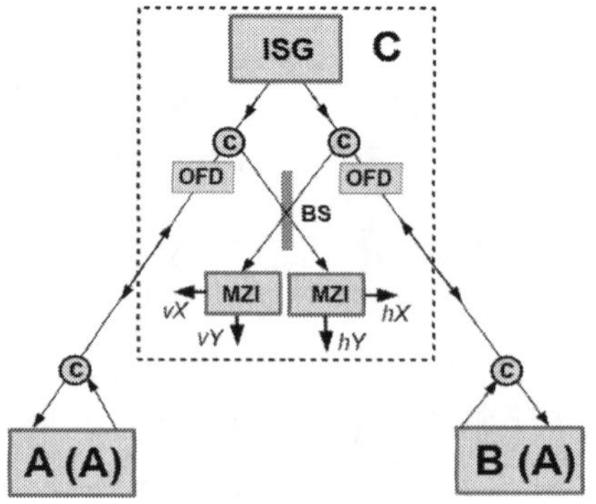

Figure 1. Basic setup of the cryptographic A-MDI-QKD system.

 https://doi.org/10.1051/epjconf/202023809002

phase shifting. Finally, in the C system there is a detection system with a Beam-Splitter (BS), two Mach-Zehnder Interferometers (MZI) and four detectors $\{hX, hY, vX, vY\}$. A and B have to use the following modal basis

$$\begin{aligned} \mathcal{B}_1 &= \{\tfrac{1}{\sqrt{2}}(|1_{aX}\rangle \pm |1_{aY}\rangle)\} = \{|1_{a\binom{P}{N}}\rangle\} \\ \mathcal{B}_2 &= \{\tfrac{1}{\sqrt{2}}(|1_{aX}\rangle \pm i|1_{aY}\rangle)\} = \{|1_{a\binom{L}{D}}\rangle\}, \quad a = h, v \end{aligned} \tag{2}$$

Note that the standard MDI protocol has states $|1_{aX}\rangle, |1_{aY}\rangle$, which are not suitable for MDI autocompensating purposes. When A and B receive the single photon state they have to choose a relative phase $\theta = \{0, \pi, \pi/2, -\pi/2\}$ in order to send a single photon state to C in base $\mathcal{B}_1$ or $\mathcal{B}_2$. When single photon states reach the C system then a BS entangles the single photon states and Bell states are produced, that is, the unitary transformation $\hat{U}_x = \exp\{i(\pi/4)\sigma_x\}$ has to be implemented, where σ_x is the first Pauli matrix. HOM states are also obtained, as for example: $|2_{hX}\rangle, |2_{vX}\rangle, ...$, however they are not used in the MDI protocol because provide ambiguous information. They can be used to calibrate the system, that is, to check that photons reach the C system at the same time. By post-selecting the events where they use the same basis, it is easy to check that the Bell states are obtained for all biphoton states: $|1_{h\alpha}1_{v\alpha'}\rangle$, with $\alpha, \alpha' = P, N$ or $\alpha, \alpha' = L, D$, that is,

$$\begin{aligned} |\Psi^-\rangle &= \tfrac{1}{\sqrt{2}}(|1_{hX}1_{vY} - |1_{vX}1_{hY}) \\ |\Phi^+\rangle &= \tfrac{1}{\sqrt{2}}(|1_{hX}1_{hY} + |1_{vX}1_{vY}) \end{aligned} \tag{3}$$

$|\Psi^-\rangle$ is obtained if $\alpha \neq \alpha'$ and $|\Phi^+\rangle$ if $\alpha = \alpha'$. The probability of detecting these states is $1/2$ in each event in the same basis. The results can be extended to polarization modes H, V and two codirectional modes 1,2 (two-core optical fiber) by the formal changes: $(X, Y) \equiv \{(H, V), (1, 2)\}$,

3 Photonic system for A-MDI-QKD

Autocompensating is obtained after one loop implemented by optical circulators (c). We present the main devices to achieve autocompensating MDI-QKD. One of them is an optical fiber delay (OFD) device placed just after the circulators in the C system. This device can be realized by a MZI with a Dove prism (DP) and an external optical path (optical fiber), as shown in figure 2. It produces a delay τ between states $|1_X\rangle$ and $|1_Y\rangle$, in order to introduce phases θ in A and B systems and obtain the different biphoton states. Next, autocompensating devices (A) are placed in A and B after the circulators. A sketch of these devices is shown figure 3. It can easily be checked that by the unitary transformation $\hat{U}_y = \exp\{i(\pi/2)\sigma_y\}$ any random unitary transformation between modes X and Y from C to A (and B) due to modal coupling by imperfections or slow temporal perturbations is compensated along the path back between A (B) and C. Transformation $\hat{U}_y$ can be implemented at A and B systems by a local loop with two cylindrical lens converters (CLC) which shift $\pm\pi/2$ the HG modes and a DP rotated $\pi/4$ as shown in figure 3. Note that when the state reaches again the ODF device the modes X and Y become synchronous again. Therefore the states reach the BS at the same time in the C system.

4 Detection photonic device

The detection system after the BS is formed by two MZIs and four detectors: hX, hY, vX, vY as shown in figure 1. The MZIs are identical to the one used inside the OFD device shown in figure 2, that is, it works like a BS depending on the spatial mode. Thus, X modes are detected in the horizontal direction and the Y modes in the vertical direction. The Bell states are identified by different coincidences results according to equation (3). The steps of the BB84 protocol can be followed to obtain QKD.

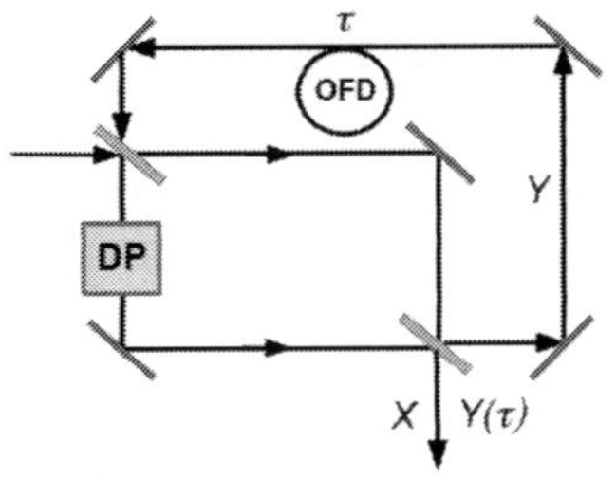

Figure 2. Basic setup of a OFD device.

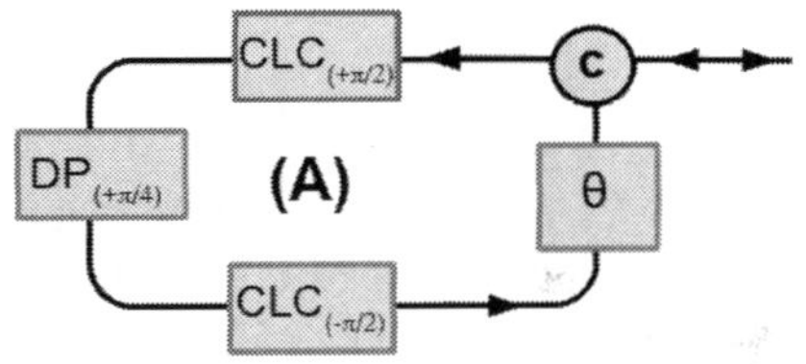

Figure 3. Basic setup of the autocompensating system.

5 Conclusions

We have proposed autocompensating MDI-QKD by using two spatial modes (Hermite-Gaussian ones) of FMF fibers and suitable biphoton quantum states. An unitary transformation (a $\pi/2$-rotation) is required, which is implemented by a local loop containing a DP between two cylindrical lens converters. The results can be extended to both polarization and codirectional modes.

Acknowledgement. Authors wish acknowledge the financial support of this work by Xunta de Galicia, Consellería de Educación, Universidades e FP, Grant GRC Nº ED431C2018/11 and a predoctoral grant (D. Balado, 2017), co-financied with the European Social Fund.

References

[1] H. Lo, M. Curty, B. Qi, Phys. Rev. Lett., **108**, 130503 (2012)
[2] D. S. Bethune, W. P. Risk, New J. Phys., **4**, 42 (2002)
[3] D. Balado, J. Liñares, X. Prieto-Blanco, J. Mod. Opt. **66**, 947-957 (2019)

Magneto-optical control and spin coupling with non-reciprocal surface waves for nanoscale thermotronics -INVITED

Svend-Age Biehs[1,*]

[1]Institut für Physik, Carl von Ossietzky Universität, D-26111 Oldenburg, Germany

Abstract. Magneto-optical materials have been proposed as promising candidates for an active control of the directionality of nanoscale heat radiation. Here, we discuss the possibility to rectify nanoscale radiative heat fluxes by means of non-reciprocal surface waves and propose a nanoscale heat flux rectifier or diode which can be controlled actively by means of externally applied fields. We furthermore, reveal the spin coupling mechanism behind the observed heat flux rectification

1 Introduction

Magneto-optical or more generally non-reciprocal materials have been proposed as promising candidates for an active control of the directionality of nanoscale heat radiation. Unexpected and very interesting effects like the thermal radiative Hall effect [1], persistent currents [2], giant magneto-resistance [3, 4],and circular heat fluxes [5] as well as persistent angular momentum and spin of thermal radiation [5] have been highlighted. We review some of the recent developments in this new direction of magneto-optical thermotronics [6]. In particular, we discuss the possibility to rectify nanoscale radiative heat fluxes by means of non-reciprocal surface waves and propose a nanoscale heat flux rectifier or diode which can be controlled actively by means of externally applied fields [7]. We furthermore, reveal the spin coupling mechanism behind the observed heat flux rectification [8].

2 Discussion

Based on fluctuational electrodynamics we derive the general expressions for mean Poynting vector and the heat exchange [7, 8] in a general configuration of N nanoparticles in dipolar approximation with temperatures $T_1, \ldots, T_N$ in an arbitrary environment at temperature T_b. These general expressions allow us for the special case $N = 2$ to evaluate the power exchanged between two InSb nanoparticles and the mean heat flux around the nanoparticle when placing both nanoparticles in close vicinity of an InSb substrate in Voigt configuration as shown in Fig. 1.

Without magnetic field the material properties of the the nanoparticles and the substrate are reciprocal and hence the heat flux in the forward and backward direction turns out to be the same. However, when turning on a magnetic field the material properties become non-reciprocal due to the Lorentz force acting on the electrons in InSb. As a consequence the

*e-mail: s.age.biehs@uol.de

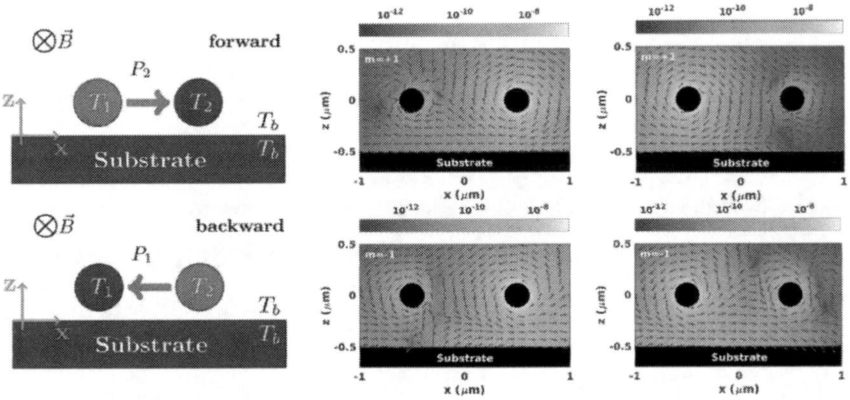

Figure 1. Directional heat flux in forward and backward direction due to the coupling of the $m = \pm 1$ resonances with the spin-momentum locked non-reciprocal surface modes for InSb.

heat flux in forward and backward direction has different values so that the interparticle heat flux becomes rectified. This effect is very strong if the particle resonances can couple to the surface mode resonances. As we will discuss in detail, this coupling is spin selective so that the particle resonance with magnetic quantum number $m = -1$ ($m = +1$) couples preferably to surface waves travelling in positive (negative) x-direction as can be seen in Fig. 1.

This coupling to the non-reciprocal surface modes can be controlled by the amplitude of the external magnetic field and can lead to a very large rectification ratio close to one. We will discuss in detail the dependence on the particle surface distance and the interparticle distance as well as the dependence on the magnetic field amplitude. Since the heat emitted by the hot particle will not only heat up the cold particle but also the surface itself there is a competition between the interparticle and the particle-surface heat flux. Therefore we study the heating dynamics in forward and backward direction to quantify the heating efficiencies.

References

[1] P. Ben-Abdallah, "Photon Thermal Hall Effect," Phys. Rev. Lett. **116**, 084301, (2016).

[2] L. Zhu and S. Fan, "Persistent Directional Current at Equilibrium in Nonreciprocal Many-Body Near Field Electromagnetic Heat Transfer," Phys. Rev. Lett. **117**, 134303 (2016).

[3] I. Latella and P. Ben-Abdallah, "Giant Thermal Magnetoresistance in Plasmonic Structures," Phys. Rev. Lett. 118, 173902, (2017).

[4] R. M. Abraham Ekeroth, P. Ben-Abdallah, J.C. Cuevas, and A. Garcia Martin, "Anisotropic Thermal Magnetoresistance for an Active Control of Radiative Heat Transfer," ACS Photonics **5**, 705 (2017).

[5] A. Ott, P. Ben-Abdallah, and S.-A. Biehs, "Circular heat and momentum flux radiated by magneto-optical nanoparticles," Phys. Rev. B **97**, 205414 (2018).

[6] A. Ott, R. Messina, P. Ben-Abdallah, S.-A. Biehs, "Magnetothermoplasmonics: from theory to applications," J. Photon. Energy **9**, 032711 (2019).

[7] A. Ott, R. Messina, P. Ben-Abdallah, S.-A. Biehs, "Radiative thermal diode driven by nonreciprocal surface waves," Appl. Phys. Lett. **114**, 163105 (2019).

[8] A. Ott and S.-A. Biehs, "Thermal rectification and spin-spin coupling of non-reciprocal localized and surface modes," Phys. Rev. B **101**, 155428 (2020).

Temperature wave-like oscillations on ultra-short and ultra-fast time scales -INVITED

Gandolfi Marco[1,2], Benetti Giulio[3], Glorieux Christ[1], Giannetti Claudio[2], Banfi Francesco[4,*]

[1]Laboratory of Soft Matter and Biophysics, Department of Physics and Astronomy, KU Leuven, Celestijnenlaan 200D, B-3001 Leuven, Belgium
[2]Interdisciplinary Laboratories for Advanced Materials Physics (I-LAMP) and Dipartimento di Matematica e Fisica, Università Cattolica del Sacro Cuore, Via Musei 41, 25121 Brescia, Italy
[3]Department of Pathology and Diagnostics – Medical Physics Unit, Azienda Ospedaliera Universitaria Integrata – Verona, P.le Stefani, 1, 37126 Verona, Italy
[4]Université de Lyon, Institut Lumière Matière (iLM), Université Lyon 1 and CNRS, 10 rue Ada Byron, 69622 Villeurbanne cedex, France

Abstract. Recent findings in the frame of temperature wave-like oscillations on the ultra-short, ultra-fast time scales in solid states devices are here reviewed. The possibility for wave-like temperature oscillations are investigated at the light of the pass-band characteristic in w-k space for the temperature scalar field. The bandpass filter characteristics are accessed in terms of the heat carriers scattering times. The concepts here reviewed are of interest for perspective design of novel thermal nano-devices.

1 Introduction

Coherent control of wave-like phenomena is driving a technological revolution in fields ranging from electronics, photonics, to phononics 1.
Although temperature has been historically taken as the paradigmatic example of an incoherent field, undergoing diffusive as opposed to wave-like propagation, on short space and time scales the Fourier law fails and the possibility for temperature wave propagation sets in (1) The ultimate goal is to devise materials exploiting the wave nature of the temperature scalar field arising in the non-Fourier heat transport regime. We here review recent progress made in this context [4-6]

2 Discussion

The Dual Phase Lag (DPL) model (2) has been introduced to correct for the instantaneous onset of the heat flux, q, upon establishment of a thermal gradient, ∇T, entailed in Fourier law. The DPL model generalizes earlier attempts – made by Cattaneo (1958) and Vernotte (1962) – to correct this pitfall by introducing causality between the cause ∇T (q), and the effect, q (∇T):

$$q(x, t+\tau_q) = -k_T \nabla T(x, t+\tau_T) \qquad (1)$$

In cases where $\tau_q < \tau_T$ the heat flux is the cause, whereas if $\tau_q > \tau_T$ the temperature gradient is the cause. Coupling the DPL with thermal energy conservation, the well-known heat diffusion equation (parabolic) is modified into a hyperbolic one:

$$\left(\frac{\tau_q}{\alpha}\right)\frac{\partial^2 T}{\partial t^2} - \frac{\partial^2 T}{\partial x^2} + \left(\frac{1}{\alpha}\right)\frac{\partial T}{\partial t} - \tau_T \frac{\partial^3 T}{\partial x^2 \partial t} = 0 \qquad (2)$$

, where $\alpha = k_T/C$ is the thermal diffusivity. The hyperbolic equation encompasses wave-terms (first two-terms) with damping (last two terms). The parabolic case is retrieved on "long" time scales when the second and third terms dominate.
Recently a user-friendly, intuitive approach [4] has been proposed to investigate under which conditions the temperature propagation safeguards coherent (wave-like) effects. This is achieved calculating the complex dispersion relation, $f(k,\omega) = 0$, for the hyperbolic equation linking the wave vectors k to the angular frequency ω. Specifically, the case $\omega \in \mathbb{C}$ and $k \in \mathbb{R}$ describes the physical situation of a temperature pulse propagating in the material, where the real part $\omega_1 = \mathcal{Re}(\omega)$ is an oscillation angular frequency, whereas the imaginary part $\omega_2 = \Im m(\omega)$ has the physical meaning of damping. Wave-like oscillations are sustained provided the quality factor $Q = \omega_1/\omega_2 > 1$. Materials sustaining temperature waves on short timescales are addressed, in the current work as Temperonic materials.
In general a material can sustain the following dynamics: (i) overdamped, $Q < 1$, (no transient temperature oscillations sustained) when q is the cause ($\tau_T/\tau_q > 1$) (ii) either over, critically or underdamped when ∇T is the cause ($\tau_T/\tau_q < 1$), the best case scenaro to observe transient temperature wave- behaviour being $\tau_T/\tau_q = 0$ (Cattaneo-Vernotte limit).
The complex-valued $\omega - k$ dispersion relation is investigated for the cases of a localised temperature pulse in space, linking the temperature wave angular frequencies ω and wave vectors k. The modal Q-factor allows discerning which temperature oscillations modes are practically accessible and allows inspecting the material as a frequency filter for the propagation of temperature oscillations. The optimal region in parameters space is discussed in a variety of systems, ranging from graphite to quantum materials [4-6].

* Corresponding author: francesco.banfi@univ-lyon1.fr

https://doi.org/10.1051/epjconf/202023810002

2 Conclusions

This presentation applies an easy-to-adopt analytical means to inspect the optimal conditions to observe temperature wave oscillations. The idea parallels the approach successfully employed in solid state physics and optics to investigate electronic wavefunction and electromagnetic wave propaga- tion in solid state devices. The calculations are applied in a variety of systems of technological interest. The present discussion is of relevance in view of engineering nanoscale thermal devices operating on ultra-fast and ultra-short time scales

References

1. M. Wegener, Science 342, 939 (2013)
2. M. Maldovan, Nature 503, 209 (2013)
3. D.Y. Tzou, Macro-to Microscale Heat Transfer: The Lagging Behavior (John Wiley & Sons, 2014)
4. M. Gandolfi, G. Benetti, C. Glorieux, C. Giannetti, and F. Banfi, Int. J. Heat Mass Transf. 143, 118553 (2019)
5. M Gandolfi, G L Celardo, F Borgonovi, G Ferrini, A Avella, F Banfi and C Giannetti, Phys. Scr. 92, 034004 (2017)
6. S. Huberman, R.A. Duncan, K. Chen, B. Song, V. Chiloyan, Z. Ding, A.A. Maznev, G. Chen, K.A. Nelson, Science 364, (2019)

Optical Resonators based on Casimir Forces -INVITED

Sol Carretero-Palacios[1,2,*], *Victoria Esteso*[1], and *Hernán Míguez*[1]

[1]Institute of Materials Science of Seville, Consejo Superior de Investigaciones Científicas (CSIC)-Universidad de Sevilla (US), Américo Vespucio 49, 41092 Seville, Spain

[2]Departamento de Física de Materiales, Universidad Autónoma de Madrid, 28049 Madrid, Spain

Abstract. The work here presented demonstrates theoretically that it is possible to create optical resonators based on levitation properties of thin films subjected to repulsive Casimir-Lifshitz forces. Our optical cavity design is made up of commonly found materials, such as silicon oxide, polystyrene or gold, with glycerol as a mediating medium, which supports high Q-factor optical modes at visible frequencies. The balance between flotation and repulsive Casimir-Lifshitz forces in the system allows the fine-tuning of the optical cavity thickness and hence its modes. Finally, we show that well-defined spectral features in the reflectivity allows by indirect means, an accurate prediction of the estimated equilibrium distance at which some part of the optical cavity arrangement levitates.

INTRODUCTION

Recently, in an elegant cutting-edge experiment [1], repulsive $F_{(C-L)}$ have been demonstrated in levitating metallic plates immersed in ethanol through a fashionable stratification of the inner dielectric media, elucidating the quantum trapping distance (d_{eq}) indirectly by spectroscopic techniques.

Here, we examine theoretically high Q-factor optical resonators formed by a levitating bilayer plate facing a coated planar substrate, all immersed in an appropriate liquid under the influence of repulsive $F_{(C-L)}$ and buoyancy [2]. We show that well-defined spectral features in the reflectivity (identified with high Q-factor optical resonators) allows an accurate prediction of d_{eq} by indirect means. Effects related to repulsive electrostatic forces are also evaluated. Specifically, optical resonators supporting high Q-factor optical modes at visible frequencies are formed by a gold (Au)-polystyrene (PS) bilayer drown in glycerol over an Au coated silicon dioxide (SiO_2) substrate (see Figure 1). The selected materials display remarkable optical quality, fine spectral reflectivity features, experimentally manageable thicknesses, and unlike alcohols, stable dissipation properties. Importantly, the relation of the dielectric permittivities gives rise to repulsive Casimir forces able to balance buoyancy.

Results and Discussion

$F_{(C-L)}$ is calculated employing the Lifshitz's theoretical formalism expressed for an arbitrary layered system [3], whereas the Poisson-Boltzmann methodology is applied to evaluate the effect of double layer forces [4,5]. Reflectance spectra are calculated employing the Transfer Matrix Method formalism [6,7]. Dielectric permittivities of all materials needed in the calculations are extracted from diverse references, in some cases employing oscillator models directly evaluated in Matsubara frequencies.

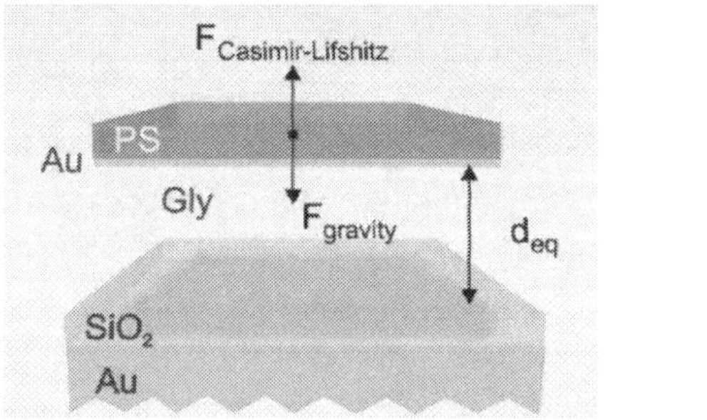

Fig. 1. Scheme of the optical cavity formed by a self-standing thin film made up of gold (Au) and polystyrene (PS) immersed in glycerol over a silica (SiO_2) coated gold substrate. Layer thicknesses range from tens to several hundreds of nanometers. The equilibrium distance at which Casimir-Lifshitz and gravity forces are balanced results in d_{eq} of few hundreds of nanometers.

Figure 2 shows a representative example of the equilibrium distances at which golden bilayer plates levitate as a function of the PS thickness building the optical resonator. For the set of parameters here considered, these can be finely tuned within $d_{eq} \in [180-375]$ nm by enhancing F_b with the PS thickness.

* Corresponding author: sol.carretero@uam.es

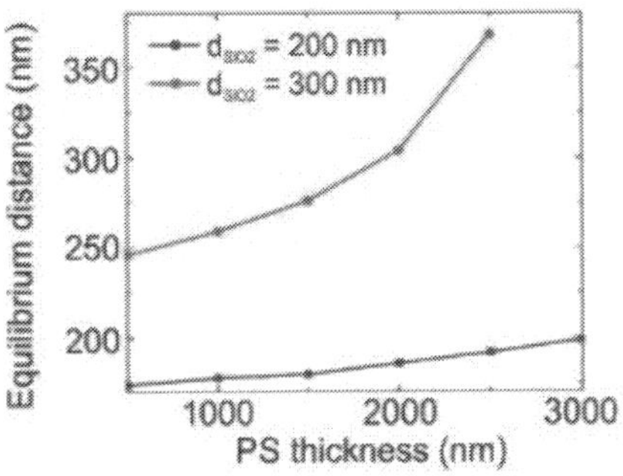

Fig. 2. Equilibrium distance as a function of the PS thickness slab in the levitating bilayer, with and d_{Au} = 30 nm fixed and setting d_{SiO_2} = 200 nm (in back) and 300 nm (in red).

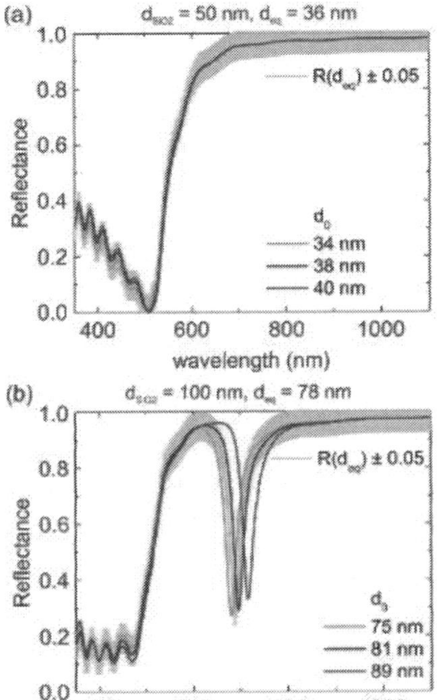

Fig. 3. (a) Reflectance of a system displaying levitation with d_{PS} = 1500 nm, d_{Au} = 30 nm, $d_{glycerol} = d_{eq}$ = 36 nm, and d_{SiO_2} = 50 nm, and an assumed error in the reflectance of ± 0.05 (in grey). For the same set of parameters, but separation distances $d_{glycerol} = d_0$ of 34, 38 and 40 nm, results are shown in orange, purple and green colors. These last systems do not yield levitation. (b) Results for the same set of parameters as in panel (a), but considering d_{SiO_2} = 100 nm, thus, $d_{glycerol} = d_{eq}$ = 78 nm, and assuming an error of ± 0.05 (in grey). Also, for separation distances of $d_{glycerol} = d_0$ of 75, 81 and 89 nm, results are shown in orange, purple and green colors. These last systems do not display levitation.

Figure 3 shows with two illustrative examples the need of having well-defined narrow spectral features to accurately estimate d_{eq}, and therefore, $F_{(C-L)}$. Both panels display high reflectance bands over a wide spectral range and, as the Q-factor of the optical cavity increases (ranging from 25 to 164 with the cavity size), narrower minima that rise in number appear at distinct wavelengths. For two SiO_2 coating thicknesses, 50 nm (a) and 100 nm (b), reflectance spectra of levitating films are shown in grey color. A deviation of ± 0.05 in the reflectance response arising from diverse error sources (such as variations in the estimation of film thicknesses, imperfect parallel alignment, surface roughness, etc) is also assumed. The spectral response of similar systems located close to the equilibrium distance (but without displaying levitation) are also depicted. Notably, for thin SiO_2 coating thicknesses (panel (a)), reflectance spectra of systems placed at d_0 = 34, 38 and 40 nm, match the estimated ± 0.05 reflectance error. This means that in a possible experiment based on this approach, any gap distance within the range $d_0 \in$ [34-40] nm could be inaccurately identified to levitation arising from just the balance of Casimir-Lifshitz forces and buoyancy, neglecting for instance the possible effect of additional forces in the system. In high Q-factor optical cavities (panel (b)), d_{eq} can be precisely determined. In this case, narrow spectral minimum corresponds to a single set of parameters, while the spectral response of bilayers located at d_0 = 75, 81, and 89 nm fall outside the assumed ± 0.05 error.

Conclusions

We have shown that high Q-factor optical resonators allows determining unambiguously, by spectroscopic techniques, the equilibrium distance of a levitating bilayer over a planar substrate due to the balance of repulsive $F_{(C-L)}$ and buoyancy. The selected choice of materials, namely SiO_2, PS, Au and glycerol, display decisive characteristics, such as repulsive $F_{(C-L)}$, adequate densities to balance such dispersion force at measurable nanoscale distances, excellent optical quality and manageable experimental fabrication. Besides, glycerol presents stable properties in contrast to alcohols suffering from evaporation.

References

1. A. Zhao, R.; Li, L.; Yang, S.; Bao, W.; Xia, Y.; Ashby, P.; Wang, Y.; Zhang, X. Science, **2019**, 364 (6444), 984–987
2. Esteso, V. ; Carretero-Palacios, S. ; Míguez, H. J. Phys. Chem. Lett. **2019**, 10, 5856–5860
3. Dzyaloshinskii, I.E.; Lifshitz, E.M.; and Pitaevskii, L.P. Adv. Phys. **1961,** 10 (38), 165-209
4. Cappella, B.; Dietler, G. Surface Science Reports, **1999**, 34, 1-104
5. Parsegian, V. A.; Gingell, D. Biophys. J., **1972**, 12, 1192
6. Thompson, R. C. Optical Waves in Layered Media. J. Mod. Opt. **1990**.
7. Hecht, E. Hecht Optics. Addison Wesley **1998**.

Nondestructive Characterization at Nanoscopic Scale by Photoacoustic and Photothermal Techniques -INVITED

Roberto Li Voti

Dipartimento di Scienze di Base ed Applicate per l'Ingegneria, Sapienza Università di Roma, Via A. Scarpa 16 Rome, 00161, Italy

Abstract. Photoacoustic and Photothermal Spectroscopy are excellent techniques for studying the optical absorption spectra of opaque and highly light-scattering substances such as nanomaterials. In this paper we review recent advances in the methodology and its novel applications at the nanoscopic scale.

One of the most challenging problems in nanomaterials research is their accurate characterization, which is fundamental for the efficient use of these technologically promising materials. Absolute absorption, quantum efficiency, thermal diffusivity, and the elastic constants are important parameters for photonic applications. Although the conventional absorption or emission techniques can provide the absorption coefficient, the determination of the absolute absorption is not a trivial measurement due to the presence of scattered light. On the contrary Photoacoustic (PAS) and Photothermal Spectroscopy are very sensitive techniques, immune to scattered or reflected light, and easy to be used to measure the absolute absorption in many different wavelength ranges.

In this summary we review recent advances in the methodology of PAS and its novel applications. In particular we highlight some works done to detect circular dichroism of intrinsic chiral materials [1] as well as extrinsic pseudo-chiral metasurfaces [2], showing how PAS can be applied to measure the selective absorption of circularly polarized light which depends on the orientation of the metasurface. PAS has been also applied to measure the resonant absorption peaks related to the guided modes of GaAs-based NW on Si in the VIS/IR range [3-5].

In disordered media PAS is confirmed to be the most appropriate technique to determine separately the absorption and the scattering coefficients allowing to determine the size of both metallic (AgNP) [6], or semiconductor nanospheres (ZnO) [7,8], or clusters of nanospheres bridged by the ligands [9], in contrast to nanolayered samples [10, 11].

On the other hand photothermal deflection spectroscopy (PDS) may be used as a complementary technique to measure the absorbance spectrum in nanostructures so to detect absorption lines, or photonic band gap in photonic crystals [12], or to evaluate the entity of the scattering phenomena in carbon nanotubes. PDS has been used in the UV/VIS/IR range from 250 to 1200 nm and by using a low modulation frequency from 1Hz to 100Hz which limits the spatial resolution so that the optical and thermal measurements are averaged in the volume.

We also underline some applications of photothermal radiometry (PTR) used to localize the internal heat sources where the pump light is absorbed, and to measure the effective thermal diffusivity. The modulation frequency obtained by an acousto-optical modulator ranges from 1 Hz up to 100 kHz, allowing the improvement of the thermal resolution till submicron range. We have applied PTR to PCM: VO_2 nanolayered samples and SiO_2/VO_2 synthetic opals [13, 14]. The pump beam is partially absorbed in the whole structure. Assuming that the heat is generated proportionally to the light intensity flowing in the structure, one may visualize the internal light propagation by performing the heat depth profile reconstruction from the photothermal radiometric data. This technique has been applied also to carbon nanotube film deposited onto a silicon substrate: in this case PTR allows to detect the thermal wave interference in the film and to measure the effective thermal diffusivity of the CNT, the thermal resistance with the substrate, and the infrared emission property of the CNT [15].

Photothermal radiometry and thermography can be also used to measure the emissivity of nanostructured materials, nanoantennas and metasurfaces which can be designed so to have specific directional thermal emittance properties [16, 17].

But in order to investigate the thermal properties at a nanoscopic scale one should necessarily work at higher frequencies in the MHz range, so to reduce the thermal diffusion length increasing the spatial resolution. One way is definitively offered by laser picosecond acoustics, by using for example a method known as optical heterodyne force microscopy (OHFM) [18, 19]. This detects photothermally induced surface vibrations in an AFM,

* Corresponding author: roberto.livoti@uniroma1.it

and can image subsurface nanoscale features through their effect on the surface displacement produced by a megahertz thermal field that is optically excited directly below the scanning AFM tip. Alternatively the thermoelastic properties of nanomaterials can be probed through imaging MHz variations in thermoreflectance in combination with picosecond ultrasonics [20, 21].

The invited talk summarizes several research topics and works done in collaboration with the Italian National Research Council, the University of Bath, the University of Dayton, the Tampere University of Technology, ICMM – CSIC in Madrid, the IOFFE Institute in S.Petersburg, and the Hokkaido University in Sapporo.

References

1. Benedetti, A., Alam, B., Esposito, M., Tasco, V, Leahu, G, Belardini, A., Li Voti, R., Passaseo, A. Sibilia, C, Scientific Reports, **7**, 5257 (2017)
2. Belardini, A., Centini, M., Leahu, G., Hooper, D.C., Li Voti, R., Fazio, E., Haus, J.W., Sarangan, A., Valev, V.K., Sibilia, C. Scientific Reports, **6,** 31796 (2016)
3. Leahu, G. Petronijevic, E., Belardini, A. Centini, M., Voti, R.L., Hakkarainen, T. Koivusalo, E., Guina, M., Sibilia, C. Scientific Reports, **7**, 2833 (2017)
4. Petronijevic, E, Leahu, G., Belardini, A., Centini, M., Li Voti, R. Hakkarainen, T., Koivusalo, E., Guina, M., Sibilia, C, Int J Thermophys, **39**, 45 (2018)
5. Petronijevic, E.,Leahu, G., Belardini, A., Centini, M., Li Voti, R., Hakkarainen, T, Koivusalo, E.,Rizzo Piton, M., Suomalainen, S.,Guina, M., Sibilia, C., Int J Thermophys **39**, 46 (2018)
6. R. Li Voti, G.Leahu, M.C. Larciprete, C. Sibilia, M. Bertolotti, I. Nefedov, I. V. Anoshkin, Int J Thermophys, **36,** 342 (2015)
7. Lamastra, F.R., Grilli, M.L., Leahu, G., Belardini, A., Li Voti, R., Sibilia, C., Salvatori, D., Cacciotti, I., Nanni, F., Nanotechnology **28**, 375704 (2017)
8. Lamastra, F.R., Grilli, M.L., Leahu, G., Belardini, A., Li Voti, R., Sibilia, C., Salvatori, D., Cacciotti, I., Nanni, F., Int J Thermophys, **39**, 110, (2018)
9. I. Fratoddi, at al., Journal of Physical Chemistry C **121,** 18110 (2017)
10. Cesarini, G., Leahu, G., Grilli, M.L., Sytchkova, A., Sibilia, C., Li Voti, R., Physica Status Solidi (C) **13**, 998 (2016)
11. Osewski, P., Belardini, A., Centini, M., Valagiannopoulos, C., Leahu, G., Li Voti, R., Tomczyk, M., Alù, A., Pawlak, D.A., Sibilia, C. Adv. Optical Mater. 1901617 (2020)
12. G.Leahu, R. Li Voti, C.Sibilia, M.Bertolotti, V.Golubev, D.A.Kurdyukov, Opt. Quant Electron **39**, 305 (2007)
13. Cesca, T., Scian, C., Petronijevic, E., Leahu, G., Li Voti, R., Cesarini, G., Macaluso, R., Mosca, M., Sibilia, C., Mattei, G., Nanoscale **12**, 851 (2020)
14. R. Li Voti, G. L. Leahu, M. C. Larciprete, C. Sibilia, M. Bertolotti, Int J Thermophys, **36,** 1004 (2015)
15. G. Leahu, R. Li Voti, M. C. Larciprete, C. Sibilia, M. Bertolotti, I. Nefedov, I. V. Anoshkin, Int J Thermophys, **36,** 1349 (2015)
16. M. Centini, M.C.Larciprete, R. Li Voti, M.Bertolotti, C.Sibilia, and M.Antezza, Opt. Express **28**, 19334 (2020)
17. M. Centini, A. Benedetti, M. C. Larciprete, A. Belardini, R. Li Voti, M. Bertolotti, and C. Sibilia, Phys. Rev. B, **92**, 205411 (2015)
18. O. Matsuda, Maria Cristina Larciprete, R. Li Voti, and O. Wright, Ultrasonics, **56**, 3, (2015)
19. M. Tomoda, O. B. Wright, R. Li Voti, Appl. Phys. Lett. **91**, 071911 (2007)
20. T. Dehoux, O.B. Wright, R. Li Voti, Ultrasonics **50**, 197, (2010)
21. M. Tomoda, R. Li Voti, O. Matsuda, and O. B. Wright, Appl. Phys. Lett. **90**, 041114 (2007)

Broadband tunable nonlinear optical response in plasmonic metamaterials -INVITED

Tiziana Cesca[1,*], *Domenico* Genchi[1], *Raul* Rangel-Rojo[2], *Jorge A.* Reyes-Esqueda[3] and *Giovanni* Mattei[1]

[1]Physics and Astronomy Department, University of Padova, via Marzolo 8, I-35131 Padova, Italy
[2]División de Física Aplicada, Centro de Investigación Científica y de Educación Superior de Ensenada, Ensenada, México
[3]Instituto de Física, Universidad National Autónoma de México (UNAM), México City, México

Abstract. Nanostructured materials with tunable nonlinear optical response are of great interest for different applications in nanophotonics. In this work we report the results of a comprehensive study on the nonlinear optical properties of two kinds of plasmonic metamaterials, i.e., honeycomb nanoprism arrays and multilayer hyperbolic metamaterials, which proved to have a very rich spectrum of parameters (as metamaterials morphology and composition, wavelength and polarization of an input beam) to exploit for controlling their nonlinear response over a broad spectral range.

1 Introduction

In recent years, nanomaterials with properly engineered nonlinear optical properties have been the object of an intense scientific research for their application in different fields in nanophotonics. Within this framework, plasmonic metamaterials have attracted considerable attention due to their strong and ultra-fast nonlinear optical (NLO) response [1] and the possibility of tailoring their nonlinear optical properties by manipulating the metamaterials nanostructure [2, 3]. Besides this, an important requirement for NLO applications is the capability to tune the material nonlinearity in a reversible way, by changing some external control parameters.

In this work, we present the results of a comprehensive study on the nonlinear optical properties of two classes of plasmonic metamaterials, namely honeycomb, Au and Ag, nanoprism arrays (HNPA) and multilayer hyperbolic metamaterials (MHM). The dependence of the NLO properties on wavelength, irradiance, and polarization of the input beam, as well as on the metamaterials morphology and composition, is investigated by using the z-scan technique [4] in the picosecond and femtosecond regimes. The temporal dynamics of the nonlinear optical response is also investigated employing the time-resolved excite-probe technique [5]. Finally, the polarization-dependent second harmonic generation (SHG) properties of the HNPA are characterized [6]. The experimental findings are compared and interpreted on the basis of the results of finite elements methods (FEM) simulations of the plasmonic properties (near-field and far-field) of the investigated nanosystems.

2 Experimental

Honeycomb Au and Ag nanoprism arrays (NPA) are fabricated on silica substrates by nanosphere lithography (NSL) and thermal evaporation. Thermal annealing treatments in air, at different temperatures, are done to modify the shape of the nanostructures (from triangular nanoprisms to hemispherical nanoparticles) without altering the array periodicity. Multilayer hyperbolic metamaterials (MHM) are produced by magnetron sputtering depositions of alternate metallic and dielectric layers. MHMs with different metallic filling fraction are obtained by changing the relative thickness of the dielectric and metallic layers.

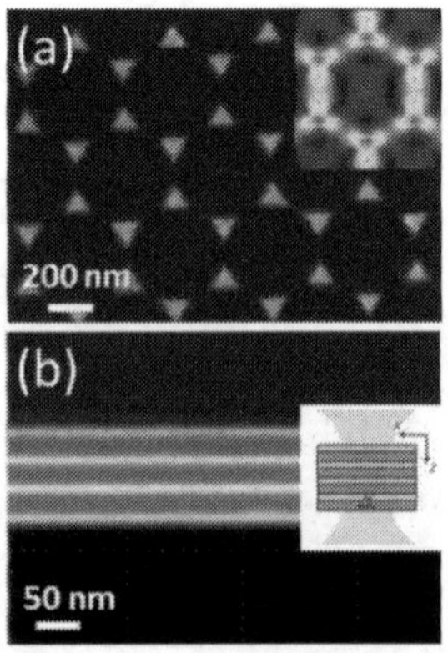

Fig. 1. (a) SEM image in plane-view of an Ag honeycomb nanoprism array (HNPA). Inset: local-field enhancement computed at the dipolar surface plasmon resonance of the HNPA. (b) SEM image in cross-section of an Au/Al_2O_3 multilayer hyperbolic metamaterial (MHM). Inset: sketch of the sample.

* Corresponding author: tiziana.cesca@unipd.it

Fig. 1 shows an example of both kinds of synthesized metamaterials: (a) Ag HNPA and (b) Au/Al_2O_3 MHM.

Finite elements method (FEM) electrodynamic simulations of the near-field and far-field optical properties of the metamaterials are carried out with the commercial software COMSOL Multiphysics.

The spectral characterization of the nonlinear optical properties of the metamaterials was performed with the z-scan technique [4] using a mode-locked ps Nd:YAG laser (pulse duration < 20 ps, 10 Hz repetition rate), coupled to an optical parametric amplifier. A fs mode-locked Ti:sapphire laser (125 fs, 100 MHz) was used to investigate the ultra-fast dynamics of the nonlinear response of the HNPA, and to characterize their second harmonic generation (SHG) properties.

3 Results and discussion

Annealing treatments in air at different temperatures are used to change the morphology of the nanostructures in the plasmonic HNPA, varying their shape from triangular nanoprisms to hemispherical nanoparticles. This allows us to control the linear optical properties of the samples, tuning the spectral position and amplitude of their surface plasmon resonance peak and, in this way, to modulate their nonlinear optical parameters (nonlinear refractive index n_2, and nonlinear absorption coefficient β). As an example, Fig. 2 shows the evolution, as a function of the annealing temperature, of n_2 for an Au HNPA. The experimental findings are well described by FEM simulations of the local-field distribution in the nanoarrays [7].

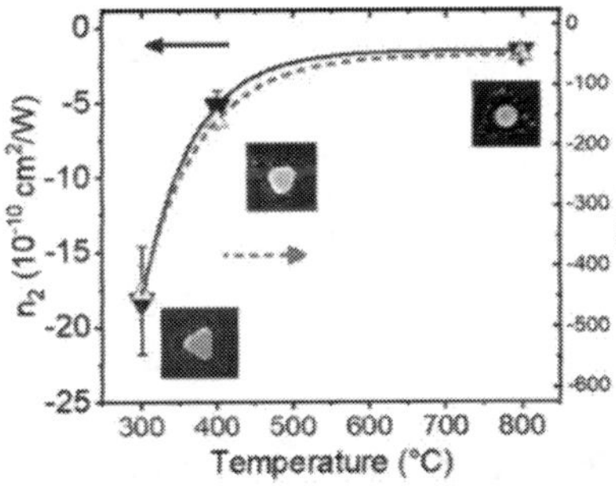

Fig. 2. Nonlinear refractive index n_2 (blue triangles, left-hand scale) of Au HNPA as a function of the annealing temperature. Open symbols (right-hand scale): imaginary component of the integrated local-field distribution in the nanoarray.

The angle of incidence and the polarization orientation of an input beam can be used instead to get a reversible and continuous modulation of the NLO response of the synthesized metamaterials. As an example, in Fig. 3(a) we report the spectral trend of the nonlinear absorption coefficient (β) of an Au/Al_2O_3 MHM (Au filling fraction 16%), at different incidence angles for TE polarization. Fig. 3(b) shows the trend of β, as a function of the linear polarization orientation of the input beam, of an Au HNPA: the trend has the 60°-symmetry of the honeycomb lattice and it is correlated to the polarization-dependent local electric-field distribution in the nanoarray, as demonstrated by FEM simulations [8].

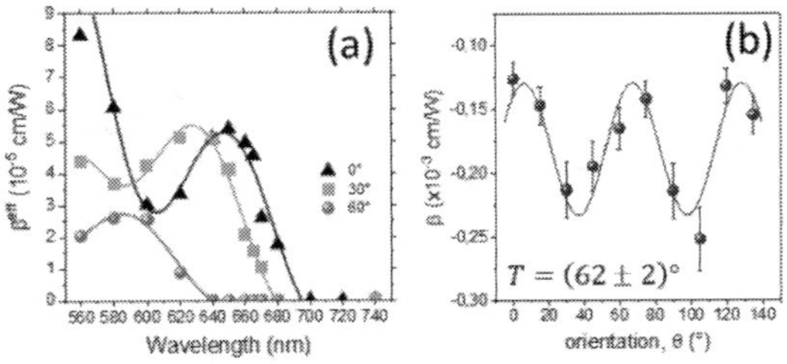

Fig. 3. (a) Spectral trend of the nonlinear absorption coefficient (β) of an Au/Al_2O_3 MHM measured at different incidence angles for TE polarization. (b) β as a function of the input linear polarization orientation of an Au HNPA.

4 Conclusions

The nonlinear optical properties of two classes of plasmonic metamaterials, namely honeycomb nanoprism arrays and multilayer hyperbolic metamaterials, have been investigated as a function of different parameters, as morphology and composition of the nanosystems or polarization and spectral range of the input beam. The results demonstrate that these metamaterials offer several degrees of freedom to tailor their nonlinear optical response, and a reversible and continuous modulation of their nonlinear optical properties can be obtained.

References

1. M. Kauranen, A. V. Zayats, Nat. Photon. **6**, 737 (2012)
2. T. Cesca, N. Michieli, B. Kalinic, I. G. Balasa, R. Rangel-Rojo, J.A. Reyes Esqueda, G. Mattei, Mat. Sci Semicon. Proc. **92**, 2 (2019)
3. T. Cesca, N. Michieli, B. Kalinic, A. Sanchez-Espinoza, M. Rattin, V. Russo, V. Mattarello, C. Scian, P. Mazzoldi, G. Mattei, Nanoscale **7**, 12411 (2015)
4. M. Sheik-Bahae, A.A. Said, T.H. Wei, D.J. Hagan, E.W. Van Stryland, IEEE Journal of Quantum Electronics **26**, 760 (1990)
5. H. Sanchez-Esquivel, K.Y. Raygoza-Sanchez, R. Rangel-Rojo, B. Kalinic, N. Michieli, T. Cesca, G. Mattei, Nanoscale **10**, 5182 (2018)
6. K.Y. Raygoza-Sanchez, I. Rocha-Mendoza, P. Segovia, A.V. Krasavin, G. Marino, T. Cesca, N. Michieli, G. Mattei, A.V. Zayats, R. Rangel-Rojo, Sci. Rep. **9**, 11514 (2019)
7. T. Cesca, M. Manca, N. Michieli, G. Mattei, Appl. Surf. Sci. **491**, 67 (2019)
8. T. Cesca, E.V. Garcia Ramirez, H. Sanchez-Esquivel, N. Michieli, B. Kalinic J.M. Gomez Cervantes, R. Rangel-Rojo, J.A. Reyes-Esqueda, G. Mattei, RSA Adv. **7**, 17741 (2017)

Engineering Local Fields in Nonlinear Plasmonic Metasurfaces -INVITED

Mikko J. Huttunen[1,*], *Saad* Bin-Alam[2], *Orad* Reshef[3], *Yaryna* Mamchur[4], *Timo* Stolt[1], *Jean-Michel* Ménard[3], *Ksenia* Dolgaleva[2,3], *Robert* W. Boyd[2,3,5] and *Martti* Kauranen[1]

[1]Photonics Laboratory, Physics Unit, Tampere University, Tampere, Finland
[2]School of Electrical Engineering and Computer Science, University of Ottawa, Ottawa, ON, K1N 6N5, Canada
[3]Department of Physics, University of Ottawa, Ottawa, ON, K1N 6N5, Canada
[4]National Technical University of Ukraine "Igor Sikorsky Kyiv Polytechnic Institute," Kyiv, Ukraine
[5]The Institute of Optics and Department of Physics and Astronomy, University of Rochester, Rochester, New York 14627, USA

Abstract. Nonlinear optical phenomena are paramount in many photonic applications ranging from frequency broadening and generation of ultrashort pulses to frequency comb-based metrology. A recent trend has been to miniaturize photonic components, resulting also in a demand for small scale nonlinear components. This demand is difficult to address by using conventional materials motivating the search for alternative approaches. Nonlinear plasmonic metasurface cavities have recently emerged as a promising platform to enable nanoscale nonlinear optics. Despite steady progress, the so far achieved conversion efficiencies have not yet rivalled conventional materials. Here, we discuss our recent work to develop more efficient nonlinear metamaterials, focusing on plasmonic metasurfaces supporting collective responses known as surface lattice resonances. These resonances can exhibit very narrow spectral features, showing potential to considerably enhance nonlinear processes via resonant interactions. We demonstrate a plasmonic metasurface operating at the telecommunications *C* band that exhibits a record-high quality factor close to 2400, demonstrating an order-of-magnitude improvement compared to existing metasurface cavities. Motivated by this experimental demonstration, we also present numerical predictions suggesting that such metasurfaces could soon answer the existing demand for miniaturized and/or flat nonlinear components.

1 Introduction

Recent investigations have shown that periodic arrays of nanoparticles, also known as metasurfaces, can act as optical cavities [1]. Major challenge in metasurface cavities has so far been their poor cavity lifetimes hindering their use in potential applications. For example, the resonances associated with individual plasmonic nanoparticles known as localized surface plasmon resonances (LSPRs) are associated with low (<10) quality factors (*Q*-factors). Already around 10-fold improvements in *Q*-factors have been recently demonstrated using dielectric nanoparticle resonators [2], while dramatically higher *Q*-factors have been predicted to be possible by utilizing collective responses known as surface lattice resonances (SLRs) [3,4]. However, currently demonstrated *Q*-factors have been so far limited to few hundreds [6].

Here, we demonstrate a plasmonic metasurface cavity design supporting an ultra-high-*Q* SLR ($Q \approx 2400$) near $\lambda = 1545$ nm [5]. Inspired by this result, we will also use nonlinear discrete-dipole approximation (NDDA) to numerically investigate second-harmonic generation (SHG) from such high-*Q* metasurfaces [7]. Over a million-fold SHG intensity enhancement is predicted.

2 Results for high-Q metasurfaces

Theory providing an intuitive and qualitative agreement with the experimental results is based on lattice sum calculations and has been described elsewhere [3,8]. Here, the lattice sum approach (LSA) was used to design a high-*Q* metasurface by arranging rectangular gold nanobars (lengths $L_y = 200$ nm, widths $L_x = 130$ nm and heights $t = 20$ nm) into a rectangular array with periodicities of $p_x = 500$ nm and $p_y = 1060$ nm, immersed in a homogeneous environment with refractive index of $n = 1.45$ (see Fig. 1a). The area of the fabricated metasurface was 600×600 μm^2, consisting of ≈1200×567 particles. Based on finite-difference time domain (FDTD) simulations the designed metasurface was expected to support a LSPR near 850 nm and a high-*Q* SLR near 1550 nm. The measured transmission spectra (see Fig. 1b), matches very well the calculated spectra. Most notably, the full width at half-maximum of the linewidth of the SLR is only $\Delta\lambda = 0.66$ nm, corresponding to a record-high quality factor of $Q = 2400$ for a plasmonic device.

3 Multiresonant high-Q metasurfaces for nonlinear optics

Then we generalize the LSA to describe nonlinear responses and finite arrays [9-11], motivated to investigate metasurfaces that could support multiple high-*Q* SLRs. By designing such metasurfaces to exhibit high-*Q* resonances, not just near the fundamental pump frequency, but also near the signal frequencies of interest, multiresonant nonlinear materials for enhanced frequency conversion could be realized [8,12]. The extension of the LSA to finite arrays and arbitrary excitation field profiles is done by using nonlinear discrete-dipole approximation (NDDA), which takes explicitly into account the scattered fields from every particle in the metasurface [7,10].

* Corresponding author: mikko.huttunen@tuni.fi

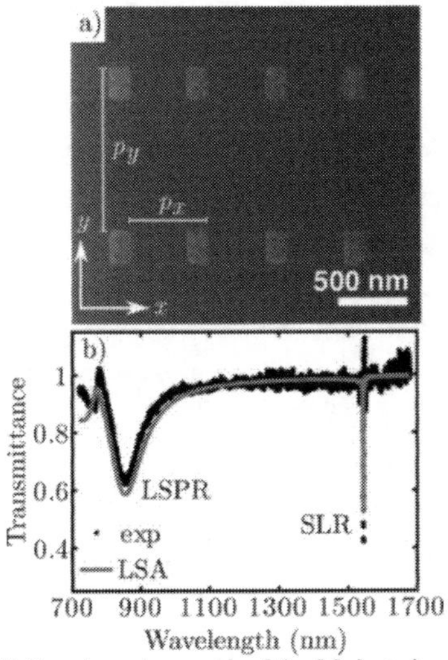

Fig. 1. a) Helium-ion micrograph of the fabricated metasurface. b) Measured (black) and calculated (red) transmission spectra showing the LSPR (850 nm) and SLR (1535 nm). The SLR peak at the telecommunications *C* band exhibits an ultra-narrow linewidth of 0.66 nm, corresponding to $Q \approx 2400$.

Here, we introduce a metasurface design consisting of asymmetric L-shaped aluminum nanoparticles with thicknesses $h = 30$ nm, left-side arm length $l_1 = 100$ nm, right-side arm length $l_2 = 90$ nm and arm width of $w = 45$ nm (Fig. 2a). Although gold is a typical plasmonic material, we were not able to find particle dimensions that would have supported LSPRs at short enough wavelengths. Subsequently, aluminium nanoparticles were used which allowed the shorter wavelength LSPR to peak near 470 nm, supporting emergence of high-Q SLRs at longer wavelengths [10]. The nanoparticles were arranged into a rectangular lattice with the periods $p_x = 395$ nm and $p_y = 793$ nm, giving rise to the two high-Q SLRs at the second-harmonic (600 nm) and the fundamental wavelengths (1200 nm), respectively (Fig. 2b). Notably, over a million-fold intensity enhancement of SHG emission was predicted to occur when the fundamental and SHG frequencies coincided spectrally with the SLRs (Fig. 2c).

4 Conclusion

We have demonstrated an SLR-based plasmonic metasurface nanocavity with a record-high Q-factor of 2400. Such high-Q metasurfaces could dramatically expand the current capabilities of metasurfaces. We have also numerically investigated multiresonant metasurfaces, which could enable applications benefiting of multiple high-Q resonances. We believe that such multiresonant high-Q metasurfaces show promise for nonlinear optical applications such as for frequency conversion.

We acknowledge the support of the Academy of Finland (Grant No. 308596) and the Flagship of Photonics Research and Innovation (PREIN) funded by the Academy of Finland (Grant No. 320165)

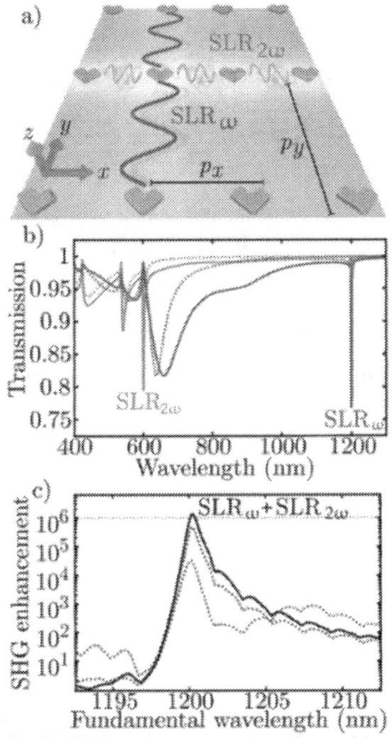

Fig. 2. a) Multiply-resonant metasurface. b) Calculated transmission spectra using the NDDA (dotted lines) and the FDTD (solid lines). Red (green) curves correspond to x-polarized (y-polarized) incident light. c) Predicted SHG intensity enhancement for a particle at the center of the metasurface. Over million-fold SHG intensity enhancement occurs for the doubly-resonant metasurface (black solid line).

References

1. N. Meinzer, W. L. Barnes, and I. R. Hooper, Nat. Photonics, 8, 889–898 (2014)
2. K. Koshelev et al., Science, **367**, 288–292 (2020)
3. M. J. Huttunen et al., Opt. Express, **24**, 28279–28289 (2016)
4. V. G. Kravets et al., Chem. Rev., **118**, 5912–5951 (2018)
5. Q. Le-Van, Quynh, et al. Adv. Opt. Mater., **7**, 1801451 (2019)
6. M. S. Bin-Alam et al., arXiv:2004.05202[physics.optics] (2020)
7. M. J. Huttunen et al., Phys. Rev. A, **97**, 053817 (2018).
8. O. Reshef et al. Nano Lett., **19**, 6429–6434 (2019)
9. L. Michaeli et al., Phys. Rev.Lett., **118**, 243904 (2017)
10. M. J. Huttunen et al., J. Opt. Soc. Am. B, **36**, E30–E35 (2019)
11. M. J. Huttunen, R. Czaplicki, and M. Kauranen, J. Nonlinear Opt.Phys. Mater., **28**, 1950001 (2019)
12. M. Celebrano et al., Nat. Nanotechnol., **10**, 412–417 (2015)

Multi-resonant optical parametric oscillator without mirrors based on 1D and 2D-PPLT nonlinear photonic crystal

Safia Mohand Ousaid[1], *Mahmoud* Chakaroun[1], *Kai-hsun* Chang[2], *Thierry* Billeton[1], *L-Han* Peng[2], and *Azzedine* Boudrioua[1]
1. Laboratoire de Physique de Lasers CNRS UMR 7538, Université Paris 13, Sorbonne Paris Cité, Villetaneuse, France
2. Graduate Institute of Photonics and Optoelectronics, GIPO, National Taiwan University, Taipei, 106, Taiwan

Abstract: In this work, we demonstrate that it is possible to obtain an optical parametric oscillator with a simultaneous amplification of multiple frequencies using only a nonlinear photonic crystal with highly polished end-faces, preventing the utilization of external cavity mirrors.

1. Introduction

Optical parametric oscillator (OPO) is an efficient tool to provide powerful parametric sources. It represents a very promising device for various technological applications. OPO is a laser oscillator where the gain depends on phase matching conditions, which link the pump wave (ω_p) and the two waves generated (signal ω_s and the complementary wave or idler ω_i) by the second order nonlinear susceptibility $\chi^{(2)}$ of the gain material. Conventional OPO consists of placing a $\chi^{(2)}$ non-linear medium in an optical cavity resonator. Most of OPOs are single resonant at either signal or idler wavelengths. Many other studies were interested in the cavity design in order to achieve a double and triple resonant OPO. Since its first demonstration in 1995 by Meyer *et al.* [1], several works have been carried-out to improve OPO efficiency and stability by optimizing the cavity mirrors. However, the use of a Fabry-Perot cavity-type with a specific design of the mirrors makes the device cumbersome and add more complexity to manage all the resonances together.

In this work, we report the demonstration of optical parametric oscillation using only a nonlinear photonic crystal polished in a particular way. The highly polished end-faces of the sample act as cavity mirrors, which prevent the utilization of external mirrors. Moreover, we show that it is possible to obtain, simultaneously, an optical parametric oscillation of multiple frequencies due to cascaded nonlinear interactions.

2. Results and discussion

The demonstration of multi-resonant optical oscillation without mirror in nonlinear photonic crystals of lithium tantalate is almost fortuitous. It appeared during a comparison study of multi-wavelength generation by cascade effect in photonic crystals 1D and 2D-PPLT. In order to improve the efficiency of this nonlinear process, we developed a specific polishing technique to obtain a very high optical quality of the sample faces. It includes lapping, softening, fast flat polishing and a final polishing to obtain a precise quality of flatness and roughness of the surface. The samples used are $LiTaO_3$ nonlinear photonic crystal of 20x10x0.5mm^3 periodically polarized by electric poling in one dimension (1D-PPLT) and two dimensions (2D-PPLT) with a period Λ=8.52μm. The 2D-PPLT is a square lattice with a circular pattern and a filling factor of 38%. The one in 1D is 50%. The filling factors are chosen to obtain the most efficient frequency conversion for these two structures [2]. For the characterization, we used the same experimental setup as depicted in reference [3]. The measurements are carried at 110°C which corresponds to the optimum conversion efficiency temperature. In these conditions, the measured out-put spectrum presents unexpected additional wavelengths as displayed in figure 1.a. (5 new peaks are observed).

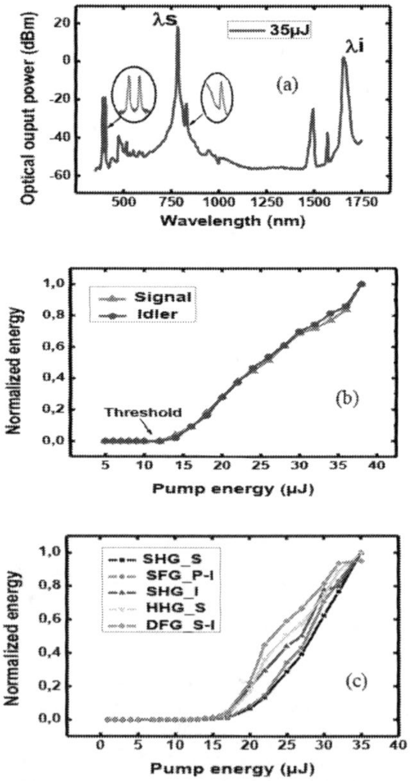

Fig.1. (a) The generated multi-wavelengths spectrum. (b) Threshold measurements of signal an idler and (c) of cascaded nonlinear processes. From 1D-PPLT nonlinear photonic crystal at T=110°C.

Safia.mohandousaid@univ-paris13.fr

As already stated in references [3, 4], it was found that the interaction of the signal λ_s at 787nm, the idler λ_i at 1641nm (in Fig.1a), and the pump can further lead to several cascaded nonlinear processes: SHG of the signal at 394nm, SHG of the idler at 820nm, SFG of the pump and the idler at 402nm, DFG of the signal and the idler at 1516nm and the half HG of the signal 1576nm. However, the phase matching conditions, alone, cannot explain the occurrence of these processes effectively. In order to explain the physical origin of this phenomenon, we suggest considering resonant oscillations inside the sample, which in turn sustain the cascaded nonlinear process. To confirm this assumption, we measured the energy threshold for each peak. Figures 1(b) and 1(c) show the obtained results for signal, idler and the other cascaded processes for the 1D crystal. It is clearly seen that these nonlinear processes are subject to a threshold phenomenon, which emphasize a resonant behaviour. In addition to that, we studied the spectrum variation of the obtained peaks at different energy pumps before and after the threshold.

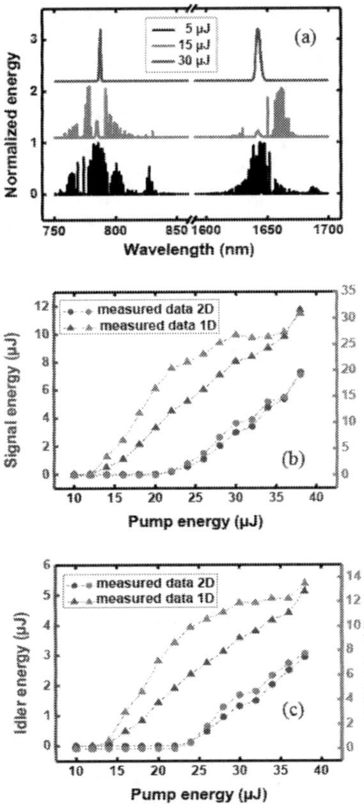

Fig.2. (a) The broad spectrum of signal and idler waves generated from a 1D-PPLT nonlinear photonic crystal before the threshold (E_p=5µJ), near and at three time of the threshold. The conversion efficiency of 1D and 2D-PPLT at 110°C (b) Signal (787nm), (c) Idler (1641nm).

The signal and idler oscillation threshold is estimated at 12µJ for 1D-PPLT and 20µJ for 2D-PPLT, that of the other waves generated by cascade effect is reached after an increase of the pump energy of 4µJ. The oscillation threshold of 2D-PPLT is relatively high compared to the 1D-PPLT, due to its low conversion efficiency. The frequencies obtained by cascade effects require a fairly powerful signal and idler and therefore greater pump energy.

As an example, the figures 2.a display the signal and the idler spectrum recorded for the 1D-PPLT before, near and at three times the threshold (E_p = 10µJ). Before the threshold, the spectrum consists of several peaks due to the multi-wavelength nature of the OPG process in the nonlinear photonic crystal PPLT. Just above the threshold, the number of peaks decrease and there two peaks at 787nm and 1641nm of low intensity appear. For a pump power equal to 3 times the oscillation threshold, we see that the 2 peaks are increased. This behavior is explained by the competition between different coupled interactions inside the nonlinear photonic crystal; only the two powerful peaks corresponding to the couple ω_s and ω_i which minimize the phase mismatch remain. These results emphasize that the cavity formed by the sample faces acts as a filter to select the most effective interaction. Note that all the previously observed peaks behave in the same way, confirming the oscillation phenomenon.

Eventually, the maximum total conversion efficiency achieved is 45% for 1D-PPLT and 27% for 2D-PPLT. That of the signal (787 nm) is 30% for 1D-PPLT, while it is 18% for the 2D-PPLT. For the idler (1641 nm), the maximum efficiency obtained in the PPLT-1D is 15% and it is 9% in the PPLT-2D. The experimental measurements obtained show that the efficiency of 1D-PPLT is 1.7 times greater than that of 2D-PPLT, which is in good agreement with the results reported in the literature [3].

3. Conclusion

In this work, we report evidence of optical parametric oscillations in 1D and 2D-PPLT nonlinear photonic crystal without using external mirrors. The oscillation threshold and the efficiency of the device should be greatly improved by using crystals with antireflection-treated sides.

4. References

[1] L. Myers, R. Eckardt, M. Fejer, R. Byer, W. Bosenberg, and J. Pierce., "*Quasi phase matched optical parametric oscillators in bulk periodically poled LiNbO₃*," JOSA B, **12**, p. 2102, (1995).
[2] Broderick, N. G. R., Ross, G. W., Offerhaus, H. L., Richardson, D. J., & Hanna, "*Hexagonally Poled Lithium Niobate: A Two-Dimensional Nonlinear Photonic Crystal*". *PRL, 84(19), 4345–4348, (2000).*
[3] S. Mohand Ousaid, J-Y. Han, K-H. Chang, M. W. Lee, M. Chakaroun, L.-H. Peng and A. Boudrioua, "*Magic crossing points of temperature dependent cascaded frequency conversions in nonlinear photonic crystals*", CLEO Munich (2019).
[4] Zohra Yellas, Min Won Lee, Régis Kremer, Kai-Hsun Chang, Mahmoud R. Beghoul, Lung-Han Peng, and Azzedine Boudrioua "*Simultaneous frequency conversions in nonlinear photonic crystals of lithium tantalite with a square lattice*", *CLEO Munich (2017)*

Safia.mohandousaid@univ-paris13.fr

Autocompensating high-dimensional quantum cryptography by phase conjugation in optical fibers

Jesús Liñares-Beiras[1,*], *Xesús* Prieto-Blanco[1], *Daniel* Balado[1,**], and *Gabriel M.* Carral[1]

[1]Quantum Materials and Photonics Research Group, Department of Applied Physics, Faculty of Physics / Faculty of Optics and Optometry, University of Santiago de Compostela, Campus Vida s/n, E-15782, Santiago de Compostela, Galicia, Spain.

Abstract. We present a system based on phase conjugation in optical fibers for autocompensating high-dimensional quantum cryptohraphy. Phase changes and coupling effects are auto-compensated by a single loop between Alice and Bob. Bob uses a source of coherent states and next Alice attenuate them up to a single photon level and thus 1-qudit states are generated for implementing a particular QKD protocol, for instance the BB84 one, together with decoy states to detect eavesdropping attacks.

1 Introduction

Quantum cryptography is based on the properties of quantum mechanics to obtain secure quantum key distribution (QKD) by using different protocols. The seminal protocol has been the so-called BB84 one in which four states define a set of two mutually unbiased basis. On the other hand, different optical fiber systems have been proposed to implement QKD cryptography. Such systems can use different kind of modes, for instance, polarization modes in monomode optical fibers, spatial modes in few-mode optical fibers and spatial codirectional modes in multicore optical fibers. One of the advantages of using optical modes is that high-dimensional QKD (HD-QKD) can be implemented, which improves the security. However, one of the most important drawbacks is that all modes need to keep stable over large distances along optical fibers. Mode instability is due to the modal coupling undergone by the modes in their propagation in optical fibers with small imperfections or slow temporal perturbations. To overcome this drawback some polarization autocompensating techniques have been proposed in cryptography with 1-qubit quantum states [1] or even with 1-qudits acquiring random relative phases [2]. In this work we propose a general autocompensating quantum cryptography technique based on phase conjugation for both few-mode optical fibers (FMF) and multicore optical fibers (MCF) where modal coupling is not negligible. Input multimode coherent states are launched and later attenuated in their path back up to a single photon level (weak coherent states), that is, 1-qudit states are produced. Moreover, decoy states have also to be generated for security purposes. Optical devices for QKD with collinear modes of FMFs can be realized by using both fiber and micro-optical components, and with codirectional modes of MCFs by using also integrated optical components.

*e-mail: suso.linares.beiras@usc.es
**e-mail: daniel.balado.souto@usc.es

2 Phase conjugation of coherent states

Let us consider a four-wave mixing (FWM) nonlinear interaction system of length L as shown in figure 1. Operator $\hat{a}_3 \equiv \hat{a}_3(0)$ is associated to the optical mode transmitted along the FWM system, and $\hat{a}_4 \equiv \hat{a}_4(L)$ is associated to the "reflected" optical mode. Likewise, there are two pump waves with very large amplitudes A_1 and A_2 and initial phases equal to zero. The nonlinear interaction strength is

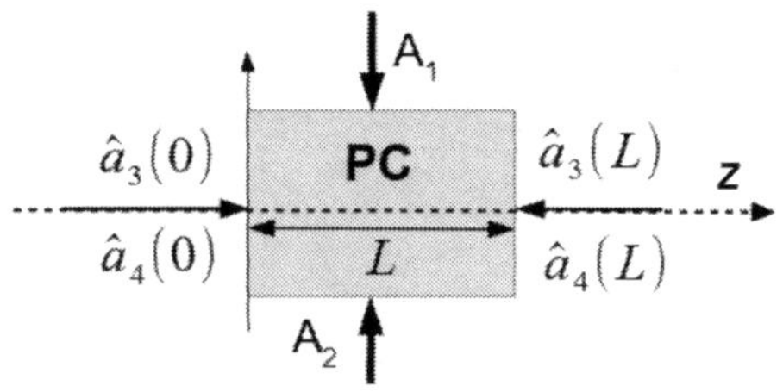

Figure 1. Four-wave mixing with two pump waves.

given by the coupling coefficient $\kappa = \chi_{eff} |A_1||A_2| L$, where χ_{eff} is an effective nonlinear susceptibility. It is wellknown that the following operator transformations are fulfilled

$$\begin{aligned} \hat{a}_3(L) \equiv \hat{a}_{o3} &= s\,\hat{a}_3 + it\,\hat{a}_4^\dagger \\ \hat{a}_4(0) \equiv \hat{a}_{o4} &= s\,\hat{a}_4 + it\,\hat{a}_3^\dagger \end{aligned} \tag{1}$$

where $s = \sec(\kappa L)$, $t = \tan(\kappa L)$. Next, let us consider a coherent state $|\alpha_3 0_4\rangle$ excited in one optical fiber mode and incident on the FWM system. The output state is given by $|L\rangle = |\alpha_3 0_4\rangle = e^{\alpha \hat{a}^\dagger_{o3} - \alpha^\star \hat{a}_{o3}} |00\rangle \rightarrow |L_c\rangle = |s\alpha\rangle\,|-it\alpha^\star\rangle$, where equation (1) has been used, that is, the FWM is a phase conjugator (PC). Note that the reflected coherent state has been conjugated. We are interested on multimode coherent states, that is, $|L\rangle = |\alpha_1 ... \alpha_N\rangle$ excited in N optical modes, therefore the output state will be

$$|L_c\rangle = (|s\alpha_1\rangle\,| - it\alpha_1^\star\rangle)...(|s\alpha_N\rangle\,| - it\alpha_N^\star\rangle \tag{2}$$

3 Phase conjugation for HD-QKD

Let us consider a FMF or MCF optical fiber supporting d modes with operators $\hat{a}_i$, $i = 1, ..., d$ in order to implement HD-QKD with 1-qudit states. For sake of simplicity we assume that polarization-maintaining optical fibers are used [3]. In that case, FMFs have negligible modal coupling, however unpredictable relative phases still happen [2]. In general, each optical mode will undergo coupling to other modes and therefore the state reaching Bob system from the Alice system will be a unpredictable 1-qudit state which prevents us to implement any protocol for QKD. Next, we show how to overcome this drawback with phase conjugation. We start from the spatial Heisenberg equation describing the modal coupling,

$$-i\hbar\frac{\partial \hat{a}_i}{\partial z} = \hbar\left(\beta_i \sum_{ij} \delta_{ij}\hat{a}_j + \sum_{ij} \kappa_{ij}\hat{a}_j\right) \equiv \sum_{ij} C_{ij}\hat{a}_j \qquad (3)$$

β_i are propagation constants and κ_{ij} are modal coupling coefficients. C_{ij} is a symmetric matrix, therefore the matrix solution $S_{ij} = \exp\{iC_{ij}z\}$ of differential equation (3) is a complex symmetric matrix, that is, $S_{ij} = S_{ji}$. On the other hand, modal coupling is an unitary transformation, therefore $S_{ij}^{-1} = S_{ij}^{\star}$. Next, let us consider q different modal couplings along z direction from a point B to point A with total matrix $M = S_1 \cdots S_q$. Therefore, if we have a PC at A we obtain $M' = S_q^{\star} \cdots S_1^{\star}$, and after the light has traveled the path back and forth, the total coupling matrix is $M'M = \mathbb{I}$, that is, the unpredictable modal coupling has been removed. Therefore if we start from a state $|L\rangle$ undergoing modal coupling, the state after the PC and travelling the path back is $|L_c\rangle$ given by equation (2). This is the key result to be used in a QKD system.

4 BB84 autocompensating system

By using the PC shown in section 2 together with the theoretical results of section 3 we design an autocompensating optical system for HD-QKD BB84 quantum cryptography in optical fibers as shown in figure 2. The first

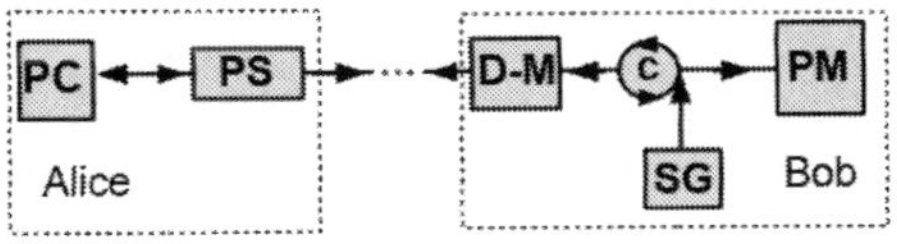

Figure 2. Basic setup of autocompensating high-dimensional cryptography by phase conjugation.

device of the system is a states generator (SG) located in Bob system, which emits a multimode coherent state $|L\rangle = |\alpha_1 ... \alpha_N\rangle$ excited in N optical modes of a FMF or MCF fiber, with $\alpha_1 = ... = \alpha_N$, in order to generate states belonging to unbiased basis. Next, one (for FMFs) or several (for MCFs) optical circulators c launches the state towards Alice. A Demultipexer-Multuplexer (D-M) device is needed to produce delays τ between modes if a FMF is used; if we use a MCF then the D-M device is not required. Next, quantum state reaches the PC, where a previous phase shifter (PS) is found in off-position. We must stress that each single mode coherent state (a core in a MCF) becomes a multimode coherent state due to modal coupling. Now, the reflected state in the PC device goes through the PS which introduces phases θ_i on the coherent states. We obtain 1-qudit states in Alice system starting from the multimode coherent states and attenuated up to single photon level,

$$|L_c\rangle = \sum_i \frac{1}{\sqrt{N}}\{e^{i\theta_1}|1_1\rangle + ... + e^{i\theta_N}|1_N\rangle\} \qquad (4)$$

These states must belong to two N-dimensional biased basis. Likewise, Alice has to produce decoy states for security purposes. When the 1-qudit again reaches the Bob system finds the D-M device, which cancels the delays τ between states $|1_j\rangle$. Finally, the circulator drives the state up to the device that implements projective measurements (PM) [2]. We show in figure 3 an integrated

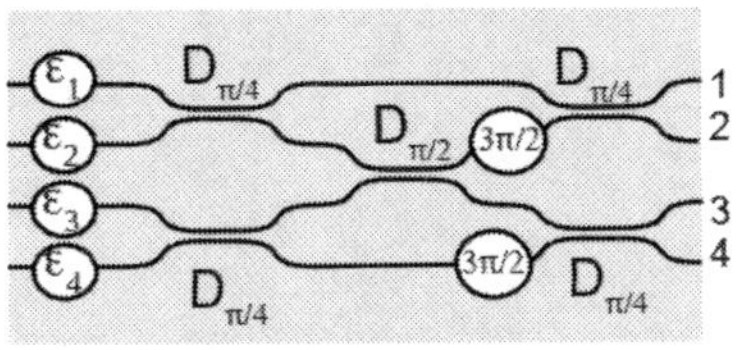

Figure 3. Sketch of an integrated quantum projector.

quantum projector (IQP) for the case of MCFs. It is formed by four directional couplers 3dB ($D_{\pi/4}$) and an inversor ($D_{\pi/2}$) together with ($3\pi/2$)-phase shifters. At the input of IQP there are four phase-shifters ϵ_i, i=1, ..., 4 which allow to choose the measurement basis. As an example, let us consider ϵ_i=0 at the input 1-quart state $|L_i\rangle = 2^{-1}(|1_{o1}\rangle - i|1_{o2}\rangle - i|1_{o3}\rangle - |1_{o4}\rangle)$. It can easily calculated that the output state is $|L_o\rangle = |1_1\rangle$, that is, it is projected on the mode 1 and so on with other unbiased orthogonal states as for example $|L_i\rangle = 2^{-1}(|1_{o1}\rangle - i|1_{o2}\rangle + i|1_{o3}\rangle + |1_{o4}\rangle)$, which is projected on $|1_2\rangle$.

5 Conclusions

We have presented an autocompensating technique based on phase conjugation for quantum cryptography in FMF and MCF optical fibers. A single loop allows to autocompensate the undesidered modal coupling, and thus a HD-QKD BB84 protocol can be implemented.

Acknowledgement. Authors wish to acknowledge the financial support of this work by Xunta de Galicia, Consellería de Educación, Universidades e FP, with a grant GRC N° ED431C2018/11 and a predoctoral grant (D. Balado, 2017), co-financied with the European Social Fund.

References

[1] D.S.Bethune,W.P.Risk, New J.Phys., **4**,42(2002)
[2] D.Balado, J.Liñares, X.Prieto-Blanco, D.Barral, JOSA B **36**, 2793-2802 (2019)
[3] L.Wang, S.LaRochelle, Opt.Lett.**40** 5846-5849 (2015)

Visible short-pulses generation by nonlinear propagation of necklace beams in capillaries

Aurora Crego[1,*], *Julio* San Roman[1], and *Enrique* Conejero Jarque[1]

[1]Grupo de Investigación en Aplicaciones del Láser y Fotónica, Departamento de Física Aplicada, University of Salamanca, E-37008. Salamanca, Spain.

Abstract. We study numerically the propagation of necklace beams through a gas-filled capillary. Ultra-short pulses in the visible (VIS) region can be obtained due to the spectral broadening of these necklace beams. This new source of few-cycle VIS pulses can be generated carrying tens of microjoules of energy using these special beams, being a valuable tool for the improvement of the standard post-compression schemes in terms of spatial stability and output energy.

1 Introduction

One of the most common post-compression techniques to obtain ultrashort pulses is the nonlinear propagation of a laser pulse as the fundamental mode of a capillary filled with gas [1]. This technique involves the spectral broadening of the laser pulse by self-phase modulation (SPM) followed by a spectral phase compensation to finally obtain the shortest possible output. Although it is possible to obtain few-cycle pulses in the near and mid-infrared [2], the possibility of obtaining short pulses in the visible region of the spectrum (400-600 nm) with this technique is still very limited. Moreover, to obtain directly an ultra-short pulse at the end of the capillary, one has to be in the anomalous dispersion region, so the pulse self-compresses during its propagation [3].

The fundamental mode of a standard capillary filled with 1 bar of argon (the EH_{11} hybrid mode) presents a normal response at 800 nm [4], which prevents the possibility of using this configuration to get self-compression. Higher order spatial modes that present anomalous dispersion [5] are therefore needed. In this work, we explore another possibility which consists in using a particular combination of high spatial modes that results in a linearly polarized beam with a spatial necklace shape [6].

2 Results

To study the nonlinear propagation dynamics in capillaries we have developed a (3+1)D (r,θ,t,z) numerical model based on the nonlinear Schrödinger equation. This model represents the complete spatio-temporal dynamics of a laser pulse propagating in a capillary, and includes dispersion, absorption, self-phase modulation, self-steepening and ionization.

We have studied the nonlinear propagation of a necklace beam with 6 beads in a capillary. This mode is a composition of the EH_{21} and EH_{41} spatial modes [4]. We propagate a 30-fs pulse with 0,2 mJ input energy at 800 nm through a capillary with 150 μm core radius filled with argon at 1 bar. We show in Fig. 1 the spatial profile of the necklace beam at the entrance of the capillary. We do not observe any spatial dynamics, like self-focusing, along the propagation through the capillary [7].

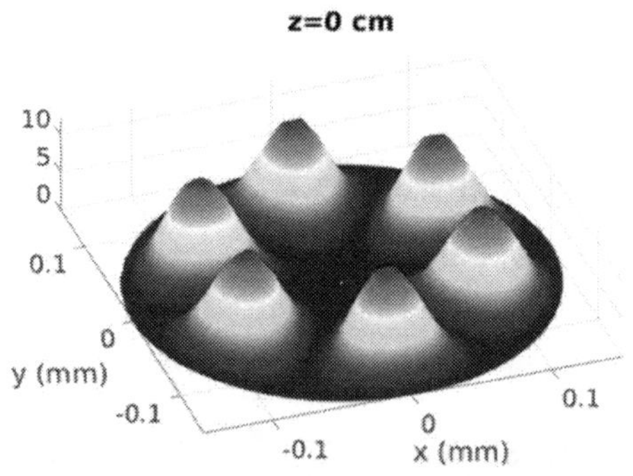

Fig. 1. Transverse profile of the input necklace beam mode with 6 beads.

For the parameters used in the simulation the zero-dispersion wavelength (ZDW) is 679 nm, so the input pulse propagates in the anomalous dispersion region. Figure 2 shows the evolution of the temporal and spectral intensity distribution at a point of maximum intensity. The self-phase modulation broadens the spectrum and this dynamic is accompanied by the self-compression towards a solitonic structure at around 70 cm inside the capillary.

* Corresponding author: acrego@usal.es

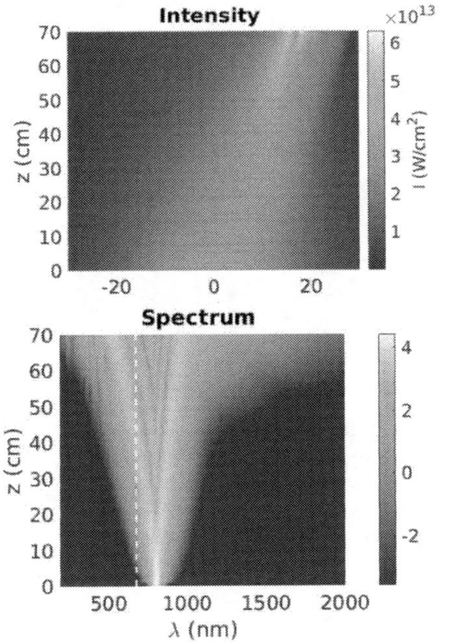

Fig. 2. Evolution of the temporal intensity (top row) and spectral intensity (bottom row, in log scale) distributions at a point of maximum intensity. The white line represents the ZDW at 679 nm.

The most notable feature of the spectrum in Fig. 2 is that it is mainly composed by two spectral bands, one which lies in the anomalous dispersion region and enables the soliton formation, and from the other one, which lies in the normal-dispersion region, we can obtain a short pulse in the visible region of the spectrum.

We have filtered the visible band of the spectrum for a pressure scan (see Fig. 3) to observe the dependence of the dynamics with the pressure. We have compared all the cases at the self-compression distance obtained at 1 bar of pressure (66 cm). From this comparison we have observed that this visible band suffers a blue-shift until 550 nm as the pressure increases.

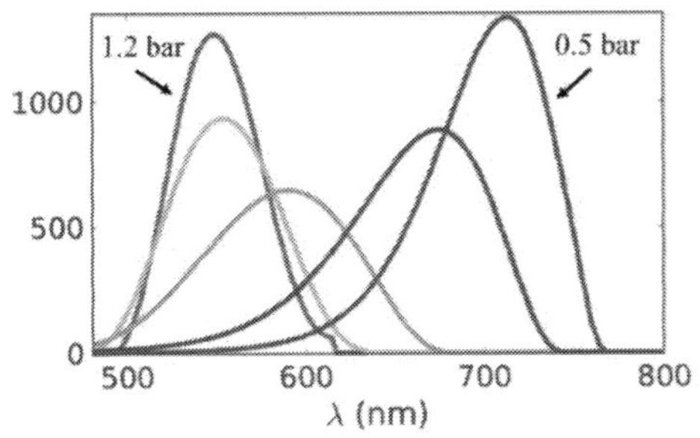

Fig. 3. Filtered visible band of the spectrum at different pressures (1.2, 1, 0.8, 0.6 and 0.5 bar from left to right) at 66 cm inside the capillary.

We have calculated the energy and the full width half maximum (FWHM) duration of the visible band along the propagation. In Fig. 4 we show the dependence with the pressure of the energy and the duration of the pulse and its Fourier limit at 66 cm inside the capillary. We can observe that the energy values are around 50 μJ in all cases and the visible pulse durations are less than 10 fs. The cases with the lowest energy (0.7-0.8 bar) have the shortest pulse duration reaching 5 fs, approximately, and could be post-compressed to a Fourier limit of 3 fs.

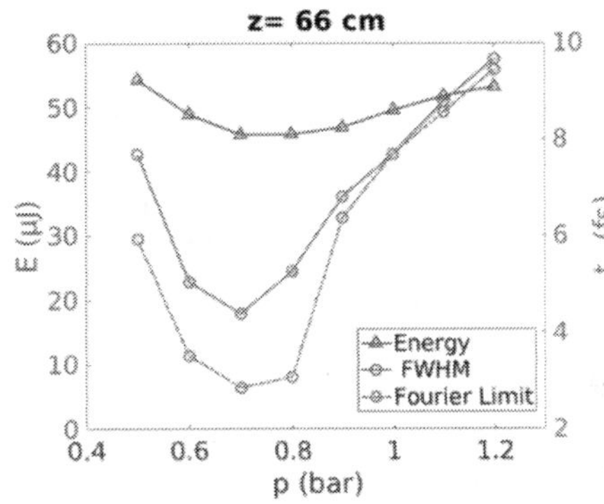

Fig. 4. Energy, FWHM and Fourier limit dependence with the pressure at 66 cm.

We have demonstrated that with the nonlinear propagation of necklace beams in capillaries we can obtain short and energetic pulses in visible wavelengths. We believe that this study paves the way to new short laser pulse sources in the visible region of the spectrum and its subsequent applications.

The authors thank support from Spanish Ministerio de Economía y Competitividad (FIS2016-75652-P) and from Junta de Castilla y León (SA287P18). A.C. is funded by the FPI grant program of Spanish Ministerio de Ciencia e Innovación and European Social Fund (BES-2017-080280).

References

1. M. Nisoli, S. De Silvestri, O. Svelto, Appl. Phys. Lett. **68**, 2793–2795, (1996)
2. E. Conejero Jarque, et al. Sci. Reports 8, 2256, (2018)
3. J. C. Travers, T. F. Grigorova, C. Brahms, F. Belli, Nat. Photonics **13**, 547–554 (2019)
4. E. A. J. Marcatili, R. A. Schmeltzer, Bell Labs Technical Journal **43**, 1783 (1964)
5. B. A. López-Zubieta, E. Conejero Jarque, I. J. Sola, J. San Roman, OSA Continuum **1**, 930–938, (2018)
6. G. Fibich, D. Shpigelman, Physica D **315,** 13–32 (2016)
7. A. Crego, E. Conejero Jarque, J. San Roman, Sci Rep **9**, 9546 (2019).

Nonreciprocal propagation in a non-Hermitian silicon photonic coupler employing graphene saturable absorption

Dimitrios Chatzidimitriou[1,*], *Alexandros* Pitilakis[1], *Traianos* Yioultsis[1], and *Emmanouil* Kriezis[1]

[1]Department of Electrical and Computer Engineering, Aristotle University of Thessaloniki, 54124 Thessaloniki, Greece

Abstract. We present the concept, analysis and design of a nonreciprocal system of coupled nonlinear waveguides, operating at the near infrared. We exploit the spatial asymmetry induced by the exceptional points of a non-Hermitian coupler which, combined with saturable absorption in the nonlinear regime, can break Lorentz reciprocity. In this work, we introduce asymmetric losses in a pair of coupled silicon waveguides by overlaying only one of them with unbiased graphene, subject to saturable absorption. The device can be used as a broadband half-duplex isolator: a two-port component offering unidirectional transmission of high-power signals.

1 Introduction

Components with nonreciprocal response are highly desirable in integrated photonics. Diodes, the elementary nonreciprocal components, exhibit unidirectional transmission and are the building blocks of advanced components such as isolators, circulators, and logic gates; these components are of fundamental importance in even more complex devices such as sources (lasers), tranceivers, and all-optical processing units. Presently, magnetic materials exhibiting Faraday rotation are mostly used for optical isolators, but they are bulky and costly to hybridly integrate in photonic platforms. Consequently, alternative ways have been proposed to break Lorentz reciprocity, such as the combination of nonlinearity and spatial asymmetry. This concept has been mainly explored using the Kerr effect in photonic resonators.

In this work, we present a concept broadband photonic diode relying on graphene saturable absorption (SA) in an asymmetric coupler. In the near infrared and for a chemical potential (μ_c) below the half photon energy, graphene's absorption is predominately due to interband transitions. This absorption is broadband, saturable, i.e., decreases with increasing intensity, has a sub-ps response time, and a much lower power threshold compared to the Kerr effect. Our concept is based on the existence of exceptional points (EPs) [1], singularities in the eigenvalue space of non-Hermitian systems. In our implementation, the EP in a lossy waveguide coupler combined with the asymmetric field overlap with the nonlinear SA material gives rise to nonreciprocal functionality, manifested as uni-directional transmission.

2 Concept, analysis and design

We consider the abstract waveguide coupler depicted in Fig. 1(a), consisting of one lossless waveguide (blue) and one nonlinear lossy waveguide (orange) whose losses are fully saturable. The waveguides are otherwise identical, i.e., their individual modes have the same propagation constant (real part). The coupler is configured as a two-port component with ports 1 and 2 denoting opposite ends of the lossy and lossless waveguides, respectively.

Assuming that the coupler's length is equal to the coupling length $L \approx L_c = \pi/2\kappa$, where κ is the coupling coefficient, then for low power input the response of the device is dictated by the operation above the EP: Guided power in the lossless waveguide is weakly coupled to the lossy one, while power guided in the lossy waveguide is greatly attenuated before being weakly coupled to the lossless waveguide [2]. For the port configuration in Fig. 1(a) this translates to very low reciprocal transmission. For high power excitation from the lossy waveguide (port 1), losses are saturated allowing full coupling to the opposite waveguide. Thus, transmission to port 2 will be high; this is the through direction. On the contrary, excitation of the lossless waveguide does not sufficiently overlap with the SA material and coupling is suppressed due to the EP, leading to low transmission to port 1; this is the isolated direction.

A coupled mode framework can be developed for the qualitative analysis and design of the device. Assuming that z is the propagation direction, $A_{1/2}$ is the E-field complex amplitude on the top/bottom waveguide normalized by the saturation amplitude ($|A_{\rm sat}|$), β is the real phase constant for both waveguides, and α is the low-power attenuation coefficient for the lossy waveguide, then the coupled equation system describing the device is

$$\frac{\partial}{\partial z}\begin{bmatrix} A_1 \\ A_2 \end{bmatrix} = \begin{bmatrix} -j\beta - \dfrac{\alpha}{1+|A_1|^2} & \kappa \\ -\kappa^* & -j\beta \end{bmatrix}\begin{bmatrix} A_1 \\ A_2 \end{bmatrix}. \quad (1)$$

Solving this system for an initial excitation at one port leads to the transmission coefficients in both directions, the forward T_{21} and backward T_{12}. Note that we assume excitation of only one port at any time. In the linear regime,

*e-mail: dcthazid@ece.auth.gr

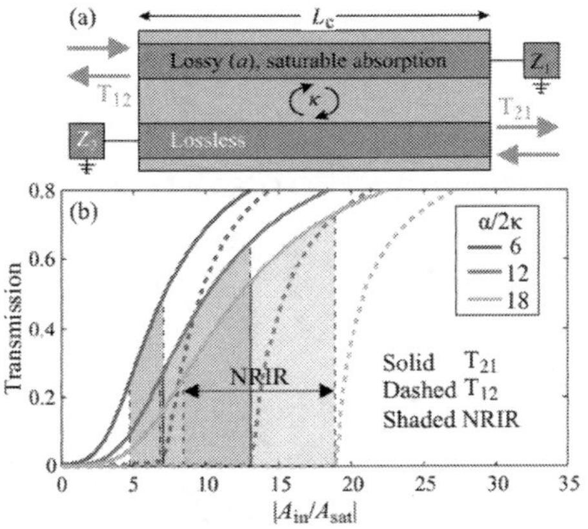

Figure 1. (a) Abstract two-port nonlinear asymmetric waveguide coupler of length $L = L_c = \pi/2\kappa$; terminals $Z_{1,2}$ denote matched loads. (b) Forward (T_{21}) and backward (T_{12}) transmission vs. normalized input amplitude; colors correspond to different values of the loss parameter $\alpha/2\kappa$.

the EP is located at $\alpha/2\kappa = 1$. In the nonlinear regime, the state of the system is power dependent, i.e., high power signals alter the state relative to the EP, which drastically changes the response. The transmission coefficients for $\alpha/2\kappa = 6, 12$ and 18 vs. the normalized input field amplitude are presented in Fig. 1(b), where the shaded regions corresponds to $T_{21} \geq -6$ dB and $T_{12} \leq -15$ dB; this is the Non-Reciprocal Intensity Range (NRIR) of this device [3]. Note that low power signals are heavily attenuated in both directions and also there exists a limiting power threshold beyond which isolation abruptly drops (T_{12} rises) and the device is transparent in both directions. The latter happens when even the low fraction of power coupled from the lossless to the lossy waveguide is able to sufficiently saturate losses to restore normal coupling operation.

The physical implementation of the coupler is shown in the inset of Fig. 2. Each Si-slot waveguide consists of two identical ribs, 360 nm wide and 180 nm tall, forming a 40 nm slot, while the gap is 640 nm leading to $L_c = 800\ \mu$m. On the top face of the left waveguide two uncoupled graphene layers are deposited, which are assumed to be unbiased ($\mu_c = 0$ eV) and possess a saturation intensity of $I_{sat} = 1\ \mathrm{MW/cm^2}$. The two layers have a total surface conductivity of 121.7 μS, leading to 0.42 dB/μm losses, assumed fully saturable.

For the validation, we firstly derive a pair of nonlinear Schrödinger equations (NLSE) [4], for the TE mode of each isolated waveguide, heuristically coupled by the weak coefficient $\kappa = \pi/2L_c$. The SA coefficient of the graphene-overlaid waveguide is formally introduced by using the overlap of the mode distribution with graphene. The simulation results are shown in Fig. 2. Although there are moderate forward losses (6 to 3 dB), the threshold CW power for nonreciprocal regime is very low (160 to 320 mW), which showcases graphene's low saturation intensity. Note that the bandwidth limiting factor is the wavelength dispersion of the coupling coefficient and mode profile, both of which are low. The results were validated by full-vector 3D beam propagation method (BPM) simulations, depicted with markers in Fig. 2; the custom BPM was implemented with finite elements in the cross-section and wide-angle Crank-Nicolson scheme in the propagation direction, iteratively stabilized for nonlinearity.

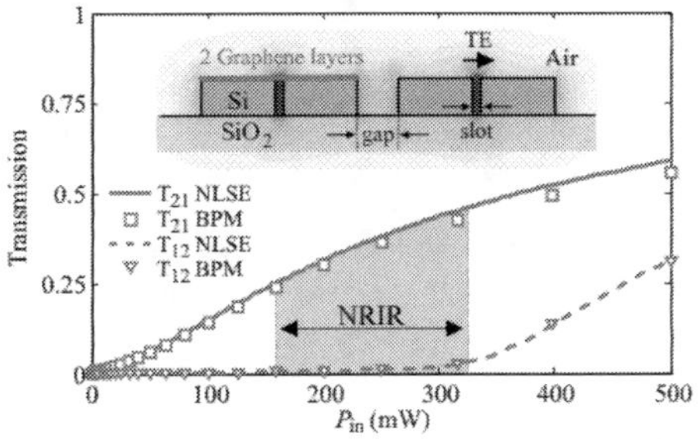

Figure 2. Forward (solid) and backward (dashed) transmission for the graphene waveguide coupler with SA. Shaded region corresponds to $T_{21} \geq -6$ dB and $T_{12} \leq -15$ dB. (Inset) Cross section (not to scale) of the silicon slot coupler with the symmetric supermode. The arrow indicates the dominant mode polarization.

3 Conclusions

Our analysis, design and evaluation show that asymmetric coupled graphene-overlaid waveguides can operated as nonlinear optical diodes, compatible with standard photonic integrated circuit technology. Our approach towards nonreciprocity provides an alternative route to the more commonly used Kerr effect, thanks to graphene's SA ultra fast response and low saturation intensity. Finally, our concept has a general scope and demonstrates how SA in conjunction with EPs can be exploited to design nonreciprocal integrated photonic components.

Acknowledgment

This research is co-financed by Greece and the European Union (European Social Fund- ESF) through the Operational Programme «Human Resources Development, Education and Lifelong Learning 2014-2020» in the context of the project "Design of nonlinear silicon devices incorporating graphene and using the Parity time symmetry concept" (MIS 5047874).

References

[1] M.A. Miri, A. Alù, Science **363** (2019)
[2] D. Chatzidimitriou, E.E. Kriezis, J. Opt. Soc. Am. B **35**, 1525 (2018)
[3] D.L. Sounas, A. Alù, Physical Review B **97** (2018)
[4] D. Chatzidimitriou, A. Pitilakis, E.E. Kriezis, Journal of Applied Physics **118**, 023105 (2015)

All-fibered high-quality 40-GHz to 200 GHz pulse sources based on nonlinear compression of besselons

Anastasiia Sheveleva, and *Christophe* Finot*

Laboratoire Interdisciplinaire Carnot de Bourgogne, 9 avenue Alain Savary, Dijon, France

Abstract. We experimentally demonstrate the generation of pulse trains with repetition rates of 40-GHz, 80 GHz and 200 GHz. Our approach is based on the nonlinear compression of besselon pulses in a highly-nonlinear fiber with normal dispersion. The pulses are of high quality, with a subpicosecond duration.

1 Introduction

Generation of very low duty-cycle high-quality optical pulse trains at repetition rate of several tens of GHz around 1.5 μm has become increasingly interesting for many scientific applications such as optical sampling or ultrahigh capacity transmission systems based on optical time division multiplexing. Unfortunately, the current bandwidth limitations of optoelectronic devices do not enable the direct generation of pulses with temporal width below a few picoseconds. Actively mode-locked fiber lasers were found to be an efficient, but onerous option to overcome this limitation. Alternative cavity-free architectures have relied on nonlinear compression in optical fibers [1] or on the use of a quadratic spectral phase to process an initial sinusoidal phase modulation applied in the temporal domain on a continuous wave [2]. Recently, we have proposed to improve this last scheme by imprinting a triangular spectral phase profile instead of a quadratic one [3], which has enabled to generate a Fourier-transform limited waveform we named besselon [4]. When the optimum depth of phase modulation is chosen, the resulting pulse train exhibits a temporal profile that is very close from a Gaussian pulse with a duty cycle of 0.18.

We demonstrate here experimentally that it is possible to further compress the waveform by taking advantage of the spectral expansion occurring upon propagation in a highly nonlinear fiber. Subpicosecond pulses at a repetition rate of 40 GHz are demonstrated. Repetition rate doubling and quintupling is also achieved thanks to line-by-line spectral processing.

2 Principle and experimental setup

Our proposed method is based on a combination of linear and nonlinear processing using phase modulations both in the temporal and spectral domains. The experimental setup we implemented is sketched in Fig. 1. A continuous wave laser at 1550 nm is first phase-modulated (PM) in the temporal domain by a sinusoidal waveform with a depth of modulation of 1.3 rad and a frequency of 40 GHz. A linear spectral shaper is then inserted to imprint a triangular spectral phase profile namely a series of $\pi/2$ spectral phase shifts between each spectral component. This converts the initial phase modulation into a train of Fourier-transform limited pulses. For our initial phase modulation, the resulting besselon waveform is close to a high-quality Gaussian pulse that is not impaired by spurious background or sidelobes [3].

The second stage of our setup relies on the nonlinear expansion of the besselon spectrum. We therefore amplify the pulse train with an erbium doped fiber amplifier (EDFA) up to an average value of 27 dBm. We have found that the nonlinear propagation in a normally dispersive fiber followed by a dispersive anomalous device [5] provides the best temporal profiles with reduced pedestals. We used a fiber with a length of 1 km, a nonlinear coefficient of 10 /km/W and a normal dispersion of –0.7 ps/nm/km. In order to avoid the deleterious effects of Brillouin backscattering, we have inserted an additional phase modulation operated at a low frequency. After the nonlinear propagation, the normal chirp of the pulses is cancelled to a large extend by imprinting a quadratic spectral phase with a programmable spectral filter [5]. By cancelling one spectral component over two or four over five, the same spectral filter can be used to double or quintuple the repetition rate of the pulse train. Results at the different stages are monitored using an optical sampling oscilloscope, an intensity autocorrelator and a high-resolution optical spectrum analyzer.

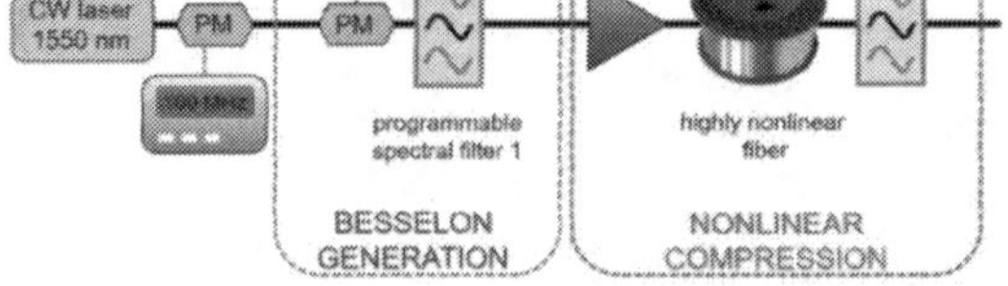

Fig. 1. Experimental setup.

* Corresponding author: christophe.finot@u-bourgogne.fr

3 Experimental results

3.1. Results at 40 GHz

The besselon temporal and spectral intensity profiles are plotted in panel (a1) and (b) respectively of Fig. 2. The temporal profile is made of 5.3 ps close to Gaussian pulses without any visible background between two pulses. The temporal intensity profile is in perfect agreement with the analytical waveform of a besselon [4] for an amplitude of the initial phase modulation of 1.3 rad. After the nonlinear propagation, the spectrum of the pulse is significantly broadened and spans over more than 1.6 THz. The nonlinear chirp is partly compensated for by applying a quadratic spectral phase of a 1.12 ps/nm. The pulse train recorded after compression is plotted in panel (c1) and demonstrates that high-quality pulses are generated with a subpicosecond duration. Additional autocorrelation measurements provide a full width duration of 850 fs (based on an assumption of a Gaussian pulse), leading to a duty cycle of 0.034. Panel (c2) recorded in the persistent mode highlights the remarkable stability of the pulse train. The experimental results are in remarkable agreement with the numerical simulations of the process based on the nonlinear Schrodinger equation.

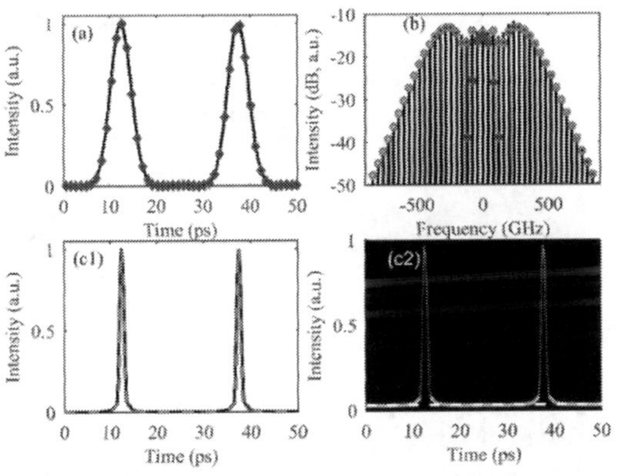

Fig. 2. (a) Temporal intensity profiles recorded on the optical sampling oscilloscope after the first linear shaping (solid line) are compared with the theoretical shape of the besselon (blue diamonds). (b) Optical spectrum after modulation (blue diamonds) and after nonlinear propagation (black lines). Results of numerical simulations are plotted with red circles. (c1) Temporal intensity profiles after the nonlinear propagation and quadratic spectral phase. Results of numerical simulation (red dashed line) are compared with experimental results (black line). Panel (c2) is similar to (c1) but recorded in the persistent mode of the optical sampling oscilloscope.

3.2. Doubling and quintupling of the repetition rate

Doubling and quintupling of the repetition rate is achieved by keeping only one spectral line over two or over five [6]. This processing is achieved thanks to the programmable spectral filter. The resulting spectra are provided in panels (a) of Fig. 3 where can note the excellent optical to noise ratio. The temporal intensity profiles recorded on the optical sampling oscilloscope (panel b and c) confirm the excellent quality of the shaped pulse train. At 80 GHz, the extinction ratio is still remarkable with a duty cycle of 0.09. Measurements made in the persistent mode also confirm the excellent stability of the source. The jitter slightly increases when operating at 200 GHz.

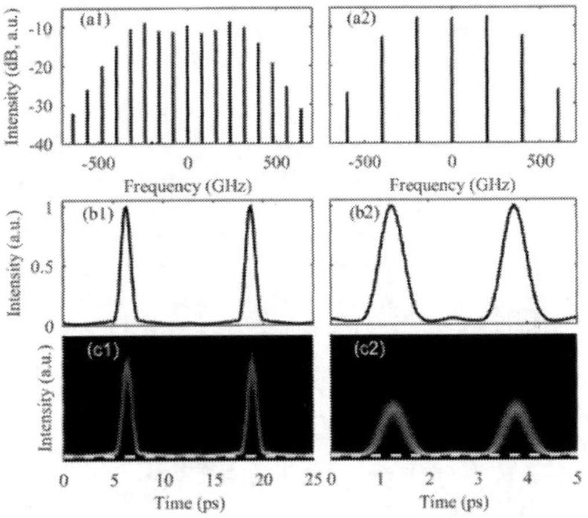

Fig. 3. (a) Spectral and (b) temporal intensity profiles after spectral components cancelation. (c) is the temporal profile recorded in the persistent mode. Panels (1) deal with a repetition rate of 80 GHz whereas panels (2) are obtained at 200 GHz. The white dotted line is the background level of the optical sampling oscilloscope.

To conclude, we have proposed to use the combination of linear and nonlinear compression schemes to achieve subpicosecond pulse trains with a very high quality and duty cycle of 0.034 at 40 GHz. Our approach is simpler than previous multi-stage architectures and only requires an initial phase modulation of 1.3 rad. The use of other operating points that are suitable for clean besselon generation will also be discussed and we will show that is possible to avoid the initial phase modulation used to prevent the Brillouin backscattering.

References

1. C. Finot, J. Fatome, S. Pitois, and G. Millot, IEEE Photon. Technol. Lett. **19**, 1711-1713 (2007).
2. T. Kobayashi, H. Yao, K. Amano, Y. Fukushima, A. Morimoto, and T. Sueta, IEEE J. Quantum Electron. **24**, 382-387 (1988).
3. U. Andral, J. Fatome, B. Kibler, and C. Finot, Opt. Lett. **44**, 4913 (2019).
4. A. Sheveleva, U. Andral, B. Kibler, S. Boscolo, and C. Finot, arXiv:2003.12630 (2020).
5. D. Grischkowsky and A. C. Balant, Appl. Phys. Lett. **41**, 1-3 (1982).
6. T. Sizer, IEEE J. Quantum Electron. **25**, 97-103 (1989).

EPJ Web of Conferences **238**, 11008 (2020) https://doi.org/10.1051/epjconf/202023811008
EOSAM 2020

Cosmology analogues in optical systems

Eric Howard[1,2,*]

[1]Department of Physics and Astronomy, Macquarie University, Sydney, Australia
[2]Centre for Quantum Dynamics, Griffith University, Brisbane, Australia

Abstract. We discuss the possibility of exploiting polariton-exciton physics as an analogue experimental tool to study challenging ideas and existing problems arising in the context of gravity theory and theoretical cosmology. We search for cosmology analogues with specific focus on simulating non-equilibrium dynamics across cosmological phase transitions in laboratory as well as employing optical analogue horizons in Bose-Einstein condensates (BECs) and signatures of white hole radiation to study gravitational and cosmological processes. Our analysis aims to uncover conceptual similarities between condensed matter systems and various phenomena in the Early Universe such as the symmetry breaking of the vacuum energy, spontaneous production of particles, false vacuum and cosmic inflation together with a number of unsolved cosmological problems.

1 Introduction

Latest theoretical and experimental advances in the physics of strong light-matter coupling, both in the classical and the quantum regimes and more recent, in the exciton-polariton research makes possible a range of new interdisciplinary forefront research topics with fundamental purposes in cosmology, gravitation and astrophysics. Analogue models of gravity [1] were inspired by various concepts from general relativity, and especially by the equivalence of optical media and space–time geometries. Such models have already being realized in recent laboratory experiments around the world. However, the simulation of cosmological scenarios and non-gravitational analogue models of large-scale cosmic processes were given much less attention in laboratories Several cosmological phenomena may be simulated in the laboratory [2], such as cosmological particle creation, caused by the expansion or contraction of the Universe, dynamical quantum phase transitions, soliton and topological defect formation (vortices, textures, domain walls or cosmic strings) that may have affected the dynamics of the Early Universe, or the presence of Hawking radiation arising from artificial event horizons, Bogoliubov-Cerenkov emission of soundwaves in bulk dilute superfluids, vacuum instability Penrose process (superradiant scattering), Unruh effect, Eardley instability, Gibbons-Hawking effect, dynamical Casimir emission and Schwinger mechanism. The standard model of cosmology describes the Early Universe as nearly homogeneous, with most inhomogeneities, including the seeds of large-scale structures as originating from the quantum vacuum fluctuations of the inflaton scalar field. The analogy between phonons in expanding BECs and quantum fields in a universe undergoing an accelerated expansion introduces an effective tool to study effects such as freezing of modes after horizon crossing and the amplification of quantum fluctuations. Such amplification mechanism was predicted to be responsible for the formation of the initial inhomogeneities as the first seeds for the formation of cosmic structures such as galaxies. A number of recent results in quantum fluids and exciton-polariton physics can be used as a test-bed capable of simulating a variety of cosmological toy models, with possibilities for recreating fundamental effects relevant for cosmological phase transitions, cosmic inflation, black holes and other gravitational scenarios in the laboratory.

2 Cosmology analogues

Quantum collective dynamics in light-matter systems and quantum fluids of light [3] such as exciton-polariton condensates are novel convenient tools to investigate phase transitions, find new universality classes for bosonic and fermionic fields, simulate solitons and topological structures occurring in the symmetry breaking phase transitions equivalent to cosmological strings or other defects in the Early Universe. Kibble-Zurek (K-Z) theory describes the breakdown of the adiabatic dynamics across a critical point, providing an insight for the preparation of ground-state phases of matter in the laboratory. Discovered in a cosmological setting, the theory was extended to quantum phase transitions such as vortices in superfluids [4] and domain walls in spin systems. While the theory described only second-order phase transitions, recent work reports the observation of a generalized K-Z mechanism across a first-order quantum phase transition in a spinor condensate. The intuitive connections between K-Z theory of topological defect production in second order phase transitions and the dynamics of the Landau-Zener transition is a centre piece

* Corresponding author: eric.howard@mq.edu.au

in understanding the cosmological phase transitions in the Early Universe. Non-equilibrium Landau-Zener tunneling in exciton-polariton condensates is a framework that may be extended to investigate dissipative cosmological dynamics occurring at the beginning of the Universe and open a new perspective toward understanding inflationary cosmology.

Exciton-polariton BECs are hybrid light-matter bosonic quantum fluids formed in a semiconductor microcavity, with the advantage of direct optical access, measurement and control of their condensate order parameters. Due to their light-matter nature, they display similar properties to normal fluids, when confined in an optical cavity, due to the effective mass of the photon and the effective photon-photon coupling. Such systems are capable of manifesting macroscopic quantum phenomena with intrinsic open-dissipative nature, enabling the investigation of universal dynamics with topologically non-trivial configurations that may describe cosmic structures. Polariton condensates embed diverse properties of lasers, atomic BECs and semiconductor physics, providing a variety of nonlinear physical phenomena, such as solitons, vortex lines in superflluids, universal scaling laws or spontaneous spin bifurcations and possible connections to cosmology, while maintaining a simple theoretical description. These features enable the simulation of different Hamiltonians with the advantage of studying effective curved spacetimes determined by the flow properties. We suggest employing characteristic phenomenological signatures in exciton-polariton BECs dynamics that may lead to observational constraints for cosmology, such as the simulation of the critical transition between cosmic inflation and the Big Bang. Such models may instigate pioneering insights into fundamental ideas and problems in standard cosmology, traditionally solved by inflation: the horizon problem, refering to the uniform temperature of causally disconnected patches of space-time, the flatness problem, refering to the extremely small spatial curvature of our universe and the monopole problem refering to the absence of Grand Unified Theory (GUT) monopoles after GUT phase transition. The spontaneous occurrence of spin vortices in polariton condensates and the associated spatial polarization patterns or spin textures are good examples of second-order phase transitions, accompanied by spontaneous symmetry breaking. Topological excitations in spinor quantum fluids, where the additional spin degree of freedom allows for mixed spin-phase topologies and the occurrence of spin vortices and hedgehogs (magnetic monopoles-like excitations) or spontaneous chiral symmetry breaking are valuable connections to large-scale non-equilibrium dynamics of cosmic structures. Cosmology toy models in exciton fluids can be made easily available as the vector order parameter can be accessed by optical means. We aim at validating such toy models and posible connections between the distribution of topological defects, determined by the scaling theory of phase transitions and signatures of universality beyond K-Z theory in the Early Universe. By combining coherence and spin properties, exciton-polariton fluids are a promising experimental platform for observing stimulated and spontaneous amplification by Hawking, Penrose, Zeldovich and superradiance effects via formation of event horizons for waves in media and rotating geometries with horizons and ergoregions. It has already been shown that the event horizon of black holes and white holes can be realized in the context of analogue gravity [5]. The dynamical equation of BECs with a sink or a vortex have the same wave equation as light in a strong curved-space. Experimental realization of black holes [6] and white holes in a nonlinear optical medium has been reported, as an analog example of spontaneous particle creation from vacuum fluctuations. Observational signatures from artificial white holes in flowing BECs may offer new predictions and open the door for further exploring the trans-plankian problem associated with Hawking radiation, gravitational backreaction and superradiance of black holes. In cosmology, the only possible white hole which by definition would radiate light, is the Big Bang. Quantum fluid analogues of white hole horizons may provide new perspectives on quantum field theory effects behind Big Bang theory. The study of signatures from sonic horizons in a moving fluid with superluminal (Bogoliubov) dispersion leading to the amplification of particle production and black hole lasing [7], may help with the understanding of spontaneous particle creation from vacuum fluctuations in quantum field theory in curved spacetime or simulating exotic forms of matter driving the cosmic inflation such as inflatons.

3 Conclusion

Properties and dynamics of quantum fluids of light, such as exciton-polariton condensates may be empoyed as a tool to simulate different cosmology and high-energy scenarios with the aim of answering existing fundamental questions and possible experimental investigation of exotic quantum effects in the Early Universe.

4 Acknowledgements

The author thanks Thomas Volz and Lorenzo Scarpelli for their expertise and useful comments.

References

1. M. Visser, Class and Quant. Grav., **15**, 1767, (1998)
2. W. H. Zurek, Nature, **317**, 505, (1985)
3. I. Carusotto, C. Ciuti, Rev Mod Phys., **85**,299, (2013)
4. K.G. Lagoudakis, M Wouters, M. Richard, A Baas, I Carusotto, R Andre, L. S. Dang, B. Deveaud-Pledran, Nature Phys., **4**, 706–710, (2008)
5. D. Gerace, I. Carusotto, Phys Rev **B86**, 144505, (2012)
6. H. S. Nguyen, D. Gerace, I. Carusotto, D. Sanvitto, E. Galopin, A. Lemaître, I. Sagnes, J. Bloch, and A. Amo, Phys. Rev. Lett. 114, 036402, (2015)
7. P. E. Larre, A. Recati, I. Carusotto and N. Pavloff, Phys. Rev. A, **85**,013621, (2012)

 https://doi.org/10.1051/epjconf/202023811009

Nonlinear temporal Fresnel diffraction induced by phase jumps in fiber optics

Anastasiia Sheveleva and *Christophe* Finot *

Laboratoire Interdisciplinaire Carnot de Bourgogne, UMR 6303, UBFC-CNRS, 9 av. A. Savary, 21000 Dijon, France

Abstract. We numerically study the temporal patterns that emerge upon the propagation of one or two π phase jumps in a nonlinear optical fiber. Both normal and anomalous regimes of dispersion are considered. The analytical derivation of linear patterns based on Fresnel's integrals provides some clues to better understand the properties of the resulting coherent structures.

1 Introduction

Diffraction is among the key effects of wave physics with applications in a broad range of technological fields. The knowledge accumulated in the field of spatial optics over the last two centuries is a valuable help to provide insights to a better understanding of the temporal dynamics of a wave experiencing the dispersive propagation. Indeed, the parabolic spectral phase induced by the dispersion of an ultrashort optical pulse is equivalent to the paraxial diffraction affecting the spatial propagation of a light beam [1]. This has been the basis of various recent applications in ultrafast optics including, for example, a temporal analogue to the Arago's spot demonstrated in optical fibers [2].

However, up to now, most the works have mainly focused to the evolution of fields that have been intensity modulated. We here discuss the temporal evolution of a continuous wave affected by one or two abrupt π-phase shifts, which is the temporal equivalent of a plane wave transmitted through a transparent plate where a localized phase jump leads to $\lambda/2$ optical path difference. We are particularly interested in the effects of the Kerr nonlinearity on the propagating wave in both the anomalous and normal regimes. We highlight the emergence of coherent nonlinear structures such as bright solitons, Peregrine solitons or black solitons.

2 Evolution of a single phase jump

We first consider the evolution of the wave affected by an ideal single π-phase shift (Fig. 1, panel a, grey line). The light propagates through a 10-km long anomalously dispersive fiber with parameters typical of the SMF-28 fiber at telecommunication wavelengths. When nonlinearity is neglected, the temporal intensity pattern can be derived analytically as [3] (see panel a, black line):

$$I(t)=\frac{1}{2}\left[C_f^2\left(\frac{\mathrm{sgn}(\beta_2)\,t}{\sqrt{\pi|\beta_2|z}}\right)+S_f^2\left(\frac{\mathrm{sgn}(\beta_2)\,t}{\sqrt{\pi|\beta_2|z}}\right)\right]$$

with t being the temporal coordinate and C_f and S_f being the Fresnel integrals, β_2 the second-order dispersion and z the propagation distance. This analytical expression can be straightforwardly interpreted by the use of the clothoid. The symmetric pattern that emerges is characterized by a central hole surrounded by two marked oscillating edges that move in opposite directions.

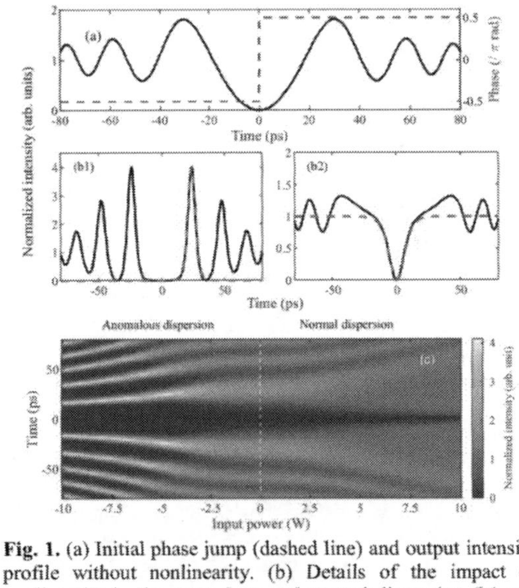

Fig. 1. (a) Initial phase jump (dashed line) and output intensity profile without nonlinearity. (b) Details of the impact of nonlinearity in the anomalous and normal dispersion (b1 and b2). Red dashed lines are fits by a bright and a black soliton. (c) Influence of the initial average power on the output pattern.

When Kerr nonlinearity impacts the propagation, the field evolution is governed by the nonlinear Schrödinger equation that is numerically solved [4]. Output profiles which are plotted on panels (b) stress that two bright fundamental solitons emerge in the anomalous regime of propagation whereas a single black soliton appears in the normal case. When nonlinearity is increased (panel c), the

* Corresponding author: christophe.finot@u-bourgogne.fr

coherent structures are subjects to the usual scaling laws of solitons.

3 Evolution of a pair of π phase jumps

We have also numerically considered the evolution of a pair of π-phase jumps spaced by $\Delta T = 40$ ps. The linear propagation (Fig. 2a) leads to a more complex pattern that can be understood quite easily in terms of interference between the temporal Arago's pattern [5] and the result of the dispersive evolution of a rectangular wave of width ΔT that is phase shifted by π. Once again, the linear profile can be analytically expressed in terms of a combination of Fresnel integrals. After 10 km of linear propagation, the dispersive pattern is made of a broad central peak surrounded by strong ripples which peak intensity has been increased compared to the previous case under discussion.

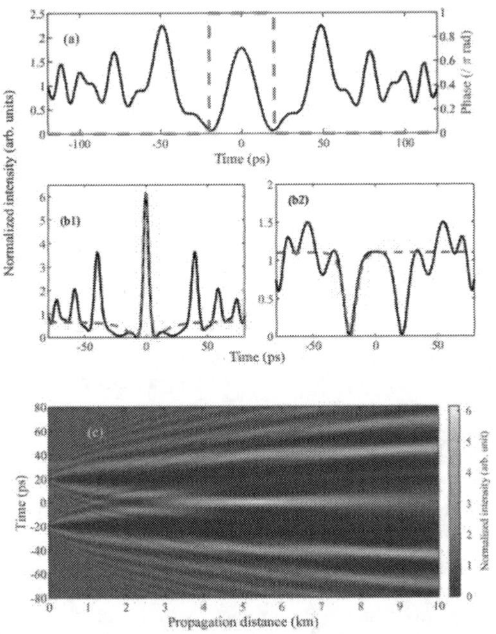

Fig. 2. (a) Initial phase profile (dashed line) and output intensity profile in the linear propagation. (b) Temporal intensity profiles in presence of nonlinearity observed after 5.37 km in a fiber with anomalous dispersion (b1) or after 10 km in a fiber with normal dispersion (b2) for an initial power of 290 mW. The red dashed lines are the profiles of a Peregrine soliton at its point of maximum focusing and of a black soliton. (c) Longitudinal evolution of the intensity profile in the anomalous regime of dispersion.

When nonlinearity is experienced, the patterns have very different dynamic according to the regime of dispersion. The anomalous regime is marked by the emergence of a strong central peak that experiences a growth and decay longitudinal evolution (see Fig. 2(c)). At the point of maximum focusing (see Fig. 2(b1)), the ultrashort structure is in close agreement with the shape of a Peregrine soliton [6], with a temporal duration below 6 ps and a peak power that is increased by a factor 6 with respect to the initial average power. This central peak may also split into a doublet of pulses [7] (Fig. 3) and is surrounded by other breathing structures that have also been identified in the nonlinear Fresnel evolution of a supergaussian pulse [8]. These observations confirm the universal impact of coherent structures linked to modulation instability that emerge not only from intensity perturbations, but also from localized phase defects.

In the normal regime of propagation, the picture gets very different. Instead of an intense central structure, the intensity profile becomes nearly constant in-between two black solitons. The position of these dips of light is fixed by position of the initial phase shifts and is not influenced by the power level of the initial continuous wave.

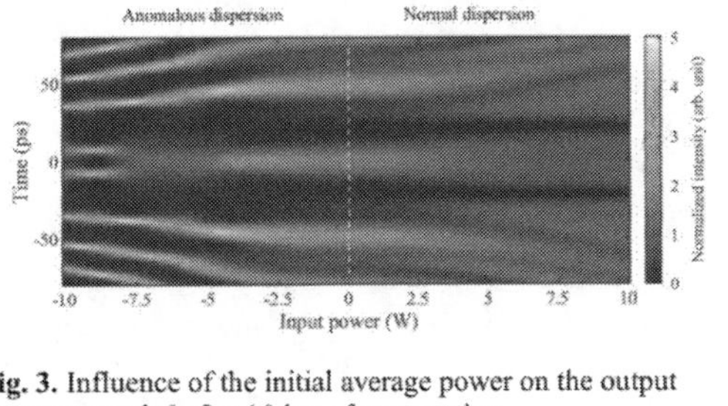

Fig. 3. Influence of the initial average power on the output pattern recorded after 10 km of propagation

To discuss the influence of the value of the initial phase offset we have carried out additional simulations. The resulting pattern can be significantly different with a loss of the temporal symmetry for the single jump case. Breathing solutions may appear in the anomalous regime of propagation and grey solitons are observed instead of black ones in the normal regime of dispersion. When non-ideal phase jumps are considered (i.e. the transition between the two phase states has a given duration), the nonlinear pattern will also be affected by this additional initial chirp.

References

1. V. Torres-Company, J. Lancis, and P. Andrés, in *Progress in Optics*, E. Wolf, ed. (Elsevier, 2011), pp. 1-80.
2. C. Finot and H. Rigneault, J. Opt **21**, 105504 (2019).
3. H. D. Betz, Appl. Opt. **8**, 1007-1013 (1969).
4. G. P. Agrawal, *Nonlinear Fiber Optics, Fourth Edition* (Academic Press, San Francisco, CA, 2006).
5. J. Komrska, Optica Acta: International Journal of Optics **14**, 127-146 (1967).
6. B. Kibler, J. Fatome, C. Finot, G. Millot, F. Dias, G. Genty, N. Akhmediev, and J. M. Dudley, Nat Phys **6**, 790-795 (2010).
7. C. Finot, Laser Phys. Lett. **17**, 025103 (2020).
8. F. Audo, B. Kibler, J. Fatome, and C. Finot, Opt. Lett. **43**, 2864-2867 (2018).

Vectorial dispersive shock waves on an incoherent landscape

Javier Nuño[1,2], *Christophe* Finot[1], *Miro* Erkintalo[3,4], and *Julien* Fatome[1,3,4,*]

[1] Laboratoire Interdisciplinaire Carnot de Bourgogne, 9 av. A. Savary, Dijon, France
[2] Departamento de Electrónica, Universidad de Alcalá, 28805 Alcalá de Henares (Madrid), Spain
[3] Department of Physics, The University of Auckland, Private Bag 92019, Auckland 1142, New Zealand
[4] The Dodd-Walls Centre for Photonic and Quantum Technologies, New Zealand

Abstract. We study the impact of temporal randomness on the formation of vectorial dispersive shock-waves that emerge due to the interaction of a partially coherent probe wave co-propagating together with an orthogonally polarized intense short pulse. Experiments carried out in a normally dispersive optical fiber demonstrate that the lack of coherence of the probe landscape acts as a strong diffusive term, which is able to hamper or inhibit the vectorial shock formation.

1 Introduction

In the last decades, shock waves have been the subject of intense studies in many areas of physics, ranging from hydrodynamics, nonlinear optics, condensates or plasmas [1]. Basically, shock waves occurring in conservative or low-dissipative systems originate from two fundamental ingredients: nonlinearity and dispersion. More specifically, in nonlinear optics, a self-defocusing regime of propagation can turn the front of an initial pulsed signal into a gradient catastrophe which is then regularized by chromatic dispersion. This regularization phenomenon then causes the onset of fast nonstationary oscillating fan structures called dispersive shock waves (DSWs) [2, 3] or undular bores [4] (in analogy with the classical "mascaret" which can be observed in specific river estuaries). Recently, we theoretically introduced and experimentally demonstrated a new type of vectorial DSW in optical fibers [5]. In contrast to previous studies, here a gradient catastrophe is generated on the phase of a CW probe through cross-phase modulation (XPM) induced by a pulsed signal owing to a vectorial configuration, leading to the equivalent of an optical axe [6]. We further explore here the nonlinear interaction occurring between the intense pulsed signal and the orthogonally polarized weak probe, especially in presence of partial incoherence. We provide here a set of experimental results carried at telecommunications wavelengths that have been closely confirmed by numerical simulations based on two coupled nonlinear Schrödinger equations corresponding to a simplified Manakov model [5].

2 Results in the coherent case

We first recall the basic principle of the vectorial interaction through XPM between a weak CW probe co-propagating in an optical fiber together with an orthogonally polarized intense short pulse (supergaussian pulse with a duration of 68 ps and a peak power of 3.8 W). The temporal intensity profiles recorded after propagation in a 13-km long normally dispersive fiber (dispersion of −124 ps/nm/km and nonlinear Kerr coefficient of 5.2 $W^{-1}.km^{-1}$) are provided in Fig. 1. The defocusing regime induces a nonlinear reshaping of the pulsed signal into subsequently parabolic, then broad and expanding sharp square pulse until the wave breaking phenomenon occurs. This increasing sharpness and temporal expansion of the pulse edges then triggers an expanding catastrophe chirp profile on the orthogonally polarized CW probe through XPM coupling. Forced by this piston effect, this nonlinear phase gradient catastrophe is then regularized by chromatic dispersion, depleting the central region of the probe while creating two repulsive shock waves with oscillating fronts onto the initial CW landscape (Fig. 1b), thus mimicking superfluid or ballistic DSWs. Regarding the pulsed signal, it progressively becomes parabolic and finally propagates self-similarly, experiencing a strong temporal broadening (Fig. 1a).

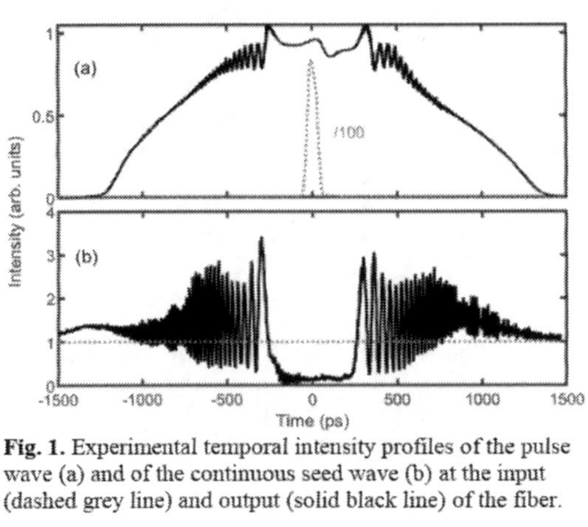

Fig. 1. Experimental temporal intensity profiles of the pulse wave (a) and of the continuous seed wave (b) at the input (dashed grey line) and output (solid black line) of the fiber.

* Corresponding author: cfinot@u-bourgogne.fr

3 Vectorial shock wave on an incoherent landscape

We replaced the coherent seed wave by the amplified spontaneous emission from an erbium-doped fiber amplifier. An optical bandpass filter with a tuneable spectral width is used to control the level of randomness of the probe. The 10-GHz temporal incoherent landscape after nonlinear propagation is visible in Fig. 2(a1) that results from the accumulation of a thousand of recorded traces. When the intense pulse is switched on, one can clearly see the typical temporal gap emerging from the fluctuations of the 10 GHz partially coherent wave. We can also make out the existence of bumps localized on each side of the temporal gap, typical of this vectorial DSW dynamics. The details of the fan structure are however completely blurred by the incoherence of the probe. Moreover, details of the deterministic pulse reported in panel (a2) show that the level of fluctuations may strongly vary along the pulse width. Indeed, significant intensity fluctuations surrounding the central part of the pulse are observed, confirming once again the XPM induced by the incoherent probe wave around the rarefaction area. In contrast, the central part and the wings of the parabolic profile remain fully coherent.

In order to further assess the statistical behavior of the waves, we have also experimentally evaluated the probability distribution function of the intensity profiles along two different temporal areas of the waves. Results are summarized in panels (b1) and (b2) for the probe and pulse waves, respectively. Whereas the rarefaction area of the probe (open blue circles, blue region in panel (a)) presents a narrow distribution centered close to zero, the bumps (red full circles, red regions in panel (a) are characterized by temporal events presenting higher peak power than the ones appearing without any interaction with the pulse (yellow crosses, yellow region in panel (a)). Regarding the pulse, the experiments confirm that the pdf of the central part (rarefaction area of the probe) is much narrower than the ones affected by the DSWs fluctuations (on the edge of the temporal gap).

Finally, we investigate in panel (c) the impact of the degree of incoherence in the probe landscape by monitoring the temporally averaged DSWs by means of a 70-GHz sampling oscilloscope triggered by means of the output pulse. For a 10- or 20-GHz incoherent landscape, we can still observe the piston effect caused by the pulse induced XPM on the probe signal and in particular the depletion of the central part. However, the fast temporal DSW fan structures tend to be damped by the randomness of the landscape. Further increasing the amount of disorder then tends to even more hamper the shock wave until its total inhibition for an incoherence higher than 40 GHz. For an incoherence of 100 GHz, the averaged recording only presents a small bump All the statistical trends observed experimentally have been closely confirmed by intensive numerical simulations of the Manakov model.

1. G. B. Whitham, *Linear and Nonlinear Waves* (Wiley, New York, 1974).
2. G. Xu, M. Conforti, A. Kudlinski, A. Mussot, and S. Trillo, Phys. Rev. Lett. **118**, 254101 (2017).
3. G. A. El and M. A. Hoefer, "Dispersive shock waves and modulation theory," Physica D **333**, 11-65 (2016).
4. J. Fatome, C. Finot, G. Millot, A. Armaroli, and S. Trillo, Phys. Rev. X **4**, 021022 (2014).
5. J. Nuno, C. Finot, G. Xu, G. Millot, M. Erkintalo, and J. Fatome, Commun. Phys. **2**, 138 (2019).
6. L. Helczynski, B. Hall, D. Anderson, M. Lisak, A. Berntson, and M. Desaix, Phys. Scr. **T84**, 81 (2000).

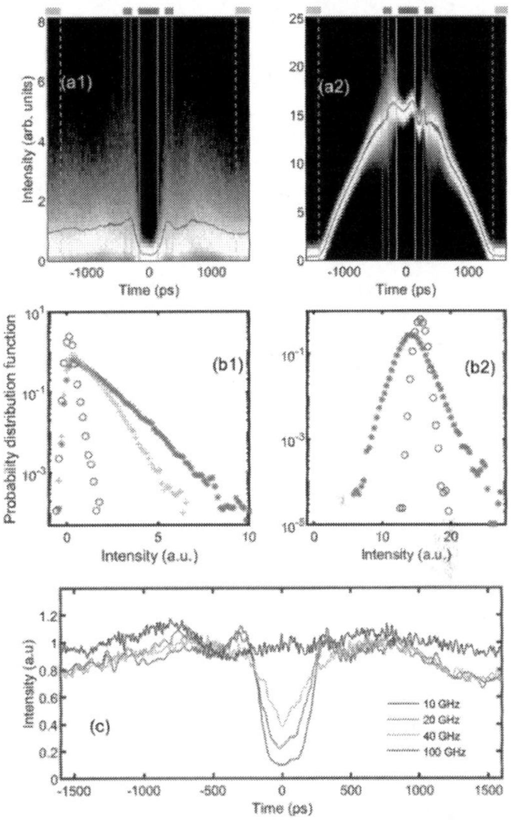

Fig. 2. Impact of the landscape partial incoherence on the vectorial DSWs. (a-b) Experimental results obtained for a pulse -peak power of 3.8 W and a spectral coherence of the probe of 10 GHz. (a) Output intensity profiles recorded for the probe and pulse waves are plotted in panels (a1) and (a2) respectively. (b) Evolution of the pdf in two different temporal sections of the probe and pulse waves (panel b1 and b2 respectively). Statistics for the central part of the probe (blue part in panels (a)) are plotted with blue open circles whereas properties in the lateral bumps (red parts in panels (a)) are plotted with red stars and with yellow crosses for the region outside the temporal pulse extend (yellow part in panels (a)). (c) Impact of the partial incoherence on the probe signal. Output temporal intensity profile of the probe wave after averaging over 1000 shots. Different levels of initial incoherence are compared: 10, 20, 40 and 100 GHz (blue, red, yellow and purple colors, respectively).

Optomechanical microwave oscillator and frequency comb generation in a full phononic bandgap 1D optomechanical crystal cavity

Laura Mércade[1,*], *Leopoldo L.* Martin[2], and *Amadeu* Griol[3] *Daniel* Navarro-Urrios[3] *Alejandro* Martínez[3]

[1]Nanophotonics Technology Center, Universitat Politècnica de València, Camino de Vera s/n, 46022 Valencia, Spain
[2]Departamento de Física, Facultad de Ciencias, Universidad de la Laguna
[3]MIND-IN2UB, Departament dEnginyeria Electrònica i Biomèdica, Facultat de Física, Universitat de Barcelona, Martí Franquès 1, 08028 Barcelona, Spain

Abstract. In this work we show that a silicon optomechanical crystal cavity can be used as an optomechanical oscillator when driven to the phonon lasing condition with a blue-detuned laser. The optomechanical cavity is designed to have a breathing like mode vibrating at $\Omega_m/2\pi$ =3.897 GHz in a full phononic bandgap. Our measurements show that the first harmonic displays a phase noise of -100 dBc/Hz at 100 kHz. Stronger blue-detuned driving leads eventually to the formation of an optomechanical frequency comb, with lines spaced by the mechanical frequency. The measured phase noise grows up with the harmonic number, as in classical harmonic mixing. We present real-time measurements of the comb waveform and show that it can be adjusted to a theoretical model recently presented. Our results suggest that silicon optomechanical cavities can play a role in integrated microwave photonics.

1 Introduction

Cavity optomechanics studies the interaction between light and sound waves simultaneously confined in a cavity [1]. In an optomechanical cavity (OMC), confined mechanical waves can coherently modulate an optical signal at MHz and even GHz frequencies via optomechanical interaction, becoming relevant in microwave photonics. Furthermore, since OMCs are nonlinear elements, multiple harmonics of the fundamental mechanical vibrations can be overimposed on the optical signal [2], a phenomenon that has been interpreted theoretically as an optomechanical frequency comb (OFC) [3].

2 Optomechanical frequency comb

The OMC that we exploit here consist in a suspended silicon nanobeam with one-dimensional (1D) periodicity [4]. The key idea is to have OM mirrors on each nanobeam side whilst optical and mechanical resonant modes are confined in the central region, having a good overlap between them. Our structure results of an optical mode at λ_r =(1522.3±0.3) nm with an optical quality factor of $Q_o = 5 \times 10^3$, presented in Fig. 1(a), and a mechanical mode at $\Omega_m/2\pi$ =3.897 GHz with an optomechanical coupling rate of $g_0/2\pi$ = (660 ± 70) kHz, depicted in Fig. 1(b). In addition, we have designed our structure in order to have a full phononic bandgap to reduce the phonon leakage. A simulation, in Fig. 1(c), of the phononic band diagram from the profile extracted from a Scanning Electron Microscopy (SEM) image, which is shown (see Fig. 1(d)), shows that the experimental mechanical modes lies into the total bandgap.

The mechanical mode can be driven to the phonon lasing regime under proper conditions. When driving the OMC with a blue-detuned laser with respect to the optical resonance at even higher power, higher-order harmonics can be observed in the detected signal. This closely resembles an optical frequency comb (OFC) of OM nature, recently analyzed theoretically in [3]. We measured up to the fifth harmonic, as can be seen in the optical spectrum of the generated OFC in Fig. 1(e). The theoretical dynamics generation[3] shows a good agreement with the acquired experimental optical traces in Fig. 1(f).

3 Optomechanical oscillator

High-quality microwave sources, which are required for a number of applications, are typically made by applying frequency multiplication to an electronic source. Amongst the different techniques to build an optoelectronic oscillator, we propose one obtained from an OM cavity when pumped with a blue-detuned laser source. In this case, since the involved mechanism is a self-sustained oscillation originated from OM interaction, from now on we will refer to it as an optomechanical oscillator (OMO). We made this experiment with the cavity proposed in the last section and we measured. For the first harmonic in Fig.2(a), the resulting noise figure evolution is presented in Fig. 2(c). For this harmonic the noise figure becomes as low as (-100±1) dBc/Hz at 100 kHz, which is a remarkable good value for an OEO oscillating at GHz frequencies. We

*e-mail: laumermo@ntc.upv.es

 https://doi.org/10.1051/epjconf/202023811011

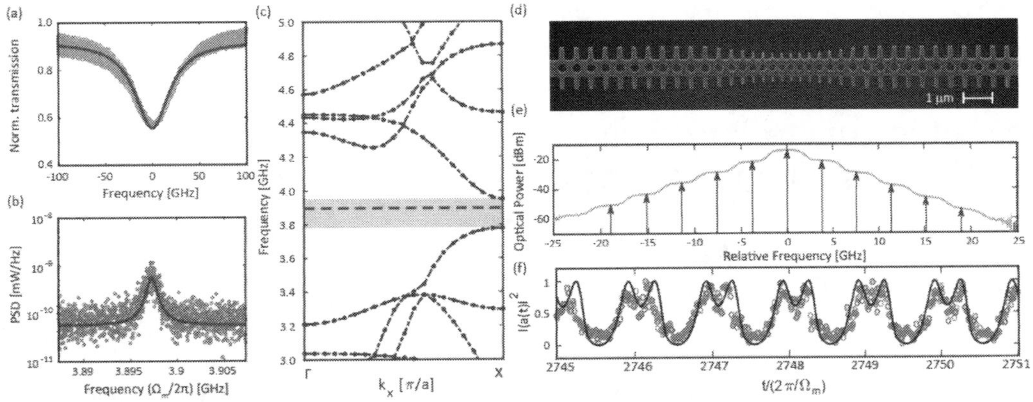

Figure 1. (a) Optical experimental resonance trace (purple) and Lorenztian fit (dark line). (b) Power spectral density of the mechanical resonance confined in the total bandgap. (c) Phononic band diagram for the real profile mirror unit cell extracted from the SEM image, including the experimental measured frequency of the mechanical mode (dashed line). (d) SEM image of the fabricated OM cavity. (e) Recorded optical spectrum in purple showing a set of peaks corresponding to different harmonics represented in blue at the expected position. (f) Comparison of the experimental (dots) and the theoretical (solid line) temporal OM-generated OFC traces.

have also measured the phase noise of the different harmonics, shown in 2(b), where the result is depicted in Fig. 2(d). In principle, the harmonic mixing process will result in an added phase noise of $20 \times log(m)$ with respect to that of the first harmonic, which is well satisfied in our device.

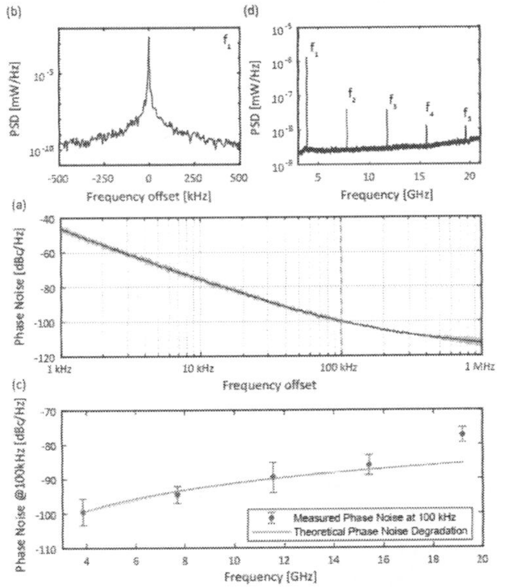

Figure 2. (a) Linear scale of the first harmonic centered at 3.87 GHz. (b) RF spectra of the generated OFC. (c) Mean phase noise of the generated microwave tone. (d) Phase noise of the different harmonics of the OFC with the theoretical phase noise degradation with respect to that of the 1st harmonic.

4 Conclusions

In summary, we have demonstrated a new silicon-chip OMC with a large OM coupling rate for a GHz mode within a full phononic bandgap, which can perform as an ultracompact OMO. Operation at cryogenic temperatures would improve the phase noise [5] as a result of the enhancement of the mechanical Q factor because of the full phononic bandgap. In addition, the preliminary demonstration of the OFC paves the way towards synthesis of microwave signals beyond the generation of pure cw tones. The main advantages of the OMC approach for microwave signal processing are its extreme compactness and low weight, highly desirable in space and satellite applications, and its compatibility with silicon electronics and photonics technology.

References

[1] M. Aspelmeyer et al., "Cavity optomechanics," Rev. Mod. Phys., vol. 86, pp. 1391–1452, 2014.

[2] M. Hossein-Zadeh and K. J. Vahala, "Photonic RF Down-Converter Based on Optomechanical Oscillation," IEEE Photonics Tech. L., vol. 20, pp. 234-236, 2008.

[3] M. A. Miri et al., "Optomechanical frequency combs," New J. Phys., vol. 20, 043013, 2018.

[4] L. Mercadé et al., "Microwave generation and frequency comb in a silicon optomechanical cavity with a full phononic bandgap," arXiv:1912.06692 [physics.optics]

[5] M. Hossein-Zadeh and K. J. Vahala, "An Optomechanical Oscillator on a Silicon Chip," IEEE J. Sel. Top. Quant., vol. 16, pp. 276–287, 2010.

Photonic generation of picosecond pulse doublets from the nonlinear splitting of super-Gaussian pulses

Christophe Finot

Laboratoire Interdisciplinaire CARNOT de Bourgogne, UMR 6303 UBFC-CNRS, 2100 Dijon, France

Abstract. We experimentally demonstrate the generation of temporal pulse doublets from the propagation of initial super-Gaussian waveforms in an optical fiber with anomalous dispersion. The nonlinear focusing dynamics leads to picosecond structures that are characterized both in amplitude and phase. Their close-to-Gaussian Fourier-transform limited shape is found in excellent agreement with numerical simulations. This single-stage reshaping scheme is energy efficient, can sustain GHz repetition rates and temporal compression factors around 10 are demonstrated.

1 Introduction

Generation of picosecond optical pulse doublets at high repetition rates remains a technological challenge. Indeed, the optoelectronics bandwidth required to generate these double pulses is much higher than the one usually required to synthesize an isolated waveform. Therefore, all-optical approaches are fully relevant to overcome the electronics bottleneck and to provide optical doublets with a duration below 10 ps. In this context, several linear solutions have relied on photonic first-order derivation [1] using long-period fiber gratings, fiber Bragg gratings or on-chip microdisk resonators. Nonlinear approaches in optical fibers can also be advantageous thanks to the combination of the dispersive and Kerr effects. Indeed, numerous examples of doublets have been reported in fiber lasers. However, their repetition rates are typically a few tens of MHz only. Cavity-free solutions that do not require mode-locking processes also offer attractive solutions and several strategies have been explored such as chirp-induced splitting of pulses [2], pulse copying thanks to an external triangular waveform [3] or the compression of higher-order solitons.

In this contribution, we introduce a new solution based on the nonlinear propagation of a super-Gaussian pulse in a fiber with anomalous dispersion. This problem, also known as the box-problem or the dam-break problem, has recently stimulated a renewed interest in optics, mainly driven by the theoretical study of the nonlinear dynamics of the temporal [4, 5] or spatial [6, 7] coherent structures that emerge upon propagation. We show here experimentally that it can find practical applications. Indeed, a pair of Gaussian-like pulses emerge during the evolution of an initial flattop pulse in a single-mode waveguide with a focusing nonlinearity [8].

2 Experimental setup

We have implemented an experiment sketched in Fig. 1 based on devices that are commercially available and typical of the telecommunication industry. An external cavity laser at 1550 nm emits a highly-coherent continuous wave that is intensity modulated with a lithium niobate modulator to obtain super-Gaussian pulses at a repetition rate of 2 GHz. The initial pulses have an initial duration of 42 ps, are highly symmetric and present a flat plateau and sharp edges so that it approximates the ideal box waveform (see dashed grey line in Fig. 3a). Pulses are amplified by an erbium-doped fiber amplifier before nonlinear propagation in a set of four identical 900m long optical SMF-28 fibers with second-order dispersion of 20.10^{-3} ps^2/m and a nonlinear Kerr coefficient of 1.1 /W/km. In order to prevent the building of unwanted stimulated Brillouin backscattering, we have inserted three optical isolators. The characterization of the output pulse properties is ensured by a complex optical spectrum analyzer (Apex Technologies) that enables us to get access to both the temporal phase and intensity profiles with a temporal resolution below 1 ps.

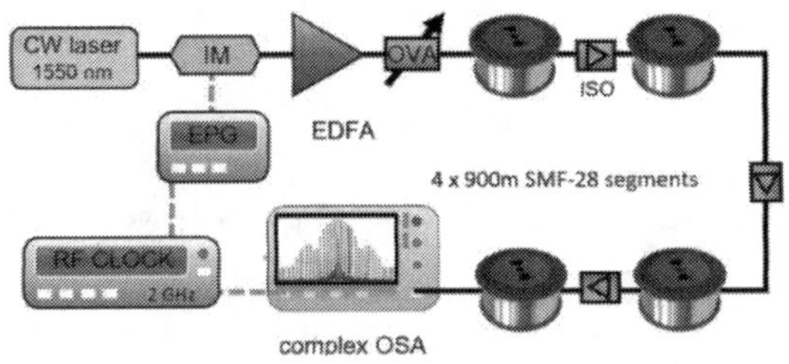

Fig. 1. Experimental setup. CW : continuous wave; IM : intensity modulator; EPG : electronic pattern generator; EDFA : erbium doped fiber amplifier; ISO : isolator ; OVA : optical variable attenuator.

* Corresponding author: christophe.finot@u-bourgogne.fr

3 Experimental results

Panels 1 of Fig. 2 summarize the experimental evolution of the intensity and phase profiles as a function of the average input power. When increasing the integrated nonlinearity in the fiber with anomalous dispersion, the initial super-Gaussian waveform reshapes into two well-separated pulses. For additional propagation or nonlinearity, a third temporal component emerges from the central part of the resulting waveform. The experimental amplitude and phase features are both reproduced closely by numerical simulations of the propagation based on the nonlinear Schrodinger equation taking into account the losses of the fiber and the exact input conditions (see panels 2).

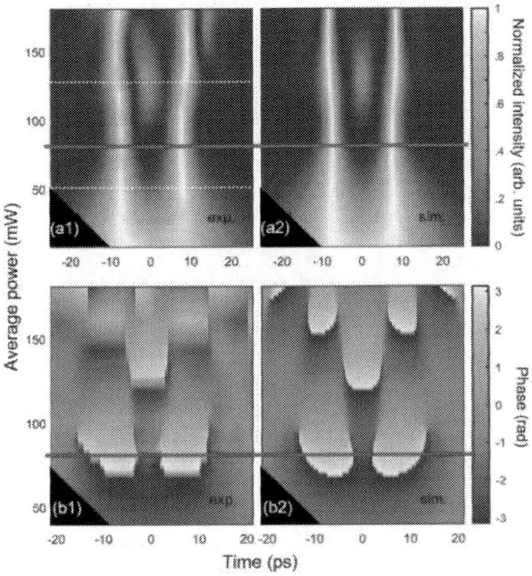

Fig. 2. Evolution of the output temporal intensity and phase profiles (panels a and b, respectively) of the waveform as a function of the initial average power for a fixed propagation length of 3.6 km. Experimental results (panels 1) are compared with numerical simulations (panels 2). Intensity profiles are normalized by the peak-power. The red horizontal line strands for the results detailed in Fig. 3.

A more detailed view of the temporal profile obtained for an average power of 80 mW (corresponding to an input peak power of 1 W and a pulse energy of 40 pJ) is proposed in Fig. 3. For this power, two identical picosecond pulses are generated and spaced by 15.2 ps. The output pulse structures have a fwhm duration of 4.3 ps (i.e a compression factor of 9.8 with respect to the input waveform) and an excellent extinction ratio is observed. Once again, the experimental results are in remarkable agreement with numerical simulations. They confirm that the temporal phase profile over the two pulses is very flat and that the two pulsed parts are in-phase. This contrasts with other methods such as optical derivation where the leading and trailing pulses are π-phase shifted [1]. A better understanding of the nonlinear dynamics will stimulate further theoretical research with more advanced tools. Using topological perspective will enable to better determine the theoretical guidelines to apply and the balance between the duration, peak power, dispersion/diffraction and nonlinearity that has to be targeted [7].

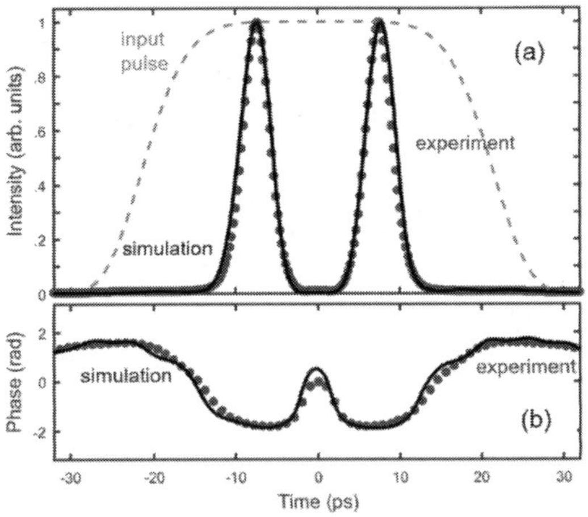

Fig. 3. Details of the temporal intensity and phase profiles (subplots a and b, respectively) obtained for an input average power of 80 mW. Experimental results (solid black lines) are compared with numerical simulations of the NLSE (blue full circles). The input intensity profile is plotted with a dashed grey line.

References

1. R. Slavík, Y. Park, M. Kulishov, R. Morandotti, and J. Azaña, Opt. Express **14**, 10699-10707 (2006).
2. L. Helczynski, B. Hall, D. Anderson, M. Lisak, A. Berntson, and M. Desaix, Phys. Scr. **T84**, 81 (2000).
3. N. Verscheure and C. Finot, Electron. Lett. **47**, 1194-1196 (2011).
4. F. Audo, B. Kibler, J. Fatome, and C. Finot, Opt. Lett. **43**, 2864-2867 (2018).
5. G. A. El, E. G. Khamis, and A. Tovbis, Nonlinearity **29**, 2798 (2016).
6. W. Wan, D. V. Dylov, C. Barsi, and J. W. Fleischer, Opt. Lett. **35**, 2819-2821 (2010).
7. G. Marcucci, D. Pierangeli, A. J. Agranat, R.-K. Lee, E. DelRe, and C. Conti, Nat. Commun. **10**, 5090 (2019).
8. C. Finot, Laser Phys. Lett. **17**, 025103 (2020).

EPJ Web of Conferences **238**, 11013 (2020) https://doi.org/10.1051/epjconf/202023811013
EOSAM 2020

Dependence of optical attenuation on radiation wavelength and waveguide geometry in copper-coated optical fibers

Pavel Cherpak[1,*], *Renat* Shaidullin[2], and *Oleg* Ryabushkin[2]

[1]Moscow Institute of Physics and Technology, Institutsky lane 9, Dolgoprudny, Moscow region, 141700, Russia
[2]Fryazino branch of Kotelnikov Institute of Radioengineering and Electronics of RAS, Fryazino, Moscow region, 141190 Russia

Abstract. We demonstrate a novel approach to the determination of optical loss coefficients in metal-coated fibers in a 0.4-1.9 μm wavelength range. It is based on measuring the change of temperature-dependent electrical resistance of the metal coating caused by laser radiation transmitted through the fiber. A number of single-mode and multimode metallized fibers were investigated using several laser sources operating in visible and near infrared ranges. The spectral dependencies of optical losses of copper-coated fibers were experimentally obtained. The region that corresponds to the minimum optical losses is located near 1 μm wavelength. The increase of radiation losses in 1.5-1.9 μm region is much steeper compared to polymer-coated fibers.

1 Introduction

Nowadays metallized silica fibers are widely used types of specialty optical fibers. A number of sensors using for laser radiation power measurement [1] or for optical intensity profile measurement [2] were developed based on metal-coated optical waveguides. Several unique high-power fiber laser sources [3, 4] were also constructed with the use of active optical fibers with the metal coating. High thermal stability of such fibers allowed to obtain kW-level fiber lasers without external active cooling. Despite its advantages, metallized fibers have significantly higher optical losses compared to optical fibers with polymer coatings. In addition, optical losses strongly depend on the radiation wavelength and fiber geometry. However, there is lack of information in literature concerning the spectral behavior of optical losses of metal-coated fibers with different waveguide geometry. Obviously, for different applications of fibers with metal coatings different parameters such as fiber geometry or source wavelength should be addressed. Therefore, accurate determination of optical loss at different wavelengths is required in order to optimize various devices based on metallized optical fibers.

2 Measurements of optical attenuation

A novel approach for the determination of optical loss coefficients of metallized fibers is based on the measurement of the change of temperature-dependent electrical resistance of metal coating conditioned by optical losses. During its propagation through the metal-coated waveguide, some part of laser radiation is scattered and eventually absorbed in the intermediate carbon layer or in the metal coating resulting in its heating. The change of metal coating resistance is measured using high precision ohmmeter. A simplified scheme of experimental setup is shown in Fig. 1. We have investigated several types of copper-coated silica fibers, further referred to as Type-A (thick multimode fiber), Type-B (thin multimode fiber), Type-C (thin single-mode fiber). Main characteristics of these fibers are listed in Table 1. Lengths of the used pieces of metallized fibers were around 0.3 m. In order to study spectral dependence of optical attenuation we have used five laser sources operating at different wavelengths (see Table 2).

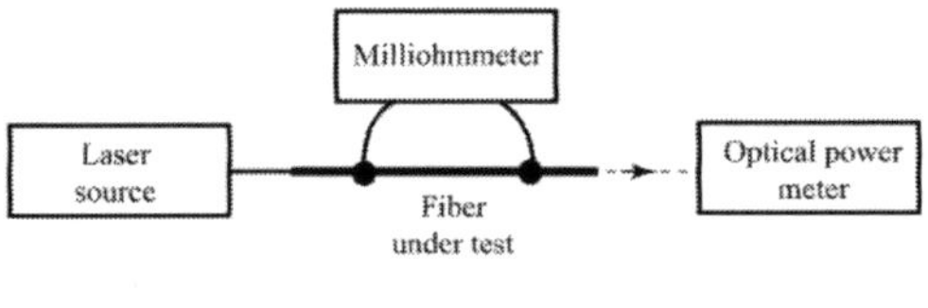

Fig. 1. A simplified scheme of the experimental setup.

Table 1. The main parameters of the studied optical fibers with a copper coating.

Notation of optical fibers	Waveguiding core diameter (μm)	Silica cladding diameter (μm)	Copper cladding diameter (μm)
Type-A	600	660	780
Type-B	113	125	160
Type-C	9	125	164

* Corresponding author: cherpak.ps@phystech.edu

Table 2. Main parameters of the employed laser radiation sources.

Laser type	Central wavelength (nm)	Maximum output power, W
Blue semiconductor laser diode	442	2.5
Second harmonic generation of Raman fiber laser	589	10
Infrared semiconductor laser diode	962	30
Yb^{3+} doped fiber laser	1070	70
Yb^{3+}/Er^{3+} doped fiber laser	1550	30
Tm^{3+} doped fiber laser	1910	40

Using the obtained experimental data together with the solution of the heat conduction problem, taking into account Newton-Richman law of convective cooling, the optical loss coefficients for the copper-coated fibers were calculated. Spectral dependence of optical losses for studied fibers is shown in Fig. 2. Dashed curves show the empirical approximation of the optical loss dependencies on the wavelength. It can be observed, that the lowest optical losses were obtained for the Type-C fiber. This can be attributed to the small diameter of the single-mode (SM) fiber core. Since diffused copper atoms tend to concentrate near the silica-metal interface, the radiation propagating in the core of SM fiber experiences lower scattering on copper atoms compared to multimode (MM) fibers. Type-A fiber demonstrated lower attenuation coefficients compared to Type-B ones. This can be attributed to the lower ratio of the copper coating thickness to the optical fiber diameter of Type-A thick multimode (MM) fiber. It is known [5], that smaller coating thickness corresponds to lower mechanical stress caused by copper layer upon silica and, therefore, to lower microbending losses. Regardless of fiber types the highest losses were observed at 1940 nm (up to 630 dB/km) and the lowest at 1070 nm (down to 1,2 dB/km). The increase of optical losses at longer wavelengths can be attributed to the growth of microbending losses [6] and absorption on molecular hydrogen bands [7]. Radiation at shorter wavelengths experiences Rayleigh scattering proportional to $1/\lambda^4$. Overall, it should be noted that optical loss in metal-coated fibers in the short-wave region is of the same order as in polymer-coated fibers, but the increase of radiation losses in 1.5-1.9 μm region is much steeper.

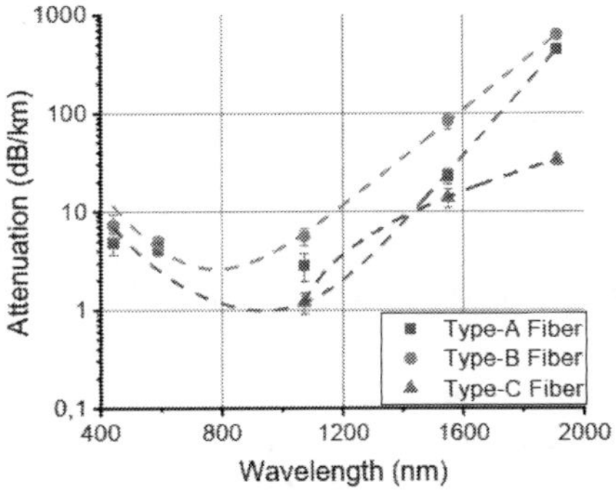

Fig. 2. Spectral dependence of optical loss for the three fibers under test (semi-logarithmic scale).

3 Conclusion

In this paper, optical attenuation coefficients of SM and MM copper-coated optical fibers at different wavelengths were determined by measuring the change of electrical resistance of the metal coating. The used method proved to be highly-sensitive and perfectly suited for measuring low optical attenuation in short pieces of metallized fiber. Investigation of spectral dependence of optical losses will help to determine optimal parameters such as fiber geometry or optical radiation wavelength for developing fiber sensors or powerful laser sources employing metal-coated fibers.

We thank the staff of the laboratory №226 of FIRE RAS for the provision of copper-coated fibers for the research.

The work was carried out within the framework of the state task.

References

1. I. Khramov, R. Shaidullin, O. Ryabushkin, Opt. Eng. **58**, 7 (2019)
2. N. Vanyushkin, N. Tereshchenko, A. Kostrov, P. Cherpak, D. Mukhankov, R. Shaidullin, O. Ryabushkin, *Seventh European Workshop on Optical Fibre Sensors*. **11199**, 116 (2019)
3. J.M.O. Daniel, N. Simakov, A. Hemming, W.A. Clarkson, J. Haub, Opt. Express. **24**, 18592 (2016)
4. C.X. Yu, O. Shatrovoy, T.Y. Fan, T.F. Taunay, Opt. Lett. **41**, 5202 (2016)
5. J.M.O. Daniel, N. Simakov, A. Hemming, W.A. Clarkson, J. Haub, *Fiber Lasers XIV: Technology and Systems*. **10083**, 100831Q (2017)
6. S. Hornung, N. J. Doran, R. Allen, Opt. Quant. Electron. **14**, 359 (1982).
7. S.L. Semjonov, V.A. Bogatyrev, A.A. Malinin, *2nd Workshop on Specialty Optical Fibers and Their Applications*. **7839**, 783912 (2010)

Pulse nonlinear optical switching in plasmonic structures

José R. Salgueiro[1,3,*] and *Albert* Ferrando[2,**]

[1]Departamento de Física Aplicada, Universidade de Vigo, E. E. Aeronáutica e do Ezpazo, As Lagoas, s/n, 32004 Ourense, Spain.
[2]Departament d'Òptica i Optometria i Ciències dela Visió, Universitat de València, Dr. Moliner, 50, Burjassot, 46100 Valencia, Spain.

Abstract. We study switching operation in a plasmonic coupler using fs-pulses. Simulations using the finite-difference time-domain method (FDTD) are carried out showing how the output changes as the pulse energy increases raising from zero to a maximum. Both cases of neglecting and realistic losses are considered in order to compare. The work is intended to explore the use of pulses for all-optical signal processing in a potentially interesting system for integrated photonics at the nanometric scale.

1 Introduction

Plasmonics has recently attracted a great interest because of the potential applications in nanophotonics, dramatically increasing the encapsulation possibilities of conventional integrated devices. In fact, integrated elements containing metals allow the light confinement to dimensions that are orders of magnitude lower than wavelength. This property together with the possibility of using nonlinear materials increases functionality allowing the design of integrated optical devices to carry out all-optical processing[1] in a nanometric scale. As nonlinear effects require high power density to trigger the suitable response it is quite common the use of pulsed sources where optical energy can be concentrated in a short time (of the order of femtoseconds). This makes necessary a specific modeling for this kind of systems as they behave generally in quite a different way respect to the case of using continuous beams since temporal and dispersive effects come into play. In the past some studies with pulses in plasmonic systems were carried out, focused on surface plasmon polariton (SPP) dynamics[2, 3] and slot waveguides[4, 5].

In this work we demonstrate the possibility of accomplishing switching in plasmonic directional couplers using short pulses, based on the total energy of the input pulse. A similar system was already studied for a CW regime[6], using in that case optical power to trigger the change of state. In the following, details of the study as it concerns modeling and results are presented.

2 System modeling

A system of two slot waveguides separated by a short distance was modeled using the FDTD method[7]. Both cores are dielectric presenting a nonlinear third-order Kerr response while claddings are supposed to be linear and metallic. Although a simple instantaneous-response model was used for the nonlinear cores, the method leaves the door open to a future inclusion of more elaborated or specific models, dependent on the pulse duration as for instance the third order Raman dispersion. Anyway, the pulsed character of the modeled beams is important as metals are strongly dispersive and have to be described by a model that properly takes into account such dispersion. In order to compare realistic results including losses to the case for which loss is absent, we modeled the metal

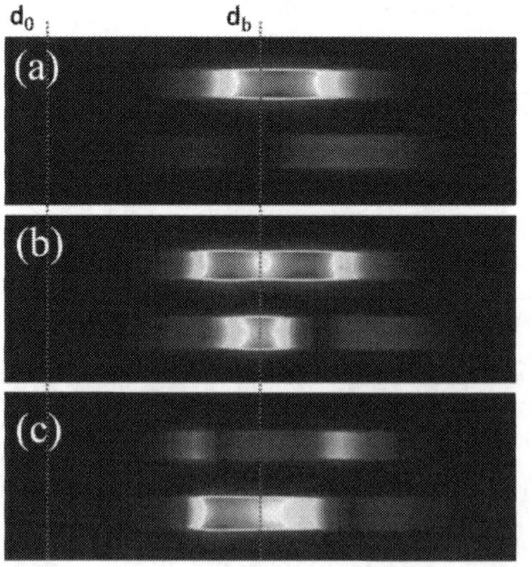

Figure 1. Three simulations for three input pulses of different energy launched into the low waveguide of the coupler and for the same propagation time. Input powers are (normalized units) $W_{in} = 0.43$ (A), $W_{in} = 3.45$ (B) and $W_{in} = 6.03$ (C). Vertical lines mark the waveguide entrance at position $d_0 = 0$ and the output located at a beat length d_b.

*e-mail: jrs@uvigo.es
**e-mail: Albert.Ferrando@uv.es

using a classical Drude method which characterizes the medium by a plasma frequency ω_p and an electronic collision frequency Γ (which is identically zero when losses are neglected)[8]. Anyway, it would be possible the use of more realistic models at optical frequencies as the modified Debye method[9], though in this case it would not be possible to isolate losses from the set of parameters and would had to be necessarily taken into account.

The calculation domain was surrounded by perfect matching layer zones (PML) to get rid of all outgoing radiation. Also, a source based on the total-field/scattered-field technique was implemented to generate a pulse at the entrance of one of the waveguides presenting a Gaussian transversal shape and also Gaussian temporal shape. The TM-polarization was used considering two electric components (E_x and E_z) in the simulation plane and a magnetic component (H_y) orthogonal to the plane.

3 Results

For the simulations we chose silver as the metal for the cladding, characterized by the parameters $\omega_p = 1.15\times10^{16}$ Hz and $\Gamma = 9\times10^{13}$ Hz at a wavelength of $\lambda_0 = 800\,\text{nm}$. Spatial coordinates were normalized by the wavenumber $k_0 = 2\pi/\lambda_0$. In such normalized units we took the width of both cores as well as the separation between them as 0.5 (corresponding to about 65 nm). Pulses where generated with the same width and a time duration of 115 fs and launched into one of the cores (first core). Power at a particular position (propagation length) was measured integrating the longitudinal (z) component of the Poynting vector ($\vec{S} = \vec{E}\times\vec{H}$) spatially on the transversal direction and also over time to obtain the total energy of the propagating pulse:

$$W(z) = \int_t\int_x [\vec{S}(x,z)\cdot\hat{z}]\,\mathrm{d}x\,\mathrm{d}t. \qquad (1)$$

The first simulation was in a linear regime, intended to determine the beat length for the coupler, i. e. the propagation distance necessary to produce the maximum transfer

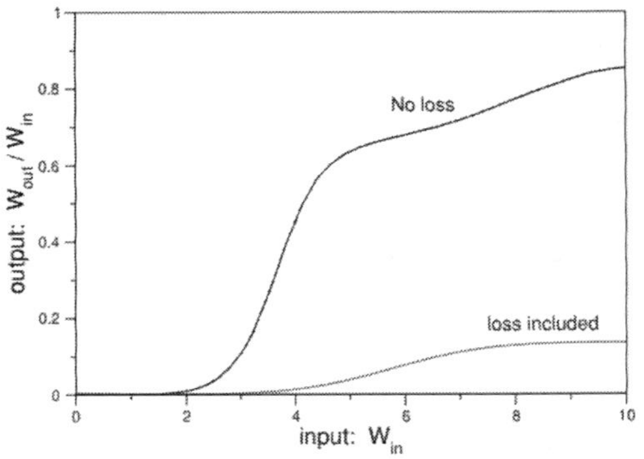

Figure 2. Switching curve: ratio of output and input pulse energy plotted against input pulse energy for both cases of neglected loss and realistic loss.

of the pulse energy to the second core. This was measured as $d_b \approx 45$ (5.73 μm). Next, a set of simulations where carried out, launching similar pulses with increasing total energy but same duration. In figure 1 it is plotted the power flux (time average of the pointing vector) for three different simulations correspondent to a linear (a), intermediate (b) and full nonlinear (c) regimes. In (a) power is completely transferred to the second core after a distance d_b producing an almost null signal at the first core output. In (b) only a fraction of energy is transferred while in (c) most of the energy remains in the first core, producing a maximum signal at the first core output.

The switching operation is illustrated in figure 2 for both cases of losses neglected and losses included. The plot represents normalized output energy against input energy. For low input signal the output is almost null as all the energy transfers to the second core. At some point there is a fast rise of the output signal (switching) to maximum values. When losses are taken into account switching is still possible but there are two important differences. On one hand output energy is quite lower (as part was converted into heat by Joule effect). On the other hand there is a loss of efficiency as the slope of the curve decreases in such a way that the rising signal extends over a double range of energy (horizontal axis) respect to the case of neglecting losses. This behavior was already observed for CW beams[6] and even solutions were proposed to improve performance[10].

The authors wish to thank the government of Spain and the local government of Galicia for the economical support under contracts FIS2017-83762-P and ED431B2018/57-GPC respectively. Financial support from Spanish MINECO through Project No. TEC2017-86102-C2-1-R is gratefully acknowledged.

References

[1] N.C. Panoiu, W.E.I. Sha, D.Y. Lei, G.C. Li, Journal of Optics **20**, 083001 (2018)
[2] Z.L. Sámson, P. Horak, K.F. MacDonald, N.I. Zheludev, Opt. Lett. **36**, 250 (2011)
[3] A. Baron, T.B. Hoang, C. Fang, M.H. Mikkelsen, D.R. Smith, Phys. Rev. B **91**, 195412 (2015)
[4] A. Pusch, I.V. Shadrivov, O. Hess, Y.S. Kivshar, Opt. Express **21**, 1121 (2013)
[5] O. Lysenko, M. Bache, N. Olivier, A.V. Zayats, A. Lavrinenko, ACS Photonics **3**, 2324 (2020)
[6] J.R. Salgueiro, Y.S. Kivshar, Appl. Phys. Lett. **97**, 081106 (2010)
[7] A. Taflove, *Computational electrodynamics, the finite difference time domain* (Artech House Publishers, Norwood, Massachussets, 2005)
[8] S.A. Cummer, IEEE Trans. Antennas Propagat. **45**, 392 (1997)
[9] H. Gai, J. Wang, Q. Tian, Appl. Opt. **46**, 2229 (2007)
[10] J.R. Salgueiro, Y. Kivshar, Opt. Expr. **20**, 9403 (2012)

3 Dimensional Motion of Photon and Its Energy

Gh. Saleh[1,*], *Reza* Alizadeh[1], *Ehsan* Dalili[1] , and *Amir* Noorbakhsh[1]

[1]Saleh Research Centre, Netherlands

Abstract. Since the electron is the photon-generator and the electron is rotating around itself and around the nucleus, the emitted photon from the electron should have the effect of these rotating motions. So, the photon has a three-dimensional motion, including a transition movement and a rotary motion. And it traverses in a helical trajectory. By using this definition, we have proved wave-particle duality at the same time and introduce a new equation for the photon motion and its energy.

1 Introduction

For a long time, the smallest object in the universe, which now is called an atom, was unknown. Until recently (a hundred years ago), no structure was imagined for that, but its structure is clear now as it could be fission. Like the atom, the photon has a world that is like an old atom for us, unfamiliar and anonymous but at the same time obvious. The photon is a child of the 20th century. In 1926, the photon so named by the physical chemist Gilbert Nor-ton Lewis. It is the 'particle of light' -or 'light quantum' - first hypothesized by Albert Einstein in 1905, and then used by him to explain the photoelectric effect.[1-3]
Saleh Research Group aims that with a new and fresh approach and with very strong reasons, reveal the unexplored depths and strengthen its connection with the beautiful photon. In this article, the behavior of photon is studied in three dimensions (instead of 2) [4]. And we present a new definition for photon's motion and new formula for its energy.

2 The photon genesis

We return to the beginning of creation of a photon from the heart of an atom, from its mother's place, which is the electron. At $T = 0 + \varepsilon$, as the electron is excited, or in other words, absorbed the given energy, it loses the extra energy or come back to the ground state by emitting an object called "photon"[5]. An important point to note is that electron rotates around itself and around its nucleus at the very high speed. So when a photon is released, it will continue to move at the speed of light and because of the very fast rotational motion of electron, it has a rotational motion too. In an obvious matter, an emitted photon from an electron has a linear and rotational motion. In fact, our rotating photon moves in at least three dimensions.

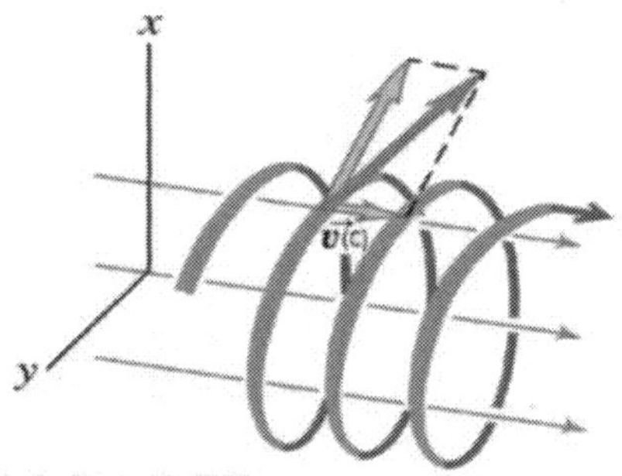

Fig. 1. 3 dimensional motion of photon [4]

3 The energy of Rotational photon

It should be noted that when a photon is emitted from an electron, it is like a spaceship, like a bullet, that has been given primary energy and thrown into space. Inevitably, when you observe it from the front, you will find it in one form and when you observe it from the side view, you will find it in the other form.[4] The front view of a photon is the image of a helical motion toward an observer, a circle with a radius of R_{mom}, and the side view of the moving photon is a sinusoidal motion.
On the other hand the equation of motion and energy of photon must show this behavior too. In fact, the emitted photon has an initial energy from which must be subtracted the Rotational kinetic energy to have the sum of the Translational kinetic energy:

Translational kinetic energy =

Initial energy − Rotational kinetic energy

Saleh Theory knows the Translational kinetic energy as the same energy that Thomas Planck had presented with his famous formula and measured empirically in the laboratory:

* Corresponding author: postmaster@saleh-theory.com

$$E = h\vartheta \quad (1)$$

The initial energy is:

$$E_i = 1/2\ mv^2 \quad (2)$$

The rotational energy is:

$$E_r = 1/2\ mr^2\ \omega^2 \quad (3)$$

So:

$$E = h\vartheta = 1/2\ mv^2 - 1/2\ mr^2\ \omega^2 \quad (4)$$

This new formula has three different parts which one part is the energy of wave-like part, and the second one is particle part, and the third part refers to the rotating energy that the color variety of the photon is related to the third part.

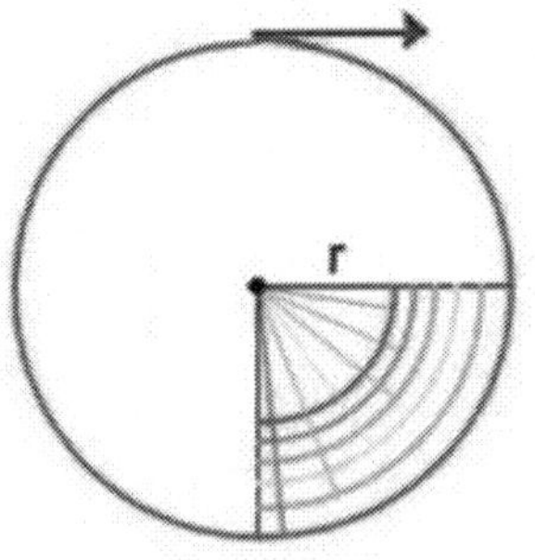

Fig. 2. Front view of photon motion of photon [4]

4 Results and Discussion

Today's all theories explain photon's motion separately. Sometimes they knew photon like wave (double-slit experiment, etc.) and sometimes like particles (photoelectric effect, etc.). But when we focused on the behavior of the generator of photon, electron, we have found that photon should have 3 dimensional motion not 2 [4]. So we presented this new motion for photon and regarding that the energy formula for that.

References

Here are some examples:

1. N. Mukunda, "The story of the photon." Resonance 5.3 (2000).
2. Pais, Abraham. Subtle is the Lord: The Science and the Life of Albert Einstein: The Science and the Life of Albert Einstein. Oxford University Press, USA, 1982.
3. J.J. Stachel,. "Einstein's miraculous year: five papers that changed the face of physics." (2005).
4. Gh. Saleh, et al. "A New Explanation for the Color Variety of Photons." MATEC Web of Conferences. Vol. 186. EDP Sciences,(2018).
5. Gh. Saleh, et al. "The Superstring Theory and the Shape of Protons and Electrons." MATTER: International Journal of Science and Technology 4.2 (2018).

EPJ Web of Conferences **238**, 12001 (2020)
EOSAM 2020
https://doi.org/10.1051/epjconf/202023812001

Transient Dynamical-Thermal-Optical System Modeling and Simulation

Luzia Hahn[1,*] and *Peter* Eberhard[1,**]

[1]Institute of Engineering and Computational Mechanics, Pfaffenwaldring 9, 70569 Stuttgart, Germany

Abstract. In this work, methods and procedures are investigated for the holistic simulation of the dynamical-thermal behavior of high-performance optics like lithography objectives. Flexible multibody systems in combination with model order reduction methods, finite element thermal analysis and optical system analyses are used for transient simulations of the dynamical-thermal behavior of optical systems at low computational cost.

1 Introduction

High resolution optical systems like astronomical telescopes or lithographic objectives consist of precise mirrors or lenses. The optical elements are mounted with high accuracy and they are very sensitive with respect to mechanical vibrations and temperature changes.

In the waver production, scanning systems use lithography objectives to project structures for a certain time onto the wafers. During the operation time, even small mechanical disturbances of the mirrors or lenses or temperature changes in the optical elements can lead to aberrated images. Sources of the vibrations can be, e.g., from influences of the wafer motion system which lead to excitations at the objective frame. The ever-growing demand of high throughput in the scanning systems demands a rise in the power of the used lasers [1]. This leads to heating of the lenses or mirrors and can therefore result in thermal aberrations. The dynamical-thermal behavior of optical systems can be described by a combination of rigid body motion, small deformations, the related stresses, material property changes, and the temperature distribution.

In order to investigate the overall dynamical-thermal-optical behavior, multi-disciplinary methods and software tools with suitable interfaces have to be used and combined. Here, a method is proposed for simulating the performance of dynamical-thermal-optical systems. Thereby, time-dependent deformations due to mechanical excitations and changes in the refraction index due to temperature gradients within the optical elements are considered. An exemplary system simulation is presented for a single lens system, shown in Fig. 1.

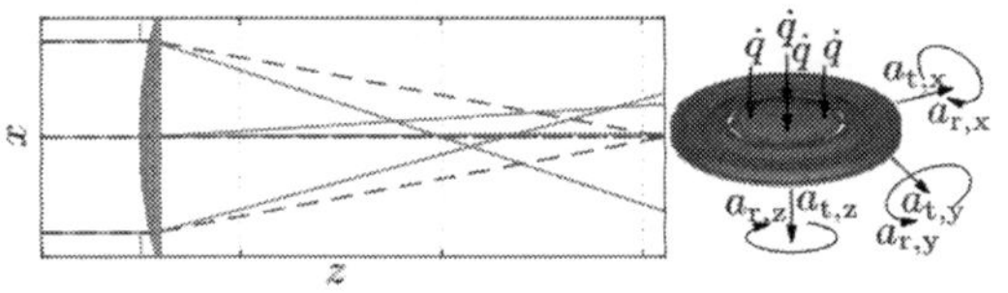

Figure 1. Single Lens optical system with all excitations (right) and the raypaths of the ideal system (blue, dashed) and the dynamical and thermal excited system (red).

2 Modeling

Dynamical-thermal-optical system analysis includes the modeling and simulation of the movement of a mechanical system, the simulation of the heat conduction in the elements and the consideration of the associated effects in the optical system simulation. Mechanical excitations lead to deformations and thus to mechanical stresses in the optical elements. Thermal changes lead to material property changes, thermal stresses, and temperature-induced displacements. A generalized workflow using cooperating software packages according to Fig. 2 can be used for holistic dynamical-thermal-optical system simulations. Therby, the mechanical deformations and stresses are considered. The thermal effects are only considered with the thermally induced refraction index changes during the optical simulation.

The first step in the mechanical modeling is, to model each optical sensitive component separately as an elastic body using a finite element (FE) software. The extraction of the node coordinates and the system matrices leads to the equations of motion which enable the calculation of the motion of each elastic body as a function of time. Additionally, the mode shapes are extracted and can be used in a model order reduction in order to reduce the number of degrees of freedom and to speed up the mechanical simulations. Afterwards, the elastic bodies and the remaining system components, modeled as rigid bodies, are assembled in a flexible multibody system [2]. The last step of the mechanical modeling is a numerical time integration which results in the movements and deformations of the dynamical-optical system dependent on time. The dynamical-optical system modeling is as also shown in [3] and [4].

*e-mail: luzia.hahn@itm.uni-stuttgart.de
**e-mail: peter.eberhard@itm.uni-stuttgart.de

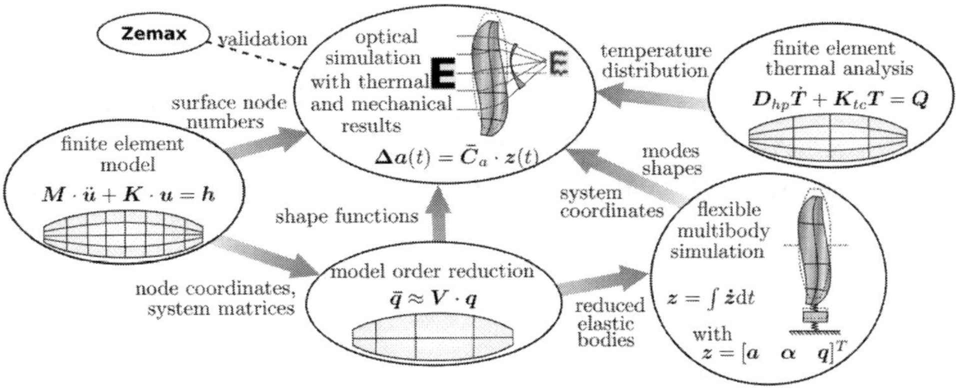

Figure 2. Workflow of the described dynamical-optical system simulation.

In order to consider the thermal effects in the system, an FE thermal analysis is applied to the thermal sensitive components. This results in the time dependent temperatures at each node of the thermal FE mesh.

After the mechanical and the thermal time simulations, sequential ray tracing which considers the thermal and the mechanical results is performed. Therefore, the rigid body motion of each optical element is transformed from the absolute description to a relative description and during the ray tracing the mode shapes, representing the deformations of the reduced bodies, are approximated accurately by polynomial functions, e.g., using a set of Zernike polynomials. Besides that, the displacements of the mechanical mesh nodes are interpolated to the thermal mesh using inverse distance weighting. The changes in the refraction index due to the temperature changes are calculated [5] and either described analytically, e.g. with cylindrical polynomials and evaluated or interpolated in the ray tracing. The optical lenses which are subject to mechanical stresses, a non uniform temperature distribution, or both have a gradient in the refraction index (GRIN). For the ray tracing through these lenses, a fourth order Runge-Kutta method is used. The rays are traces stepwise through the GRIN lenses and in each spatial step the changes in the refraction index due to stress and temperature are calculated and considered. For the optical simulation considering the mechanical and thermal effects, the program OM-Sim is developed. Besides the ray tracing, it can provide the associated wavefront aberrations and an image simulation.

3 Numerical Example and Conclusion

As an example of the performance of an integrated dynamical-thermal-optical system simulation a single lens system is presented in Fig. 1. The dashed lines show exemplary four ray paths for the ideal system. The lens is dynamically excited with a predefined movement in all space directions and an analytical, radial temperature gradient around the points of impact of the rays to the lens is applied. Stresses are not considered in this example. The resulting ray paths for the mechanical and thermal excited system are presented in Fig. 1 in red. The mechanical deformation and the thermal gradient was scaled during the ray tracing such that the aberration effects get visible. Due to the radial temperature gradient, the focal length decreases as it can be seen in pink in Fig. 3. Due to the mechanical deformations the focus point is shifted away from the optical axis, as shown in Fig. 3 with the green dashed rays. So, the dynamical and thermal effects influence the imaging performance of the optical system and both should be considered in a holistic optical system simulation.

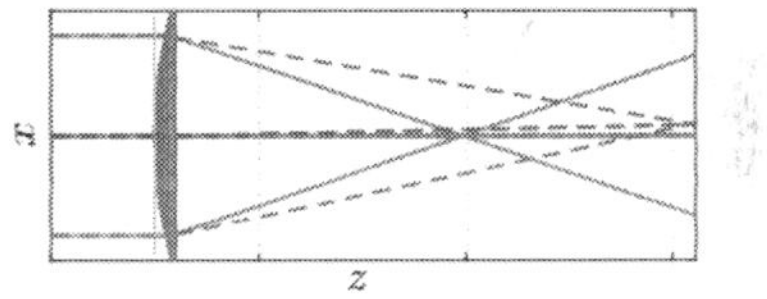

Figure 3. Raypaths of the thermal excited system (pink) and the dynamical excited system (green, dashed).

References

[1] L. Zhao, L. Dong, X. Yu, P. Li, P. Qiao,Active lens for thermal aberration compensation in lithography lens, Appl. Opt. **57(29)**, 8654-8663 (2018)

[2] R. Schwertassek, O. Wallrapp, *Dynamik flexibler Mehrkörpersysteme. (in German)* (Vieweg, Braunschweig, 1999)

[3] N. Wengert, *Gekoppelte dynamisch-optische Simulation von Hochleistungsobjektiven. (in German)* (Dissertation, Schriften aus dem Institut für Technische und Numerische Mechanik der Universität Stuttgart **40**, Shaker Verlag, Aachen, 2015)

[4] J. Störkle, *Dynamic simulation and control of optical systems* (Dissertation, Schriften aus dem Institut für Technische und Numerische Mechanik der Universität Stuttgart **58**, Shaker Verlag, Aachen 2018)

[5] N.N., Temperature Coefficient of the Refractive Index, SCHOTT Technical Information **TIE-19**, 2 (2016).

https://doi.org/10.1051/epjconf/202023812002

Integration of Arrayed Waveguide Grating on Strip-Loaded Slot Waveguide for Sensing Applications

Janvit Tippinit*, *Markku* Kuittinen, and *Matthieu* Roussey

Institute of Photonics, University of Eastern Finland, P.O. Box 111, 80101 Joensuu, Finland

Abstract. We demonstrate the integration of arrayed waveguide grating (AWG) de-multiplexer with a low loss and low index contrast platform, i.e., a striped-loaded slot waveguide (SLWG). The SLSW-AWG is designed to operate around the wavelength λ = 1550 nm with a four-output channel. The device experiences excellent output parameters: channel width $\delta\lambda$ = 1.39 nm at -3dB, channel spacing $\Delta\lambda$ = 2.43 nm and extinction ratio ER = 9.87 dB, useful for sensing applications.

1 Introduction

Arrayed waveguide grating (AWG) is one of the most important integrated photonic devices in telecom and sensing systems. This device can be used as wavelength multiplexers and de-multiplexers, combining several wavelengths or separating them to a single wavelength. The major problem of AWGs, however, is optical insertion losses, originating mainly from the field mismatch due to the high refractive index difference between the waveguide structures and the surrounding. This is enhanced by the sidewall roughness, as it can be seen with, e.g., silicon (Si) and silicon nitride (Si_3N_4) on silicon dioxide (SiO_2) waveguides [1-2]. To overcome the loss due to the sidewall roughness, a very near-infrared (~900 nm) Si_3N_4 AWG has been demonstrated using a double etch step, which applied at the access waveguides near star couplers and the bends in the arrayed waveguides, reducing the index contrast [3].

We propose another approach, which uses a novel waveguide structure, so-called striped-loaded slot waveguide (SLSW). This waveguide is constituted of a horizontal slot waveguide, a layer of low index material (n_L) between two layers of higher index material (n_H), and a polymer strip on the top, inducing the refractive index difference locally in the slot waveguide. The strong confinement of the electric field of the mode occurs in the n_L region, located underneath the etched polymer strip. This makes the waveguide perfect for integrated devices since the structures can be patterned directly on the polymer and not on the guiding layer. There is a very limited interaction of the optical field and the sidewall roughness of the strip. SLSW has been, moreover, integrated with some nano- and microstructures such as Y-junctions, ring resonators, and photonic-crystal-based structures [4-5]. In this paper, we design an AWG on SLSW structure at the center wavelength of 1550 nm for sensing application, and to the best of our knowledge, no SLSW-AWG has been reported yet.

2 Methodology and results

The optical loss is one of the main concerns in photonic devices, and SLSW has a great advantage in reducing the loss due to the sidewall roughness. For operation λ = 1550 nm, the waveguide consist of two TiO_2 slab waveguides (n_{TiO2} = 2.27) with the height of h_1 = 200 nm and h_3 = 180 nm, and a SiO_2 slab waveguide (n_{SiO2} = 1.44) with the height h_2 = 80 nm in between. On the top of these layers, polymer ($n_{Polymer}$ = 1.60) is deposited with a height of 250 nm, and the polymer strip waveguide is patterned with a width of 1.2 μm. This structure provides the strong optical confinement for TM polarization in the SiO_2 layer [6], as shown in figure 1.

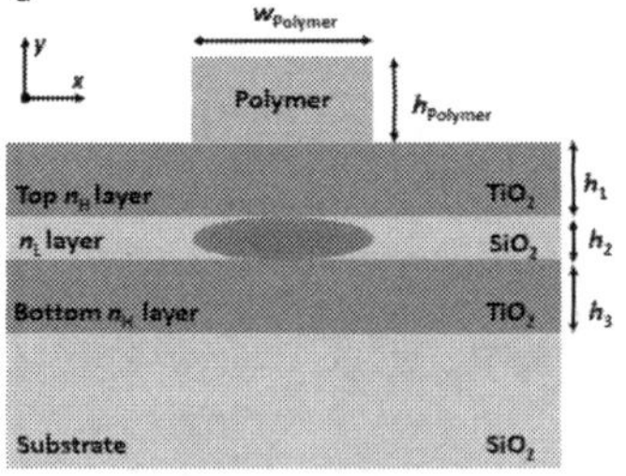

Fig. 1. Schematic of cross-section of a SLSW.

Table 1. Design parameters.

Parameter	Notation	Value
Number of array waveguides	N	22
Diffraction order	m	102
Output waveguide separation	D	6.48μm
Array waveguide separation	Λ	6.48μm
Focal length	f	352.62μm
Length increment	ΔL	77.114 μm

* Corresponding author: janvit@uef.fi

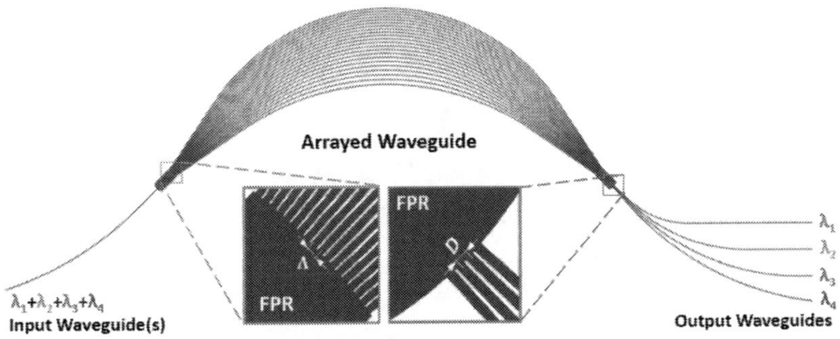

Fig. 2. Schematic of the SLSW-based AWG device.

We employ a beam propagation method (BPM) to design the SLSW-AWG device. The polymer strip waveguide on the top of the sandwich layer of TiO_2 and SiO_2 causes slightly higher in the effective index of the core ($n_{eff,core}$= 2.03) than of the cladding ($n_{eff,clad}$= 1.99). We design the device with the parameters listed in table 1, and the entire geometry is depicted in Fig. 2.

Figure 3 shows the normalized power spectrum at the output of each channel. We obtain a channel width $\delta\lambda$ = 1.39 nm at -3dB and a channel spacing $\Delta\lambda$ = 2.43 nm. The extinction ratio, ER = 9.87 dB, is sufficient to obtain well-separated channels. The proposed device is capable to perform a four-wavelength de-multiplexer with a wide channel spacing as well as with the excellent extinction ratio. This makes the device suitable for sensing applications.

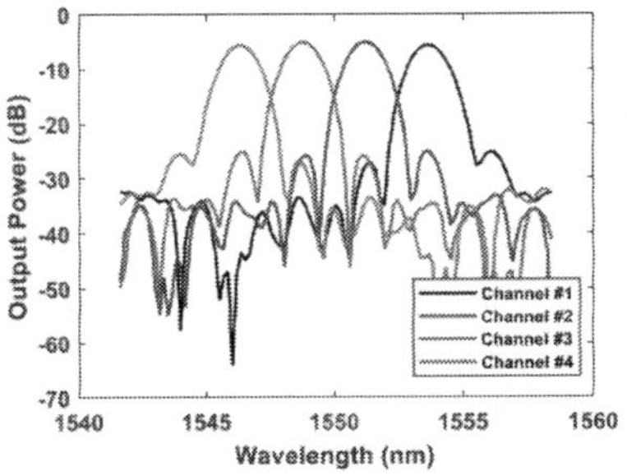

Fig. 3. Spectrum at each output of the device.

4 Conclusions

We have succeeded to design an SLSW-based AWG device, which operates at the center wavelength of 1550 nm. The device is capable of de-multiplexing four wavelengths with a separation of 2.43 nm. We obtain the channel width at 3dB of 1.39 nm and well-separated output spectra with the extinction ratio of 9.87 dB.

References

1. D. Dai, Z. Wang, J.F. Bauters, M.-C. Tien, M. J. R. Heck, D.J. Blumenthal, J.E. Bowers, Opt. Express **19**, 14130-14136 (2011)
2. P. Dumon, W. Bogaerts, D.V. Thourhout, D. Taillaert, R. Baets, J. Wouters, S. Beckx, and P. Jaenen, Opt. Express **14**, 664-669 (2006)
3. D. Martens et al., in IEEE Photonics Technology Letters, **27**(2), 137-140 (2015)
4. M. Roussey, L. Ahmadi, S. Pélisset, M. Häyrinen, A. Bera, V. Kontturi, J. Laukkanen, I. Vartiainen, S. Honkanen, M. Kuittinen, Opt. Lett. **42**, 211-214 (2017)
5. S. Pélisset, M. Roussey, J. Eur. Opt. Soc. - Rapid Publ. **15**, 6 (2019).
6. S. Pélisset, J. Laukkanen, M. Kuittinen, et al, J. Eur. Opt. Soc. - Rapid Publ. **13**, 37 (2017)

Cavity-enhanced detection of transient absorption signals

Fernanda C. Rodrigues-Machado[1*], *Pauline* Pestre[1], *Liam* Scanlon[1], *Shirin A.* Enger[2], *Jack C.* Sankey[1], and *Lilian I.* Childress[1**]

[1]Department of Physics, McGill University, Montreal, Canada
[2]Medical Physics Unit, McGill University, Montreal, Canada

Abstract. We present a simple, high-duty-cycle, cavity-enhanced optical absorption measurement technique based on delay-limited Pound-Drever-Hall (PDH) sideband locking. The chosen circuit naturally provides real-time readout of the amplitude quadrature of the PDH error signal, which can be mapped onto the cavity's internal loss rate while using the phase quadrature to lock sideband frequency to the cavity mode. Our proof-of-concept device comprises a 5-cm-long Fabry-Perot cavity with a 450 kHz bandwidth (finesse 6800, 350 ns power ringdown), and a feedback bandwidth of several MHz, limited primarily by the group delay of our electronics. This technique could readily be applied to other optical resonators such as fiber cavities, with potential applications in radiation dosimetry.

1 Introduction

We work on the development of a time resolved cavity-enhanced absorption sensing technique based on Pound-Drever-Hall (PDH) sideband locking.

PDH locking [1] relies on phase modulation of a laser beam and posterior demodulation of the power reflected from an optical cavity by mixing with a local oscillator at the same modulation frequency. With the correct choice of phase for the local oscillator (corresponding to the phase quadrature of the power), the resulting mixed-down voltage provides a derivative-like signal that can be used to lock the cavity on resonance to an incoming laser beam. Here, we lock the cavity to one of the sidebands generated by phase modulation, which allows a high locking bandwidth (~MHz) [2].

In addition to creating the locking error signal (phase quadrature), we use a $\pi/2$-shifted signal and a second mixer to get the PDH amplitude quadrature. The amplitude quadrature has a Lorentzian shape about the detuning of the sideband (at modulation frequency Ω), and provides a non-zero signal when the sideband is locked to the cavity. This signal carries real-time information from the cavity and can be mapped onto the cavity's internal losses.

2 Experimental apparatus

We use a 5-cm-long, single-sided Fabry-Perot cavity with finesse 6800 to demonstrate the technique (Fig. 1.a). A 1550 nm laser beam is phase-modulated by an electro-optical modulator (EOM), which receives an RF signal ($\Omega \sim$ 1GHz) from a voltage-controlled oscillator (VCO). The laser beam reaches the cavity and the reflected light is collected by a fast photodiode. Half the signal coming from the photodiode is used for PDH feedback (V_y) [1, 2], while the other half is phase-shifted by $\pi/2$ and mixed with the VCO signal to generate the amplitude quadrature (V_x). When the cavity is locked to one of the sidebands, the V_x signal slowly fluctuates about a DC offset, and varies linearly with cavity lifetime.

With the high bandwidth provided by sideband PDH locking, we can keep the cavity locked on resonance for several hours, allowing the detection of transient signals.

3 Results

We measure the sensitivity noise floor with power spectral densities and translate the amplitude quadrature voltage signal to absorption (units of cm^{-1}) by calibrating the apparatus with swept ring-down measurements [3].

The expected shot-noise limit in units of $cm^{-1}/Hz^{1/2}$ is calculated by

$$\langle S_a \rangle = a\sqrt{2\, e\, \eta\, P_{PD}\, Z^2\, \xi} \quad (1)$$

where a is a conversion factor determined by calibration (units of cm^{-1}/V), e is the electron charge, η is the photodiode responsivity (in A/W), P_{PD} is the total power reaching the photodiode, Z is the diode transimpedance (in V/A) and ξ accounts for the unitless circuit components power gains.

Figure 1.b shows our sensitivity floor. The locked (blue) curve is the signal we use to monitor cavity absorption. Our measurements are currently limited by laser classical noise, achieving sensitivities on the order of $10^{-10}cm^{-1}/Hz^{1/2}$, within a factor of <30 from shot-noise limit (red curve, confirmed by results from equation 1 - green curve). These results are comparable to the most sensitive absorption measurements reported to date [4, 5].

*e-mail: fernanda.rodriguesmachado@mcgill.ca
**e-mail: lilian.childress@mcgill.ca

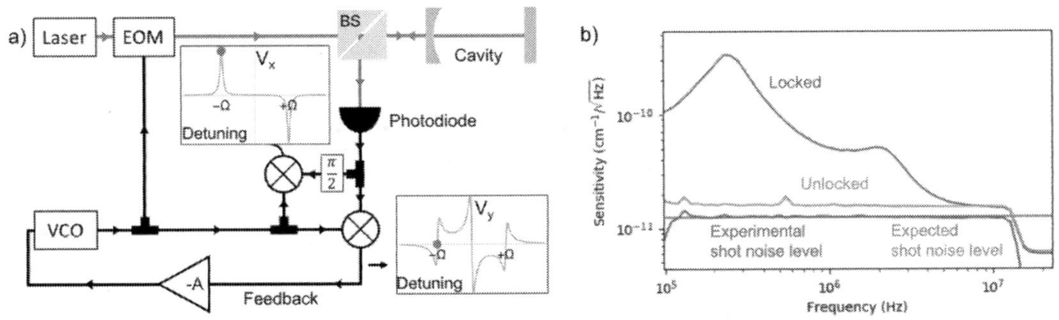

Figure 1. a) Experimental setup (simplified) for sideband PDH locking and intra-cavity absorption sensing. EOM: electro-optical modulator. BS: non-polarizing beamsplitter. VCO: voltage-controlled oscillator. -A: feedback controller. $\pi/2$ indicates a phase shift. Insets show PDH amplitude quadrature (V_x) and phase quadrature (V_y). Red dots indicate the lock point on the lower sideband. b) Setup noise power spectral densities converted to units of absorption. Blue: noise floor of the system while the laser is locked to the cavity. Orange: unlocked laser noise (quantum limited) and electronics noise. Red: laser shot noise level (electronics subtracted). Green: shot noise limit calculated with equation 1.

4 Outlook

We are currently looking into phase-detection techniques (e.g. with amplitude modulation of the laser beam) to acquire direct absorption, calibration-free measurements. We are also exploring noise cancellation techniques to subtract the classical noise seen in our locked measurements (Fig. 1.b - blue) and achieve shot-noise-limited measurements, even at lower frequencies.

Our described technique can be applied to other optical resonators such as fiber cavities, which would provide micrometer spatial resolution and GHz bandwidth. In the near future, we aim to employ such fiber cavities in studies of radiation dosimetry. The continuous monitoring of the cavity potentially allows the detection of solvated electrons, short lived entities generated in water when radiation is incident. This highly sensitive technique would allow the real-time detection of low radiation doses (~ 1mGy) against an absorptive background.

References

[1] E.D. Black, American journal of physics **69**, 79 (2001)
[2] C. Reinhardt, T. Müller, J.C. Sankey, Optics express **25**, 1582 (2017)
[3] Y. He, B. Orr, Applied Physics B **75**, 267 (2002)
[4] J. Ye, L.S. Ma, J.L. Hall, JOSA B **15**, 6 (1998)
[5] L.S. Ma, J. Ye, P. Dubé, J.L. Hall, JOSA B **16**, 2255 (1999)

Crystallization of HfO2 thin films and their influence on laser induced damage

Zoltán Balogh-Michels[1,*], *Igor* Stevanovic[1], *Ruggero* Frison[2], *Andreas* Bächli[1], *Daniel* Schachtler[1], *Thomas* Gischkat[1], *Antonia* Neels[2], *Alexander* Stuck[1], and *Roelene* Botha[1]

[1]RhySearch, Optical Coatings Department, 9471 Buchs SG, Switzerland
[2]Empa, Center for X-ray Analytics, 8600 Dübendorf ZH, Switzerland

Abstract. We present our investigation on the crystallization of IBS HfO2 on (0001) SiO2. The crystallization was studied by in-situ XRD. The activation energy was 2.6±0.5 eV. The growth follows a two-dimensional mode. LIDT measurements (5000-on-1) with 10 ns pulses at 355 nm on 3QWT HfO_2 layers shows that the crystallization leads to increase of the laser irradiation resistance. The 0%-LIDT of the as coated sample was 3.1 J/cm2 and increased to 3.7 J/cm2 after 5h @ 500°C.

1 Introduction

Thin-film interference coatings, like antireflection or high reflectance coatings as well as wavelength and polarization selective coatings, are key elements of optical components as they allow to tune or even radically alter their optical properties [1,2]. Current high-end optical coatings in the VIS/NIR range are mainly based on ion beam sputtered (IBS) multilayers. Thermal annealing of a single layer or multilayer stacks is a well-established way to improve the properties like optical losses or laser irradiation resistance [3]. Above a temperature T_{cryst} the films starts to crystallize which can be wanted or unwanted. Thus, the understanding of the crystallization process, the dependence on time and temperature and the activation energy are essential for defined controlling the thermal treatment and tuning the properties.
In the case of laser applications, HfO_2 is standard high index material. Investigations show that high temperature annealing leads to the crystallization and the laser-induced damage behaviour improves [4]. In this work, investigate the crystallization of thin HfO_2 films by means of in-situ x-ray diffractometry and its effect on the laser-induced damage threshold.

2 Experimental setup and results

2.1. Crystallization

For the investigation of the crystallization, we deposited thin HfO_2-single layers with thickness of 10, 15, 20 and, 50 nm. The coating was done by means of reactive ion beam sputtering from a metallic Hf-target (Plasmaterials with 3N purity) using a Veeco Spector 1.5 Dual Ion Beam Sputter (DIBS) tool placed in a clean-room environment (ISO 5-7). In order to achieve complete oxidization the process O_2 flux during deposition was 31 sccm. The temperature during film growth was 100°C. As substrates, we used epi-polished 10x10 mm² SiO_2 single crystals with 0001 orientation from CrysTec. Prior coating, the samples were in-situ treated with energetic O_2 ions (1 keV) for 10 minutes using the second plasma source of the Veeco DIBS machine.
The in-situ XRD thermal annealing experiments were carried out using a Bruker D8 Discovery DaVinci diffractometer equipped with an Anton Paar DHS1100 high temperature stage. Since polycrystalline films were expected we selected a grazing incidence geometry, with $\omega = 2°$ as incoming angle. The 2θ range was 28 - 35°, which contained the expected main peaks of both the cubic and monoclinic HfO_2 [5]. Isochronal annealing on 10 nm thick HfO_2-samples in steps of 50°C and a holding time of 30 min was carried out to gain an overall view of the transformation. Based on this information, we performed isothermal scans on the 15, 20, and 50 nm thick samples with temperatures between 550 and 650°C. The period between each single scan was 6 minutes. Figure 1 shows the results of the isothermal scans of the 50 nm thick sample at a temperature of 550°C. The fitted integral value of the peaks in Fig. 1a gives the crystalline fraction for each scan, which is depicted in Fig. 1b as a function of time. As can be seen in the small inset in a logarithmic scale the measurement points can be clearly described by a linear fit function. From this the crystallization times t_c and dimensionality of the growth n' were calculated [6]. The calculated n' was between 1.85 and 2.15 for the different thicknesses and temperatures indicating 2D-growth. The corresponding Arrhenius-plots of crystallization times gives an activation energy for crystallization of 2.6±0.5 eV.

* Corresponding author: zoltan.balogh@rhysearch.ch

 https://doi.org/10.1051/epjconf/202023812004

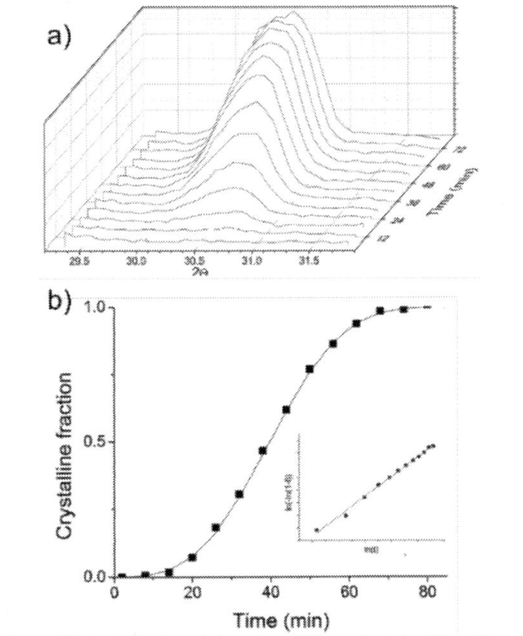

Fig. 1. Time evolution of the cubic HfO_2 111 peak of the 50 nm thick layer during the 550°C annealing a) as well as the relating fitted peak integral as a function of annealing time b). The inset shows the curve in logarithmic scale.

2.2 Laser damage testing

Applying the same coating procedure, we deposited 127 nm (3QWT @ 355 nm) thick HfO_2-single layers for the investigation of the LIDT. Therefore we used fused silica samples (P4-polished, rms roughness of about 3 Å, scratch-dig 20-10) provided by WZW Optics AG. Prior to the deposition, the samples were precision cleaned using a wet bench placed in the cleanroom. The detailed cleaning process is described in Ref. [7]. After deposition, the samples were annealed at 500°C for 1 h and 5 h, respectively. Temperature and time were chosen regarding the results of the Arrhenius-plots. Laser-damage testing was performed at 355 nm using the LIDT setup at RhySearch [7]. Pulse duration was 10 ns and the spot size (1/e) at the surface was 230 µm.

Fig. 2 shows the damage onsets as well as the 0%-LIDT values after 100 and after 5000 pulses for the different samples. The damage onset and the 0%-LIDT in case of the as-deposited (x-ray amorphous) layer are about 3.0 and 4.1 J/cm^2, respectively. The partially crystallized sample after 1h @ 500°C shows a decreased fluence for the onset and the 0%-LIDT. In contrast, the full-crystallized sample that was thermally treated for 5h @ 500°C shows a slightly higher damage onset and 0%-LIDT compared to the as-deposited sample.

3 Conclusions

Controlled nucleation and crystallization is a key for the defined and controlled improvement of optical properties of thin films like laser damage resistance. Therefore, the knowledge of growth mode, nucleation time as well as activation energy is essential. In our investigation on crystallization of HfO_2 thin films we found that an activation energy of 2.6±0.5 eV.

In accordance with the investigations of other groups [3.4] the full crystallized layer shows higher laser-induced damage resistance compared to the amorphous sample. In contrast, it seems that the intermedia state of the partially crystallized film has a significantly lower damage resistance compared to the amorphous one.

The crystallization behaviour, as well as the dependence on the optical properties, as well as the laser-induced damage resistance, will be investigate in depth in the next years.

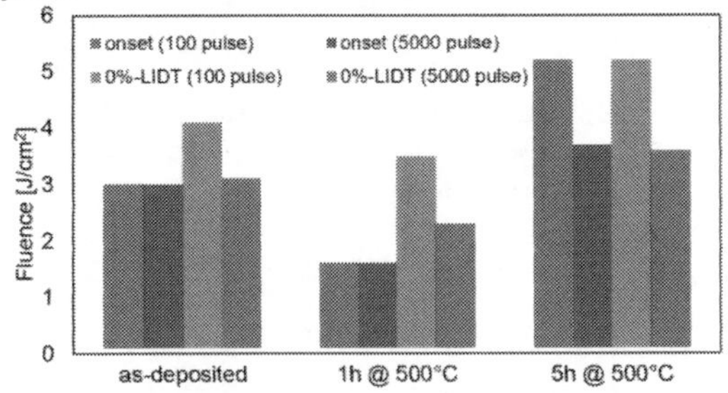

Fig. 2. Damage onsets as well as the 0%-LIDT values after 100 and after 5000 pulses for the different samples.

We would like to thank the canton of St-Gallen and the Principality of Liechtenstein for supporting RhySearch.

References

1. N. Kaiser, H.K. Pulker, Optical Interference Coatings (Springer Verlag, Berlin, 2003)
2. D. Ristau, *Laser-induced damage in optical materials*, (CRC Press, 2014)
3. C. Wang, Y. Jin, D. Zhang, J. Shao, Z. Fan, Opt. Laser Technol, **41**, 570 (2009)
4. G. Abromavicius, S. Kicas, and R. Buzelis, Opt. Mater., **95**, 109245 (2019)
5. Y. Xie, Z. Ma, Y. Su, Y. Liu, L. Liu, H. Zhao, J. Zhou, Z. Zhang, J. Li, and E. Xie, J. Mater Res., **26**, 50 (2011)
6. M. Avrami, J. Chem. Phys., **7**, 1103 (1939)
7. Th. Gischkat, D. Schachtler, Z. Balogh-Michels, R. Botha, A. Mocker, B. Eiermann, S. Günther, Proc. SPIE 11173, Laser-induced Damage in Optical Materials 2019, 1117317.

Correlation between polychromatic image quality metrics to predict visual acuity in pseudophakic patients

Jesús Armengol,[1,*] *Nuria* Garzón,[2] *Fidel* Vega,[1] *Irene* Altemir,[3] and *Maria S.* Millán.[1]

[1]Departament d'Òptica i Optometria. Universitat Politècnica de Catalunya. Barcelona Tech. 08222. Terrassa. Spain.
[2]Miranza IOA. 28003 Madrid, Spain and Optometry and Vision Department. UCM University. 28037 Madrid, Spain.
[3]Miguel Servet University Hospital. Ophthalmology Department. 50009 Zaragoza, Spain.

Abstract. The optical quality of three multifocal intraocular lenses (MIOLs) was measured in vitro using the area under the modulation transfer function (MTFa) and energy efficiency (EE). We evaluated those metrics through-focus (TF) by axially scanning the image space with three wavelengths (R, G, B). High linear correlation between MTFa and EE was found for all cases. Correlations between optical in vitro results (polychromatic TF-MTFa and TF-EE) and clinical visual acuity (VA) defocus curves for the three MIOLs allowed us to derive mathematical expressions. Both metrics proved to be equivalent to predict postoperative VA.

1 Introduction

Cataract surgery is evolving from a solution to visual impairment to a refractive solution. The use of multifocal intraocular lenses (MIOLs) is increasingly widespread to improve the refractive performance.

By analyzing the through focus optical quality of MIOLs, Marsack et al. [1] concluded that MTFa and EE -through the light-in-the-bucket (LIB) estimate- were among the six PSF-based metrics that accounted for over 70% of the variance in postoperative LogMar VA.

2 Method

To obtain in vitro the images produced by the MIOLs, we used the on-bench eye model described in detail in [2]. The system met the standards of the International Organization for Standardization, ISO 11979-2: 2014. We have studied three MIOLs: a FineVision PODF (Physiol), a Versario 3F (Bausch+Lomb), and a Tecnis Symfony (J&J Vision), all of them with a diffractive-based design and 20D base power. These lenses have an addition between + 3D and + 3.5D for the near focus and between + 1.5D and + 1.75D for the intermediate focus. We measured MTFa and EE to study the trough-focus MIOL performance. MTFa is the area under the MTF curve for the range of frequencies of interest in vision. We measured the MTF in two perpendicular directions by placing a cross-line object of 10 microns slit width. We integrated the average MTF up to the 50cyc/mm (≈15 cyc/degree). We imaged a pinhole and measured EE from the LIB estimation as reported in detail elsewhere [3]. Each MIOL image space was TF scanned sequentially with blue (B) 455nm, green (G) 530nm and red (R) 650 nm wavelengths.

The clinical data were obtained from 102 eyes of 52 patients recruited in two clinical trials carried out at two ophthalmology centers. Both studies were prospective, consecutive and non-randomized and followed the tenets of the declaration of Helsinki. The patients underwent bilateral and symmetrical cataract surgery followed by MIOL implantation into the capsular bag. Previously, they had been fully informed about the study and signed a consent form. The local ethics committee approved the corresponding trial. Monocular defocus VA curves from −5.00 D to + 3.00 D, with the patients wearing their best distance correction, were measured during the last postoperative follow up.

3 Results

Figure 1 shows the TF-MTFa (A) and TF-EE (B) curves for a MIOL (FineVision) with 3mm pupil. Both groups of curves appear to be very similar for each wavelength. The polychromatic TF curves (black line) are computed from linear combination of the RGB curves weighted by their photometric coefficients.

Figure 2 depicts, for every MIOL and wavelength, the (MTFa, EE)x pairs at each defocus position x(D). The figure reveals very good linear correlation between MTFa and EE ($R^2 > 0.96$).

Following the methodology described in closely related studies [4,5], Figure 3 shows in blue dots the clinical VA(x) outcomes versus a polychromatic (poly) function for every defocus or image vergence (x): (A) MTFa_poly(x) and (B) EE_poly(x). We fit the points of Figures 3 (A, B) to either exponential functions (green solid line) (Eqs. 1, 2).

* Corresponding author: jesus.armengol@upc.edu

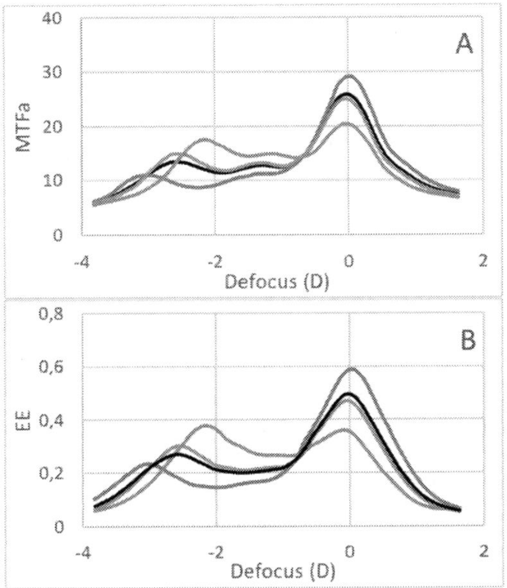

Fig. 1. Monochromatic R, G and B MTFa (A) and EE (B) versus defocus (at spectacle plane) obtained with 3mm pupil for the FineVision MIOL. Polychromatic functions in black.

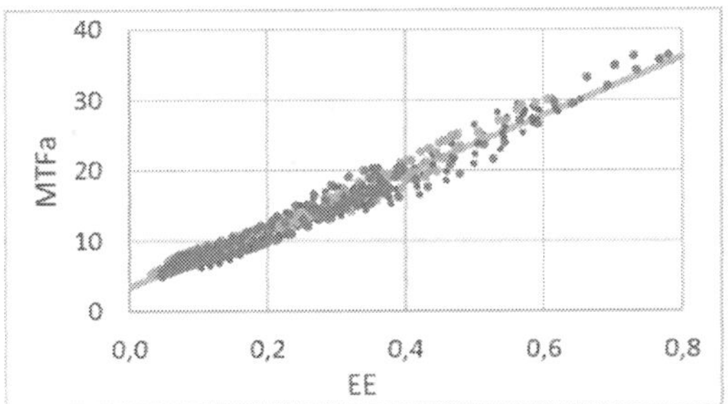

Fig. 1. MTFa vs. EE for the three MIOLs with 3mm pupil and three wavelengths (R, G, B dots). Best linear fit (yellow).

The mathematical expressions and their correlation coefficients (R^2) are

$$VA = 1.828 * e^{-0.230*MTFa_{poly}} + 0.014 \quad R^2 = 0.94 \quad (1)$$

$$VA = 0.707 * e^{-7.746*EE_{poly}} \quad R^2 = 0.90 \quad (2)$$

4 Discussion

MTFa and EE are objective metrics to assess in-vitro the optical image quality of MIOLs; the larger the value of any of these two metrics, the better the optical quality.

Our experimental results (Fig. 2) show a clear linear correlation between MTFa and EE values ($R^2 > 0.96$). Prior research [4-7] showed the potential of MTFa as a preclinical metric for predicting average postoperative VA at different defocus values. In this work we have proved the ability of the EE metric to predict clinical VA outcomes of patients implanted with three different designs of MIOLs.

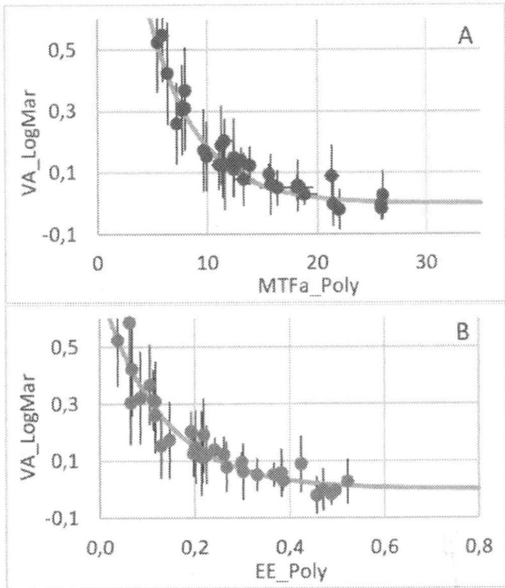

Fig. 3. Clinical Visual Acuity versus polychromatic (A) MTFa_Poly and (B) EE_Poly for 3mm pupil (blue dots). Exponential fits (green line).

The results show, in agreement with former works [4,5,7], non-linear relationships between VA and both, MTFa_poly and EE_poly. The mathematical functions that fit the data of Figure 3 with high correlation coefficients, $R^2 = 0.94$ for VA vs MTFa_poly (Eq. 1) and $R^2 = 0.90$ for VA vs EE_poly (Eq. 2), clear a path to predict average VA of pseudophakic patients from objective on-bench quality metrics. In the conditions of our study, TF-EE is a very useful optical metric that can replace MTFa as a preclinical quality parameter.

References

1. J. D. Marsack, L. N. Thibos, and R. A. Applegate. J. Vis. **4**(4), 322–328 (2004).
2. F. Vega, F. Alba-Bueno, and M. S. Millan. J. Eur. Opt. Soc. Rapid Publ. **9**, 14002 (2014).
3. M. S. Millán, F. Vega, and I. Ríos-López. Investig. Ophthalmol. Vis. Sci. **57**(4), 2021–2028 (2016).
4. A. Alarcon, C. Canovas, R. Rosen, H. Weeber, L. Tsai, K. Hileman, and P. Piers. Biomed. Opt. Express **7**(5), 1877 (2016).
5. F. Vega, M. S. Millán, N. Garzón, I. Altemir, F. Poyales, and J. M. Larrosa. Biomed. Opt. Express **9**(10), 4893 (2018).
6. J. Fernández, M. Rodríguez-Vallejo, J. Martínez, N. Burguera, and D. P. Piñero. J. Refract. Surg. **35**(12), 789–795 (2019).
7. G. Cardona, F. Vega, M. A. Gil, C. Varón, J. A. Buil, and M. S. Millán. Eur. J. Ophthalmol. **28**(1), 36–41 (2018).

Funding

Spanish Ministerio de Economía y Competitividad y fondos FEDER (DPI2016-76019-R).

FDTD Modeling of double ultrashort pulse propagation in nonlinear absorbing media.

J.D. Pisonero, *O.* Varela, E. García, *I.* Hernández, *J.* Ajates, *J.L.* Sagredo, *M.* Olivar, *C.* Albarran, L. Roso and *C.* Méndez

Centro de Láseres pulsados (CLPU), Parque Científico. C/Adaja, 8. 37185 Villamayor, Salamanca, Spain.

Abstract. An approach based on the finite-difference time-domain (FDTD) method is developed for simulating the dynamics of two ultrashort laser pulses inside a saturable absorbing media. This work discusses the results obtained using this numerical model for the prediction of the nonlinear absorbing media behaviour as well as how it affects the final double pulse combination. These results can be used to improve contrast cleaning conditions for high power laser chains and for synchronization studies, this last application was checked in the VEGA facility lab as a code validation.

1 Introduction

Since the FDTD algorithm was first developed by Yee in 1966 [1], this method has been successfully applied to a large number of electromagnetic problems. In particular, an explicit scheme of this method was applied to the analysis of wave propagation in saturable absorbers and gain lasers [2]. These works showed that for small signal intensities and slowly varying pulses, the method was able to give same results as frequency-domain methods using the linear susceptibility function.

On the other hand, saturable absorbers are elements that are now included in front-end parts of high-intensity laser chains in order to enhance the temporal contrast of these laser pulses for laser-matter interaction experiments [3,4].

Modelling short laser pulse propagation in these kind of materials is therefore an interesting topic to improve ultrashort pulse characteristics and management. In particular, we present results corresponding to the behaviour of two laser pulses interactions in these kind of media.

1.1 Simulations

1.1.1 FDTD formulation

A detailed description of the used method can be found in [2], where atomic rate equations describing the time evolution of atomic energy level populations under the influence of applied signals is included. In this work, the formulation used all over that reference is followed.

1.1.2 FDTD results.

The code was used to analyze the behaviour of two equal ultrashort laser pulses travelling inside the saturable absorber RG850. The data used to model this media are the same ones used at reference [2]. The steps values of the mesh were chosen to be Δt=0.018fs and Δz=0.002µm. The resonance frequency was taken to be ω_a=2.38fs^{-1} and the total decay rate $\Delta\omega_a$=0.025fs^{-1}. Finally, the lifetime of the atoms was chosen to be τ_{21}=10ns so the value of the factor κ was κ=7.89x10^{-8}.

Different simulations were done in order to see how one pulse affects the other depending on their initial time when both have an electric field value close to but lower than the one that is needed to saturate the absorber (E_0=8V/µm). Main results are shown in Figure 1.

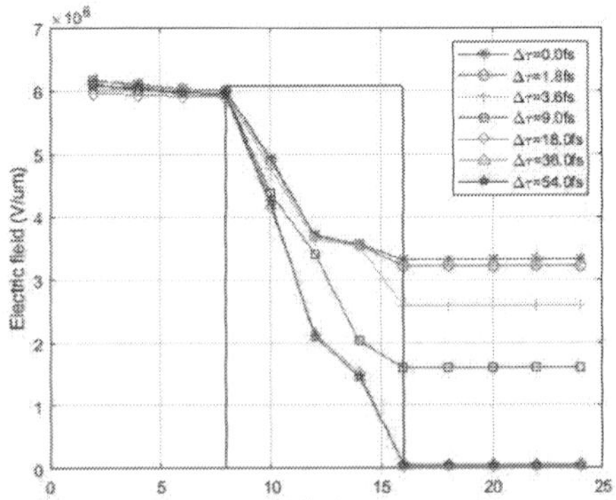

Fig. 1. Electric field during the propagation of two pulses through a SA at different delay times between pulses. SA is shown as a solid square in the graphic.

In figure 1, both pulses arrive to the SA front face after 7 microns propagation. SA width is chosen to be 8 microns. We can see that when delay between two pulses is larger than 18fs, them both are completely absorbed in the media but the closer they become, the larger the pulse that is leaving the absorber results.

* Corresponding author: jdpisonero@clpu.es

2 Experimental results

VEGA system is a petawatt (PW) laser user facility belonging to the Centro de Láseres Pulsados in Salamanca that allows developing experiments for exploring the physics of intense lasers interactions with matter. The Ti:Sapphire Chirped Pulse Amplification (CPA) based laser chain has three common front-end outputs (VEGA 1-2-3) at 1 PW, 150 TW and 20 TW respectively.

Preliminary experimental validation of the FDTD code was done in the VEGA facility lab with a leak of VEGA 2 beamline (5 J, 30 fs, 10 Hz). The leak from VEGA2 is introduced in a Michelson interferometer that divides the initial pulse in two beams with the same energy. One of the mirror is connected to a linear stage to change the delay between both beams. After that both pulses pass through the SA and their spectrum is measured to analyze the absorption in the media.

2.1 Results

To analyze the results, spectrums were taken and integrated in a spectrometer at different delays between pulses. Those data are shown in Figure 2 in which there are plotted the measures of both pulses taken independently and both pulses passing together through the SA. In this figure data show that the addition of pulses spectrum taken individually is not matching with the spectrum taken with both pulses arriving at the same time, showing a quite larger transmission in this last case.

This results match with the theoretical simulations shown in Figure 3. In this figure, theoretical and experimental measurements are shown at different delays. The experimental data were taken using a separation of 10 microns between them, what is equivalent to 33fs of delay. On the other hand, simulations were made with more resolution specially when the pulses are very close to each other.

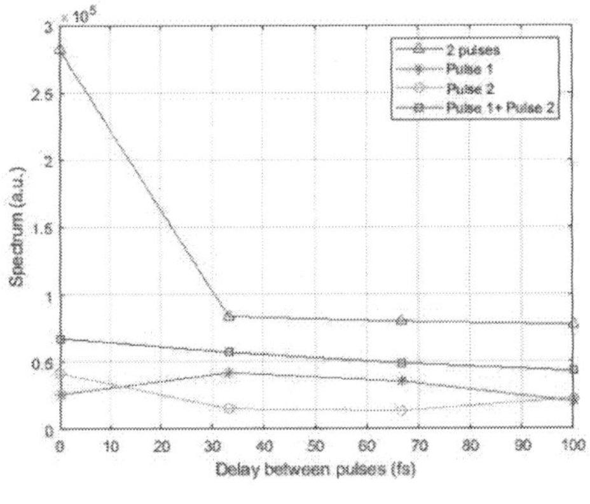

Fig. 2. Measurements of integrated spectrums with SA.

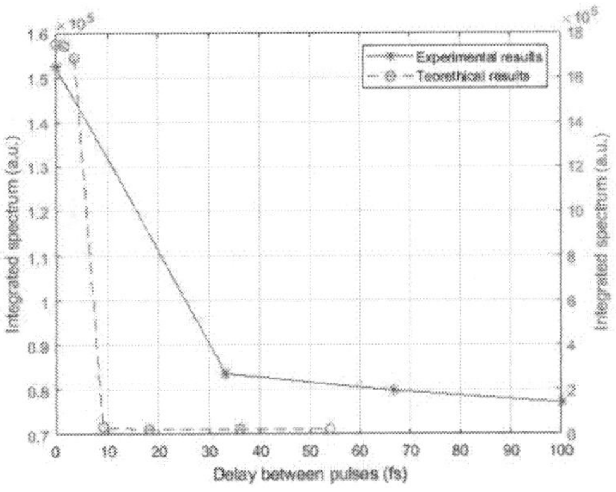

Fig. 3. Comparative between theoretical and experimental results for integrated output spectrum at different delay conditions.

The data shown in those figures prove the difference between the cases in which the pulses pass through the SA with 0 delay and those with more delay, so the less delay the pulses have, the more spectrum is measured at the output. This result is the same result obtained with the FDTD simulation.

3 Conclusions

In this work we have tested an implementation of an FDTD code. Simulations of pulses propagation in saturable absorbers have been developed and tested in the lab. This tool will allow future studies for short pulses synchronization with fs resolution as well as high power laser pulses contrast designs.

References

1. K.S. Yee, IEEE Trans. Antenn. Propag. **14**(3), 302-307 (1966)
2. Amit S. Nagra and Robert A. York, IEEE Trans. Antenn. Propag. **46**, 3 (1998)
3. Jae Hee Sung, Seong Ku Lee, Tae Moon Jeong, Chang Hee Nam, Appl. Phys. B (2014) 116:287-292
4. S. Fourmaux et alt. Optics Express **19**, 9, 8486 (2001)

Optimizing the CMOS Sensor-Mode for Extreme Linear Dynamic Range MEMS-based CAOS Smart Camera Imaging

Nabeel A. Riza [1,*], *Nazim* Ashraf [1] and *Mohsin* Mazhar [1]

[1]School of Engineering, University College Cork, Cork, Ireland

Abstract. Experimentally highlighted are the limitations & optimizations of the CMOS sensor-mode engaged for the CAOS smart camera. Demonstrated is an optimized method for finding the Camera Response Function (CRF) using a calibrated target.

1 Introduction

Recently introduced and demonstrated is the CAOS smart camera that not only operates over a very wide optical band (e.g., 350 nm to 2700 nm), it most importantly provides an extreme linear Dynamic Range (DR) reaching 177 dB [1]. This Pixels of Interest (POI) CAOS smart camera operates by first engaging a multi-pixel sensor (e.g., CMOS/CCD/FPA sensor) for initial lower limited linear DR scene capture that in-turn provides intelligence for the CAOS-mode POI scene capture to produce a net improved extreme linear DR image via image fusion methods.

Linear High Dynamic Range (HDR) imaging is critical for deciphering low contrast targets within HDR scenes, including enhancing multispectral (e.g., color) imaging. Furthermore, accurate and reliable image capture is important for mission critical applications where incorrect image data can lead to catastrophic system failure. The CAOS smart camera inherently meets these stringent requirements and forms a robust machine vision system that can impact several scientific, industrial, and consumer applications. Given many visible light applications, the lower cost CMOS HDR sensor is chosen to design and test the CAOS smart camera. Initial experiments using the 87 dB HDR sCMOS Quantalux sensor camera model Si 2100-M from Thorlabs has shown the limitations of CMOS sensor cameras versus the CAOS camera when subjected to imaging tests using calibrated HDR test targets with low contrast stepped zones [2]. Current CMOS sensors with specified HDR performance inherently produce a non-linear CRF over the HDR, in particular in the low light and bright light regions. Experiments indeed show such a non-linear CRF prevents a CMOS camera from registering differential output signals with adequate SNR for capture of low contrast targets within a HDR [2]. A classic approach used to enhance digital sensor camera DR is to use multiple un-calibrated synthesized or real images of different exposure times engaged with an algorithm to first generate the CRF and then deploy this full range CRF with real capture multi-exposure images to generate an HDR image that was otherwise not possible using a single exposure limited DR image. It has recently been shown in Ref.3 that by using known calibrated HDR targets to experimentally determine the best estimate of the true CRF, some leading prior-art multi-exposure algorithms produce non-robust HDR images [3]. Part of this non-robustness comes from the inaccuracy of the deployed CRF as well the fact that these algorithms engage the full CRF range data that fails to maintain linearity required for ideal multi-exposure HDR image generation operations. Ref.3 also introduced a calibration empowered HDR image generation method that restricts data processing operations to an optimal smaller DR linear CRF range and engages unweighted multi-exposure processing to produce improved robustness image data. This paper highlights further limitations of the deployed CMOS sensor technology and demonstrates additional optimizations for experimental CRF generation, leading to improved accuracy and reliability of the HDR image when compared to HDR image generation using prior-art algorithms.

2 CMOS Camera CRF Generation and Multi-Exposure HDR Imaging Experiments

Under bright light 60 KLux illumination conditions using the Image Engineering (Germany) LG3 light box illuminating a 16 patch HDR calibration target with ND filter optimized attenuations, the Thorlabs 16-bit v_p output signal CMOS camera (using a C-mount GMZ18108 lens) for the brightest (or 0 dB) patch starts to trigger pixel saturations (i.e., $v_p=2^{16}-1=65535$) for 0.93% of pixels in the patch zone for the used exposure time while giving a computed mean v_p called $v_p(avg)$ of 39218. Ideally, this $v_p(avg)$ should be 65535 as the CMOS sensor received a uniform patch illumination, so all pixels in patch should have produced a $v_p=65535$. To counter this experimental anomaly, the raw acquired CRF v_p data in Ref.3 was

* Corresponding author: n.riza@ucc.ie

scaled by a factor of 1.67, as the v_p(avg)=39218 at the hint of pixel saturation is expected to be at the saturation value of 65535. In reality, such scaling ignores the highly nonlinear behaviour of the deployed CMOS sensor for bright light conditions where the onset of saturated pixel signal outputs has started. Because of this saturated pixel triggering behaviour observed for the deployed CMOS sensor that maybe common place for CMOS & other digital sensor technologies, this paper proposes an improved CRF generation technique based on pixel v_p output signal histogram analysis and calibration target knowledge. Specifically, demonstrated is CRF generation using a 90 dB 16-patch HDR target similar to ref.3 target design with patch attenuation values in dB of 0, 8, 14, 18, 28, 32, 36, 40, 44, 52, 58, 64, 70, 78, 84, 90. The camera exposure time is set to 3.703 ms as it gives the 0 dB brightest patch v_p(avg) of 64537 which is very near the 16-bit v_p limit. Next pixel v_p value histogram analysis for 16 patches is done that shows that the 28 dB patch is the first patch of the brighter patches to show no saturated pixels, i.e., no v_p values of 65535.

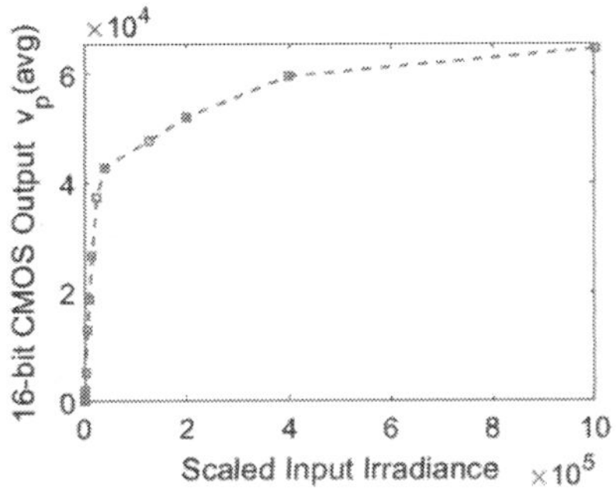

Fig. 1. Experimental CRF.

Let us assume that the scaled input irradiance value I_s for the zero attenuation or 0 dB patch is $I_s=10^6$ for the Fig.1 CRF plot construction. Note that any sufficiently large arbitrary I_s should be used to allow accurate slope computations between adjacent data points in plot. The no light patch black region, i.e., for scaled $I_s = 0$ measures a v_p(avg) of 191. Slope analysis for all data points shows that near continuous linear CRF behaviour is between the 32 dB patch with a v_p(avg)=37486 and 84 dB patch with a v_p(avg)=389. Note that for any v_p> 64537, an $I_s=10^6$ is assigned. Similarly, for any v_p<191, an I_s=0 is assigned. Thus per our ref.3 method, the experimental camera linear dynamic range or LDR_E = 20log(37486/389) = 39.66 dB implying that for a near 80 dB HDR target recovery, only 2 multi-exposure images are needed with a 100X factor between exposure times. To test the Experimental (E) CRF and the ref.3 method, a 78 dB 16 patch HDR target is observed using specific exposure times of 59 µs and 5.9 ms that are picked based on the v_p(avg) max/min selection method in ref.3. Table 1 E1 row shows the measured patch DR values. Unlike our ref.3 method, prior-art methods do not specify the number and increasing time factors for the multi-exposure images, so picked are 4 images starting with the 59 µs shortest exposure image and 3 other images with 1.5, 15, and 100 increase in exposure time. The Fig.1 CRF is used for all the ref.3 stated 4 prior-art methods, although these methods propose to use un-calibrated target CRFs which would further enhance uncertainty of the HDR image. Table 1 E2 row shows data generated using the 4 images with our ref.3 method.

Design (dB)	E1	E2	MP	DM	MN	RBS
0	0	0	0	0	0	0
8	10.1	9.7	11.7	12.7	3.7	14.8
14	15.7	15.7	17.5	17.5	10.9	20.5
20	21.9	22.1	24.9	24.2	26.4	28.9
26	26.5	26.1	29	29	32.1	31.2
32	35.2	34.9	37.5	35.2	37.2	35.6
36	40.2	39.4	42.2	38.9	41.5	39.9
40	41.8	42.3	45.9	43.1	46.5	44.8
44	43.7	45.1	49.7	46.8	50.3	48.7
50	53.5	53.6	61.6	57.6	60.6	58.6
56	56.5	56.8	65.1	61.3	64	61.5
60	61.1	62.5	69.9	67.3	70.2	66.2
64	64.1	66	74.2	71.3	74.3	69.1
68	68	68.1	80.8	75.7	77.4	73
74	73.4	73.5	86.5	82.8	84	78.5
78	77.4	77.4	89.3	89.9	90.8	82.4

Table 1. Part 1: CMOS Camera Imaged E1 & E2 Data using Proposed Method versus 4 Prior-Arts.

3. Conclusion

The bright light CRF region of the tested s-CMOS camera shows a highly non-linear response and non-uniform unreliable individual sensor pixel 16-bit voltage output signals despite uniformity of the illumination. Compared to leading prior-art multi-exposure HDR methods, the proposed calibration image deployed CRF generation technique using pixel outputs histogram analysis along with the ref.3 proposed restricted CRF linear zone selection and extraction multi-exposure method experimentally shows higher robustness to non-linearities in CMOS sensor and multi-exposure image fusion techniques

References

1. N. A. Riza and M. A. Mazhar, "177 dB Linear Dynamic Range Pixels of Interest DSLR CAOS Camera," IEEE Photonics Journal (open access), Volume 11, Number 2, April 2, 2019.
2. N. A. Riza and M. A. Mazhar, "Robust Low Contrast Image Recovery over 90 dB Scene Linear Dynamic Range using CAOS Camera," IS & T Electronic Imaging Conf. Proc. Paper, Jan.26, 2020
3. N. A. Riza and N. Ashraf, "Calibration Empowered Minimalistic Multi-Exposure Image Processing Technique for Camera Linear Dynamic Range Extension," IS & T Elect. Imaging Conf. Proc., 2020

MEMS-based CAOS Smart Camera 177 dB Linear Extreme Dynamic Range Imaging Tests with Calibrated and Controlled Incoherent White Light and Laser Light Targets

Nabeel A. Riza [1,*], and *Mohsin* Mazhar [1]

[1]School of Engineering, University College Cork, Cork, Ireland

Abstract. Highlighted are the linear extreme dynamic range calibrated target imaging tests of the CAOS smart camera showing a 177 dB linear dynamic range operation with an optical attenuation controlled bright laser target

1 Introduction

Linear Extreme Dynamic Range (XDR) imaging is required in applications where objects of interest occur in a scene with extreme brightness to weak light zones with low contrast targets of interest embedded within these extreme zones. Recently introduced and demonstrated is the Coded Access Optical Sensor (CAOS) smart camera that has the fundamentals to provide an linear XDR along with a wide spectral response, e.g., 350 nm to 2700 nm [1-2]. This smart XDR camera with a linear camera response function is designed on the principles of the XDR RF multi-access wireless mobile network and operates with a variety of image access modes to suit the pixel extraction environment [3]. This camera engages the classic operations CMOS/CCD/FPA sensor to provide a "DC" mode of operation initial image and CAOS time-frequency modulated imaging modes representing the "AC" modes of operations for XDR capture. The CAOS smart camera is an AC+DC camera optimized based on imaging conditions and the application. In fact, such optimizations can take place using a variety of computational processing methods, including machine learning. This paper highlights updated processed Dynamic Range (DR) data from the experimental set-up in ref.4 where the CAOS smart camera underwent linear XDR imaging tests with calibrated and controlled incoherent white light and laser light targets [4].

2 CAOS Smart Camera Linear Extreme Dynamic Range Imaging Tests

The details of the CAOS smart camera test conditions are described in ref.4. Highlighted here are the updated processed linear DR imaging results presented as tables and a plot. The LG3 Image Engineering (Germany) light-box is used as the white light illumination source. A custom 36 patches target with different designed patch DR values giving an overall 160 dB test scene XDR is used to check the designed (D) linear DR values in dB versus the measured (M) numbers in dB observed using the CAOS smart camera. Table 1 shows the DR for the CMOS-mode of the camera showing accurate recovery till a DR of 32 dB. Table 2 shows the improved linear imaging DR via the CDMA-mode of the camera showing accurate recovery till a DR of 45.8 dB.

Table 1. CMOS-Mode White Light Imaging DR Data.

Patch #	D (dB)	M (dB)	SNR	Patch #	D (dB)	M (dB)	SNR
1	0	-	545	11	45.8	42.28	4.2
2	4.6	4.69	317	12	50.2	45.27	3
3	9.2	9.19	189	13	54.8	48.26	2.1
4	13.8	12.99	122	14	59.4	50.35	1.7
5	18.2	17.41	73	15	64.0	51.72	1.4
6	22.8	21.94	43.5	16	68.6	52.87	1.24
7	27.4	26.32	26.33	17	73.2	54.05	1.08
8	32.0	30.41	16.4	18	77.8	54.63	1.012
9	36.6	33.72	11.2	19	82.2	54.64	1.009
10	41.2	37.70	7.1	20	86.8	54.72	1.0001

Table 2. CDMA-Mode White Light Imaging DR Data.

Patch #	D (dB)	M (dB)	SNR	Patch #	D (dB)	M (dB)	SNR
1	0	-	1802	8	32.0	32.78	41
2	4.6	4.21	1110	9	36.6	33.65	37
3	9.2	8.20	701	10	41.2	40.27	18
4	13.8	13.77	370	11	45.8	43.8	10.6
5	18.2	18.28	220	12	50.2	44.61	5.8
6	22.8	22.19	140	13	54.8	49.82	3.5
7	27.4	25.85	92				

Table 3 shows the additional improvement in linear DR via the FM-TDMA mode of the camera showing accurate recovery till a DR of 73.2 dB. Experiments indicate that

* Corresponding author: n.riza@ucc.ie

https://doi.org/10.1051/epjconf/202023812008

inter-patch crosstalk at the image plane such as due to camera front lens optics is limiting the higher DR (weaker patch) recovery. Hence the CAOS FM-TDMA mode is re-engaged while blocking target plane light from all patches except the patch under measurement.

Table 3. FM-TDMA Mode White Light Imaging DR Data.

Patch #	D (dB)	M (dB)	SNR	Patch #	D (dB)	M (dB)	SNR
1	0	-	4620	10	41.2	41.10	41
2	4.6	4.07	2891	11	45.8	44.11	29
3	9.2	9.10	1621	12	50.2	49.35	16
4	13.8	13.83	940	13	54.8	53.02	10.3
5	18.2	18.09	576	14	59.4	59.78	4.7
6	22.8	22.25	357	15	64.0	61.05	4.1
7	27.4	25.23	253	16	68.6	67.46	2
8	32.0	32.71	107	17	73.2	70.02	1.5
9	36.6	34.51	87	18	77.8	72.15	1.1

Table 4. FM-TDMA Mode XDR Data with patch blocking.

Patch #	D (dB)	M (dB)	SNR	Patch #	D (dB)	M (dB)	SNR
1	0	-	100	15	64.0	62	97.8
2	4.6	4.8	98.3	16	68.6	66.4	97.5
3	9.2	9.3	98.9	17	73.2	71.6	96
4	13.8	13.9	97.6	18	77.8	71.1	93
5	18.2	14.8	95.6	19	82.2	80.9	88
6	22.8	19.2	98.3	20	86.8	85.7	71
7	27.4	23.6	98.2	21	91.4	91.1	50
8	32.0	28.6	98.2	22	96.0	95.9	34
9	36.6	32.3	98.1	23	100.6	101.3	20
10	41.2	39.2	98.2	24	105.2	106	8.2
11	45.8	42.1	97.9	25	109.8	111	13.3
12	50.2	47.2	97.7	26	114.2	115.2	8.4
13	54.8	51.4	97.8	27	118.8	124.4	7.4
14	59.4	57.2	98.4	28	123.4	125.3	9.4

Table 5. FM-mode XDR Data with a laser patch target.

Attenuation (OD)	D (dB)	M (dB)	SNR	Attenuation (OD)	D (dB)	M (dB)	SNR
0	0	-	269.3	6.5	130	128.1	8.4
1	20	19.9	268.7	7.4	148	147.3	5.4
2	40	39.3	268.7	7.7	154	152.5	6.8
3	60	57.9	268.3	7.8	156	154.7	5
4	80	80	161.3	8.4	168	163.9	17.3
5	100	97.4	36.06	8.8	176	173	6.3
5.5	110	107.6	10.74	9	180	177	4

Table 4 shows the further improvement in DR recovery to 123.4 dB. Table 5 shows DR data taken using a laser beam as an Optical Density (OD) controlled single patch up-to a 180 dB XDR value. These readings are taken using the FM-mode, showing a recovery up-to a 177 dB XDR. Fig.1 plot shows the linear 177 dB XDR response of the demonstrated CAOS smart camera.

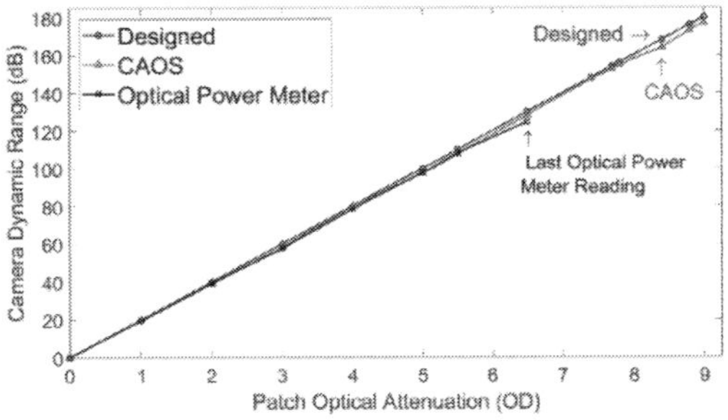

Fig. 1. Linear 177 dB XDR response using Table 5 Data.

3. Conclusion

Calibrated target image testing is a robust method to ascertain the linear DR imaging capability of a camera. Hence highlighted are such initial tests for the CAOS smart camera that both point to the limitations of the testing infrastructure and the deployed camera operations. For example, brightness limits on a pixel basis of the white light LG3 source-based XDR test allowed image recovery to 123.4 dB DR. The brighter visible laser single patch OD-controlled target allowed linear XDR recovery to 177 dB.

References

1. N. A. Riza, Coded Access Optical Sensor (CAOS), USA Patent 10,356,392 B2, July 16, 2019.
2. N. A. Riza, "The CAOS Camera Platform – Ushering in a Paradigm Change in Extreme Dynamic Range Imager Design," Invited Paper SPIE OPTO, Vol. 10117, Photonics West, Jan. 30, 2017.
3. N. A. Riza and M. A. Mazhar, "Laser beam imaging via multiple mode operations of the extreme dynamic range CAOS camera," Appl. Opt., Vol. 57, No.E20, 2018
4. N. A. Riza and M. A. Mazhar, "177 dB Linear Dynamic Range Pixels of Interest DSLR CAOS Camera," IEEE Photonics Journal (open access), Volume 11, Number 2, April 2, 2019.

Micro-force measurement with pre-curvature long-period fiber grating-based sensor

Walter S. J. Ferreira[1,2], *Paulo* S. S. dos Santos[3,5], *Paulo* Caldas[3, 4*], *Pedro* A. S. Jorge[3,5], *João* M. S. Sakamoto[2,1]

[1]Technological Institute of Aeronautics [Instituto Tecnológico de Aeronáutica - ITA], Praça Mal. Eduardo Gomes, 50, São José dos Campos-SP, 12228-900, Brazil.
[2]Institute for Advanced Studies [Instituto de Estudos Avançados - IEAv], Trevo Cel. Av. José A. A. do Amarante, 01, São José dos Campos- SP, 12228-001, Brazil.
[3]INESC TEC - Institute for Systems and Computer Engineering, Technology and Science, 4200-465 Porto, Portugal.
[4]Escola Superior de Tecnologia e Gestão, Instituto Politécnico de Viana do Castelo Rua Escola Industrial e Comercial de Nun'Álvares, n.º34 4900-347 Viana do Castelo
[5]Departamento de Física e Astronomia da Faculdade de Ciências da Universidade do Porto, Rua do Campo Alegre 687, 4169-007 Porto, Portugal

Abstract. In this work, a long-period fiber grating (LPG) based sensor was evaluated as a sensing device for micro-force measurement, in the order of micro Newtons. It was used an LPG fabricated by arc-inducted technique in a SMF-28 standard optical fiber. The optical fiber was fixed between two clamps with a separation of 150 mm with the middle of the LPG located at the center. Characterizations were performed in terms of temperature, curvature and strain. The grating was then used as a micro-force sensor by means of both curvature and strain, induced by a hung mass in a stretched fiber. Furthermore, the evaluation of a pre-curvature LPG was performed to assess if an increase of sensitivity is achieved. Micro-force sensitivity achieved with the stretched LPG was 1.41 nm/mN and it was demonstrated that its sensitivity can be enhanced to 5.14 nm/mN with a pre-curvature of 2.2 m^{-1} applied to the LPG, achieving a spectral resolution of at least 15.6 μN.

1 . Introduction

The research field in optical fiber sensors (OFS) has been steadily growing, mainly because OFS presents interesting characteristics, such as resistance to most common chemicals, lightweight, inherently small form factors, and capability to withstand high temperatures, as well as their immunity to electromagnetic interference. Some of these OFS already have established a firm position as industry solutions, used in hard-to-reach areas, performing, for example, corrosion sensing, or in aeronautics and even space industry where fiber Bragg grating (FBG) has been already passed the proof-of-principle stage and is now an established technology [1]–[4].

Apart from FBG, another kind of grating, with well-known characteristics is long-period fiber grating (LPG). This kind of grating is easy to fabricate in cost-effective ways, using, for example, electric-arc discharges. These gratings present some advantages over FBGs, such as refractive index (RI) sensing or low back-reflection. The light is coupled from the fundamental forwarding propagating LP_{01} core mode to higher-order forward propagating cladding modes [5]–[7] by applying periodic modulations on the fiber core, using phase masks or point-by-point markings (with femtosecond lasers, CO_2 lasers, argon ion lasers, or induced electric-arc discharges). The LPG spectral behavior consists of high attenuation rejection bands with a full width at half maximum (FWHM) of several nanometers. The spectral characteristics depend on the temperature, strain, curvature, and RI of the surrounding medium [3].

This multi-parameter sensing capability must be considered when they are to be used in the design of sensing devices. For the specific application of this work, temperature and RI variations are undesirable. However, if the medium surrounding the LPG does not change, the induced RI wavelength shifting does not present a relevant issue when working in an environment like a laboratory without significant temperature and humidity changes. On the other hand, temperature effects must be considered since the LPG temperature coefficients are around 0.154 nm/ºC, and can have an impact on the final measurements [1]. Bhatia [1] showed that lower temperature sensitivities are achieved for longer grating periods, contrasting with higher sensitivities for curvature, RI and strain for shorter grating periods, so there should be suitable parameters for producing a LPG tuned for having higher sensitivities for curvature and strain sensing while diminishing the effects of temperature fluctuations [8].

At present, both strain and curvature sensors have been developed and vastly reported with sensitivities around 5.285 nm/m^{-1} and 49.3 pm/m^{-1} for curvature and up to 2 pm/με and 1.2 pm/με for strain, for LPGs and FBGs, respectively [9]–[13]. In this paper, it is reported and studied a simple mechanism to take advantage of the LPG curvature and strain sensitivity to perform high sensitivity micro-force measurements. A body of calibrated mass induces the curvature and strain, and the response of a stretched LPG and a pre-curvature LPG written in SMF-28 optical fiber to measure micro-force is studied, in order to assess if an increase of sensitivity is achieved for the pre-curvature LPG.

2 Materials and methods

The electric-arc LPGs fabrication setup consists of a broadband light source (BBLS) connected at one end of the Corning SMF-28 optical fiber, and the other end connected to an Optical Spectrum Analyzer (OSA) to monitor the evolution of the LPG modes while in production. A DC discharge unit from a BICC AFS3100 optical fiber splice machine produces the markings, while a linear stage controller controls the grating period.

The produced LPG tracked mode chosen was the LP_{16} with a period of 395 μm, which provided a 1550 nm peak, monitored during a temperature sweep from 35.0ºC to 99.8ºC. Its sensitivity to temperature was characterized with a connection to a BBLS (Thorlabs fiber-coupled SLD Source 2.5 mW) and an OSA (Yokogawa AQ6370), while the grating was inserted inside an incubator (ESCO Isotherm incubation chamber) with a resolution of ±0.1°C, in a fixed position by two clamps, assuring no curvature induced changes. Thus, the temperature sensitivity of 109 pm/°C (in wavelength shifting) and –0.07 dB/°C (in amplitude variation) was measured.

Typical strain and curvature measurements were made using the setup presented in Fig. 1. This configuration allows the LPG to bend when the linear translation stage (LTS) is moved towards the fixed stage (FS), achieving curvature measurement. Strain measurement can be performed when the LTS is moved far away from the FS, stretching the optical fiber. A simple geometrical approach was used to calculate the curvature radius, R, only knowing its width (w) and height (h), as shown in Fig. 1, by using the following equation:

$$R = \frac{h}{2} + \frac{w^2}{8h} \qquad (1)$$

Thus, the curvature, *C*, can be calculated by *C* = *1/R*. To calculate the strain, the following equation can be used:

$$\Delta\varepsilon = \frac{\Delta L}{L} \qquad (2)$$

where *ΔL* is the change in length, and *L* is the original length.

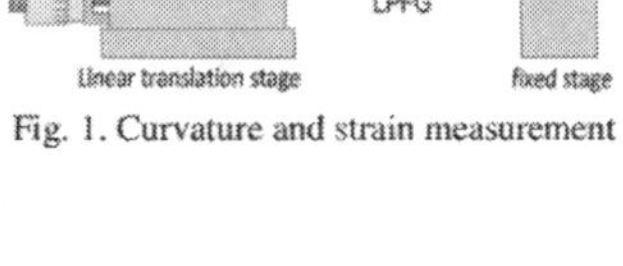

Fig. 1. Curvature and strain measurement setup.

* Corresponding author: pcaldas@estg.ipvc.pt

The initial spacing between the LTS and FS was 150 mm, with the optical fiber stretched. For measuring curvature, displacement steps of 50 μm were applied (moving the LTS towards the FS), while for measuring strain, displacement steps of 5 μm were applied (moving the LTS away from the FS).

3 Results and discussion

Employing the LPG proposed in this work, curvature and strain sensitivities were measured to be 12.64 nm/m^{-1} [Fig. 2(a)] and 0.47 pm/με [Fig. 2(b)], respectively. Force measurements were then performed by placing different masses on the center of the LPG, which was initially stretched (without pre-curvature). Although the contact with the mass induced a change in RI response towards lower wavelengths, these changes only resulted in a constant spectral offset, therefore not affecting the experiment. The weights of the masses produced forces (considering the earth gravitational acceleration constant of 9.806 m/s^2), which, in turn, induced both curvature and strain on the fiber, making these contributions to produce an effective wavelength shift increase (and both in the same direction). A set of masses ranging from 10 mg to 500 mg were hooked to the LPG, yielding the results presented in Fig. 2(c), with a sensitivity of 1.41 nm/mN.

In the curvature measurements [Fig. 2(a)], it is possible to see a linear behavior, and the middle of the studied range is around 2.2 m^{-1}. This value was thus chosen as an operation point on the micro-force measurement with pre-curvature. Therefore, to test the LPG response, it was previously created a pre-curvature in it of 2.2 m^{-1}, and then forces were applied by using different calibrated masses from 10 mg to 300 mg, as presented in the experimental setup of Fig. 3. The results obtained are presented in Fig. 4(a), which was best fitted as a quadratic response (green curve). Regarding the linear range between 98 μN and 686 μN (red curve), the sensitivity obtained was 5.14 nm/mN, showing an enhancement of the pre-curvature LPG in comparison with the stretched LPG.

The spectral resolution, r, of this setup, can be estimated considering the values obtained from two measurements linked with two different forces, with the following expression [14]:

$$r = \frac{2\sigma}{S_m} \qquad (3)$$

where σ is the highest standard deviation between the two steps, and S_m is the sensitivity to mass variations. Regarding the stretched LPG, the resolution measurement is shown in Fig. 4(b).

The step of Fig. 4(b), between approximately 1524.30 nm and 1524.44 nm, was obtained using two different forces (9.8 μN and 49 μN) and the results were acquired using the OSA set to its limiting resolution of 20 pm. The worst obtained standard deviation was 11 pm, the sensitivity (S_m) was 1.41 nm/mN, resulting in a spectral resolution of 15.6 μN, for the worst-case (stretched LPG).

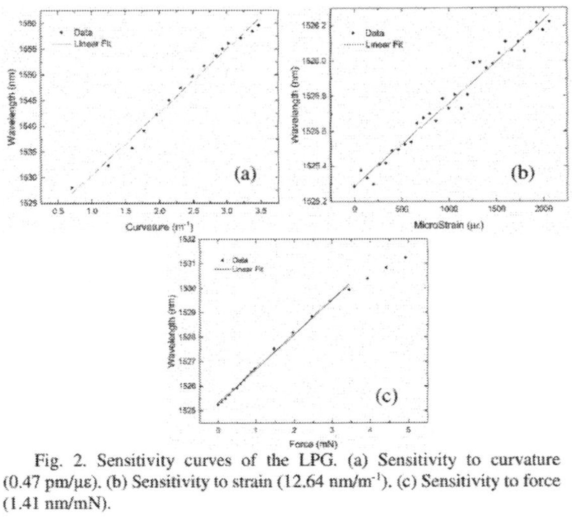

Fig. 2. Sensitivity curves of the LPG. (a) Sensitivity to curvature (0.47 pm/με). (b) Sensitivity to strain (12.64 nm/m^{-1}). (c) Sensitivity to force (1.41 nm/mN).

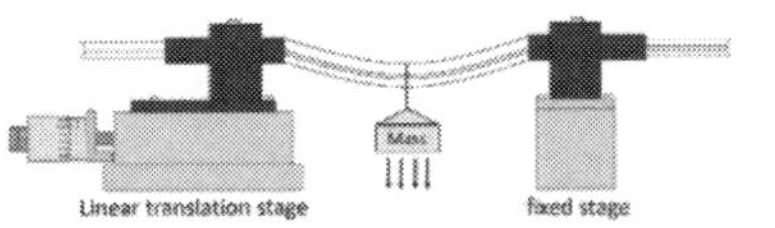

Fig. 3. Experimental setup to micro-force measurement with a pre-curvature LPG.

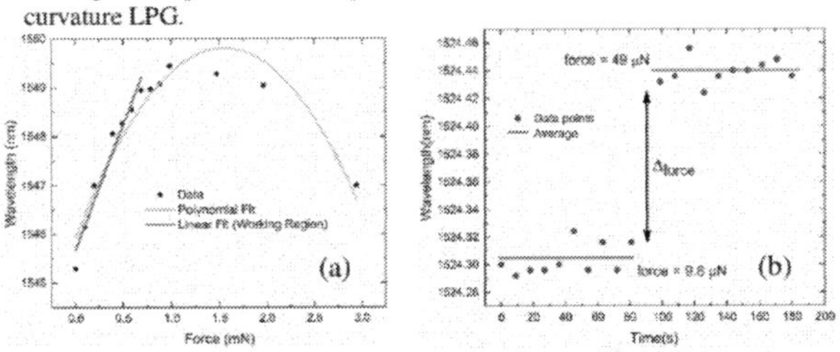

Fig. 4. Results of force measurement. (a) Sensitivity to force (5.14 nm/mN) for pre-curvature LPG (pre-curvature of 2.2 m^{-1}). (b) Resolution measurement for forces of 9.8 μN and 49 μN.

4 Conclusions

The LPG sensitivities to curvature, strain, and weight were measured to be 0.47 pm/με, 12.64 nm/m^{-1}, and 1.41 nm/mN, respectively. Regarding the micro-force measurement, we obtained an enhancement from a sensitivity of 1.41 nm/mN (stretched LPG) to 5.14 nm/mN (pre-curvature LPG), a factor higher than 3. Thereofore, with the proposed technique, based on using a pre-curvature LPG on the center of its linear range, one should be able to enhance the sensitivity of an LPG as a micro-force sensor.

Acknowledgments

The result presented in this paper is supported by grant #2017/20128-2 and grant #2019/04986-4, São Paulo Research Foundation (FAPESP) and financed in part by Coordenação de Aperfeiçoamento de Pessoal de Nível Superior - Brasil (CAPES) Finance Code 001, and is part of the Project "On Chip Whispering Gallery Mode Optical Microcavities For Emerging Microcontaminant Determination In Waters" - SAFE WATER, supported and co-funded by the European Commission, Directorate-General Communications Networks, Content and Technology (DG CONNECT) under the ERA-NET Cofund scheme - Horizon 2020 "Horizon 2020 – the Framework Programme for Research and Innovation (2014-2020).

References

[1] V. Bhatia, "Applications of long-period gratings to single and multi-parameter sensing," Opt. Express, vol. 4, no. 11, p. 457, 1999.

[2] X.-W. Zhao and Q. Wang, "Mini review: Recent advances in long period fiber grating biological and chemical sensors," Instrum. Sci. Technol., pp. 1–30, Sep. 2018.

[3] V. Bhatia and M. Vengsarkar, "Optical fiber long-period grating sensors.," Opt. Lett., vol. 21, no. 9, pp. 692–4, 1996.

[4] R. Pirich and K. D'Ambrosio, "Fiber optics for harsh environments," in 2011 IEEE Long Island Systems, Applications and Technology Conference, 2011, pp. 1–4.

[5] A. M. Vengsarkar, P. J. Lemaire, J. B. Judkins, V. Bhatia, T. Erdogan, and J. E. Sipe, "Long-period fiber gratings as band-rejection filters," J. Light. Technol., vol. 14, no. 1, pp. 58–64, 1996.

[6] G. Rego, "Arc-Induced Long Period Fiber Gratings," J. Sensors, vol. 2016, 2016.

[7] T. Allsop et al., "Comparison between femtosecond laser and fusion-arc inscribed long period gratings in photonic crystal fibre," vol. 7357, p. 73570J, 2009.

[8] S. Chaubey, P. Joshi, M. Kumar, R. Arya, A. K. Nath, and S. Kher, "Design and development of long-period grating sensors for temperature monitoring," Sadhana - Acad. Proc. Eng. Sci., vol. 32, no. 5, pp. 513–519, 2007.

[9] Q. Wang and Y. Liu, "Review of optical fiber bending/curvature sensor," Measurement, vol. 130, pp. 161–176, Dec. 2018.

[10] J. Kong, X. Ouyang, A. Zhou, H. Yu, and L. Yuan, "Pure Directional Bending Measurement With a Fiber Bragg Grating at the Connection Joint of Eccentric-Core and Single-Mode Fibers," J. Light. Technol., vol. 34, no. 14, pp. 3288–3292, Jul. 2016.

[11] P. Geng et al., "Two-dimensional bending vector sensing based on spatial cascaded orthogonal long period fiber," Opt. Express, vol. 20, no. 27, p. 28557, Dec. 2012.

[12] R. Guyard, D. Leduc, C. Lupi, and Y. Lecieux, "Global overview of the sensitivity of long period gratings to strain," Opt. Laser Technol., vol. 79, pp. 62–73, May 2016.

[13] R. J. Black, D. Zare, L. Oblea, Y.-L. Park, B. Moslehi, and C. Neslen, "ON THE GAGE FACTOR FOR OPTICAL FIBER GRATING STRAIN GAGES."

[14] F. Chiavaioli et al., "Towards a Uniform Metrological Assessment of Grating-Based Optical Fiber Sensors: From Refractometers to Biosensors," Biosensors, vol. 7, no. 4, p. 23, Jun. 2017.

Optical design of a variable angle irradiation system for skin cancer laser phototherapy

Diogo Garcia[1*], *João* M. P. Coelho[2,3], and *Pedro* Vieira[1]

[1]Departamento de Física, Faculdade de Ciências e Tecnologia, Universidade Nova de Lisboa, Portugal
[2]Laboratório de Óptica, Lasers e Sistemas, Faculdade de Ciências, Universidade de Lisboa, Portugal
[3]Instituto de Biofísica e Engenharia Biomédica, Faculdade de Ciências, Universidade de Lisboa, Campo Grande, Lisboa, Portugal

Abstract. Photothermal therapy is a developing therapeutic technic mainly regarding the usage of nanoparticles considering their proneness to assemble around malignant tissue and induce their destruction by converting electromagnetic radiation into thermal radiation. In order to optimize this process, controlling the area of incidence in a way that it can fit the tumour and prevent damage of healthy cells is of great importance. This study was conducted by designing an optical system based on zoom systems for manipulating an initial input beam and adjust it to the most common skin cancer sizes. Two solutions are proposed, accomplishing spot sizes at 150 mm from the last lens within the approximate range of 2 mm to 76 mm, with beam divergences lower than 3 mrad.

1 Introduction

Photothermal therapy (PTT) is a minimally invasive therapeutic that can be used for cancer treatment. Recent advances using laser radiation have introduced nanoparticles as a solution to increase the volume of the affected malignant tissue and reduce damaging adjacent healthy cells [1]. In particular, the use of the surface plasmon resonance (SPR) effect by functionalized nanoparticles bounded to the tumour allows to maximize the conversion of laser radiation into thermal energy [2].

However, even considering that the SPR effect minimizes the interaction with healthy cells (where there are no nanoparticles) to reduce the energy being applied and thus minimizing costs, adjusting the laser beam on the tumour should be important. In general, lasers used in the considered PTT technique emit in the near-infrared and their beam is delivered by an optical fibre, which means having a divergent beam. Thus, the irradiation area can be adapted to the tumour size just by adjusting the distance. However, and considering the possible sizes of tumours, this solution is somehow limited.

Regarding more superficial tumours, such as those derived from skin cancer, they are commonly categorized into two types: melanoma and non-melanoma skin cancer. Non-melanoma skin cancer main forms are basal cell carcinoma characterized by an average diameter of 12.2 mm, ranging from 2.0 to 53.0 mm [3] and squamous cell carcinoma with an average tumour size of 24.9×20.3 mm^2 [4]. Accordingly with [5], melanomas can be characterized by an average diameter of 14.4 mm, ranging from 2.0 to 76.0 mm. Thus, we stablished the goal of designing an optical system capable of dynamically manipulate the size of the laser beam to the size of the tumour in the range 2 mm to 76 mm, keeping the distance of system to the tumour constant.

2 Methodologies

The starting point for the considered design was that of a simple zoom lens [6], a mechanical assembly of optical elements (lenses) for which the angle of irradiance (and focal length) can be varied. One of its most basic assemblies consists in having two positive lenses with the same focal lengths and one negative in the middle with a focal length, in absolute value, of less than half that of that of the positive lenses. **Figure 1** shows a schematic of this principle as implemented in this study, illustrating the main parameters to be considered: the input (D_i) and output (D_o) beam diameters, the distance between lens vertices (t_1 for L_1-L_2, and t_2 for L_2-L_3), t the total length of the system, and d the distance from the last lens to the plane of irradiation. Although the designs of professional zoom systems are typically more complex, our application does not require complete collimation over all the range of output beam sizes and some divergence/convergence was considered acceptable. Nevertheless, the lowest divergence was always pursued.

Zemax™ optical design software was used, and it was first assumed the design of custom-made lenses, and after the use of catalogue lenses to have a cheaper solution.

The input beam was assumed as collimated and having 50 mm diameter, a solution achievable using a collimator associated to the output of the fibre delivery system of a diode laser. The wavelength of the radiation was 808 nm and the distance between the output lens (L_3) and the tumour was fixed as 150 mm. The latter value was set based on our experience with *in vivo* animal testing.

* Corresponding author: dmm.garcia@campus.fct.unl.pt

https://doi.org/10.1051/epjconf/202023812010

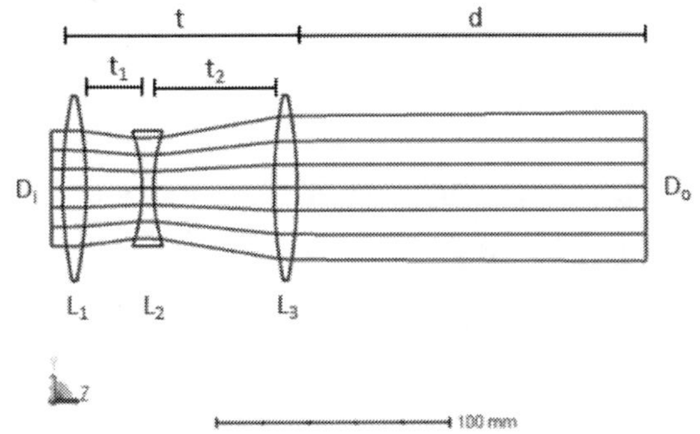

Fig. 1. Schematic of the basic zoom lens system considered in this work.

3 Results and analysis

The study revealed two solutions considered to solve the problem, following the principle mentioned before and using Zemax™ design tools, and **Table 1** resumes their performance.

The first system was designed with three custom-made lenses. The focal length of the positive lenses (80 mm diameter) was 195.85 mm and that of the negative lens (50 mm diameter) -78.34 mm. At the irradiation plane, D_o ranged from 1.7 mm to 75.4 mm by changing t_1 and t_2. The corresponding values can be shown in **Figure 2**.

Table 1. Main parameters of the designed systems. (*Maximum divergence at the range where *t* is constant)

System	D_o (mm)	t (mm)	Max. divergence* (mrad)
Custom-made	1.7 to 75.4	101 to 154	1.4
Catalogue	1.8 to 87.5	114 to 134	3.0

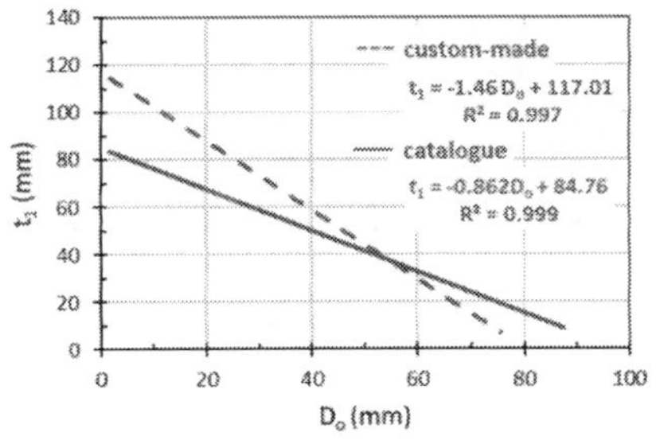

Fig. 2. Relation between the output beam diameter (D_o) and t_1. t_2 changes are ruled by: (1) custom-made version - if $t_1 \geq 64$ mm ($D_o \leq 34.5$ mm) then $t_2 = 12$ mm, or else $t_2 = 76 - t_1$; (2) catalogue version - if $t_1 \geq 65$ mm ($D_o \leq 22.0$ mm) then $t_2 = 10$ mm, or else $t_2 = 75 - t_1$.

Following the first design, a more cost-efficient approach was followed resorting to catalogue lenses. The solution was obtained using two L-BCX269 (Ross Optical) positive lenses (150.60 mm focal length and 85.50 mm diameter) and a 017-0425 (OptoSigma) negative lens (-59.54 mm focal length and 45.72 mm diameter). With this design the values of D_o can be made to change between 1.81 mm and 87.52 (**Figure 2**) but having higher values of maximal divergence (**Table 1**).

Other relevant parameter is the total system length, *t*. The chosen lenses were characterized by its small f/# which allowed to obtain reduced total system lengths (**Table 1**) to facilitate future clinical implementation. Therefore, by manufacturing limitations, the diameter of the negative lenses (both solutions) is smaller than that of the positive ones. Another different aspect from traditional zoom lens systems is the maintenance of a constant value *t* for a certain range. Outside this range, only t_1 changes (increase), and the beam starts to converge. This situation only occurs if one needs smaller spots, that is, when targeting smaller tumours.

4 Conclusions

Two solutions were proposed where a zoom-type lens system can dynamically change its output beam diameter according to the most common sizes of skin cancer. They were designed targeting smaller divergence values, shorter total system lengths and simpler future mechanical design.

None of the designs completely satisfy all the targeted parameters, in particular the low collimation along all the zoom range, and the total length of the system will be affected by the laser beam expander that was not included in this study. Nevertheless, these results point to a simple solution that can dynamically help laser phototherapy mediated by functionalized nanoparticles.

Future work will consist in testing the designs using laboratorial setups and in the development of a prototype for animal *in vivo* tests.

The authors would like to acknowledge the support by the Portuguese Society for Optics and Photonics, SPOF, and the funding by Fundação para a Ciência e a Tecnologia (FCT) through the project UIDB/00645/2020.

References

[1] J. Fong, Y. Ng, S. Ng, Chapter 7 in *Fullerenes, Graphenes and Nanotubes: A Pharmaceutical Approach* (William Andrew Applied Science Publishers 2018).
[2] C. Oliveira Silva, P. Rijo, J. Molpeceres, L. Ascensão, A. Roberto, A. Fernandes, R. Gomes, J. Pinto Coelho, A. Gabriel, P. Vieira, C. Pinto Reis, *Ther. Deliv.*, **7**, 287-304 (2016).
[3] E. Cigna, M. Tarallo, M. Maruccia, V. Sorvillo, A. Pollastrini, N. Scuderi, *J. Skin Cancer*, **2011**, 476362 (2011).
[4] S. Lee, W. Hwang, K. Kim, H. Kim, J. Hwang, K. S. Kim, Arch. Plast. Surg., 43, 538-543 (2016).
[5] R. de Oliveira Filho, D. de Oliveira, M. Souza, M. da Silva, M. Brandão, *Einstein (Sao Paulo)*, **13**, 506–509 (2015).
[6] K. Kim, S.-H. Choe, J.-M. Ryu, H. Choi, *Mathematics*, **8**, 581 (2020).

https://doi.org/10.1051/epjconf/202023812011

Surface Plasmon Resonance sensor based on a D-shaped photonic crystal fiber with a bimetallic layer

Markos P. Cardoso [1,*], *Anderson O.* Silva [2], *Amanda F.* Romeiro [1], *M. Thereza* R. Giraldi[3], *João C.W.A.* Costa[1,**], *José. L.* Santos[5,6] *José. M.* Baptista[4,6] and *Ariel* Guerreiro[5,6]

[1]Federal University of Pará, Applied Electromagnetism Laboratory, Rua Augusto Corrêa, 01, 66075-110, Belém, Pará, Brazil
[2]Federal Center for Technological Education Celso Suckow da Fonseca, Av Maracana, 20271-110 – Rio de Janeiro, Brazil
[3]Military Institute of Engineering, Laboratory of Photonics, Praça Gen. Tibúrcio, 80,22290-270, Rio de Janeiro, Brazil
[4]Faculdade de Ciências Exatas e da Engenharia, Univerisidade da Madeira, Campus da Penteada, 9020-105 Funchal, Portugal
[5]Faculadade de Ciências da Universidade do Porto, Rua do Campo Alegre s/n, 4169-007 Porto, Portugal
[6]INESC TEC, Rua Dr Roberto Frias, 4200-465 Porto, Portugal

Abstract. The investigation of a D-shaped photonic crystal fiber sensor with a bimetallic layer for operation at the visible and infrared spectra is presented. The bimetallic layer is composed by silver and gold slabs deposited adjacently on the flat face of the fiber. It is shown that this architecture allows the excitation of two sharply distinguished plasmon resonance, which suggest potential applications for multiparameter sensing.

1 Introduction

Surface plasmon resonance (SPR) sensors have attracted much attention due to their technical features as highly sensitive characteristic and wide operating spectral range [1]. Sensors based on SPR have many potential applications such as biological detection [2]. Owing to advantages such as low loss and large mode area, photonic crystal fibers (PCF) have been used as a suitable platform for exciting surface plasmon resonances. Besides, guiding modes can be easily manipulated by varying the PCF's structural parameters [3]. A large variety of SPR sensors constructed with PCFs have been reported as: refractive index sensors [1], temperature sensors [4] and chemical sensors [5].

A sensing platform based on a D-shaped photonic crystal fiber with a bimetallic layer is theoretically investigated in this work. This is a new approach of a multiplasmonic sensing configuration, that is not based on coupling between plasmons in overlaid metallic films. Rather, it considers distinct and independent plasmons, with different responses that may facilitate simultaneous measurements of more than one physical parameter, as demonstrated for another design in Ref. [6]. In this study we consider a simpler design, with a bimetallic layer formed by two adjacent slabs of silver and gold, deposited on the flat surface of the PCF. Numerical computations have shown that a sensitivity of 3010 nm/RIU is achieved for the measurement of a single parameter (refractive index).

2 Designed Structure

The scheme of the structure proposed is depicted in Figure 1. It is obtained by a side polished PCF with a hexagonal array of air holes on a matrix of fused silica.

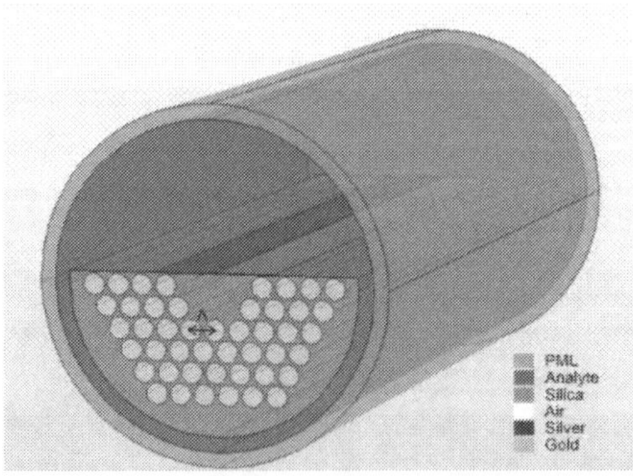

Figure 1. Schematic design of the D-shaped photonic crystal fiber with a bimetallic layer.

The diameter of the fiber is D = 24 μm and the diameters of the air holes is designated by d. The distance between two adjacent air holes is kept fixed at $\Lambda = 2\mu m$ while the ratio among d and Λ is $d/\Lambda=0.88$. The Finite Element based software COMSOL Multiphysics [7] was used to carry out numerical calculation. To adequately evaluate the sensing properties of the D-shaped fiber, we applied a perfectly matched layer (PML) boundary condition with 10% of the fiber diameter for the truncation of the computational domain.

*e-mail: markosdenardi@gmail.br
**e-mail: jweyl@ufpa.br

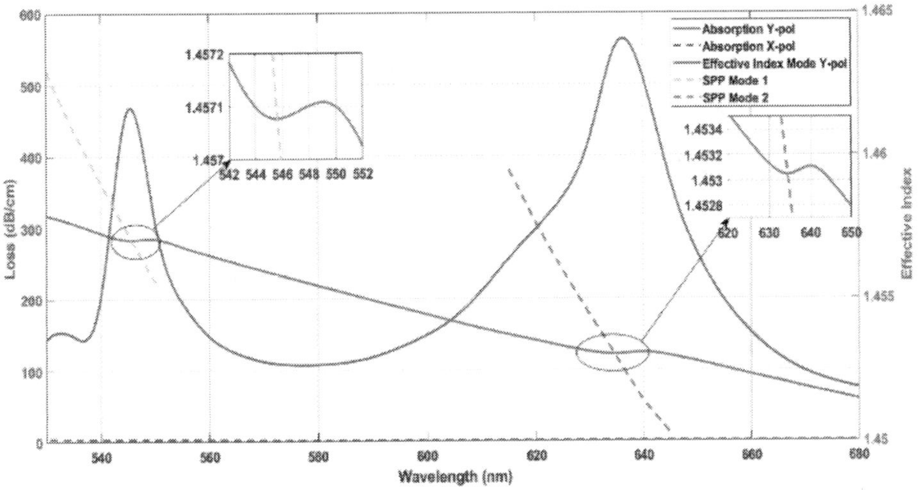

Figure 2. Dispersion and absorption curves of the D-shaped PCF with a bimetallic layer.

3 Results and Discussion

The dispersion curves for the SPP modes and the fundamental fiber mode together with the absorption curve are plotted in Figure 2 for $n_{analyte}$=1.36. The resonant coupling is reached at the phase matching between the SPP mode and the fiber guided mode, resulting in maximum confinement losses. The computed sensitivity is accounted for the variation of the resonant wavelength of maximum absorption for a variation of 10^{-2} in the refractive index of the analyte. Preliminary results are presented in Table 1, where a sensitivity of 3010 nm/RIU and resolution of 3.32x10^{-5} RIU are achieved considering an experimental spectral variation detection of 0.1 nm.

Table 1. Sensitivity and Resolution for the variation of analyte refractive index.

RI	λ_{peak} [nm]		S_λ [nm/RIU]		R [RIU]	
	Ag	Au	Ag	Au	Ag	Au
1.28	430.6	545.9	970	700	1.03x 10^{-4}	1.42x 10^{-4}
1.30	450.3	560.4	1180	800	8.47x 10^{-5}	1.25x 10^{-4}
1.32	474.6	577.9	1480	1080	6.76x 10^{-5}	9.25x 10^{-5}
1.34	505.1	600.8	1980	1580	5.05x 10^{-5}	6.32x 10^{-5}
1.36	545.5	636.1	3010	2630	3.32x 10^{-5}	3.80x 10^{-5}

4 Conclusion

A refractive index SPR sensor based on a D-shaped photonic crystal fiber with a bimetallic layer is analyzed in this work. This approach of a multiplasmonic sensing configuration is not based on coupling between plasmons, but on the generation of two distinct and independent plasmons, with distinct responses with potential interest for multiparameter sensing. Preliminary results of a single parameter (refractive index) show a sensitivity of 3010 nm/RIU and a resolution of 3.32x10^{-5} RIU.

*e-mail: markosdenardi@gmail.br
**e-mail: jweyl@ufpa.br

This study was financed in part by the Coordenação de Aperfeiçoamento de Pessoal de Nível Superior – Brasil (CAPES) – Finance code 001 and Conselho Nacional de Desenvolvimento Científico e Tecnologico (CNPq). It was also financed by the ERDF—European Regional Development Fund through the Operational Program for Competitiveness and Internationalization—COMPETE 2020 Program and by National Funds through the Portuguese funding agency, FCT—Fundação para a Ciência e a Tecnologia within project "GreenNanoSensing" POCI-01-0145-FEDER-032257.

References

1. JunJun Wu, Shuguang Li, Xinyu Wang, Min Shi, Xinxing Feng and Yundong Liu, Applied Optics, Vol 57, No 15, 2018.
2. S. Jabbari, B. Dabirmanesh, S. S. Arab, M. Amanlou, and S. Daneshjou, Sens. Actuators B 240,519–527, 2017.
3. Md. Rabiul Hasan, S. Akter, A.A. Rifat, S. Rana, K. Ahmed, R. Ahmed, H. Subbaraman, D. Abbott, Biosens. IEEE Sens. J. 18 (1) (2017) 133–140F. De Lillo, F. Cecconi, G. Lacorata, A. Vulpiani, EPL, 84, 2008.
4. Wang, Yong & Huang, Qing & Zhu, Wenjie & Yang, Minghong & Lewis, Elfed, Optics Express. 26.1910, 2018.
5. Thenmozhi, Harikesavan & Mani Rajan, Senthil & Ahmed, Kawsar, Optik - International Journal for Light and Electron Optics. 180. 264-270. 2019.
6. D. F. Santos, A. Guerreiro, J. M. Baptista, Optik - International Journal for Light and Electron Optics. IEEE Sensors Journal, 17, 8, 15, 2017.

EPJ Web of Conferences **238**, 12012 (2020) https://doi.org/10.1051/epjconf/202023812012
EOSAM 2020

Superhydrophobic Asphalt Pavements: Surface Improvement

Iran Rocha Segundo[1,*], *Salmon Landi Jr.* [2,3], *Elisabete Freitas*[1], *Verônica Castelo Branco*[4], *Manuel F. M. Costa*[2], *Joaquim Carneiro*[2]
[1]Civil Engineering Department, University of Minho, Azurém Campus, Guimarães, Portugal
[2]Physics Department, University of Minho, Braga & Guimarães, Portugal
[3]Federal Institute Goiano, Rio Verde – GO, Brazil
[4]Transportation Engineering Department, Federal University of Ceará, Fortaleza, Brazil

Abstract. The most adverse weather condition for road safety happens when there is water, snow, or ice on the road surface because their presence highly decreases friction. Therefore, it is essential to drain or repel them quickly. If the water drops are repelled from the surface or the ice/snow formation is avoided with the application of superhydrophobic coatings, roads become safer. In order to functionalize the asphalt mixtures used in road pavements, nano/micromaterials, such as Polytetrafluoroethylene (PTFE), TiO_2, and SiO_2, among others have been applied by spraying coating. The mixes are usually characterized by the water contact angle, and the surface roughness is typically assessed by optical and electron analysis. This research work aims to present a brief overview of superhydrophobic asphalt mixtures.

1 Introduction

The water, snow, and ice over the surface of the road pavements cause time waste, economic loss, polluting the environment and daggering passenger safety, and traffic congestion. Also, due to the presence of ice/snow over airport roads, flights are cancelled, causing personal problems for travellers and millions of losses for the airline companies and airports [1]. The conventional method to melt the ice/snow over the pavement's surface is spreading salt (mainly chloride salts, for example, NaCl and $CaCl_2$) and chemical compounds, which affect the asphalt pavement performance and oxidize vehicles and special structures (i.e., bridges), and, at long-term period, it causes environmental problems [2].

One of the main functional characteristics of road pavements is friction. It expressively reduces when there is water or ice on the road surface. Thus, it is essential to drain or repel the surface water quickly. Superhydrophobic materials can do it fast, and, by their self-cleaning effect, the dirt particles are removed [1,3]. This research work aims to present an overview of superhydrophobic capability applied to asphalt mixtures.

2 Superhydrophobic Asphalt Mixtures Using Nano/microparticles

Wetting is the ability of a liquid (mainly water) to maintain contact with a solid surface, resulting from intermolecular interactions. As reported by Wenzel (1936), wettability is described by a thermodynamic process. Regarding the hydrophilic surfaces, the process is spontaneous since the wet interface free energy is inferior to that of the dry interface. Whereas for hydrophobic surfaces, the process is non-spontaneous because the dry interface free energy is lower than that of the wet interface [1,3,4]. Furthermore, the wettability is also directly related to the surface roughness. Increases on roughness raise the contact angle (CA) between the (water) drop and the material surface; that is, it can improve the hydrophobicity [4].

The superhydrophobic capability is achieved when the CA is higher than 150° [5]. While, according to some authors, CA should be lower than 5 or 10° to have a superhydrophilic, or super-wettable surface [5].

Superhydrophobic surfaces are not wetted by water, presenting low surface energy. They are usually designed by two methods: i) applying surface treatments with low-surface-energy, and ii) creating a rough surface [6]. The benefits of the promotion of this capability for the materials are diverse, providing functions such as water-resistant, anti-icing, antibacterial, contaminant-free, self-cleaning, and anticorrosive [5].

3 Application Method

Some nano/micromaterials (with a grain size between 6 nm and 45 μm) have been applied on asphalt mixtures: TiO_2, ZnO [3], modified-SiO_2 [7], modified fluorine-containing polymer with nano-CaO [4], Magnesium–Aluminium layered double hydroxides (Mg-Al LDHs) [2], and PTFE [1]. Usually, solutions (or dispersions) are prepared containing the nano/microparticles. They are applied by spraying over the asphalt mixture surfaces. In some cases, there is just one spraying process or even two. For the case of two spraying processes, the binder layer is applied, then the dispersed particles are sprayed over the binder layer [1,3].

* Corresponding author: iran_gomes@hotmail.com

https://doi.org/10.1051/epjconf/202023812012

4 Analysis

The superhydrophobic capability is mostly evaluated by optical and electron analysis. The main method is by measuring the CA of the water and the surface. As mentioned before, CA must be higher than 150° to guarantee the superhydrophobic surface.

As previously described, the surface should be rough, being essential to the analysis of the microtopography of the surface. The surface roughness is usually analysed by optical methods, for example, by optical triangulation based-microtopography (Fig. 1). In this case, surface parameters can be assessed in order to characterize it and compare materials and treatments.

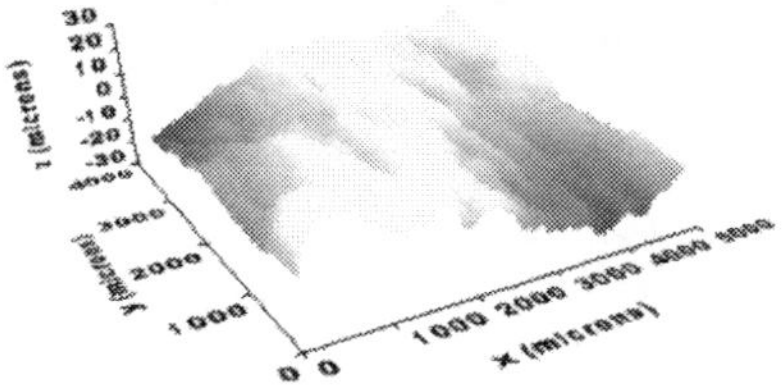

Fig.1. Microtopography of an asphalt mixture.

Also, the images by Scanning Electron Microscope (SEM) are essential to understand the microtopographic structure of the surface. In Fig. 2, the self-cleaning effect is showed by water drops carrying the dirt particles that were previously deposited, as the effect of the Lotus Flower [8]. A biomimetic approach is developed to provide a self-cleaning effect.

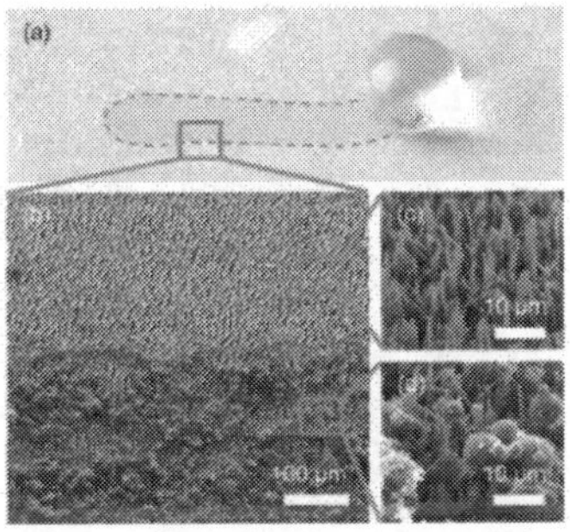

Fig. 2: a) Water drop carrying dirt. Adapted from [8].

Tests to check the feasibility are also conducted: friction [1,9], water drop freezing [2], and durability [2,4].

5 Results on Literature

Table 1 summarizes the main results in the literature. Most of the research work has used nanoparticles. The CA ranges from 150 to 166°, depending on the material and application method. The highest CA was achieved by using microparticles of PTFE [1]. The superhydrophobic asphalt mixtures can present rough surface [2,4], delay water freezing [2,7], decreasing the ice adhesion force [2]. Besides, friction was slightly influenced by water and ice [9]. Durability should be further explored.

Table 1. Superhydrophobic Asphalt Mixtures.

Material	Grain size (nm)	Dispersion Medium	CA (°)	Ref.
TiO_2	23 – 28	Water	150	[3]
Modified-SiO_2	20	Silanes, Water, Ethanol, and Ammonia	151	[7]
TiO_2 ZnO	23 – 28, 45000	Water	155	[3]
Mg-Al LDHs	100	Naphtha	152	[2]
*	*	*	157	[9]
Fluorine polymer with nano-CaO	6 – 20	Water	163	[4]
PTFE	12000	Acetone**	166	[1]

*Not described by the authors

**One binder layer composed of epoxy resin dissolved in xylene was sprayed before the application of the particles.

Conclusions

Superhydrophobic coatings containing PTFE, TiO_2, TiO_2 and ZnO, SiO_2, Mg-Al LDHs, and fluorine polymer with nano-CaO nano/microparticles have been applied over asphalt mixtures to improve safety. They are tested under optical and electron analysis (CA, surface roughness, and SEM images), friction, water drop freezing, and durability. Relevant literature results indicate that road can be safer with superhydrophobic asphalt mixtures.

References

1. A. Arabzadeh, H. Ceylan, S. Kim, K. Gopalakrishnan, A. Sassani, Transp Res Rec J Transp Res Board **2551**, 10–17 (2016)
2. C. Peng, H. Zhang, Z. You, F. Xu, G. Jiang, S. Lv, R. Zhang, Constr Build Mater **189**, 227–235 (2018)
3. I. Rocha Segundo, C. Ferreira, E.F. Freitas, J.O. Carneiro, F. Fernandes, S. Landi Júnior, M.F. Costa, Constr Build Mater **166**, 36–44 (2018)
4. J.H.O. Nascimento, P. Pereira, E. Freitas, F Fernandes, 7th Int. Conf. Maint. Rehabil. Pavements Technol. Control, At Auckland, New Zealand (2012)
5. H. Zhu, Z. Guo, W. Liu, Chem Commun **50**, 30, 3900–3913 (2014)
6. S. Heinonen, E. Huttunen-Saarivirta, J.P. Nikkanen, M. Raulio, O. Priha, J. Laakso, E. Storgårds, E. Levänen, Colloids Surfaces A: Physicochem Eng Asp **453**,1, 149–161 (2014)
7. Y. Gao, L. Qu, B. He, K. Dai, Z. Fang, R. Zhu, Constr Build Mater **191**, 270–280 (2018)
8. J. Lee, R.S. Fearing, Langmuir **28**, 43, 15372-15377 (2012)
9. E. Eriskin, S. Karahancer, S. Terzi, M. Saltan, Procedia Eng **187**, 532–7 (2017)

Sensor Based on Multiple Fiber Bragg Gratings for Diameter Measurement

Victor Cardoso[1,4,7], *Paulo* Caldas[4,5], *M. Thereza* R. Giraldi[2], *Orlando* Frazão[3,4], *Claudio* Carvalho[6], *João* Costa[1], and *José* L. Santos[3,4]

[1]Federal University of Pará, Applied Electromagnetism Laboratory, Rua Augusto Corrêa, 01, 66075-110, Belém, Pará, Brazil
[2]Military Institute of Engineering, Laboratory of Photonics, Praça Gen. Tibúrcio, 80,22290-270, Rio de Janeiro, Brazil
[3]Department of Physics and Astronomy, Faculty of Sciences of University of Porto, Rua do Campo Alegre, 687, 4169-007 Porto, Portugal
[4]Institute for Systems and Computer Engineering, Technology and Science, Rua do Campo Alegre, 687, 4169-007 Porto, Portugal
[5]Polytechnic Institute of Viana do Castelo, Rua Escola Industrial e Comercial de Nun'Álvares, 4900-347, Viana do Castelo, Portugal
[6]Brazilian Agricultural Research Corporation
[7]victorcard@ufpa.br

Abstract. A method based on multiple Fiber Bragg Gratings (FBG) and a structure fabricated using a 3D printer for diameter measurement is proposed and experimentally demonstrated. The sensor is easily developed inserting the FBG on the structure and fixing the FBG in two points. Due to this arrangement, when the diameter is reduced a force is applied on the sensor and this causes a wavelength shift. The study is divided in two steps using one experimental setup. At first, each FBG response is independently evaluated concerning the diameter variation and after that, the FBGs are evaluated together. The results demonstrated that the structure can be used as a way to monitor the diameter variation in some applications.

1 Introduction

In several areas the diameter monitoring presents an important function because it will influence the production and consequently generate financial losses and in some cases can also lead to social problems. For example, in the gaseous hydrogen transportation infrastructure the pressure into the pipeline is high, causing deformation and/or fatigue and consequently a disruption [1].Another example is the expansion and contraction monitoring of the tree stem that is directly related to irrigation, since it depends on the soil water deficit [2–4]. The use of optical sensors for these cases is interesting because they present characteristics such as immunity to electromagnetic interference, compact size, resistance to pressure, heat and corrosion, and the possibility of transmission over long distances. The Fiber Bragg Grating (FBG) is used extensively in several applications, mainly, for strain measurement since this parameter is associated with the change in the wavelength of the reflected signal, which is accurately determined and catalogued for each grating [5].

In this work, we investigate a method for diameter monitoring using a sensor based on multiple Fiber Bragg Gratings. We use an experimental setup that uses the principle of the strain gauge based on FBG for diameter monitoring. We develop a structure in a 3D printer so that it was possible to perform the analysis of the diameter variation of different structures. This work presents preliminary results using only two FBGs

*This study was financed in part by the Coordenação de Aperfeiçoamento de Pessoal de Nível Superior – Brasil (CAPES) – Finance code 001, the Conselho Nacional de Desenvolvimento Cientifico e Tecnologico (CNPq), Institute for Systems and Computer Engineering, Technology and Science (INESC TEC) and the Department of Physics and Astronomy, Faculty of Sciences of University of Porto from Portugal)

2 Preliminary Experimental Setup and Sensor Operation

We proposed a configuration that can be observed in Fig. 1(a). We developed the structures using a 3D printer for the FBGs supporting and to simulate the diameter to be monitored, as shown in Fig. 1(b). We used two pieces for this experiment, one for each FBG. The diameter variation is made using a linear moving translation stage. The FBGs were fixed in the piece in two points so that when the diameter changes a force will tension or compress the sensor, as shown Fig. 1(b). The FBGs were used in series. Light from a broadband source was injected in the circulator and after crossing it, the optical beam was introduced in the sensing head formed by the two FBGs. The reflection spectrum of the sensing head was monitored with an Optical Spectrum Analyzer (Yokogame AQ6370C), which is illustrated in Fig. 1(a). The FBGs were fabricated in single-mode fiber with a reflection wavelength of 1558 nm each one. The pieces were created so that it was possible to do an analysis of circular structures like a pipeline, the stem of a tree and other structures which need to have their diameter variation monitored. A strain is applied on the optical fiber by the variation of the diameter of the structure (Δd) and therefore it was possible to obtain the sensor response. In order to obtain the sensitivity of the sensors, in this experiment we analyzed each FBG separately and after that the FBGs were tested together.

3 Results and Discussion

The results obtained from the experiments are depicted in Fig. 2. It is possible to observe that when reducing the

 https://doi.org/10.1051/epjconf/202023812013

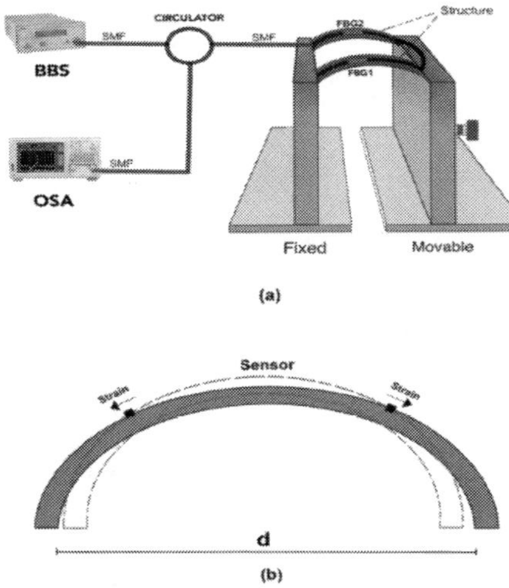

Figure 1. (a) Schematic of the experimental setup. (b) Mechanism of operation of the structure that demonstrates the diameter reduction using the proposed structure developed with a 3D Printer.

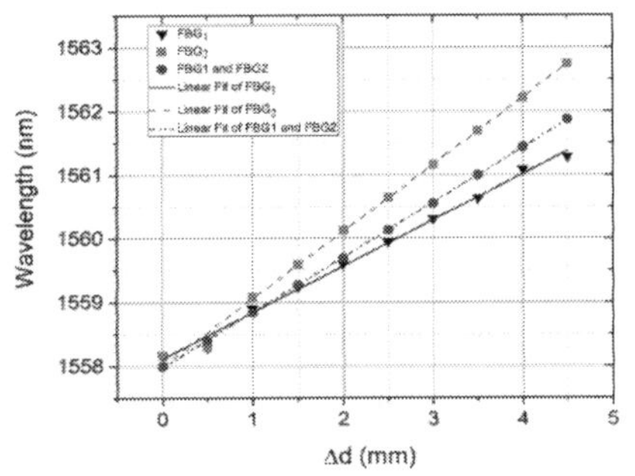

Figure 2. Wavelength shift caused by the diameter reduction.

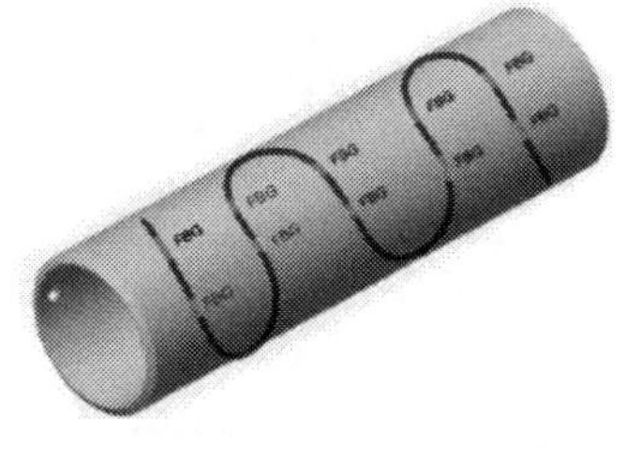

Figure 3. Schematic of the experimental support under construction for Multiple FBGs in a cylindrical structure.

diameter, the results presented a linear behavior when using just one FBG or both of them. When just one FBG is under test, it presented a wavelength shift due to the applied force, while the other one remained centered at 1558 nm since it was not tensioned. For this case, FBG1 and FBG2 presented good linear responses with $R^2 = 0.9940$ and $R^2 = 0.9959$, respectively. The sensitivity obtained was 0.7261 nm/mm and 1.05 nm/mm, respectively.

When we analyzed the FBGs acting together, the same force was applied in each piece. It was possible to observed a wavelength shift due to the two FBGs together, superimposed. For this case, the sensor presented a good linear response with $R^2 = 0.9994$ and a sensitivity of the 0.8577 nm/mm. This study presents preliminary results for two FBGs. The next step will be to perform measurements on cylindrical structures using multiple FBGs, as shown in Fig. 3, applying non-uniform forces and consequently variations in the cylindrical structure. Therefore, it will be possible to measure the eccentricity variation around and/or along the cylindrical structure.

4 Conclusion

In conclusion, we proposed an experimental investigation on a method for diameter monitoring using a sensor based on multiple Fiber Bragg Gratings. We developed an experimental setup that is simple and repeatable. The results using the proposed method demonstrate that it is possible to monitor the diameter of some applications such as pipelines, expansion and contraction of the tree stem, Structure Health Monitoring (SHM) and so on. The use of multiple FBGs enable to measure the eccentricity variation around and/or along a cylindrical structure. With the printed piece used in this work, it is possible to use the FBG to monitor the diameter variation from 0 to 4.5 mm. Diameter variations of these sizes in some stem trees permit to obtain the irrigation levels of these vegetations. This study presents preliminary results and the next step will be to perform measurements on cylindrical structures using multiple FBGs.

References

[1] R.L. Amaro, E.S. Drexler, A.J. Slifka, International Journal of Fatigue **62**, 249 (2014)

[2] M. Cohen, D.A. Goldhamer, E. Fereres, J. Girona, M. Mata, The Journal of Horticultural Science and Biotechnology pp. 55–60 (2001)

[3] M. Martín-Palomo, M. Corell, I. Girón, L. Andreu, E. Trigo, Y. López-Moreno, A. Torrecillas, A. Centeno, D. Pérez-López, A. Moriana, Agricultural Water Management **218**, 115 (2019)

[4] R. Çakir, Field Crops Research (2004)

[5] H. lin Liu, Z. wei Zhu, Y. Zheng, B. Liu, F. Xiao, Optical Fiber Technology (2018)

Three-dimensional laser writing inside silicon using THz-repetition-rate trains of ultrashort pulses

Andong Wang[1,*], *Amlan* Das[1], and *David* Grojo[1,*]

[1]Aix-Marseille Univ., CNRS, LP3 UMR 7341, 13009 Marseille, France

Abstract. Three-dimensional laser writing inside silicon remains today inaccessible with the shortest infrared light pulses unless complex schemes are used to circumvent screening propagation nonlinearities. Here, we explore a new approach irradiating silicon with trains of femtosecond laser pulses at repetition rates up to 5.6 THz. This extremely high repetition rate is faster than laser energy dissipation from microvolume inside silicon, thus enabling unique capabilities for pulse-to-pulse accumulation of free carriers generated by nonlinear ionization, as well as progressive thermal bandgap closure before any diffusion process comes into play. By space-resolved measurements of energy delivery inside silicon, we evidence a net increase on the level of space-time energy localization. The improvement is also supported by experiments demonstrating an apparent decrease of the energy threshold for modification and drastic improvements on the repeatability, uniformity, and symmetricity of the produced features. The unique benefits of THz bursts can provide a new route to meet the challenge of 3D inscription inside narrow bandgap materials.

1 Introduction

Ultrafast laser 3D inscription inside materials is today used in some of the most advanced technologies from 3D quantum photonics and optical communication to nanosurgery [1,2]. While this would be highly beneficial for the semiconductor industry, they however remain so far limited to transparent materials including dielectrics and some polymers. We have recently established that narrow bandgap semiconductors make hardly achievable local material manipulation in the bulk because of strong propagation nonlinearities experienced by ultrafast infrared light in these materials [3-5]. Although many progresses have been made recently, they are still inaccessible with the shortest pulses unless complex schemes are implemented. For example, we have demonstrated the first bulk modification inside silicon with femtosecond laser pulses by applying a solid-immersion focusing strategy [4]. However, the contact nature of the technique creates severe difficulties for implementations. In this manuscript, we describe a new solution which holds promises for application because of a possibility for implementation with a simple arrangement of crystals in the beam path [6].

2 Methods

The solution proposed in this paper relies on a multi-time scale optimization of the interactions. It is inspired by the new ultrafast GHz-burst mode of irradiation that has been recently introduced for unprecedented performances in femtosecond laser surface machining [7]. However, ours require pulse trains at even higher repetition rates so that accumulation effects can give their full potential before dissipation processes (including carrier and temperature diffusions [8]) come into play in silicon. We use a femtosecond laser system (Pharos) to pump an optical parameter amplifier delivering pulses of duration <200fs at 1550-nm wavelength. The output energy was controlled by the combination of a half-waveplate and polarizer. A parallel grating pairs is built to stretch the pulses up to 21 ps.

To generate the pulse trains, we rely on an engineered arrangement of birefringent crystals to split a pulse in delayed replica. Trains of 2, 4, 8, 16, 32, and 64 pulses are produced by combining 6 crystals of different pre-defined thicknesses. The shortest inter-pulse delay (the one applicable to the 64-pulse trains) of 180 fs between pulses corresponds to an irradiation at a repetition rate of ~ 5.6 THz. The obtained trains of pulses for all crystal combinations are systematically verified by using a long-scan auto-correlator. For irradiation studies, we tightly focus the beam deep inside silicon samples (intrinsic (100), focusing depth exceeding 300µm) with a microscope objective of 0.85NA. Under these focusing conditions, strong single pulse develop a plasma in pre-focal region that limits the intensity and energy deposition near the focus, which restrains modification from occurring [3,4]. With the THz-repetition-rate train of weaker pulses, one can accumulate carrier at the focus. The duration of the generated trains exceeding electron-phonon coupling time (~ps) but below electron and temperature diffusion times (~ns). Then, we can expect

* Corresponding author: andong.wang@univ-amu.fr; david.grojo@univ-amu.fr

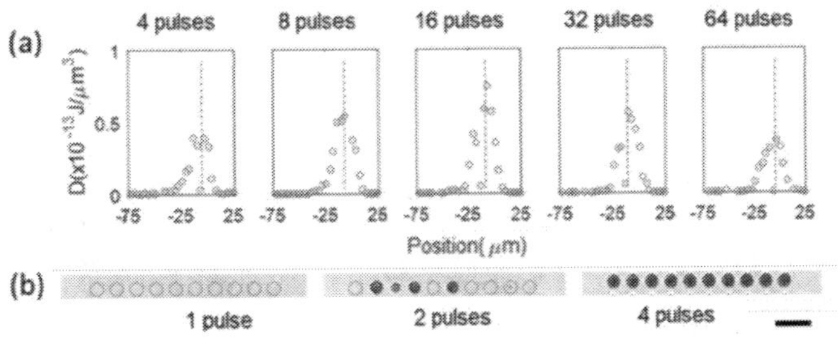

Fig. 1. Accumulating energy density with pulse trains. (a) Estimated absorbed energy densities (peak) along the propagation distance (geometrical focus at zero-position) with trains of femtosecond pulses (5.6 THz) focused with NA=0.45 (b) Top-view infrared images of modifications produced with trains of 4-ps pulses focused (NA=0.85) at 300μm below the surface. All irradiation energies are the same (2.1μJ/train) repeated 10 times for direct comparisons of damage probabilities). The scale bar is 50 mm.

local heating gradually increasing absorption during the train.

To quantify the conditions applied to silicon near focus, we measure the fluence distributions by a Z-scanning microscopy approach similar to the one used in refs. [4]. While the measurement is not possible for the highest NA, we make this analysis for a beam focused with NA=0.45. Then, we estimate from the measurements the absorbed energy density by simple derivation of energy with respect to the position on the optical axis. We also systematically confirm the enhanced conditions in matter by damage test studies for different energy levels, number of pulses in the trains and separation between pulses.

3 Results and Discussions

The validity of the THz burst mode of irradiation to enhance the conditions in semiconductors is demonstrated by comparisons on the level of spatio-temporal light energy localization measured under various THz-burst irradiation conditions. As shown in figure 1a, measurements reveal for given incoming energy that it exists an optimum number of femtosecond pulses (around 16) to apply to reach the highest level of absorbed energy density. This shows nonlinear delocalization the beam [9] persist with train of pulses and the new control parameters are used for appropriate excitation build-up and limited retro-reaction.

While this level of improvement does not allow to reach the threshold for permanence modification at NA=0.85, the benefit of burst mode to limit delocalization of the beam is supported by a range of experiments demonstrating improved 3D writing performances with pulse duration of ~5-7 ps. As shown in figure 1b, while single pulses of 2.1 μJ does not allow to modify Si, splitting the energy between 2 pulses, makes possible to cross the threshold for modification despite non-reproducible modification (damage probability ~ 40%). Additional splitting to deliver the same energy with a train of 4 pulses lead to highly repeatable features indirectly showing increased absorbed energy densities in Si with the number of pulses in the trains.

The THz repetition rate is so high in the tested pulse trains that an important question may arise on the difference on the material response between trains and stretched pulses. While we could not make systematic comparisons to study in details this question, there are already a striking difference revealed with this study. It is the absence of modification with THz trains of 64-pulses of 200 fs corresponding to an apparent duration of interaction of about 12 ps while single Gaussian pulses of duration exceeding 7 ps can readily produce permanent modifications. This is a direct evidence that the detailed temporal characteristics of irradiation play a major role in these situations. This reveals that the detailed temporal characteristics of irradiation should not be overlooked for the development of ultrafast 3D laser writing technologies applicable to narrow bandgap materials.

This research has received funding from the European Research Council (ERC) under the European Union's Horizon 2020 research and innovation program (Grant Agreement No. 724480).

References

1. R. R.. Gattass and E. Mazur. Nat. Photonics **2**, 219–225 (2008).
2. L. Jiang, A.-D. Wang, B. Li, T.-H. Cui, and Y.-F. Lu, Light: Science and Applications **7**, 17134 (2018).
3. A. Mouskeftaras, et al. Appl. Phys. Lett. **105**, 191103 (2014).
4. M. Chanal, et al. Nat. Commun. **8**, 773 (2017).
5. V.Y. Fedorov et al. Phys. Rev. Lett. **117**, 043902 (2016).
6. A. Wang, A. Das, D. Grojo. Research **2020**, 8149764, (2020).
7. C. Kerse, et al. Nature **537**, 84–88 (2016).
8. A. Mouskeftaras, et al. Appl. Phys. Lett. **108**, 041107 (2016).
9. M. Chambonneau, et al. Phys. Rev. Appl. **12**, 024009 (2019).

 https://doi.org/10.1051/epjconf/202023812015

Crystalline lens imaging during accommodation with a slit-scanning tomography system: preliminary results

Jessica Gomes and Sandra Franco

Centre of Physics, University of Minho, 4710-057, Braga, Portugal

Abstract. A slit-scanning tomography system previously developed in the Centre of Physics of University of Minho allows to capture images of the eye and posteriorly calculus of the biometric (thickness and curvature) and optical (wavefront aberrations) properties of the corneal and crystalline lens surfaces. It is intended to optimise the system for an accurate computing of these properties of the crystalline lens in the relaxed state and during ocular accommodation. In this study, preliminary experimental results of the crystalline lens imaging of a subject during accommodation are presented.

1 Introduction

The human eye is an optical system with different refractive components where the cornea and crystalline lens are the most important and determine the optical quality of the eye. [1]

In order to obtain a clear image of objects, at different distances, on the retina, it is necessary that the lens changes its position, shape, and thickness, called this process ocular accommodation. [2] These alterations cause changes in the optical quality of the eye. [3]

Previous studies claimed that the ocular aberrations have impact on determining the accommodative direction and influence the precision of the accommodative response. [4] However, there is no in-depth knowledge about the impact of the biometric and optical changes of the lens in the ocular accommodation.

The study of the changes in the lens during accommodation and their consequent influence on the characteristics of the accommodative response can provide important information of the accommodative process, in healthy eyes and in dysfunctional accommodative responses. Moreover, it can be helpful to evaluate the results of intraocular lenses implantation and its efficiency.

Over the last years, an optical tomography system was developed in the centre of physics of University of Minho, which allow to obtain images of the anterior ocular segment and do its 3D reconstruction. It is possible to calculate its thickness, curvature and the wavefront aberrations. This system was later also applied to the lens in relaxed state. [5-9]

The purpose of this work was to optimize the system to obtain accurate information of the crystalline lens, both in the relaxed state and during ocular accommodation. It is intended to make the system able to study the changes in the lens changes during accommodation.

2 Material and Methods

2.1. Slit-scanning tomography system

This system (figure 1.A) contains two systems; a lighting system, which projects a beam of light in the anterior segment of the eye, scanning and allowing different optical sections to be obtained; and an optical system, which obtains the images of the optical sections.

An algorithm developed for this purpose (figure 1.B) allows analysing these images and calculating the biometric parameters, such as the thickness and curvature of the cornea and lens, and the optical properties, i.e., the wave front aberrations.

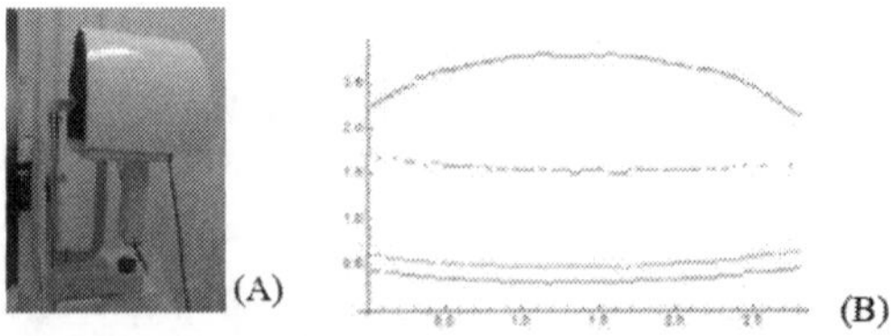

Fig. 1. (A) Slit-scanning tomography system. (B) Example of a mathematical adjustment for the calculation of biometric and optical properties of the anterior (blue) and posterior (orange) corneal surfaces and anterior (green) and posterior (red) crystalline lens surfaces.

2.2. Data collection procedure

The study adhered to the tenets of the Declaration of Helsinki and was approved by the Ethical Subcommission of Life and Health Science of University of Minho. After the explanation of the procedures, an informed consent was obtained from the subject.

In this preliminary study, measurements were taken on a subject without mydriatic medication.

* Corresponding author: sfranco@fisica.uminho.pt

The subject was asked to fix a distant point with the left eye, maintaining his accommodation relaxed, while the images were captured in the right eye. After measurements in the relaxed state, the accommodation was stimulated in 4.00 D, placing a negative spherical lens in front of the left eye, while images continued to be captured in the contralateral eye.
The images of the eye's subject with and without accommodation were analysed in the previously developed algorithm and the curvature radius and power of the anterior surface of the crystalline lens in both conditions were calculated.

3 Results

3.1. Imaging with the slit-scanning tomography system

The images of the subject's eye captured by the slit-scanning tomography system in the relaxed state and with 4.00 D of accommodative stimulus are shown in figure 2. It was possible to identify the anterior and posterior surfaces of the cornea and the anterior surface of the lens.

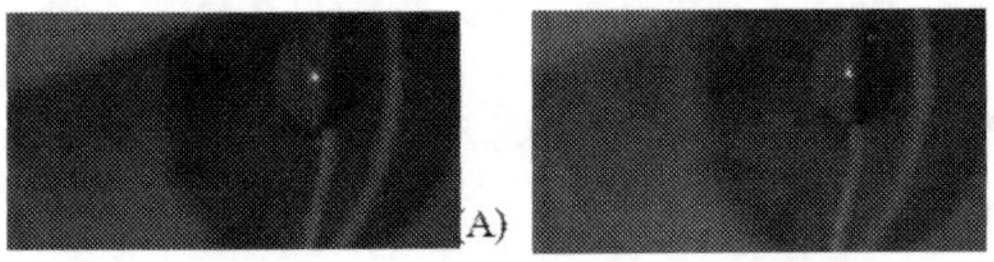

(A)

Fig. 2. Image captured with the tomography system of the subject's eye unaccommodated (A) and accommodated (B).

(B)

3.2. Image processing of the crystalline lens

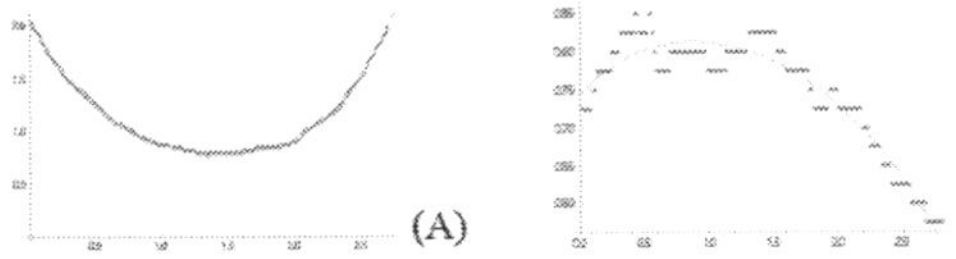

(A)

Fig. 3. Edge detection and curve fitting of the anterior surface of the lens unaccommodated (A) and accommodated (B).

(B)

3.3. Changes in properties of the crystalline lens with accommodation

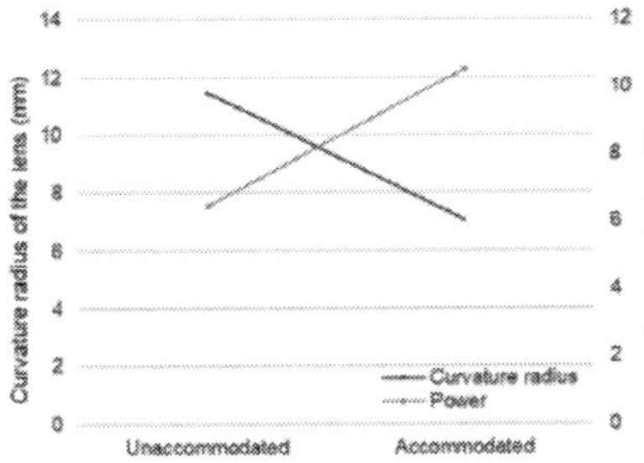

Fig. 4. Curvature radius and power of the lens in relaxed state and with an accommodative stimulus of 4.00 D.

The central curvature radius and power of the anterior surface of the lens were obtained for the unaccommodated and accommodated state (Fig 4). For an accommodative stimulus of 4.00D, the subject accommodated 4.05 D, reducing the curvature radius of the lens from 11.46 mm to 7.04 mm.

4 Conclusions

The slit-scanning tomography system allows the detection of the changes with accommodation of the biometric parameters of the crystalline lens, such as curvature radius and power. As it was excepted, in this preliminary study the anterior surface of the lens presented smaller curvature radius and greater power when the accommodation was being stimulated.
It is intended to use this system to perform a 3D reconstruction of the lens and obtain the changes with accommodation of other properties, such as the thickness and wavefront aberrations.

Acknowledges

This work was supported by the Portuguese Foundation for Science and Technology (FCT) in the framework of the Strategic Funding UID/FIS/04650/2019 and by the project PTDC/FIS-OTI/31486/2017 and POCI-01-0145-FEDER-031486.

References

1. P. Artal, A. Benito, J. Tabarnero. The human eye is an example of robust optical design. J Vis **6**, 1 (2006)
2. Y. Wang, Y. Shao, Y. Yuan. Simultaneously measuring ocular aberrations and anterior segment biometry during accommodation. J Innov Opt Heal Sci **8**, 1550005 (2015)
3. J. He, S. Burns, S. Marcos. Monochromatic aberrations in the accommodated human eye. Vision Res **40**, 41-48 (2000)
4. E. Fernández, P. Artal. Study of the effects of monochromatic aberrations in the accommodation response by using adaptive optics. J Opt Soc Am A **22**, 1732-1738 (2005)
5. S. Franco, J. Almeida, M. Parafita. Optical corneal tomography. In mastering the techniques of IOL power calculations, ed. A Garg, J E Hoyos, D Dementiev, 21-23. New Delhi: Jaypee Brothers Medical Publishers (P) Ltd (2009)
6. S. Franco, J. Almeida, M. Parafita. Corneal thickness and elevation maps computed from optical rotary scans. J Refract Surg **20**, S576-S580 (2004)
7. S. Franco, J. Almeida, M. Parafita. Measuring corneal thickness with a rotary scanning system. J Refract Surg **18**, S630-S633 (2002)
8. C. Oliveira, J. Almeida, S. Franco. Crystalline lens imaging with a slit-scanning system. In Technology and Medical Sciences 93-96 (2011)
9. C. Oliveira, J. Almeida, S. Franco. Imaging of aphakic intraocular lens with a slit-scanning tomography system. In International Conference on AOP – International Society for Optics and Photonics **8001**, 800135 (2011)

https://doi.org/10.1051/epjconf/202023812016

The influence of age, sex and visual dysfunction on reading ability in school-children

António M. G. Baptista[1,*], *Pedro Serra*[2], *Carlos Silva*[3], *Sandra Franco*[1], *James Gilchrist*[4], *Brendan Barrett*[5]

[1]University of Minho, Centre of Physics, Rua da Universidade 4710 – 057 Braga, Portugal
[2]Instituto Superior de Educação e Ciência, Alameda das Linhas de Torres 179, 1750-142 Lisboa, Portugal
[3]Clinical Practice, Braga, Portugal
[4]Senior Researcher, UK
[5]University of Bradford, Optometry Department, Richmond Rd, Bradford BD7 1DP, UK

Abstract. Binocular and accommodative dysfunctions affect the efficiency and coordination of the eyes, and this may impair reading performance. Reading can be evaluated using accuracy and speed measures. In this study, we searched for evidence that dysfunctions of accommodation (i.e. accommodation insufficiency, IA) and binocular vision (i.e. convergence insufficiency, CI) and also hyperopia have impact on reading ability. A sample of 292 children read a text following vision screening. The sample was composed of control children and children with the conditions: AI; CI and hyperopia. Multilinear regression analysis was used to evaluate the influence of variables sex, age and visual condition (control, AI, CI and hyperopia) on reading accuracy and speed. The results showed that ~7% of the variation in reading accuracy was explained by a model where age ($p=0.001$), and presence of AI ($p<0.001$) and hyperopia ($p=0.019$) were statistically significant. On the other hand, ~38% of the variation in the index of speed was explained by age alone ($p<0.001$). These results confirm the importance of age on children's ability to read, and suggests that AI and hyperopia, but not CI may affect reading accuracy. Reading speed does not appear to be affected by the presence of binocular or accommodative dysfunction.

1 Introduction

Binocular and accommodative dysfunctions affect the efficiency and coordination of the eyes, and this may impair reading performance, a critical skill for learning. Reading performance can be evaluated using accuracy and speed measures. Accuracy (1) reflects the percentage of words correctly read [1], and speed (2) corresponds to the number of words correctly read per minute [1]. In this study, we searched for evidence that the most prevalent dysfunctions of accommodation (i.e. accommodation insufficiency, AI) and binocular vision (i.e. convergence insufficiency, CI) and also hyperopia have impact on reading ability.

$$\text{Accuracy}=\frac{\text{Total number of words read-Total number of errors}}{\text{Total number of words read}} \quad (1)$$

$$\text{Speed}=\frac{\text{Total number of words read-Total number of errors}}{\text{Reading time}}\times 60 \quad (2)$$

2 Methods

Accommodation insufficiency was defined as monocular accommodative amplitude less than 12.5 spherical dioptres. Convergence insufficiency was defined as [near point of convergence equal or greater than 6 cm {AND} near exophoria at least four prism dioptres greater than distance exophoria {AND} (failing Sheard's criterion {OR} positive fusional vergence equal or less than 15 prism dioptres base-out)]. Hyperopia was defined as being positive on the blur test (+2.00 spherical dioptres) in either eyes. A sample of 292 children read a text of a traditional, Portuguese tale "O rei vai nu" [1] following vision screening tested for: autorefractor; monocular visual acuity with habitual refraction; cover test; near point of convergence; blur test +2.00 spherical dioptres; distance phoria; stereopsis (TNO); monocular near visual acuity; monocular accommodative amplitude. All testing took place in the school (29 schools were visited). The sample was composed of control children (absence of AI, CI and hyperopia as defined above, stereopsis equal or better than 120 arc sec and monocular visual acuity equal or better than 0.8 decimal) and children with the conditions: AI; CI and hyperopia. Multilinear regression analysis was used to evaluate the influence of variables sex (binary variable: male=0 and female=1), age and visual condition (nominal variable: control, AI, CI and hyperopia; e.g. when AI=1 then control=0 and CI=0 and hyperopia=0) on reading accuracy and speed.

This research was conducted according to the guidelines outlined in the Declaration of Helsinki. Written informed consent was obtained from children parents. This study

* Corresponding author: abaptista@fisica.uminho.pt

received ethical approval by the Ethics Subcommittee for Life and Health Sciences of Minho University.

3 Results

Number of control children: male: n=91, 9.7±1.4 years; female: n=85, 9.6±1.3 years. Number of children with the conditions AI: male: n=48, 10.1±1.4 years; female: n=36, 10.8±1.6 years. Condition CI: male: n=5, 9.5±1.8 years; female: n=6: 10.0±2.0 years. And hyperopia: male: n=5, 8.9±2.2 years; female: n=15, 9.5±1.5 years. The results showed that ~7% of the variation in reading accuracy was explained by a model where age (p=0.001), and presence of AI (p<0.001) and hyperopia (p=0.019) were statistically significant but not sex (p=0.954) or CI (p=0.344). On the other hand, ~38% of the variation in the index of speed was explained by age alone (p<0.001), i.e. the presence or absence of hyperopia (p=0.598), AI (p=0.294) or CI (p=0.565) or sex (p=0.384) did not affect speed.

4 Conclusions

The results confirm the importance of age on children's ability to read, and suggests that accommodation insufficiency and hyperopia, but not convergence insufficiency may affect reading accuracy. Reading speed does not appear to be affected by the presence of binocular or accommodative dysfunction.

Acknowledges

This work was supported by the Portuguese Foundation for Science and Technology (FCT) in the framework of the Strategic Funding UID/FIS/04650/2013.

References

1. Anabela O.D.C. Carvalho, Teste de avaliação da fluência e precisão de Leitura : O Rei (EDIPSICO, 2010)

The angular momentum of vectorial non-paraxial fields and the role of radial charges in orbit-spin coupling

Omar El Gawhary[1,2,*] and *Paul* Urbach[2]

[1]VSL Dutch Metrology Institute, Thijsseweg 11 2629 JA Delft, Netherlands
[2]Optics Research Group, Delft University of Technology, Lorentzweg 1 2628 CJ Delft, Netherlands

Abstract. Electromagnetic fields carry a linear and an angular momentum, the first being responsible for the existence of the radiation pressure and the second for the transfer of torque from electromagnetic radiation to matter. The angular momentum is considered to have two components, one due to the polarization state of the field, usually called Spin Angular Momentum (SAM), and one due to existence of topological azimuthal charges in the field phase profile, which leads to the Orbital Angular Momentum (OAM). For non-paraxial fields these two contributions are not independent of each other, something which is described as spin-orbit coupling. It has been recently proved that electromagnetic fields necessarily carry also invariant radial charges that, as discussed in this work, play a key role in the angular momentum. Here we show that the total angular momentum consists in fact of three components: one component only dependent on the spin of the field, another dependent on the azimuthal charges carried by the field and a third component dependent on the spin and the radial charges contained in the field. By properly controlling the number and coupling among these radial charges it is possible to design electromagnetic fields with a desired total angular momentum. In this way it is also possible to discover fields with no orbital angular momentum and a spin angular momentum typical of spin-3/2 objects, irrespective of the fact that photons are spin-1 particles.

1 Introduction

Electromagnetic fields carry an angular momentum which is responsible for the transfer of torque from light to matter. After the work by Allen et. al [1] the interest on the angular momentum of light, its origin and effect on matter, has gradually grown through the years and currently represents an active area of investigation, especially in view of possible applications to high-capacity optical communications, novel metrology techniques, classical and quantum imaging [2–5]. There are essentially two sources the angular momentum: one due to the polarization state of the field (spin angular momentum or SAM, for short) and the other due to the presence of a helical wavefront in the electromagnetic field (orbital angular momentum or OAM)[6–15]. In this work we discuss an insightful expression of the total angular momentum (TAM) in terms of the s and p components of the angular spectra of the field. We then use that expression of the TAM and combine it with the recently introduced representation of a field in terms of Helmholtz Natural Modes which allows us to identify two types of topological

*e-mail: o.elgawhary@tudelft.nl

charges carried by a field: an azimuthal charge and a radial charge. The existence of azimuthal charges has been known for long as they are linked to the orbital angular momentum of a field. On the other hand, the existence of topological radial charges seemed to have remained unnoticed to the scientific community. Interestingly, as we are going to present in this work, radial topological charges are responsible for the orbit-spin coupling in non-paraxial fields. Once the main mechanism behind the generation of the Total Angular Momentum (TAM) of a field is clear, we can show the existence of fields endowed with only a SAM contribution, which however corresponds to spin-3/2 particles, something surprising when dealing with electromagnetic fields. In this work we will consider the time-averaged TAM, which can be written in the following way

$$\boldsymbol{J} \;=\; \frac{\varepsilon_0\mu_0}{2}\,\mathrm{Re}\left\{\int \left[\mathbf{r}\times\left(\mathbf{E}\times\mathbf{H}^\star\right)\right]d\mathbf{r}\right\} \tag{1}$$

where the $\star$ symbol stands for complex conjugation. We will show that for a beam propagating along the z-direction, the z-component of the TAM per unit length can be expressed in Fourier space as

$$\mathfrak{J}_z = \mathrm{Re}\left\{\frac{\varepsilon_0}{2i\omega}\int\left[\left(A_s^\star\frac{\partial A_s}{\partial\varphi}+A_p^\star\frac{\partial A_p}{\partial\varphi}\right)-\frac{(2\pi)^2}{k}\frac{\rho^2}{k_z}A_s^\star A_p\right]\rho d\rho d\varphi\right\} \tag{2}$$

Eq.2 shows that TAM can in fact be expressed in terms of the sum of a s-polarized contribution (which we will call S-TAM), a p-polarized contribution (which we will denote as P-TAM) and a s-p cross-term (which we can denote as SP-TAM). A lot can be learned on the physics governing the TAM by expressing any field in terms of vectorial Helmholtz Natural Modes (HNMs) [16, 17]. As we will show in this paper, this allows us to write the angular momentum per unit energy as made of three parts (n denotes the azimuthal charge carried by the field and m the radial one)

$$\gamma \;=\; \frac{\sigma_z}{\omega}+\frac{1}{\omega}\frac{\mathrm{Re}\left\{\sum_{m,n}\frac{4}{3}n|c_{m,n}|^2+\sum_{\substack{m,m',n\\ m\neq m'}}nc^\star_{m',n}c_{m,n}f(m,m')\right\}}{\sum_{m,n}\frac{4}{3}|c_{m,n}|^2+\sum_{\substack{m,m',n\\ m\neq m'}}c^\star_{m',n}c_{m,n}f(m,m')}+$$

$$+\;\frac{\sigma_z}{\omega}\frac{\mathrm{Re}\left\{\sum_{m,n}\frac{2}{3}|c_{m,n}|^2-\sum_{\substack{m,m',n\\ m\neq m'}}c^\star_{m',n}c_{m,n}f(m,m')\right\}}{\sum_{m,n}\frac{4}{3}|c_{m,n}|^2+\sum_{\substack{m,m',n\\ m\neq m'}}c^\star_{m',n}c_{m,n}f(m,m')} \tag{3}$$

Eq.3 tells us that the TAM has three main contributions: a first term only dependent on the spin σ_z, a second one dependent on the azimuthal charges n present in the field (the orbital part) and a last term which depends on the spin σ_z and the radial charges m contained in the field. $\mathfrak{J}_z/W$ can be interpreted as if each photon in the field contributes to the angular momentum through a term $\sigma_z\hbar$ due to its spin, a term $n\hbar$ due to its orbital angular momentum and a third term, again proportional to the spin σ_z and function of the radial charges carried by the field. The last contribution, which is absent within the paraxial regime, is a special trait of the electromagnetic theory. In the talk we will discuss on the coupling among radial charges, identify cases when such coupling is absent and the interesting case where no azimuthal charge is present. [18].

References

[1] L. Allen, M.W. Beijersbergen, R.J.C., Spreeuw and J.P. Woerdman,"Orbital angular momentum of light and the transformation of Laguerre-Gaussian laser modes" Phys. Rev. A, **45**, 8185, 1992.

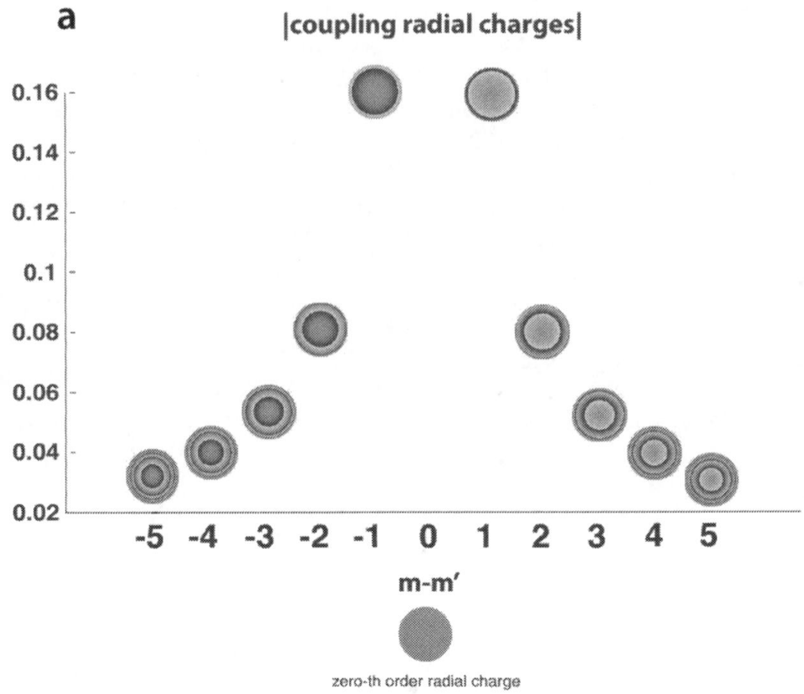

Figure 1. Panel **a**: Absolute value of the coupling function $f(m, m') = 1/[2i\pi(m-m')]+1/[2\pi^2(m-m')^2]$ among neighbouring radial modes. The curve refers to the coupling of the 0-th order radial charge (shown close to the origin of the x-axis in the graph) with the neighbouring radial charges, of order $m \in [-5, 5]$. On the x-axis the order of each charge is reported, while the phase profile of each charge is used as marker on the curve. Only radial charges corresponding to HNMs of same azimuthal charge n can couple to each other.

[2] *Quantum Imaging*, M. Kolobov Editor, Springer Singapore, 2007.

[3] *Quantum metrology, Imaging and Communication*, D.S. Simon, G. Jaeger, A.V. Sergienko Editors, Springer Cham, Switzerland, 2017.

[4] Y. Yan, G. Xie, M. P. J. Lavery, H. Huang, N. Ahmed, C. Bao, Y. Ren, Y. Cao, L. Li, Z. Zhao, A. F. Molisch, M. Tur, M. J. Padgett, A. E. Willner, "High-capacity millimetre-wave communications with orbital angular momentum multiplexing", Nature Comm., **5**, 4876, 2014.

[5] T. Van Mechelen and Z. Jacob, "Universal spin-momentum locking of evanescent waves", Optica, **3**, 118 – 126, 2016.

[6] L. Allen, M.J.Padgett and M. Babiker, "The orbital angular momentum of light", in Progress in Optics XXXIX, Elsevier, Amsterdam, Netherlands, E. Wolf Editor, 291-372, 1999.

[7] C. Cohen-Tannoudji, J. Dupont-Roc, G. Grynberg, *Photons & Atoms*, Wiley-VHC Weinheim, Germany, 2004.

[8] S. M. Barnett, L. Allen, "Orbital angular momentum and non paraxial light beams", Optics Comm., **110**, 670-678, 1994.

[9] K. Y. Bliokh, M. A. Alonso, E. A. Ostrovskaya, and A. Aiello, "Angular momenta and spin-orbit interaction of nonparaxial light in free space", PRA, **82**, 063825, 2010.

[10] S.J. van Enk, G. Nienhuis, "Commutation rules and eigenvalues of spin and orbital angular momentum of radiation fields", J. Modern Opt. **41**, 963-977, 1994.

[11] K. Y. Bliokh, F. Nori, "Transverse and longitudinal angular momenta of light", Physics Reports, **592**, 1-38, (2015).

[12] M. S. Soskin, V. N. Gorshkov, M. V. Vasnetsov, J. T. Malos, and N. R. Heckenberg, "Topological charge and angular momentum of light beams carrying optical vortices", Phys. Rev. A **56**, 4064, 1997.

[13] S. M. Barnett, L. Allen, R. P. Cameron, C. R. Gilson, M. J. Padgett, F. C. Speirits and A. M. Yao, "On the natures of the spin and orbital parts of optical angular momentum", Journal of Optics, **18**, 064004, 2016.

[14] S. M. Barnett, M. Babiker, M. J. Padgett, "Optical orbital angular momentum", Phil. Trans. R. Soc. A 375, 20150444, 2017.

[15] K. Y. Bliokh, F. J. Rodriguez-Fortuno, F. Nori, A. Zayats, "Spin-orbit interactions of light", Nature Phot.,**9**, 796-808, 2015.

[16] O. El Gawhary, "Helmholtz Natural Modes: the universal and discrete spatial fabric of electromagnetic wavefields ", New J. Phys. **19**, 013021, 2017.

[17] O. El Gawhary, "On a propagation-invariant, orthogonal modal expansion on the unit disk: going beyond Nijboer–Zernike theory of aberrations", Opt. Lett. **40**, 2626-2629, 2015.

[18] O. El Gawhary, T. van Mechelen, H.P. Urbach, "Role of Radial Charges on the Angular Momentum of Electromagnetic Fields: Spin-3/2 Light", Phys. Rev. Lett., **121**, 123202, 2018.

Optical Simulation and Design of Spatial Heterodyne Spectrometers for Remote Sensing Applications

C.-A. Stöckling[1,*], *W.* Pöschel[2,**], and *S. A.* Schulz[1,***]

[1]Hamburg University of Applied Sciences, Faculty of Engineering and Computer Science, Berliner Tor 5, 20099 Hamburg
[2]SIOS Meßtechnik GmbH, Am Vogelherd 46, 98693 Ilmenau

Abstract. We present the optical simulation and design concept of a spatial heterodyne spectrometer (SHS) for mobile applications. A framework using Python and Zemax OpticStudio was developed for the optical system design and automated tolerance analysis from the incoming light, the spectrometer and the imaging lens system to the 2D detector. The spectrometer design and the fabrication methods were validated using a test setup in the VIS spectral range for a future SHS fabrication in the LWIR spectral band operating on small unmanned aerial vehicles (SUAV) for remote sensing applications.

1 Introduction

In recent years, the demand for small and robust sensors for mobile applications has increased due to the further development of field robotics. Spectrometers will be an important part of these future solutions. Simple spectroscopic measurements can be solved by the usage of transmission filters for specific wavelength ranges or by systems based on continuously variable filters. Higher spectral resolution in the IR spectral bands usually requires interferometers like Michelson-type sensors or Fourier-Transform-Spectrometers (FTS). For hyperspectral imaging applications these systems combine spatial data with spectral signatures. The miniaturization of hyperspectral sensors enables the integration on small autonomous robotic systems and allows new applications in remote sensing.

For applications in field robotics and processing industry, optical spectrometers will operate in harsh environments under extreme environmental conditions. Rapid change in temperature, high vibration levels and alternating radiant flux are caused by the exploration and investigation of dangerous areas, i.e. in the oil and mining industry. The reliable operation on mobile robots demands small, lightweight and ruggedized hyperspectral sensors. At present, filter-based spectral cameras fulfill many criteria, but with the lack of spectral fingerprint capability because of their low spectral resolution, robust spectrometer designs are required. A spectrometer with no moving parts is essential for such applications. A stationary Fourier-Transform Spectrometer design in a push-broom configuration combines high optomechanical stability, large entrance aperture and high spectral resolution for mobile remote sensing applications.

*e-mail: christian.stoeckling@haw-hamburg.de
**e-mail: poeschel@sios.de
***e-mail: stephan.schulz@haw-hamburg.de

2 Fundamentals of a SHS spectrometer

A SHS with a focal plane array (FPA) to record an entire Fizeau fringe pattern at once in the LWIR spectral range was presented by Harlander and Roesler in 1990 [1], although the first designs of these spectrometers called SISAM were realized by Connes in the late 1950's. The SHS adapts a Michelson configuration with the two end mirrors replaced by diffraction gratings which are identical tilted about the angle Θ to the optical axis which is known as Littrow configuration [3].

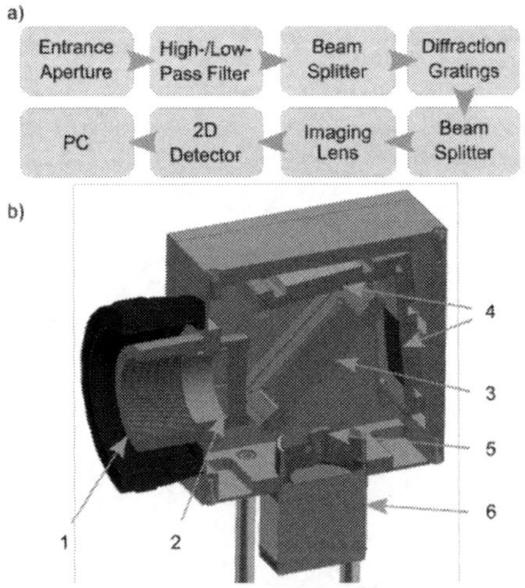

Figure 1: (a) Configuration of the SHS and (b) 3D assembly view: entrance aperture (1), filter (2), beam splitter (3), gratings (4), imaging lens (5), and detector (6).

A SHS is heterodyning the signal, so the spatial frequency of the fringe pattern is the same for every wavelength with the same distance to the Littrow wavelength. A short or long pass filter has to be used to get an unambiguous result and an extended spectral range can be achieved by field widening. In a basic principle of operation (Fig. 1), the input wave front is divided at the beam splitter, and a wavelength dependent wave front phase shift is induced by the gratings. Both opposed phase shifted wave fronts form the Fizeau fringe pattern with a characteristic spatial frequency. This 2D interferogram on the detector (FPA) can be converted via Fourier Transformation (FT) to its equivalent spectrum.

3 Python Framework for SHS Simulation

For the exact and reliable SHS design an object-based Python framework was developed. The framework is capable of simulating the whole chain of generating a interferogram from a single wavelength or given spectrum and transform the interferogram to a spectrum for an ideal SHS (Fig. 2). The SHS parameters are defined in a SHS object to generate the interferogram and also improved spectra. For further improvement methods like apodization with different windowing are implemented.

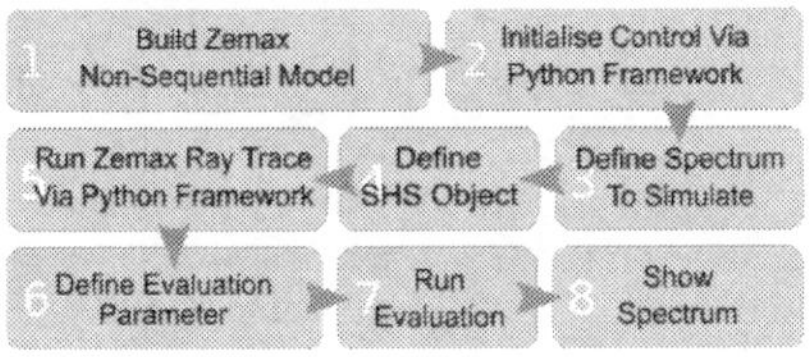

Figure 2: Flow chart of a typical SHS simulation

The python framework simulates a SHS without imperfections. This allows the investigation of SHS parameter variation and its impact of the system performance of an undisturbed system. Disturbation of the system on purpose and simulating coatings, filters and other obstacles in the optical path to study the effects is implemented. To investigate the phenomena induced by more complex elements like lenses to the SHS system, the framework is coupled with Zemax OpticStudio for ray tracing. The python framework controls the operation of Zemax for closed-loop optimization to generate a interferogram, which is evaluated by the framework. Fig. 2 shows a typical simulation run with the framework and Zemax OpticStudio combined.

4 Simulation and Experimental Results

To test the python framework a SHS for the visual spectral range was simulated and built. Figure 3 shows the neon spectra for the simulation and the experimental data with a neon spectral lamp (Oriel 6032) in good agreement. Deviations are caused by the uncollimated light source and due to standard single lens imaging which impose wavefront aberrations.

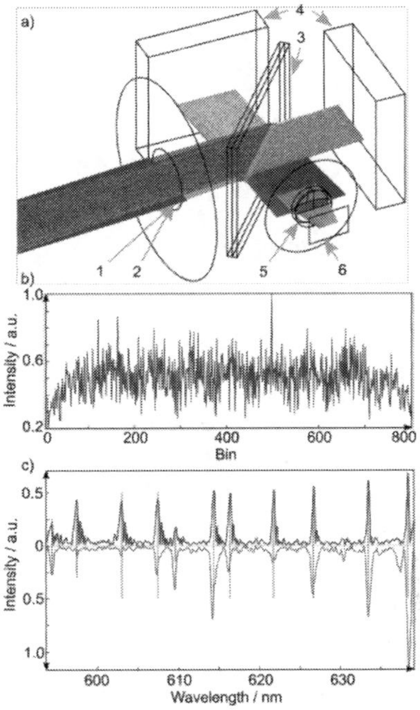

Figure 3: Simulation results: (a) Zemax model of a VIS-SHS, (b) interferogram of a neon spectral lamp, (c) measured (blue) data, simulated (green) data with NIST reference [5] (red).

5 Conclusion and outlook

A SHS for the visible spetral range was designed, simulated and built based on a developed Python simulation framework and combined with the optics simulation ZEMAX software. We have shown that the Python framework is well suited to predict the performance of a SHS by experimental demonstration of the simulation by a reference SHS build and characterization in the visible spectral range. The Python framework allows the closed-loop optimization for future SHS designs in the LWIR spectral range for remote sensing applications.

The authors acknowledge the financial support by the Federal Ministry of Education and Research of Germany (BMBF) in the project ATHMOS (13N14762).

References

[1] J. Harlander, F. L. Roesler, SPIE **1344**, 120-131 (1990)

[2] P. Connes, Journal de Physique et le Radium **19**, 215-222 (1958)

[3] C. Palmer, E. Loewin, *Diffraction grating handbook* (Newport Corporation, New York, 2005)

[4] C. R. Englert, D. D. Babcock, J. M. Harlander, Optical Engineering 48 (10), 105602, (2009)

[5] A. Kramida, Y. Ralchenko, J. Reader, NIST ASD Team, *NIST Atomic Spectra Database (ver. 5.7.1).*

 https://doi.org/10.1051/epjconf/202023812019

Phase-coherent THz-wave demodulation with a photomixer

Florin Lucian Constantin[1,*]

[1]Laboratoire PhLAM, CNRS UMR 8523, 59655 Villeneuve d'Ascq, France

Abstract. The nonlinear electrical response from a low temperature grown GaAs photomixer driven with the optical beat of two continuous-wave lasers allows frequency mixing between radiofrequency and THz-waves. Phase-sensitive demodulation of the signal from a modulated THz-wave is theoretically and experimentally demonstrated using a photomixer electrically driven with an alternative voltage at the modulation frequency.

1 Introduction

The Terahertz photonics research has applications in various domains including biological imaging, remote sensing, high-speed telecommunications, and security screening [1]. The optical frequency down-conversion (photomixing) was exploited for continuous-wave (cw) THz generation using the optical beat of two cw lasers and an ultrafast photoconductive antenna [2]. The optical frequency control enabled compact, room-temperature operated, cost-effective and low-power consuming THz photomixing systems with narrow spectral linewidth, wide frequency tunability, high dynamic range and high signal-to-noise ratio. The modulation of the THz waves, critical parameter for high-speed wireless telecommunications, was demonstrated through various approaches [3]. Particularly, the nonlinear electrical conduction in a low temperature grown GaAs (LTG-GaAs) photomixer was exploited for direct current modulation of the photomixing signal [4]. This contribution addresses theoretically and experimentally the demodulation of a THz-wave using the photomixing approach. The LTG-GaAs photomixer was exploited for the detection of a modulated THz-wave [5]. When the photomixer is electrically driven with an alternative voltage oscillating at the modulation frequency of the THz-wave, the phase-sensitive DC response signal of the photomixer indicates the demodulation of the THz-wave.

2 Results

The electrical conduction in the photomixer is modeled by unidimensional carrier transport in a thin LTG-GaAs layer. Two laser beams, carrying the same optical power P are detuned by an angular frequency ω around the mean angular frequency ϖ, are spatially superposed and enter at normal incidence in the photoconductor. The current density is expressed as:

$$j(t)=\frac{2e\eta P}{\hbar\varpi Ad}\left[\begin{array}{l}\tau_n(E)v_n(E)\left(1+\dfrac{\omega\tau_n(E)sin(\omega t)+cos(\omega t)}{1+(\omega\tau_n(E))^2}\right)\\ +\tau_p(E)v_p(E)\left(1+\dfrac{\omega\tau_p(E)sin(\omega t)+cos(\omega t)}{1+(\omega\tau_p(E))^2}\right)\end{array}\right] \quad (1)$$

The superposed laser beams are focused on a photoconductive area A, and absorbed within a depth d with a quantum efficiency η. The carrier lifetimes $\tau_{n,p}(E)$ depend on the applied electric field [6]. The dependences are described here with linear functions that take into account the carrier heating:

$$\tau_{n,p}(E)\approx\tau_{0n,0p}[1+3\lambda eE/(2k_BT)] \quad (2)$$

in function of the Boltzmann constant k_B, the electron charge e, the room temperature T=300 K, the low-field carrier lifetimes $\tau_{0n,0p}$ and the distance λ between the Coulomb barriers in the LTG-GaAs photoconductor. The carrier drift velocities have a nonlinear dependence on the electrical field [7,8], expressed as:

$$v_{n,p}(E)=v_{sn,sp}/[1+v_{sn,sp}/(\mu_{0n,0p}E)] \quad (3)$$

with the saturation velocities $v_{sn,sp}$ and the low-field mobilities $\mu_{0n,0p}$ of the electrons and holes.

The total electric field in the photoconductor may be expressed as:

$$\begin{array}{l}E(t)=E_{DC}+E_{RF}\cos(\omega_{RF}t+\phi_{RF})\\ +\sum_n E_{THz,n}\cos((\omega_{THz}+n\omega_{RF})t+\phi_{THz,n})\end{array} \quad (4)$$

the sum between a bias electric field E_{DC} and a radiofrequency electric field (amplitude E_{RF}, angular frequency ω_{RF}, phase ϕ_{RF}) from an alternative voltage applied on the photomixer electrodes. The last terms are the spectral components of a THz electric field (amplitudes $E_{THz,n}$ angular frequencies $\omega_{THz}+n\omega_{RF}$,

* Corresponding author: FL.Constantin@univ-lille1.f

phases $\phi_{THz,n}$) that is induced by a THz-wave modulated at ω_{RF} which is detected with the photomixer. The electrical nonlinearity arising from $v_{n,p}(E)$ and $\tau_{n,p}(E)$ dependences enables frequency mixings between the THz-wave spectral components and the radiofrequency in the LTG-GaAs photoconductor. The amplitudes and phases of the spectral components of the current density are analytically calculated with eq. (1-4), using the photomixer parameters d=2 µm, A=12.6 µm^2, $\hbar\varpi/e$=1.5 V, η=0.68, the LTG-GaAs photoconductor parameters from refs. [6-8] and the modulation parameters. Particularly, the modulated THz-wave generates a spectral component of the current density at ω_{RF} that can be demodulated by coherent mixing with the radiofrequency signal at ω_{RF} applied on the phomixer electrodes.

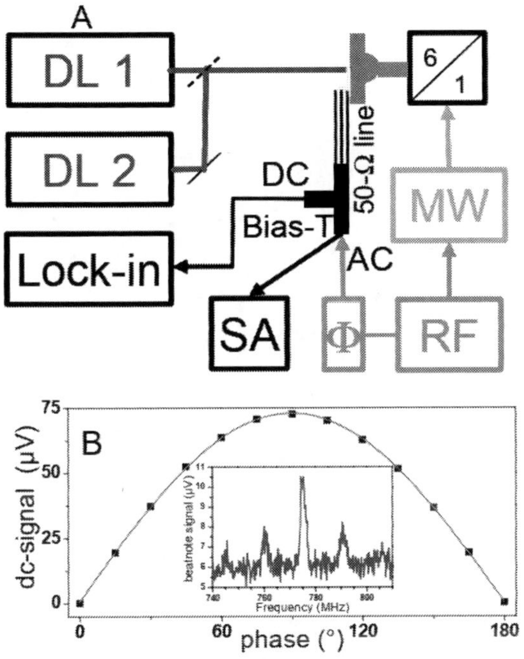

Fig. 1. A. The experimental setup. The signal from the radiofrequency synthesizer RF is used to modulate the THz-wave and, after application of an adjustable phase offset, is coupled to the photomixer. B. Phase-sensitive dependence of the demodulated signal. Inset : Spectrum of the heterodyne detection signal of a THz-wave at 102.072 GHz modulated at 15 MHz. Optical beat power 2×14 mW. RBW=1 MHz, sweep time 4 ms, 16 video averages.

The experiment (Fig. 1.A) is based on a LTG-GaAs photomixer with a log-spiral antenna driven by the optical beat of two extended cavity diode lasers (DL1 and DL2) emitting at 820 nm. The extremities of the antenna are connected to a coaxial microwave line addressed with a bias-T. The DC path is used to measure the response from the photomixer using a lock-in detection technique (Lock-in). The AC path is used to record the heterodyne down-conversion signals with a microwave spectrum analyser (SA) or, alternatively, to apply direct-current modulation on the photomixer with a RF synthesizer (RF). A 75-110 GHz electronic frequency multiplier driven by a microwave synthesizer (MW) is used to generate a THz-wave which is focused on the photomixer antenna with a silicon lens. All synthesisers and the spectrum analyser are phase-referenced on a 10 MHz GPS-disciplined quartz oscillator.

A multifrequency THz-wave is generated by modulating MW with a 15 MHz signal provided by RF. The spectrum of the THz-wave is down-converted using the heterodyne detection scheme with the photomixer and recorded with SA. The spectral components, spaced by the modulation frequency, are shown in the inset of Fig. 1.B. The next step is to demodulate the THz waveform by exploiting a phase-delayed signal from RF (amplitude 1.5 V, frequency 15 MHz) that is capacitively coupled to the photomixer. For sensitive detection, RF is 100% amplitude modulated at 1 kHz and the dc-signal in the photomixer circuit is recovered with a lock-in amplifier. The Fig. 1.B presents the dependence of the dc-signal in function of the relative phase between the THz-wave modulation signal and signal applied to the photomixer. The sinusoidal dependence demonstrates demodulation of the THz-wave with the photomixer. The photomixing approach enables integration of the functions of the optoelectronic detection and the electrically-driven demodulation in a single photonic circuit for THz signal processing.

References

1. S.S. Dhillon *et al.*, J. Phys. D **50**, 043001 (2017)
2. E.R. Brown *et al.*, Appl. Phys. Lett. **66**, 285-287 (1995)
3. Z.T. Ma *et al.*, Research Vol. **2019**, 6482975 (2019)
4. F.L. Constantin, Proc. of SPIE Vol. **9894**, 98941K (2016)
5. F.L. Constantin, IEEE J. Quantum Electr. **47**, 1458-1462 (2011)
6. N. Zamdmer and Q. Hu, Appl. Phys. Lett. **75**, 2313-2315 (1999)
7. E.R. Brown, Appl. Phys. Lett. **75**, 769-771 (1999)
8. E.R. Brown *et al.*, Appl. Phys. Lett. **64**, 3311-3313 (1994)

AUTHOR INDEX

AUTHOR INDEX

AUTHOR INDEX

AUTHOR INDEX

AUTHOR INDEX

AUTHOR INDEX

EDP Sciences – Web of Conferences
17, Avenue du Hoggar
Parc d'Activité de Courtabouf
BP 112
F-91944 Les Ulis Cedex A
France

ISBN 978-1-7138-1795-6

CURRAN ASSOCIATES INC.
proceedings.com

International Conference on Computing and Precision Engineering 2015 (ICCPE 2015)

MATEC Web of Conferences Volume 71 (2016)

Sun-Moon Lake, Taiwan
27 - 29 November 2015

edp sciences